Bioactive Compounds in Foods

Bioactive Compounds in Foods

Edited by

John Gilbert
Central Science Laboratory, Sand Hutton, York, UK

Hamide Z. Şenyuva
*Scientific and Technological Research Council of Turkey (TÜBİTAK),
Ankara Test and Analysis Laboratory (ATAL), Ankara, Turkey*

Blackwell
Publishing

This edition first published 2008
© 2008 by Blackwell Publishing Ltd

Blackwell Publishing was acquired by John Wiley & Sons in February 2007. Blackwell's publishing programme has been merged with Wiley's global Scientific, Technical, and Medical business to form Wiley-Blackwell.

Registered office
John Wiley & Sons Ltd, The Atrium, Southern Gate, Chichester, West Sussex, PO19 8SQ, United Kingdom

Editorial office
9600 Garsington Road, Oxford, OX4 2DQ, United Kingdom

For details of our global editorial offices, for customer services and for information about how to apply for permission to reuse the copyright material in this book please see our website at www.wiley.com/wiley-blackwell.

Library of Congress Cataloging-in-Publication Data

Bioactive compounds in foods/edited by John Gilbert, Hamide Z. Şenyuva.
    p.   cm.
  Includes bibliographical references and index.
  ISBN-13: 978-1-4051-5875-6 (hardback : alk. paper)
  ISBN-10: 1-4051-5875-1 (hardback : alk. paper)
  1. Food–Analysis.   2. Food–Toxicology.   3. Food contamination.
4. Bioactive compounds   I. Gilbert, John.   II. Senyuva, Hamide Z.
  TX545.B56   2008
  664′.07–dc22
                                                              2007038437

A catalogue record for this book is available from the British Library.

Set in 10/12 pt, Times by Aptara Inc., New Delhi, India
Printed and bound in Singapore by Markono Print Media Pte Ltd

1   2008

For the cover:
Brown mushrooms photo © Alan Dunlop-Walters – Fotolia.com
French fries photo © Anette Linnea Ras mussen – Fotolia.com
Golden maize photo © AGphotographer – Fotolia.com
Mixed vegetables photo © Michelle Marsan – Fotolia.com
Mussels photo © Alison Bowden – Fotolia.com

Contents

Color plate section between pages 224 and 225

Contributors

I. Blank
Nestle Product
Technology Center, Orbe,
Switzerland

L. Cano-Lerida
Johnson Matthey Catalysis,
Chilton, Billingham,
UK

D. Clarke
Central Science Laboratory, Sand Hutton,
York, UK

C. Crews
Central Science Laboratory, Sand Hutton,
York, UK

M. Dekker
Department of Agrotechnology and
Food Science, Wageningen University,
Wageningen,
The Netherlands

J. Gilbert
Central Science Laboratory, Sand Hutton,
York, UK

V. Gökmen
Food Engineering Department,
Hacettepe University,
Ankara, Turkey

J. Hajslová
Institute of Chemical Technology,
Prague, Czech Republic

C. Hamlet
RHM Technology Ltd, The Lord Rank
Centre, High Wycombe, Buckinghamshire,
UK

F. Hannah
University of London, University Marine
Biological Station, Millport, Isle of
Cumbrae, Scotland, UK

T. Herraiz
Spanish Council for Scientific Research,
Juan de la Cierva, Madrid,
Spain

M. Knize
University of California, Lawrence
Livermore Program, National Laboratory,
Biology & Biotechnology Research
Livermore, CA, USA

R. Krska
Christian Doppler Laboratory for
Mycotoxin Research, Center for Analytical
Chemistry, Department for
Agrobiotechnology (IFA-Tulln), University
of Natural Resources and Applied Life
Sciences, Konrad Lorenz,
Tulln, Austria

J. W. Leftley
Scottish Association for Marine Science,
Oban, Argyll, Scotland, UK
and Integrin Advanced Biosystems Ltd,
Marine Resource Centre, Barcaldine,
Argyll, Scotland, UK

M. Reinik
Estonian Health Protection Inspectorate,
Tartu Laboratory, Tartu, Estonia

M. Roasto
Department of Food Science and Hygiene,
Estonian University of Life Sciences,
Kreutzwaldi, Tartu, Estonia

M. Rose
Central Science Laboratory, Sand Hutton,
York, UK

V. Schulzova
Institute of Chemical Technology.,
Prague, Czech Republic

K. Scudamore
KAS Mycotoxins, Taplow, Maidenhead,
Berks, UK

H. Şenyuva
Scientific and Technological Research
Council of Turkey (TÜBİTAK),
Ankara Test and Analysis Laboratory
(ATAL), Ankara, Turkey

T. Tamme
Department of Food Science and
Hygiene, Estonian University of Life
Sciences, Kreutzwaldi, Tartu,
Estonia

R. Verkerk
Department of Agrotechnology and
Food Science, Wageningen
University, Wageningen, The
Netherlands

P. Walton
Department of Chemistry, University of
York, Heslington,
York, UK

Part One
Natural toxicants

1 Introduction

John Gilbert and Hamide Z. Şenyuva

1.1 DIFFERENT PERCEPTIONS OF CHEMICALS IN FOOD

There are several different ways in which non-essential substances in foods can be categorised. They might be thought of as 'naturally occurring' and therefore as intrinsic components of the foods. These substances are thus difficult to avoid without stopping consumption of the food in question. Alternatively these non-essential chemicals might be 'man-made' in terms of their source and therefore their presence in food might be deemed avoidable. Another simplistic way of looking at these substances could be as being either 'chemical' or 'biological' in origin. To the non-scientist the notion of something present in food, which is described as being chemical (man-made) tends to be thought of in emotive terms and deemed as being highly undesirable. In contrast and quite irrationally the notion of anything 'naturally-occurring' in foods is thought of as being benign, simply because it is 'natural'. Thus, by this logic substances of biological origin are felt to be safer than those of chemical origin. It is this superficial categorisation which results in public wariness concerning pesticides and additives in foods (thought of as man-made chemicals), and yet less concern, for example, regarding natural toxicants such as mycotoxins thought of as natural being of biological origin.

There are also other dimensions which compound the public misunderstandings when trying to gauge the true significance of these non-essential substances in foods. The fact that it is not the consumption of any chemical substance which is important, but it is the total amount consumed which is critical, is a simple concept but one often ignored and sometimes difficult to communicate. Additionally, substances have different biological effects depending on the dose, some being beneficial at perhaps low levels of intake but harmful at higher exposure levels. Some substances might be benign for some sectors of the population and harmful for others (e.g. those suffering from intolerance or allergenic reactions).

All food additives must have a demonstrated essential function and necessarily undergo a rigorous scientific safety evaluation before they can be approved for use. Assessments are based on reviews of all available toxicological data in both humans and animal models. From the available data, the maximum level of additive that has no demonstrable effect is determined. Additives are used to carry out a variety of functions which are often taken for granted. Foods are inevitably subjected to many environmental conditions, such as temperature changes, oxidation and exposure to micro-organisms, which can change their original composition. Food additives play a key role in maintaining the food qualities and characteristics that consumers demand, keeping food safe, wholesome and appealing from farm-to-fork.

There are also examples of where increased consumption of a food item is desirable from a beneficial dietary perspective (e.g. increased consumption of oily fish) but exposure to contaminants contained in the fish need to be minimised. How do you balance the beneficial effects in reducing heart disease from high fish consumption against the negative effects of dioxin/PCB exposure, and decide on an optimum level of consumption of fish. Unfortunately, the science is not yet sufficiently well enough developed to answer these questions and a pragmatic approach is therefore needed. There is thus an apparent complexity in terms of the significance of chemicals in foods. This complexity has proved very difficult to communicate to the general public, and has been one of the issues underlying the many and various food scares of recent years.

1.2   RESIDUES AND CONTAMINANTS IN FOODS

'Man-made' or chemical contaminants can be those that occur as a result of deliberate treatment of the food supply, e.g. pesticides and veterinary drugs. In these two cases these chemicals are described as 'residues' implying that the detectable amounts are a remaining proportion (albeit small) of that used in the treatment process, i.e. 'residues'. These residues are well understood toxicologically and are well regulated, with such low-level residues as occurring in foods, generally being kept to a minimum. Authorisation of the use of pesticides and veterinary drugs is based on good practice (e.g. good agricultural practice) limited to maximum residue limits (MRLs) to levels which are unavoidable and toxicologically in-significant. Thus, if good agricultural practice (GAP) is followed, minimal levels of pesticide residues occur in fruit and vegetables. Similarly, minimal levels of veterinary drug residues should be detectable if correct withdrawal times are observed between treatment of animals with permitted drugs and the eventual slaughter of the animal.

Other man-made chemicals in foods might be derived from environmental sources or might result from adventitious occurrence during processing or might result from transfer from food packaging materials. In these cases the substances are described as chemical contaminants rather than residues. Or in the case of contamination by migration from packaging they are described (in the USA) as 'indirect food additives'. At least in principle these sources of contamination can be eliminated or controlled.

Environmental contaminants are probably the more difficult to control and efforts have to be focussed on reducing the sources, i.e. reducing emissions into the environment, which can take many years to achieve. Thus, many years after PCBs were banned, these contaminants are still found in the food chain as a result of their long-term persistence. Some metals in the environment are the result of release into water supplies through industrial activity such as mining, and some chemicals such as dioxins and furans are a result of waste disposal through incineration or through uncontrolled industrial activity. In general, as countries become more developed and industrialised there is a greater use of a wider variety of chemicals in all walks of life, with greater possibilities of discharge into the environment. Polybrominated dioxins and furans (substances used as flame retardants), perfluorinated carboxylic acids such as perfluorooctanoic acid (PFOA) and perfluorooctane sulfonate (PFOS) used as industrial surfactants, fabric protectors and in teflon coatings are just a few of a growing list of 'new' chemicals that have gained widespread use and eventually are being found in biological samples and appearing in the food chain. Many of these contaminants (e.g. phthalates), which are still in widespread use, have become ubiquitous making it almost impossible to find biological samples free of contamination.

Contamination during food production and processing can occur from a variety of sources from the point of harvest of an agricultural product, through transport, storage and during subsequent food preparation and processing operations. This type of contamination can be very difficult to predict, as every point in the food chain represents a potential contamination source, thus should be thought as critical control point (CCP). The hazard analysis critical control point (HACCP) approach adapted from use for microbiological prevention is now frequently used by the food industry in an analogous way to control chemical, physical and biological contamination.

Food packaging contamination (or migration as it is known) of low-molecular-weight compounds is an important problem of plastics packaging and other materials intended to come into contact with food products. Migration itself is controllable and depends on the type of packaging material, the intimacy of contact, temperature and duration of contact. This is a highly regulated area, where only substances which are authorised, are permitted to be used as monomers or starting materials for the manufacture of polymers or as additives in polymer systems for food packaging. In the EU, this positive list is supplemented by limits on maximum residual levels in the packaging as well as migration limits applied to the foods (or food simulants). In the USA, a slightly different approach is applied using 'packaging usage factors' and 'food consumption factors' but the result is the same tight system of regulatory approval. These regulations are being progressively extended to cover all other types of food contact materials such as can coatings, paper and board, cookware etc. although despite these controls there are nevertheless sudden 'surprises' in terms of contaminants like 2-isopropyl-thioxanthone (ITX) which provides an example of a typical unanticipated 'food scare'.

Together these sources of contamination can result in several hundred possible chemicals that might enter the food supply and all these need to be understood, monitored and controlled.

1.3 NATURAL TOXICANTS IN FOODS

Whilst there is no problem in categorising residues in foods and no problem in categorising those chemicals identified from the above sources as being contaminants, there are a group of substances which although thought of as undesirable might also be regarded as 'naturally occurring'. In this 'natural' category of chemicals in foods are those which are 'inherent constituents' and these relate primarily to plants and fungi (mushrooms). Sometimes these substances are referred to as 'natural toxins' or 'natural toxicants'. Some of these toxic constituents might be detoxified during processing, e.g. during preparation or cooking. Practices have evolved over time to ensure that only suitable plants are selected for human consumption that are free of such toxicants or they are detoxified when suitably prepared. The presence of lectins (also known as hemagglutinins) in various beans is a good example where detoxification occurs when the beans are heated at above 90°C for a short time. The lectins are proteins which are heat-denatured, but can otherwise cause agglutination of red blood cells and may negatively influence nutrient absorption. Eating these beans uncooked or after 'slow cooking' at only low temperatures can lead to toxic effects. Similarly for fungi, only those free of harmful components have over time been identified and thus selected for human consumption.

There is also another category of substances, which are natural in food in the sense that they are an inevitable consequence of preparing the food and are thus unavoidable. Some of these substances are not present in the food in its raw state and are only formed during processing, frequently as a result of heat treatment. Some of these substances are simply formed during normal high temperature cooking (e.g. furan and acrylamide), and are products of browning reactions in the food. Non-enzymic browning or Maillard browning is the reaction which leads to the development of highly desirable colour and flavours in cooked foods, as well as minor amounts of undesirable substances. Other substances such as polycyclic aromatic hydrocarbons (PAHs) are a result of processes like pyrolysis and can result during cooking such as barbequing, or can result from deliberate smoking of foods to develop desirable flavours.

1.4   DEVELOPMENTS IN ANALYTICAL METHODOLOGY

Much of the concern with the presence of chemicals in foods whether residues, contaminants or 'natural toxicants' has arisen in recent years because of our improved ability to detect and identify trace levels in foods. Unfortunately this improved analytical capability has not been matched by our ability to interpret these results in terms of potential harm to human health.

Over the past 25 years there have been astonishing developments in analytical techniques which have enabled monitoring of a significantly greater number of trace levels of chemicals in foods, at lower and lower levels with an increasing degree of confidence both in accuracy and precision. Combined gas chromatography–mass spectrometry (GC/MS) is now routinely used for the analysis of volatile and semi-volatile residues and contaminants. Bench-top GC/MS instruments have become significantly more reliable, fast scanning and data acquisition have improved as has sensitivity. Multi-residue GC/MS of more than 100 pesticide residues at ng/g (ppb) sensitivity in a single chromatographic run is now routine.

Although not yet quite at the same stage of widespread accessibility, combined liquid chromatography–mass spectrometry (LC/MS) has also developed into being a sensitive and robust technique. Developments in LC/MS have had an even greater impact in the food safety area opening up the possibilities of determining previously intractable substances which lack chromophores and previously precluded detection by HPLC.

This trend is set to continue in the future with the introduction of a new generation of fast scanning, highly sensitive, accurate mass, time-of-flight mass spectrometers coupled to either GC or HPLC (GC/TOFMS or LC/TOFMS). These instruments offer new possibilities of rapid screening of crude food extracts against constructed databases of residues or contaminants enabling identification of target compounds with a high degree of confidence. Tentative identification of novel contaminants will also be possible based on accurate mass information which can be more rigorously confirmed by resorting to more sophisticated quadrupole time-of-flight mass spectrometer (QTOF-MS) tandem instruments.

The widespread use of GC/MS and LC/MS and the constant push towards measuring lower and lower levels of residues and contaminants has meant that chemicals previously below limits of detection are now being determined in foods, and new previously unrecognised contaminants are being found. Unfortunately our ability to assess the significance of these very low levels of contaminants in the food supply has not advanced at the same pace as the analytical techniques. The significance of long-term low-level exposure to trace levels of chemicals from foods is generally unknown, except where there is clear evidence of carcinogenicity. The effects of complex mixtures of chemicals (cocktail effects) are even more difficult to assess. There is generally a tendency to error on the side of caution and apply the 'precautionary principle' in risk assessment. However, this approach is not particularly helpful when the chemical in question is either an unavoidable inherent component or is intrinsically linked to flavour formation in cooked foods.

1.5   EMERGING RISKS

Recognising the damage which has been done over the past 20 years not only in economic terms with problems like bovine spongiform encephalopathy (BSE) in beef, but also in terms of public confidence in the food supply, enormous efforts have been made to avoid similar problems in the future. The notion of 'emerging risks' has been coined and work has been undertaken to try to identify risks in advance, as a means of anticipating future food safety problems. The definition of 'emerging' covers known risks that previously occurred, subsided and are now re-occurring; risks that are of growing concern as exposure patterns change; risks that are now only just being seen because of developments in science; and risks that emerge from changes in food and agricultural production. In a European Food Safety Authority (EFSA) project EMRISK a number of case studies were examined to try to see what lessons could be learnt from previous experiences. In this project the release of unauthorised genetically modified organism (GMO) material into the food chain (the Starlink™ incident in the USA), intoxication from increased consumption of botanical products (e.g. star anise tea), illegal use of antibiotics in production of shrimps and the use of 'natural' pesticides provided a range of case studies each with different lessons to be learnt.

An EU FP6 project called Go-Global. (http://www.goglobalnetwork.eu/default.aspx) was started in 2007. This project aims to develop a global food safety network and amongst other tasks will focus upon the issue of emerging risks in the food chain. The project aims to collect information from a variety of sources about changes of any sort that can be anticipated that might have an impact on the food chain. The project is looking particularly at trends and priorities in research activities as well as at economic and other drivers. From publicly available sources the intention is to examine how research funding is being spent and to look and see whether such funding is likely to lead to changes in agricultural production, changes in food handling, changes in food processing, changes in food preservation etc. The initial collection of information will provide some mapping of the areas of responsibility within the public sector (Government) and the sources of funding within each geographical area using websites, research strategies and any key contacts, thus providing a virtual network and can be used in the future for updating. Some of the research areas might provide indications of any obvious trends, e.g. any trend towards increased funding in any particular area or might identify any obvious drivers (e.g. increased costs need to reduce waste) that are significant in influencing research priorities. Some of the areas that might be considered are as follows:

1. Changes in the environment – trends in environmental pollution (increased/reduced), changes in climate (temperature/rainfall) that might impact on types of agricultural production – new diseases, new crops
2. Changes in agricultural production methods – more/less use of agrochemicals – changes in use of agrochemicals
3. Changes in type of crops and livestock being produced – trends in production
4. Changes in patterns of export/import of agricultural products and processed foods
5. Changes in handling practices for agricultural products
6. Changes in food processing methods – new food preservation techniques
7. Changes in food packaging – new packaging technologies

As an example across the EU there have been moves to reduce pesticide usage, and to increase organic production, which avoids or largely excludes the use of synthetic fertilisers and pesticides, plant growth regulators, and livestock feed additives. There has been increased production of certain crops like rapeseed due to subsidies but increased dependence on imported foods from outside the EU. The food industry has responded to consumer demands for 'fresher', more natural foods by reducing the use of food additives, e.g. chemical preservatives, and moving towards shorter shelf-life and minimally processed foods. Consumers are being more adventurous in their tastes and habits and new food products and new styles are being introduced. Growing concern with health has led to interest in functional foods and totally new products are being introduced, together with increased consumption of herbal remedies (botanicals).

All of these changes can potentially have an impact in terms of levels of contaminants, natural toxicants and process contaminants in foods. New varieties of crops or growing crops under different climatic conditions might affect the levels of glucosinolates or phytoestrogens. Climate change can mean that mycotoxins previously only seen in tropical regions of the world start to infect crops in temperate climates, e.g. aflatoxins were found in 2005 in maize in Italy leading to elevated aflatoxin M_1 levels in milk. Changes in sea temperatures impact on populations of dinoflagellates leading to increased incidence of phycotoxins and identification of novel toxins in seafood products. Changes in food processing conditions may lead to increased levels of processing contaminants like acrylamide or indeed formation of novel biologically active compounds not previously recognised. Ant 'horizon scanning' which needs to be undertaken on a global scale might provide insights into changes and help to anticipate these 'emerging risks' so that remedial action can be taken before they develop into a food scare.

1.6 BIOACTIVE COMPOUNDS IN FOODS

This book was originally intended as a second edition of 'Natural Toxicants in Food' edited by David Watson in 1998. However, when the scope of the book was discussed it became apparent that there was a need to produce more than just an update. We did decide to retain and update chapters on pyrrolizidine alkaloids, and glucosinolates, where there have been analytical advances, which have impacted in these areas. Similarly, there have been substantial advances both in the mycotoxin and phycotoxin areas in identifying new toxins and in studying increased occurrence, which justified an update of these two chapters. These areas have been complemented by a new chapter on 'mushroom toxins', which is of increasing interest with growing demand for wild fungi and increased consumption of raw mushrooms. We have added a completely new chapter on phytoestrogens in foods reflecting the growing interest in the new sources of phytoestrogens and the interest in health benefits of consumption. We have also added a chapter on the natural occurrence of nitrates and nitrites in foods, which represents an important dietary source of naturally occurring undesirable substances but frequently overlooked. It would be impossible to cover all aspects of natural toxicants in a single book and inevitably we have had to be selective. We have not covered the important area of glycoalkaloids nor cyanogens from plant sources. There are also a whole range of antinutritional factors in foods such as enzyme inhibitors, antivitamins, mineral binding agents and goitrogenic compounds which would need to be included in a comprehensive treatise. Similarly other plant toxins such as coumarins, furocoumarins, saponins, lupin alkaloids, oxalates, methylxanthines and glycyrrizin have not been covered.

The area of carboline alkaloids in foods is less well known than many other areas of processing contaminants and a new chapter on this subject has been included in this book. These

THβC and βC alkaloids occur in foods under a wide range of concentrations and distribution patterns. They exhibit a broad range of biological activities including antimicrobial, antiviral, antiparasitic, antioxidant, neuroactive, citotoxic and neurotoxic actions. Consequently, their presence in foods and the diet is of considerable interest. These alkaloids form under mild conditions in foods resulting from a non-enzymatic Pictet-Spengler cyclisation during food production, processing and storage. Their formation depends on an array of biological, chemical and technological factors, and although in this book we have characterised them as 'naturally occurring', they equally in some respects might be thought of as processing contaminants. Heterocyclic amines in foods again have tended to be under-recognised in terms of importance as processing contaminants and this deficiency has been rectified by devoting a chapter to this topic. Although polycyclic aromatic hydrocarbons (PAHs) have been known for many years, as contaminants derived from pyrolysis, there has been a renewed interest in terms of the extent of PAH contamination of foods in general. Regulatory limits set by the EU and developments in methodology justified inclusion of a chapter on PAHs in this book. 3-Monochloropropane-diol has been long recognised as a processing contaminant, originally the focus being on its origin from hydrolysed vegetable protein. However, in recent years other foods and other mechanisms of formation have been identified as have a growing list of chloropropanol and chloroesters in foods. By adding separate up-to-date reviews on acrylamide and on furan in foods, which are both relatively new topics of interest, for the first time in we have linked together 'inherent toxicants' and 'heat processing contaminants' together in one book. This links together both inherent constituents and processing derived chemicals recognising both classes as being 'bioactive' but not distinguishing beneficial or detrimental bioactivity. This drawing together of the two areas also recognises (as discussed in the sections above) that they are both in a sense 'natural' and the difficulty in both instances of prevention of occurrence in foods.

2 Pyrrolizidine Alkaloids

Colin Crews and Rudolf Krska

Summary

Pyrrolizidine alkaloids occur as natural secondary metabolites of a very large number of plants found in a range of climates. Very many compounds share the pyrrolizidine skeleton and are toxic to varying degrees to animals and to humans, causing liver disease and in some cases cancers. The chemistry of the formation and particularly the metabolism of pyrrolizidine alkaloids has been studied in some detail. The alkaloids appear to have a function of protecting the plant from predators although some insects are able to ingest the toxins, without harm, for their own defense.

The exposure of humans to pyrrolizidine alkaloids is of great interest as it can take many forms, ranging from mass poisonings by contaminated grain through low-level intake in honey made from pyrrolizidine-containing plants to the deliberate consumption of herbal teas and medicines.

A growing awareness of this threat to human health and our greater understanding of the metabolic processes have been made possible by major technological advances in methods of analysis, particularly in the combination of high performance liquid chromatography and mass spectrometry. This highly sensitive and specific technique, supported by improved extraction procedures and complementary methods, has allowed us to improve our knowledge of human exposure and is gradually leading to the implementation of advice and regulations which will help protect future generations from the effects of these widespread toxins.

2.1 INTRODUCTION

Among the many chemical compounds ingested when plants are consumed, the alkaloids perhaps have the most potent action on bodily function. These effects are often harmful, but sometimes can also be beneficial or desired, leading to the intentional ingestion of many alkaloid-containing plants as medicines or intoxicants. Consumption of toxic alkaloids in some plants elicits a response that may be delayed by years and in this circumstance the toxicity of the plant might be unobserved or denied by the subject.

Alkaloids are basic nitrogenous compounds in which the nitrogen is usually contained within a heterocyclic ring system. Their functions in the plant are relatively poorly understood but are possibly linked to a protective function as they appear to act as feeding deterrents to a wide range of animals, from insects to herbivorous mammals. The potent pharmacological action of the alkaloids has led to intensive study and a wealth of literature has been established;

recent reviews include those of Cooper-Driver (1983), Cheeke (1989), Rizk (1991a), Hartmann (1991), and Huxtable (1992).

The alkaloids are widely distributed, occurring in some 20% of flowering plants. However, this chapter will concentrate on the group of alkaloids that is perhaps of greatest significance to humans, the pyrrolizidine alkaloids, which occur in a wide variety of food plants. With the exception of the ubiquitous caffeine these are the naturally occurring alkaloids most likely to be ingested in foodstuffs.

2.2   THE PYRROLIZIDINE ALKALOIDS

Pyrrolizidine alkaloids are toxic secondary metabolites of a wide variety of plants found in various environments throughout the world, from the colder temperate climates to sparsely vegetated hot dry regions. Plants containing pyrrolizidine alkaloids have been responsible for numerous outbreaks of poisoning of livestock on several continents and continue to cause serious economic damage. Although most of the respective pasture-contaminating plants are unpalatable to grazing animals, they may be foraged in times of food shortage or ingested via contaminated silage. In recent years, pyrrolizidine alkaloids have been identified as causing human deaths in less developed countries as a result of contamination of cereal crops and harvested seed, and they have been suspected of causing illness following intentional ingestion as vegetables and in the form of herbal remedies. Several authors have described specific aspects of pyrrolizidine chemistry and toxicity or have given overviews: some of the more comprehensive reviews have been presented by Bull *et al.* (1968), Peterson and Culvenor (1983), Mattocks (1986), World Health Organisation (1988), Rizk (1991b), Stegelmeier *et al.* (1999), and Fu *et al.* (2001).

The pyrrolizidine structure is based upon two fused five-membered rings that share a bridgehead nitrogen atom, forming a tertiary alkaloid. The nitrogen atom is very often present as the oxide. In nature the rings are most frequently substituted with a hydroxymethylene group at position C-1 and a simple hydroxyl group at position C-7, forming a structure known as a necine base. The bases most commonly encountered are heliotridine, retronecine, supinidine, and otonecine; the structures of which are shown in Fig. 2.1.

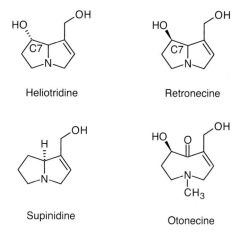

Fig. **2.1**   The major bases of the pyrrolizidine alkaloids.

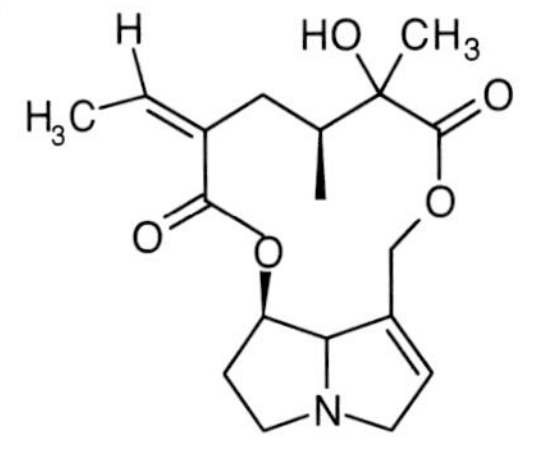

Senecionine

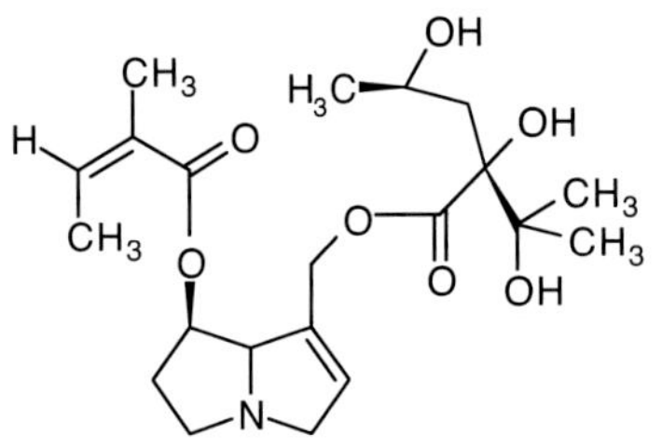

Echimidine

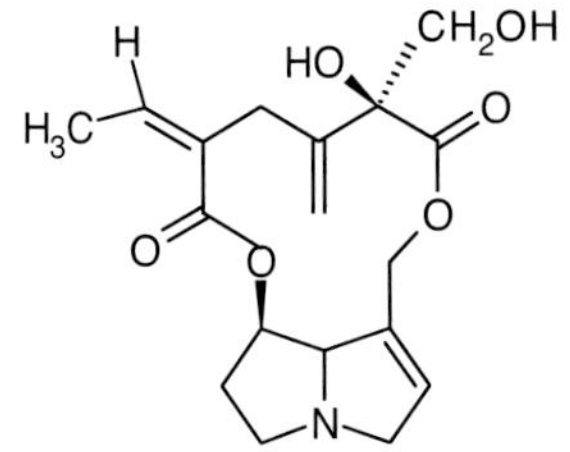

Riddelliine

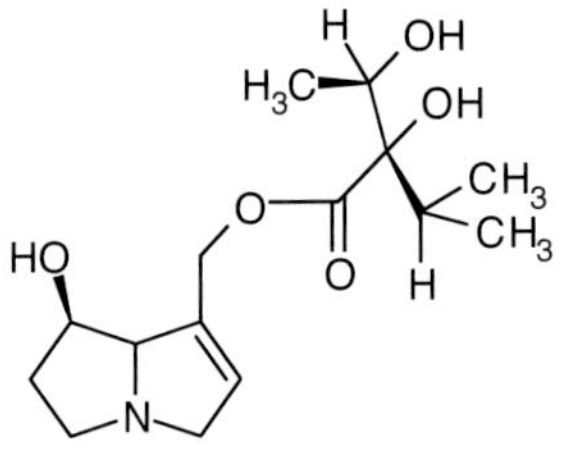

Lycopsamine

Fig. 2.2 Typical pyrrolizidine alkaloids.

The necine base alcohols are normally esterified with any of a series of characteristic (necic) acids to form pyrrolizidine alkaloids. The esterification may be in the form of C-1 mono-esters, open-chain diesters or, more frequently, a macrocyclic diester. The substituting acids are mostly highly branched chains of five to ten carbon atoms substituted with methyl, methylene, hydroxyl and/or keto groups, producing several relatively complex alkaloids. Summaries of the structures of many of the pyrrolizidine alkaloids, especially those most commonly associated with human toxicity, have been assembled by Rizk (1991b), Hartmann and Witte (1995), and Roeder (1995). Over 350 pyrrolizidine alkaloids have so far been identified and characterized, not including the N-oxide forms. Examples of some important representatives are shown in Fig. 2.2.

2.3   OCCURRENCE

Pyrrolizidine alkaloids are found mainly in the families Compositae (Asteraceae), Boraginacea, and Leguminosae, but also in Apocyanacae, Ranunculacae, and Scrophulariacae. The genera *Senecio, Eupatorium, Symphytum, Cynoglossum, Heliotropium*, and *Crotalaria* contain the species most frequently associated with human illness. These genera are widely distributed throughout different climates and their pyrrolizidine alkaloid-containing species could comprise as much as 3% of the world's flowering plants (Smith and Culvenor, 1981). Comprehensive lists of plants containing unsaturated pyrrolizidine alkaloids have been published (Smith and Culvenor, 1981; Mattocks, 1986; Rizk, 1991b; Hartmann and Witte, 1995).

Each species contains an unusually characteristic range of pyrrolizidine alkaloids and a specific ratio of free base to N-oxide (Molyneux and James, 1990). Some species may contain essentially only a single pyrrolizidine alkaloid, notably riddelliine in *Senecio riddelli*, but most contain over five.

Alkaloid content varies between species, plant organ, site, and season, and can be up to several percent of the plant's dry weight. In general, levels are considerably higher in roots than in leaves and they are higher in buds, inflorescences and young leaves than in older leaves, and lower still in stems. In some species, notably *Crotalaria*, high levels (up to 5% dry weight) are often found in the seed (Johnson and Molyneux, 1984; Johnson *et al.*, 1985), presenting a serious threat to health in cases of contamination of grain intended for human consumption. Several necine bases lack the 1,2-unsaturation and their alkaloids are relatively nontoxic. These saturated pyrrolizidine alkaloids may be found associated with their toxic counterparts in particular species, but they are much less widespread.

2.3.1   Formation and function

The biosynthetic routes of some of the major pyrrolizidine alkaloids have been studied in detail (Hartmann and Ober, 2000). In the biosynthesis of the base retronecine, depending on plant species, either L-ornithine or L-arginine or both are combined to form two molecules of putrescine, and the biosynthesis proceeds via homospermidine.

The necic acids are synthesized from the α-amino acids L-valine, L-leucine, L-isoleucine, and L-threonine by a route compiled by Roeder (1995). These reactions occur in the roots of the plant where the primary product in most species studied is the N-oxide (Hartmann and Toppel, 1987; van Dam *et al.*, 1994). The oxide is much more water-soluble than the base and can be transported within the plant. The mechanism and purpose of pyrrolizidine alkaloid transport are not yet well understood, but the alkaloids are specifically channeled via the phloem to the younger leaves and flowering parts of the plant, where they accumulate. The purpose of this is possibly to provide important tissues with a deterrent towards herbivores. Many grazing animals avoid eating plants that contain pyrrolizidine alkaloids, although sheep and goats have some tolerance. Some insects, particularly certain moths and butterflies, accumulate the toxins, which they can then use as a defense against predators or as intermediates in the production of substituted carbonyl pyrrolizidines that act as pheromones. Concise and informative reviews have been published describing the varied functions and uses of pyrrolizidine alkaloids by insects (Boppré, 1990; Hartmann and Witte, 1995).

2.4   EXPOSURE

The two most significant sources of exposure of humans to pyrrolizidine alkaloids are the accidental contamination of foodstuffs and the intentional ingestion of plants containing the alkaloids in the form of culinary vegetables or herbal medicines. The incidence of pyrrolizidine poisoning of humans has probably been underestimated owing to the lack of association between plants and disease, poor recognition of chronic effects, and the time lag between ingestion and the appearance of symptoms in subacute poisoning (Roitman, 1983). Many plants that contain pyrrolizidine alkaloids are deliberately consumed as food or herbal remedies in all parts of the world and reports of toxins in materials are increasing in number (Mattocks, 1986; Hirono, 1993; Roeder, 1995, 2000; Bertram *et al.*, 2001; Fu *et al.*, 2002b).

2.4.1 Contamination of foods

Direct accidental contamination of grain with seed from pyrrolizidine alkaloid-containing plants occasionally leads to a major incident. The plants usually responsible are *Heliotropium lasiocarpum*, *H. popovii* and *H. europaeum* which grow well with wheat (Prakash *et al.*, 1999). In 1976 in Afghanistan over 20% of a population of 7200 villagers who had consumed wheat contaminated with the seed of *Heliotropium* showed signs of liver disease (Mohabbat *et al.*, 1976). The wheat had been consumed over a period of 2 years with an estimated minimum intake of 1.5 g of the alkaloid in the form of the N-oxide. In the previous year, contamination of local grain heavily contaminated with seeds of *Crotalaria* species was responsible for the death of 28 patients from 67 people affected in four villages in India (Tandon *et al.*, 1976). A more recent outbreak (1992–1993) was precipitated by a famine and a delay in the wheat harvest which allowed growth of *Heliotropium* within the wheat crop. It has been documented in some detail (Chauvin *et al.*, 1994).

The use of plants containing pyrrolizidine alkaloids as foods is limited mostly to Japan where *Petasites*, *Tussilago*, *Symphytum*, and *Farfugium* are consumed (Hirono, 1993); however, there is a paucity of information on the exposure to the alkaloids from these sources.

Contamination of foods on a smaller scale occurs where there is an intermediate agent between the plant source and the foodstuff. Examples of this are the transfer of pyrrolizidine alkaloids from plants into milk by herbivores, into eggs via chickens, and into honey by bees.

Contamination of milk and its effects on suckling animals has been reviewed by Panter and James (1990), most studies not showing serious effects on the offspring. The proportion of pyrrolizidine alkaloids passed into milk from goats fed *Senecio jacobaea* (ragwort) was determined to be about 0.1% of that ingested (Deinzer *et al.*, 1982); however, there is evidence that water-soluble metabolites (Eastman *et al.*, 1982) and pyrrolizidine N-oxides (Candrian *et al.*, 1991) are present. With the pooling of milk samples being widespread practice, any risk to health is probably confined to the consumption of milk from individual animals and particularly species such as goats, which are comparatively willing to eat pyrrolizidine-containing plants (Molyneux and James, 1990).

In a report of contamination in hens' eggs, levels of a number of pyrrolizidines in eggs laid by hens accidentally poisoned with seeds of *Heliotropum* and other pyrrolizidine-containing plants reached about 40 μg per egg (Edgar and Smith, 2000).

Relatively high concentrations (up to 4 mg/kg) of pyrrolizidines have been measured in honey produced from *S. jacobaea* (Deinzer and Thompson, 1977; Crews *et al.*, 1997) and *Echium plantagineum* (Culvenor *et al.*, 1981). Pioneering studies into this problem have been carried out in Australia at the Commonwealth Scientific and Industrial Research Organisation (CSIRO) where the occurrence of a range pyrrolizidine alkaloids from *Heliotropium* and *Echium* have recently been confirmed in Australian honeys at levels of about 2 mg/kg (Beales *et al.*, 2004; Betteridge *et al.*, 2005). In cases where honey is contaminated, the plant responsible is usually the predominant nectar source, during its flowering season, in the locality of the hives (Fig. 2.3). There is therefore considerable potential for the localized contamination of honey. Transfer has been associated with the nectar but pollen, which is actively collected by bees and is a constituent of honey, has also been shown to be a major site of pyrrolizidine alkaloid storage in the plant (Boppré *et al.*, 2005). Honeys produced in study hives situated adjacent to sites in the UK where *Senecio jacobaea* flowered were found to contain *Senecio* pollen, and alkaloids at concentrations ranging from 0.01 to 0.06 mg/kg (Crews *et al.*, 1997).

Human exposure to pyrrolizidine alkaloids from contaminated honey is considered to be a problem and has been reviewed in detail (Edgar *et al.*, 2002). A wide range of plants containing pyrrolizidine alkaloids make a significant contribution to honey production

Fig. 2.3 Wild plants of *Echium vulgare* in New Zealand. A rich monofloral source of nectar and pyrrolizidine alkaloids. Photo courtesy of Barrie Wills. For a color version of this figure, please see Plate 1 of the color plate section that falls between pages 224 and 225. (Reprinted with permission from Betteridge *et al.* in *Journal of Agricultural and Food Chemistry*, **53**, 1894–1902, Figure 4. Copyright 2005 with permission from the American Chemical Society.)

world-wide. Europeans eat an average of about 1 g of honey per capita per day but some consumers, including infants in the UK (Ministry of Agriculture, Fisheries and Food, 1995), and in Australia (Edgar *et al.*, 2002), can consume far more than this.

Another potential dietary source of pyrrolizidine alkaloids is oil obtained from borage seed, *Borago officinalis*, which is popular in Europe on account of its high content of the beneficial gamma-linolenic acid. However, the pyrrolizidine content of this plant is very low (Larson *et al.*, 1984; Lüthy *et al.*, 1984) and that of its oil even lower (Wretensjo and Karlberg, 2003). The European borage should not be confused with the various and more toxic colloquially named borages derived from *Echium* species found in New Zealand.

Numerous reports of poisonings directly related to herbal teas and similar products have been published (Weston *et al.*, 1981; Kumana *et al.*, 1985; Margalith *et al.*, 1985; Ridker *et al.*, 1985; Culvenor *et al.*, 1986; Ridker and McDermott, 1989). The practice in Jamaica of brewing teas from uncultivated plants for medicinal purposes (bush teas) has been associated with epidemics of pyrrolizidine poisoning (Bras *et al.*, 1954). However, government campaigns there have led to a reduction in the frequency of these incidents.

2.4.2 Pyrrolizidines in herbal preparations

Increasing interest in "alternative" therapies and herbal medicines in Europe and the USA has led to preparations of pyrrolizidine alkaloid-containing plants being made widely

Fig. 2.4 Herbal products sold in the UK derived from pyrrolizidine-containing plants.

available commercially and publicized for their health-giving properties. Comfrey (*Symphytum officinale*), in particular, has long been a popular herb in Europe and the USA. Preparations of comfrey in the form of dried leaves, dried root, and root powder tablets and capsules, often mixed with other herbs, have been sold with active promotion of the plant's supposed healing and digestive properties. Some examples of herbal products sold in the UK which are derived from plants likely to contain pyrrolizidine alkaloids are shown in Fig. 2.4. They include tinctures and flowers from coltsfoot (*Tussilago*) and teas, leaf, and root material from comfrey, with the latter bearing a suitable health warning.

Herbal preparations of *Symphytum*, *Tussilago*, *Borago*, and *Eupatorium*, in the form of leaf, root powders, tablets and root extract tinctures, sold in the UK in 1994 were surveyed for pyrrolizidine alkaloid content (Ministry of Agriculture, Fisheries and Food, 1994). Comfrey (*Symphytum*) tablets contained up to 5000 mg/kg and root powders up to 8300 mg/kg of pyrrolizidine alkaloids, giving estimated potential intakes in excess of 35 mg/day. Comfrey and borage (*Borago*) leaf preparations intended for consumption as teas contained less than 100 mg/kg total pyrrolizidine alkaloids. About 50% of the total acetyllycopsamine and symphytine but only about 5% of the lycopsamine were extracted into the water on brewing comfrey leaf teas, possibly due to binding of the more polar lycopsamine to the plant tissue.

A survey of comfrey leaf and root products sold in the USA in 1989 showed them to contain up to 1200 mg/kg of pyrrolizidine alkaloids (Betz *et al.*, 1994). Teas prepared from comfrey root and leaf showed preferential extraction of acetyllycopsamine and acetylintermedine over lycopsamine and intermedine. Later studies confirmed that pyrrolizidine alkaloids were present in many comfrey preparations sold in the USA (Altamirano *et al.*, 2005). In one case where the N-oxides were reduced prior to determination the level of symphytine measured increased from 0.1 to 1 mg/L (Oberlies *et al.*, 2004).

More recently, political changes in China have made aspects of Chinese culture more accessible to the West and this has raised the popularity of traditional Chinese remedies. This activity has been followed by studies of the composition of such medicines which has revealed that of the wide range of herbal plants used in China, about 50 have so far been identified as containing toxic pyrrolizidine alkaloids (Zhao *et al.*, 1989; Roeder, 2000; Fu *et al.*, 2002b).

The misidentification of plants is an additional risk to consumers of vegetables and herbs. This is particularly likely to occur where the intended plant is closely related to a more toxic species. For example, Sperl *et al.* (1995) reported a poisoning case in which *Adenostyles* had been gathered and consumed in place of the less toxic *Tussilago*. Similar and additional problems of herbal products have been described by Huxtable (1990a).

2.5 REGULATIONS

Regulations and recommendations have been introduced in some countries in attempts to limit human exposure. Regulations introduced in Germany by the Federal Health Bureau (Germany Federal Health Bureau, 1992) limit the tolerable pyrrolizidines in herbal medicines to levels providing less than 1 µg per day orally based on an assessment of genotoxic carcinogenicity, reduced to 0.1 µg per day when used for over 6 weeks. The regulations are notable for the fact that the limits can easily be exceeded by consumption of contaminated foods such as eggs and honey, nevertheless similar limits are likely to be adopted across Europe (Roeder, 2000). Food Standards Australia New Zealand (FSANZ) has set a provisional exposure level of 1 µg/kg body weight per day and advised heavy consumers not to eat honey from *Echium plantagineum* every day (FSANZ, 2004). Herbal preparations containing comfrey root were removed voluntarily from the market in the UK (Ministry of Agriculture, Fisheries and Food, 1994) and in the USA (FDA).

2.6 TOXICITY AND METABOLISM

2.6.1 General toxicity

The toxicity and metabolism of pyrrolizidine alkaloids have been studied and reviewed (McLean, 1970; Mattocks, 1986; Winter and Segall, 1989; Huxtable, 1990b; Segall *et al.*, 1991; Cheeke and Huan, 1995). The main details are given here.

Ingestion of pyrrolizidine alkaloids has long been associated with severe damage to the liver. The hepatic veins become blocked by a build up of connective tissue, a condition known as veno-occlusive disease. Giant cells, which are dysfunctional, develop in the liver through the intense antimitotic activity of pyrrolizidine alkaloids preventing the completion of cell division. The condition frequently leads to death. The symptoms have been shown to be identical to those of Budd–Chiari syndrome, an illness characterized by thrombosis of the hepatic veins that leads to enlargement of the liver, ascites, and portal hypertension (Bull *et al.*, 1968; McLean, 1970).

Long-term, sub-lethal doses of pyrrolizidine alkaloids are usually associated with cattle exposed during grazing to plants containing them or with humans who habitually take herbal medicines. Long-term exposure to low levels of pyrrolizidine alkaloids may cause cumulative damage to body organs and cancer, but even exposure to high levels is not necessarily fatal.

Toxicity is associated with unsaturation of the pyrroline ring, the presence of one or two hydroxyl groups on the pyrroline ring, esterification of one or both of these groups, and branching of the esterifying acid chain (Prakash *et al.*, 1999).

2.6.2   Metabolism

Bioactivation of pyrrolizidine alkaloids is required to produce toxicity, and this is effected by P450 enzymes. Activation of toxins in the liver results from the formation of polar compounds that are conjugated to glutathione. This produces hydrophilic compounds that can be readily excreted, the intended effect being detoxification. In the case of pyrrolizidine alkaloids some of these metabolites are highly reactive and toxic.

The metabolic reactions are hydrolysis to the parent base and acids, oxidation by microsomal oxygenases to form the N-oxide, and dehydrogenation to form pyrroles. This is represented diagrammatically in Fig. 2.5. Hydrolysis forms necines and necic acids with no toxicity. Dehydrogenation forms pyrrole esters which are highly reactive and toxic to the liver, and which can also be hydrolyzed to alcoholic pyrroles having mutagenic and carcinogenic activity (Prakash *et al.*, 1999). Pyrroles can bind to tissue nucleophiles such as glutathione and DNA.

The relative toxicity of pyrrolizidine alkaloids towards different animals varies dramatically between animal species. Guinea pigs, hamsters, gerbils, and quails show considerable resistance (Cheeke and Pierson-Goeger, 1983; Cheeke, 1994). Differences in available esterase and oxygenase enzymes are believed to be related to the variability in toxicity towards different animal species and sexes. Experiments showing that metabolic detoxification pathways vary according to the animal species and the alkaloid involved have been summarized by Cheeke (1994) and Prakash *et al.* (1999). These variations are most likely to be due to differences between the animals' P450 enzyme systems, which catalyze the pyrrole production reactions and the conjugation reactions at different relative rates in different species (Cheeke and Huan, 1995).

Pyrrolizidine N-oxides have the same toxicity in principle as their bases but as these are considerably more hydrophilic it would be expected that their rapid excretion would make them far less important as toxins. However, it has been shown recently that their carcinogenicity on oral administration equals that of the parent alkaloid (Chou *et al.*, 2003).

2.6.3   Carcinogenicity and mutagenicity

There is considerable evidence from animal studies using plants and isolated compounds, and from determination of DNA adducts, that pyrrolizidine alkaloids have carcinogenetic activity (Cook *et al.*, 1950; Schoental, 1968; International Agency for Research on Cancer, 1983). An extensive investigation has been made showing the genotoxicity and carcinogenicity of a typical pyrrolizidine alkaloid, riddelliine (Chan, 1993), and the subject has been summarized in reviews by Fu *et al.* (2001, 2002a). Considerable progress has been made towards understanding the mechanisms in more detail. For example, adducts identified from the reaction of the toxic metabolite dehydroretronecine (derived from riddelliine) with calf thymus DNA have been isolated and separated by high performance liquid chromatography (HPLC),

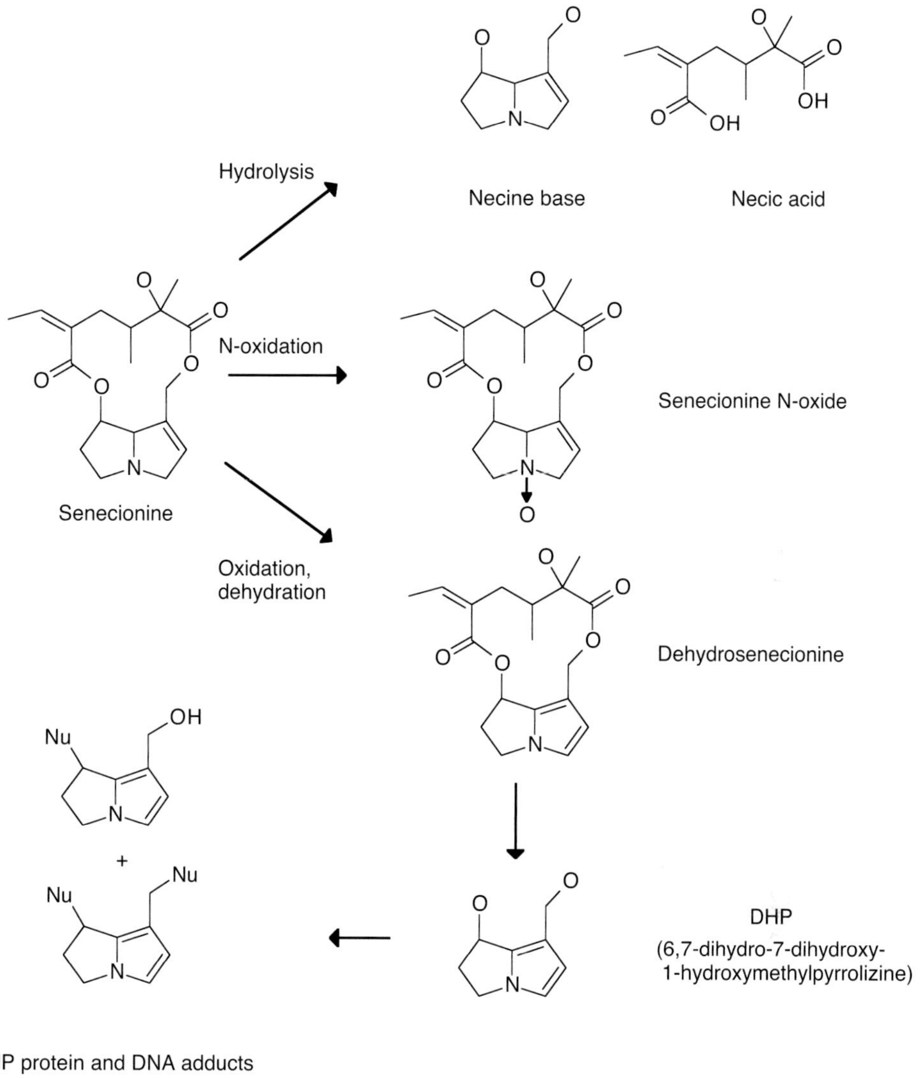

Necine base   Necic acid

Hydrolysis

N-oxidation

Senecionine N-oxide

Senecionine

Oxidation,
dehydration

Dehydrosenecionine

DHP
(6,7-dihydro-7-dihydroxy-
1-hydroxymethylpyrrolizine)

DHP protein and DNA adducts

Fig. 2.5   Metabolic pathways for the pyrrolizidine alkaloid senicionine. Adapted from Stegelmeier et al. (1999), Prakash et al. (1999), and Fu et al. (2002b).

characterized, and identified as identical to those from rats treated with riddelliine (Fu *et al.*, 2001; Yang *et al.*, 2001).

2.7   ANALYTICAL METHODS

Several authors have reviewed analytical methodology for pyrrolizidine alkaloids with concentration on their determination in certain sample types, e.g. plants (Roeder, 1999), honey (Edgar *et al.*, 2002), and human liquids and tissues (Stewart and Steenkamp, 2001).

The methods include visualization by color reaction, thin layer chromatography (TLC), gas chromatography (GC), and HPLC. In recent years, HPLC with tandem mass spectrometric detection has come to prominence in this as in many other fields. These, and some other extraction and determination methods are described below. Quantitative analysis has long been compromised by a lack of commercially available pure standards. Only monocrotaline, retrorsine, and senecionine are readily available commercially and preparative isolation techniques, which have been based mainly on countercurrent chromatography (Kim *et al.*, 2001) or multiple development TLC (Mroczek *et al.*, 2006), remain rather cumbersome.

2.7.1 Extraction

Extraction of pyrrolizidine alkaloids from plant materials, medicines, biological materials, and food has usually followed two routes, based on the use of mid-polarity solvents such as chloroform, or the use of acidified alcohol or water. In the former the material is extracted using a Soxhlet apparatus or by soaking and/or boiling, and the organic solution obtained can be reduced in volume easily by evaporation. Even so, subsequent solution in acid has often been used to enable reduction of the N-oxides to free bases by addition of zinc dust. The acid solution obtained by either approach can be washed with nonpolar organic solvent to remove lipids, and made strongly alkaline with ammonia to enable partition of the alkaloids back into organic solution for analysis.

Solid phase extraction (SPE) columns have been used to purify plant and food extracts containing pyrrolizidine alkaloids. Those based on diatomaceous earth have been applied to plants (Witte *et al.*, 1993; Mossoba *et al.*, 1994) and to honey (Crews *et al.*, 1997). Ergosil, a surface-treated silica gel, preferentially binds certain classes of alkaloids and allows isolation and concentration of free base pyrrolizidine alkaloids while pigments and interfering peaks are easily washed out (Gray *et al.*, 2004). SPE cartridges containing strong cation exchange medium have been applied to the extraction of pyrrolizidine alkaloids from plant tissues (Chizzola, 1994; Mroczek *et al.*, 2002), and from honey (Beales *et al.*, 2004; Betteridge *et al.*, 2005). Pyrrolizidine alkaloids are recovered from these columns with methanol containing dissolved ammonia. The cation-exchange SPE columns offer a major advantage when combined with HPLC separation in that the alkaloid free bases and N-oxides can be isolated and chromatographed directly in a single run without prior reduction of the oxides.

Where reduction techniques are preferred, zinc dust, which requires several hours to complete may decompose and parts of the N-oxides (Stelljes *et al.*, 1991) can be replaced by sodium dithionite as the salt (Crews *et al.*, 1997) or bound to a resin (Chizzola, 1994; Colegate *et al.*, 2005).

2.7.2 Gas chromatography

Pyrrolizidine alkaloids can be determined directly by GC of organic solutions of the free bases or their derivatives (Lüthy *et al.*, 1981; Deinzer *et al.*, 1982; Bicchi *et al.*, 1985, 1989a; Kelley and Seiber, 1992; Witte *et al.*, 1993; Conradie *et al.*, 2005). Witte *et al.* (1993) demonstrated that about 100 pyrrolizidine alkaloids can be separated and identified without derivatization using GC with mass spectrometry (GC/MS) in combination with retention time data on a low polarity methylpolysiloxane stationary phase (OV-1, DB-1). More polar stationary phases such as 50% phenylmethylpolysiloxane (DB-17), have also been applied to the separation of pyrrolizidine alkaloids (Stelljes *et al.*, 1991).

An alternative approach in which total levels of pyrrolizidine alkaloids are determined is to subject the alkaloids to mild hydrolysis and determine the retronecine base by gas chromatography after derivatization (Hovermale and Craig, 1998).

GC/MS with tandem MS detection has been used less, and mainly in the examination of metabolites (Winter *et al.*, 1988; Schoch *et al.*, 2000). Chemical ionization (CI) techniques have similarly found limited application (Lüthy *et al.*, 1981; Bicchi *et al.*, 1989a; Mossoba *et al.*, 1994). As an alternative to MS for retrieving molecular-specific information, GC has been coupled to matrix isolation/Fourier transform infrared spectroscopy, which uses the characteristic fingerprint region of the FTIR spectra for compound identification (Bicchi *et al.*, 1989b; Mossoba *et al.*, 1994). Still, unequivocal identification was only obtained either when reference standards were available, or when GC/MS was used for confirmation.

The alkaloids are usually analyzed without derivatization but peak shape and resolution are improved by derivatization, which also reduces decomposition in the injector. The most frequently used derivatives are the trimethylsilyl ethers formed by reaction with bis-(trimethylsilyl)trifluoroacetamide (BSTFA) but reaction with methane or butane boronic acids is also very useful, alkyl boronates readily forming stable cyclic derivatives across pairs of hydroxyl groups that are in reasonably close proximity. Winter *et al.* (1988) presented a method for the simultaneous determination of parent alkaloids, N-oxides, and hydrolytic metabolites by analyzing their trimethylsilyl derivatives with tandem MS and GC/MS.

2.7.3   High performance liquid chromatography

Analytical methods for pyrrolizidine alkaloids based on HPLC separation offer several advantages over GC, including the possibility to simultaneously determine the free bases, the thermally labile N-oxides and pyrrolizidine alkaloid metabolites. HPLC methods are used for both analysis and semi-preparative isolation of pyrrolizidine alkaloids. For these purposes, various reversed-phase (RP) column materials, e.g. C_{18} (Lin *et al.*, 1998; Cui and Lin, 2000; Mroczek *et al.*, 2004a), C_8 (Parker *et al.*, 1990), phenyl (Kedzierski and Buhler, 1986; Crews *et al.*, 1997), and cyano-bonded (Brown *et al.*, 1994) packing materials have been used, usually with either a basic binary mobile phase comprising ammonium hydroxide and acetonitrile, or an acidic mobile phase consisting of formic/acetic acid or their ammonium salts in methanol or acetonitrile. Gray *et al.* (2004) systematically investigated the performance of some C_{18}, C_8, and phenyl packing materials under different mobile phase compositions with regard to peak shape and resolution. In order to overcome the problem of peak asymmetry with RP columns, Mroczek *et al.* (2002) successfully applied ion-pairing to the separation of pyrrolizidine alkaloids with various structures (N-oxides, free bases, otonecine-based pyrrolizidine alkaloids).

Several HPLC methods have been reported for pyrrolizidine alkaloid analysis (Kedzierski and Buhler, 1986; Brown *et al.*, 1994; Mroczek *et al.*, 2002), but conventional UV detectors lack the requisite sensitivity and specificity. An alternative to UV detection for pyrrolizidine alkaloids not having chromophores in their structure was presented by Schaneberg *et al.* (2004) who used evaporative light scattering detection (ELSD) for the simultaneous determination of nine pyrrolizidine alkaloids and N-oxides including the poorly UV-sensitive heliotrine. However, MS is the detector of choice in terms of specificity and sensitivity.

Various interfaces have been used to couple LC to MS such as thermospray (TSP) (Ndjoko *et al.*, 1999), atmospheric pressure chemical ionization (APCI) (Crews *et al.*, 1997; Beales *et al.*, 2004; Gray *et al.*, 2004; Mroczek *et al.*, 2004a, 2006), electrospray ionization (ESI) (Lin *et al.*, 1998; Altamirano *et al.*, 2005; Betteridge *et al.*, 2005; Boppré *et al.*, 2005;

Colegate *et al.*, 2005), and thermabeam (TB) (Mroczek *et al.*, 2004b). A helpful review has been published by Sherma (2003), who summarized the state-of-the-art in HPLC/MS techniques currently in use for the analysis of botanical medicines and dietary supplements.

Currently, atmospheric pressure ionization (API) techniques, i.e. ESI and APCI, are pre-eminent. These are both "soft" ionization techniques that provide molecular weight but no structural information as the degree of fragmentation is low. Structural information may be obtained with API interfaces by collision-induced dissociation (CID), which can be either accomplished in the ion source of single-stage MS (in-source CID) or, more elegantly, with multiple-stage or tandem MS using triple quadrupole (TQ) or ion trap (IT) instruments.

The TB interface also provides structural information by creating a desolvated analyte "particle beam" from the LC effluent that may subsequently be ionized by electron impact (EI) ionization, however, at much less sensitivity than e.g. TQ instruments. Mroczek *et al.* (2004b) used LC/EI-MS with a TB interface to screen plant extracts for toxic pyrrolizidine alkaloids. Lin *et al.* (1998) used a quadrupole instrument to obtain structural information of pyrrolizidine alkaloids in blood samples from rats by performing both in-source CID (CID-HPLC/MS) and fragmentation of parent ions in the collision cell (HPLC/MS/MS).

The coupling of HPLC with IT-MS and with MS/MS via API interfaces has found wide application in analysis on account of the many experimental possibilities this type of instrumentation offers (Beales *et al.*, 2004; Wuilloud *et al.*, 2004; Mroczek *et al.*, 2004a, 2006; Boppré *et al.*, 2005; Colegate *et al.*, 2005), and fragmentation patterns have been described for a large number of pyrrolizidine alkaloids in the cited publications. An example of HPLC combined with MS applied to the analysis of pyrrolizidine alkaloids in honey is shown in Fig. 2.6.

2.7.4   Other methods

Nuclear magnetic resonance (NMR) has long been used for the determination and confirmation of the structure of pyrrolizidine alkaloids. Comprehensive collections of ^{1}H NMR (Logie *et al.*, 1994) and ^{13}C NMR (Molyneux *et al.*, 1982; Roeder, 1990) data derived from studies of a large number of pyrrolizidine alkaloids have been published. Both approaches have been used quantitatively (Pieters and Vlietinck, 1985, 1987; Pieters *et al.*, 1989a), and they have been compared with GC and HPLC methods by Pieters *et al.* (1989b) and Pieters and Vlietinck (1986). The proton NMR spectra of plant extracts are often complex and quantification difficult, therefore, ^{1}H NMR is better suited for the determination of total pyrrolizidine alkaloid levels while ^{13}C NMR, with its better resolving power, can be used for quantitative analysis of individual pyrrolizidine alkaloids. Quantitative results of chromatographic and NMR methods show good agreement, but low sensitivity and long accumulation times of NMR measurements pose main disadvantages and limit its application to sample extracts with high pyrrolizidine alkaloid content and rather large sample sizes of about 10 mg. An interesting method using direct online coupling between HPLC, MS, and ^{1}H NMR was developed for the detection and identification of pyrrolizidine alkaloids in crude plant extracts where HPLC/^{1}H NMR analysis delivered complementary information to MS that helps to distinguish isomeric structures (Ndjoko *et al.*, 1999). However, due to low sensitivity, higher amounts of extract (mg instead of μg) need to be injected to obtain good quality onflow HPLC/^{1}H NMR spectra.

During the past decade, enzyme-linked immunosorbent assays (ELISA) have become generally accepted and used for pyrrolizidine alkaloid analysis (Roeder, 1999; Than *et al.*, 2005). Various ELISAs have been developed for screening different matrices for a range of

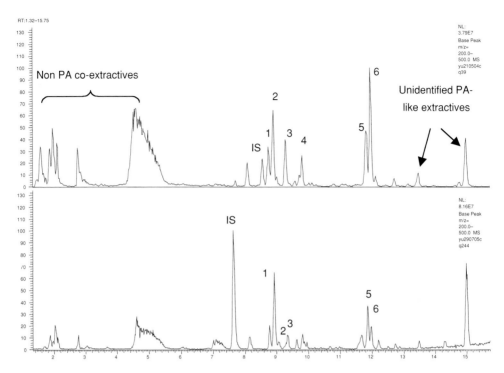

Fig. 2.6 HPLC ESI MS chromatogram showing about 2500 µg of pyrrolizidine alkaloids (and their N-oxides) per kg in two *Echium vulgare* honey samples. IS, internal standard (senecionine for the top chromatogram and heliotrine for the bottom). The alkaloids identified included (1) echimidine, (2) echimidine-N-oxide, (3) vulgarine-N-oxide, (4) acetyl echimidine-N-oxide, (5) echivulgarine, and (6) echivulgarine-N-oxide.

pyrrolizidine alkaloids, with the aim of achieving high sensitivity (detection limits in the ng and pg range), high sample throughput, simplicity, and reliability. Pyrrolizidine alkaloid molecules, however, are generally too small to react as immunogens. They have to be linked to larger complexes by forming *N*-hydroxysuccinimide derivatives via the alkaloids' hydroxyl groups and subsequent protein conjugates via the succinic acid ester group in order to create a reasonable immune response (Zuendorf *et al.*, 1998; Lee *et al.*, 2001). Roseman *et al.* (1996) developed a different approach by using quaternary pyrrolizidinium salts as immunogens. Most ELISAs are competitive assays using polyclonal antibodies (Roeder and Pflueger, 1995; Roseman *et al.*, 1996; Lee *et al.*, 2001), although some work has been published on monoclonal antibodies (Zuendorf *et al.*, 1998). As the pyrrolizidine alkaloid antibodies usually do not effectively cross-react with the respective N-oxides an additional reduction step during sample preparation is required to enable determination of the total pyrrolizidine alkaloid content.

A relatively large number of color reactions can indicate the presence of pyrrolizidine alkaloids. The simpler reactions are not specific, detecting all alkaloids or tertiary ones. Some more complex procedures have much improved specificity. The techniques have been well summarized by Roeder (1999). Quantitative determination of pyrrolizidine alkaloids by spectrophotometry commonly applies a color reaction that involves pyrrole formation and subsequent reaction with *Ehrlich* or *Mattocks* reagent (4-dimethylaminobenzaldehyde) and allows the determination of pyrrolizidine alkaloids at concentrations levels down to 1 μg/g (Mattocks, 1967a; Bartkowski *et al.*, 1997). The same color reactions are used to visualize pyrrolizidine alkaloids on TLC plates (Mattocks, 1967b; Huizing and Malingre, 1981; Chizzola, 1994). Alternatively, *Dragendorff's* iodoplatinate reagent can be used which shows less specificity but ten times higher sensitivity than *Mattocks* reagent (Parvais *et al.*, 1994). Various TLC methods have been published, applying different plates and solvent systems (Roeder, 1999). However, separation efficiency of TLC is low, thus limiting its application mainly to screening purposes. For unambiguous determination of multi-component extracts that contain more than six alkaloids, more efficient separation methods such as GC and HPLC are necessary. Although sensitivity and specificity of photometric and TLC techniques cannot compete with that of GC and HPLC, color reactions offer an easy and fast screening method that indicate the presence of pyrrolizidine alkaloids in a sample down to low μg per kg levels.

Conclusions

Despite our increasing knowledge of their chemistry and toxicity, pyrrolizidine alkaloids continue to be a significant threat to human health. Their plant sources flourish through their diversity and lack of palatability to grazing animals. Education of people in the developing countries has reduced their exposure to harmful traditional herbal medicines. Conversely, a revival of interest in natural and alternative therapies has led to the commercial sale of toxic plant preparations in the more developed countries. The scale of the threat caused by the consumption of pyrrolizidine-containing herbs and vegetables will only be realized fully once the genotoxic action and chronic low-dose toxicity of the major alkaloids are understood more thoroughly. The risk of large-scale poisoning through cereal contamination, however, clearly remains serious in view of the continuing practice of consuming poor quality grain in times of drought and famine.

Investigations into the metabolism of pyrrolizidine alkaloids, the mechanism of their toxicity and their genotoxic activity proceed apace, driven by an increased awareness of their potential for causing disease and the availability of ever more sophisticated techniques for studying the action of toxic chemicals at the cellular and sub-cellular level. Because of the great variety of pyrrolizidine alkaloid structures and the diversity of their effects in different species such studies will continue for a long time.

Such studies have benefited greatly in recent years by substantial increases in the performance of analytical instrumentation, particularly in the fields of NMR and HPLC/MS. The recent advances should make a substantial contribution to our understanding of the magnitude of the pyrrolizidine problem and guide the way to the setting and enforcement of sensible regulations governing human and animal exposure to these important toxicants.

References

Altamirano, J.C., Gratz, S.R. and Wolnik, K.A. (2005) Investigation of pyrrolizidine alkaloids and their N-oxides in commercial comfrey-containing products and botanical materials by liquid chromatography electrospray ionization mass spectrometry. *Journal of AOAC International*, **88**(2), 406–412.

Bartkowski, J.P.B., Wiedenfeld, H. and Roeder, E. (1997) Quantitative photometric determination of senkirkine in *Farfarae folium. Phytochemical Analysis*, **8**(1), 1–4.

Beales, K.A., Betteridge, K., Colegate, S.M. and Edgar, J.A. (2004) Solid-phase extraction and LC-MS analysis of pyrrolizidine alkaloids in honeys. *Journal of Agricultural and Food Chemistry*, **52**(21), 6664–6672.

Bertram, B., Hemm, I. and Tang, W. (2001) Mutagenic and carcinogenic constituents of medicinal herbs used in Europe or in the USA. *Pharmazie*, **56**(2), 99–120.

Betteridge, K., Cao, Y. and Colegate, S.M. (2005) Improved method for extraction and LC-MS analysis of pyrrolizidine alkaloids and their N-oxides in honey: Application to *Echium vulgare* honeys. *Journal of Agricultural and Food Chemistry*, **53**(6), 1894–1902.

Betz, J.M., Eppley, R.M., Taylor, W.C. and Andrzejewski, D. (1994) Determination of pyrrolizidine alkaloids in commercial comfrey products (*Symphytum* sp.). *Journal of Pharmaceutical Sciences*, **83**(5), 649–653.

Bicchi, C., D'Amato, A. and Cappelletti, E. (1985) Determination of pyrrolizidine alkaloids in *Senecio inaequidens* D.C. by capillary gas chromatography. *Journal of Chromatography*, **349**(1), 23–29.

Bicchi, C., Caniato, R., Tabacchi, R. and Tsoupras, G. (1989a) Capillary gas-chromatography positive and negative-ion chemical ionization mass-spectrometry on pyrrolizidine alkaloids of *Senecio inaequidens* using ammonia and hydroxyl ions as the reagent species. *Journal of Natural Products*, **52**(1), 32–41.

Bicchi, C., Rubiolo, P. and Frattini, C. (1989b) Capillary gas chromatography-Fourier transform infrared spectroscopy of pyrrolizidine alkaloids of *Senecio inaequidens* DC. *Journal of Chromatography*, **473**(1), 161–170.

Boppré, M. (1990) Lepidoptera and pyrrolizidine alkaloids, exemplification of complexity in chemical ecology. *Journal of Chemical Ecology*, **16**(1), 165–185.

Boppré, M., Colegate, S.M. and Edgar, J.A. (2005) Pyrrolizidine alkaloids of *Echium vulgare* honey found in pure pollen. *Journal of Agricultural and Food Chemistry*, **53**(3), 594–600.

Bras, G., Jellife, D.B. and Stuart, K.L. (1954) Veno-occlusive disease of the liver with non-portal type of cirrhosis, occurring in Jamaica. *Archives of Pathology*, **57**, 285–300.

Brown, M.S., Molyneux, R.J. and Roitman, J.N. (1994) A general method for high-performance liquid chromatography of pyrrolizidine alkaloid free bases and N-oxides. *Phytochemical Analysis*, **5**(5), 251–255.

Bull, L.B., Culvenor, C.C.J. and Dick, A.T. (1968) *The Pyrrolizidine Alkaloids: Their Chemistry, Pathogenicity and other Biological Properties*. North-Holland, Amsterdam.

Candrian, U., Zweifel, U., Lüthy, J. and Schlatter, C. (1991) Transfer of orally-administered [H-3] seneciphylline into cows milk. *Journal of Agricultural and Food Chemistry*, **39**(5), 930–933.

Chan, P.C. (1993) NTP Technical Report on toxicity studies of riddelliine, administered by gavage to F344/N rats and B6C3F$_1$ mice. NIH publication No. 94-3350.

Chauvin, P., Dillon, J.C. and Moren, A. (1994) An outbreak of Heliotrope food poisoning, Tadjikistan, November 1992–March 1993. *Cahiers Santé*, **4**, 263–268.

Cheeke, P.R. (1989) *Toxicants of Plant Origin. Vol. 1: Alkaloids*. CRC Press, Boca Raton, FL.

Cheeke, P.R. (1994) A review of the functional and evolutionary roles of the liver in the detoxification of poisonous plants, with special reference to pyrrolizidine alkaloids. *Veterinary and Human Toxicology*, **36**(3), 240–247.

Cheeke, P.R. and Pierson-Goeger, M.L. (1983) Toxicity of *Senecio jacobaea* and pyrrolizidine alkaloids in various laboratory animals and avian species. *Toxicological Letters*, **18**(3), 343–350.

Cheeke, P.R. and Huan, J. (1995) Metabolism and toxicity of pyrrolizidine alkaloids, in *Phytochemicals and Health. Current Topics in Plant Physiology*, Vol. 15 (eds. D.L. Gustine and H.E. Flores). American Society of Plant Physiologists, Rockville, MD, pp. 155–164.

Chizzola, R. (1994) Rapid sample preparation technique for the determination of pyrrolizidine alkaloids in plant extracts. *Journal of Chromatography A*, **668**(2), 427–433.

Chou, M.W., Wang, Y.-P., Yan, J., Yang, J.C., Berger, R.D., Williams, L.D., Doerge, D.R. and Fu, P.P. (2003) Riddelliine N-oxide is a phytochemical and mammalian metabolite with genotoxic activity that is comparable to the parent pyrrolizidine alkaloid riddelliine. *Toxicology Letters*, **145**(3), 239–247.

Colegate, S.M., Edgar, J.A., Knill, A.M. and Lee, S.T. (2005) Solid-phase extraction and HPLC-MS profiling of pyrrolizidine alkaloids and their N-oxides: A case study of *Echium plantagineum*. *Phytochemical Analysis*, **16**(2), 108–119.

Conradie, J., Stewart, M.J. and Steenkamp, V. (2005) GC/MS identification of toxic pyrrolizidine alkaloids in traditional remedies given to two sets of twins. *Annals of Clinical Biochemistry*, **42**, 141–144.

Cook, J.W., Duffy, E. and Schoental, R. (1950) Primary liver tumours in rats following feeding with alkaloids of *Senecio jacobaea*. *British Journal of Cancer*, **4**, 405–410.

Cooper-Driver, G.A. (1983) Chemical substances in plants toxic to animals, in *Handbook of Naturally Occurring Food Toxicants* (ed. M. Rechcigl). CRC Press, Boca Raton, FL, pp. 213–240.

Crews, C., Startin, J.R.S. and Clarke, P.A. (1997) Determination of pyrrolizidine alkaloids in honey from selected sites by solid phase extraction and HPLC-MS. *Food Additives and Contaminants*, **14**(5), 419–428.

Cui, Y.Y. and Lin, G. (2000) Simultaneous analysis of clivorine and its four microsomal metabolites by high-performance liquid chromatography. *Journal of Chromatography A*, **903**(1–2), 85–92.

Culvenor, C.C.J., Edgar, J.A. and Smith, L.W. (1981) Pyrrolizidine alkaloids in honey from *Echium plantagineum* L. *Journal of Agricultural and Food Chemistry*, **29**(5), 958–960.

Culvenor, C.C.J., Edgar, J.A. and Smith, L.W. (1986) *Heliotropium lasiocarpum* Fisch and Mey identified as cause of veno-occlusive disease due to a herbal tea. *Lancet*, **1**(8487), 978–978.

Deinzer, M.L. and Thompson, P.A. (1977) Pyrrolizidine alkaloids: Their occurrence in honey from Tansy Ragwort (*Senecio jacobaea* L.). *Science*, **195**, 497–499.

Deinzer, M.L., Arbogast, B.L. and Buhler, D.R. (1982) Gas chromatographic determination of pyrrolizidine alkaloids in goat's milk. *Analytical Chemistry*, **54**(11), 1811–1814.

Eastman, D.F., Dimenna, G.P. and Segall, H.J. (1982) Covalent binding of two pyrrolizidine alkaloids, senecionine and seneciphylline, to hepatic macromolecules and their distribution, excretion, and transfer into milk of lactating mice. *Drug Metabolism and Disposition*, **10**, 236–240.

Edgar, J.A. and Smith, L.W. (2000) Transfer of pyrrolizidine alkaloids into eggs: Food safety implications. *Natural and Selected Synthetic Toxins*, **745**, 118–128.

Edgar, J.A., Roeder, E.L. and Molyneux, R.J. (2002) Honey from plants containing pyrrolizidine alkaloids: A potential threat to health. *Journal of Agricultural and Food Chemistry*, **50**(10), 2719–2730.

FSANZ (2004) *Pyrrolizidine Alkaloids in Food: A Toxicological Review and Risk Assessment*. Technical Report Series No. 2. 2001. Available from: http://www.foodstandards.gov.au/mediareleasespublications/technicalreportseries/index.cfm

Fu, P.P., Chou, M.W., Xia, Q., Yang, Y.C., Yan, J., Doerge, D.R. and Chan, P.C. (2001) Genotoxic pyrrolizidine alkaloids and pyrrolizidine alkaloid N-oxides – Mechanisms leading to DNA adduct formation and tumorigenicity. *Journal of Environmental Science and Health Part C – Environmental Carcinogenesis and Ecotoxicology Reviews*, **19**(2), 353–385.

Fu, P.P., Xia, Q., Lin, G. and Chou, M.W. (2002a) Genotoxic pyrrolizidine alkaloids – mechanisms leading to DNA adduct formation and tumorigenicity. *International Journal of Molecular Sciences*, **3**, 948–964.

Fu, P.P., Yang, Y.C., Xia, Q.S., Chou, M.W., Cui, Y.Y. and Lin, G. (2002b) Pyrrolizidine alkaloids – Tumorigenic components in Chinese herbal medicines and dietary supplements. *Journal of Food and Drug Analysis*, **10**(4), 198–211.

Germany Federal Health Bureau (Bundesgesundheitsamt) (1992) Bundesanzeiger, 111, 4805. *Deutsche Apotheker-Zeitung*,**132**, 1406–1408.

Gray, D.E., Porter, A., O'Neill, T., Harris, R.K. and Rottinghaus, G.E. (2004) A rapid cleanup method for the isolation and concentration of pyrrolizidine alkaloids in comfrey root. *Journal of AOAC International*, **87**(5), 1049–1057.

Hartmann, T. (1991) Alkaloids, in *Herbivores, Their Interactions with Secondary Plant Metabolites. Vol. 1: The Chemical Participants*, 2nd edn (eds. G.A. Rosenthal and M.R. Berenbaum). Academic Press, San Diego, pp. 79–121.

Hartmann, T. and Toppel, G. (1987) Senecionine N-oxide, the primary product of pyrrolizidine alkaloid biosynthesis in root cultures of *Senecio vulgaris*. *Phytochemistry*, **26**(6), 1639–1643.

Hartmann, T. and Witte, L. (1995) Chemistry, biology and chemoecology of the pyrrolizidine alkaloids, in *Alkaloids: Chemical and Biological Perspectives*, Vol. 9 (ed. S.W. Pelletier). Pergamon Press, Oxford, pp. 156–233.

Hartmann, T. and Ober, D. (2000) Biosynthesis and metabolism of pyrrolizidine alkaloids in plants and specialized insect herbivores, in *Topics in Current Chemistry: Biosynthesis – Aromatic Polyketides, Isoprenoids, Alkaloids*, Vol. 209 (eds. F.J. Leeper and J.C. Vederas). Springer, Berlin, pp. 207–243.

Hirono, I. (1993) Edible plants containing naturally-occurring carcinogens in Japan. *Japanese Journal of Cancer Research*, **84**(10), 997–1006.

Hovermale, J.T. and Craig, A.M. (1998) A routine method for the determination of retronecine. *Fresenius Journal of Analytical Chemistry*, **361**(2), 201–206.

Huizing, H.J. and Malingre, T.M. (1981) Ion-pair adsorption chromatography of pyrrolizidine alkaloids. *Journal of Chromatography*, **205**(1), 218–222.

Huxtable, R.J. (1990a) The harmful potential of herbal and other plant products. *Drug Safety*, **5**(suppl 1), 126–136.

Huxtable, R.J. (1990b) Activation and pulmonary toxicity of pyrrolizidine alkaloids. *Pharmacological Therapeutics*, **47**(3), 371–389.

Huxtable, R.J. (1992) The toxicology of alkaloids in food and herbs, in *Handbook of Natural Toxins. Vol. 7: Food Poisoning* (ed. A.T. Tu). Marcel Dekker, New York, pp. 237–262.

International Agency for Research on Cancer (1983) *IARC Monographs on the Evaluation of the Carcinogenic Risk to Humans. Vol. 31: Some Food Additives, Feed Additives and Naturally Occurring Substances*, IARC, Lyon, pp. 207–246.

Johnson, A.E. and Molyneux, R.J. (1984) Variation in pyrrolizidine alkaloid content of plants, associated with site, stage of growth and environmental conditions. *Plant Toxicology, Proceedings of the Australia–USA Poisonous Plants Symposium,* Brisbane, May 14–18, 1984, pp. 209–218.

Johnson, A.E., Molyneux, R.J. and Merrill, G.B. (1985) Chemistry of toxic range plants. Variation in pyrrolizidine alkaloid content of *Senecio, Amsinckia* and *Crotalaria* species. *Journal of Agricultural and Food Chemistry*, **33**(1), 50–57.

Kedzierski, B. and Buhler, D.R. (1986) Method for the determination of pyrrolizidine alkaloids and their metabolites by high performance liquid chromatography. *Analytical Biochemistry*, **152**(1), 59–65.

Kelley, R.B. and Seiber, J.N. (1992) Pyrrolizidine alkaloid chemosystematics in *Amsinckia. Phytochemistry*, **31**(7), 2369–2387.

Kim, N.C., Oberlies, N.H., Brine, D.R., Handy, R.W., Wani, M.C. and Wall, M.E. (2001) Isolation of sym-landine from the roots of common comfrey (*Symphytum officinale*) using countercurrent chromatography. *Journal of Natural Products*, **64**(2), 251–253.

Kumana, C.R., Ng, M., Lin, H.J., Ko, W., Wu, P.-C. and Todd, D. (1985) Herbal tea induced hepatic veno-occlusive disease: Quantification of toxic alkaloid exposure in adults. *Gut*, **26**, 101–104.

Larson, K.M., Roby, M.R. and Stermitz, F.R. (1984) Unsaturated pyrrolizidines from borage (*Borago officinalis*), a common garden herb. *Journal of Natural Products*, **47**(4), 747–748.

Lee, S.T., Schoch, T.K., Stegelmeier, B.L., Gardner, D.R., Than, K.A. and Molyneux, R.J. (2001) Development of enzyme-linked immunosorbent assays for the hepatotoxic alkaloids riddelliine and riddelliine N-oxide. *Journal of Agricultural and Food Chemistry*, **49**(8), 4144–4151.

Lin, G., Zhou, K.-Y., Zhao, X.-G., Wang, Z.-T. and But, P.P.H. (1998) Determination of hepatotoxic pyrrolizidine alkaloids by online high performance liquid chromatography mass spectrometry with an electrospray interface. *Rapid Communications in Mass Spectrometry*, **12**(20), 1445–1456.

Logie, C.G., Grue, M.R. and Liddell, J.R. (1994) Proton NMR-spectroscopy of pyrrolizidine alkaloids. *Phytochemistry*, **37**(1), 43–109.

Lüthy, J., Zweifel, U., Karlhuber, B. and Schlatter, C. (1981) Pyrrolizidine alkaloids of *Senecio alpinus* L. and their detection in feedstuffs. *Journal of Agricultural and Food Chemistry*, **29**(2), 302–305.

Lüthy, J., Brauchli, J., Zweifel, U., Schmid, P. and Schlatter, CH. (1984) Pyrrolizidine alkaloid in arzneipflanzen der Boraginaceen *Borago officinalis* L. and *Pulmonaria officinalis* L (in German). *Pharmaceutica Acta Helvetiae*, **59**, 242–246.

Margalith, D., Heraief, E., Schindler, A.M., Birchler, R., Mosimann, F., Aladjem, D. and Gonvers, J.J. (1985) Veno-occlusive disease of the liver due to the use of tea made from Senecio plants. A report of two cases. *Journal of Hepatology*, **1**(suppl 2), S280.

Mattocks, A.R. (1967a) Spectrophotometric determination of unsaturated pyrrolizidine alkaloids. *Analytical Chemistry*, **39**, 443–447.

Mattocks, A.R. (1967b) Detection of pyrrolizidine alkaloids on thin-layer chromatograms. *Journal of Chromatography A*, **27**, 505–508.

Mattocks, A.R. (1986) *Chemistry and Toxicology of Pyrrolizidine Alkaloids*. Academic Press, London, New York.

McLean, E.K. (1970) The toxic actions of pyrrolizidine (*Senecio*) alkaloids. *Pharmacological Reviews*, **22**(4), 429–483.

Ministry of Agriculture, Fisheries and Food (1994) *Naturally Occurring Toxicants in Food*. Food Surveillance Paper No. 42. HMSO, London.

Ministry of Agriculture, Fisheries and Food (1995) *Surveillance for Pyrrolizidine Alkaloids in Honey*. Food Surveillance Information Sheet No. 52. HMSO, London.

Mohabbat, O., Younos, S.M., Merzad, A.A., Srivastava, R.N., Sediq, G.G. and Aram, G.N. (August 7, 1976) An outbreak of hepatic veno-occlusive disease in north-western Afghanistan. *Lancet*, **2**, 269–271.

Molyneux, R.J., Roitman, J.N., Benson, M. and Lundin, R.E. (1982) ^{13}C NMR-spectroscopy of pyrrolizidine alkaloids. *Phytochemistry*, **21**(2), 439–443.

Molyneux, R.J. and James, L.F. (1990) Pyrrolizidine alkaloids in milk: Thresholds of intoxication. *Veterinary and Human Toxicology*, **32**(suppl), 94–103.

Mossoba, M.M., Lin, H.S., Andrzejewski, D., Sphon, J.A., Betz, J.M., Miller, L.J., Eppley, R.M., Trucksess, M.W. and Page, S.W. (1994) Application of gas chromatography/matrix isolation/Fourier transform infrared spectroscopy to the identification of pyrrolizidine alkaloids from comfrey root (*Symphytum officinale* L.). *Journal of AOAC International*, **77**(5), 1167–1174.

Mroczek, T., Glowniak, K. and Wlaszczyk, A. (2002) Simultaneous determination of N-oxides and free bases of pyrrolizidine alkaloids by cation-exchange solid-phase extraction and ion-pair high-performance liquid chromatography. *Journal of Chromatography A*, **949**(1–2), 249–262.

Mroczek, T., Ndjoko, K., Glowniak, K. and Hostettmann, K. (2004a) On-line structure characterization of pyrrolizidine alkaloids in *Onosma stellulatum* and *Emilia coccinea* by liquid chromatography-ion-trap mass spectrometry. *Journal of Chromatography A*, **1056**(1–2), 91–97.

Mroczek, T., Baj, S., Chrobok, A. and Glowniak, K. (2004b) Screening for pyrrolizidine alkaloids in plant materials by electron ionization RP-HPLC-MS with thermabeam interface. *Biomedical Chromatography*, **18**(9), 745–751.

Mroczek, T., Ndjoko-Ioset, K., Glowniak, K., Miekiewicz-Capala, A. and Hostettmann, K. (2006) Investigation of *Symphytum cordatum* alkaloids by liquid–liquid partitioning, thin-layer chromatography and liquid chromatography-ion-trap mass spectrometry. *Analytica Chimica Acta*, **566**(2), 157–166.

Ndjoko, K., Wolfender, J.L., Roder, E. and Hostettmann, K. (1999) Determination of pyrrolizidine alkaloids in *Senecio* species by liquid chromatography/thermospray-mass spectrometry and liquid chromatography/ nuclear magnetic resonance spectroscopy. *Planta Medica*, **65**(6), 562–566.

Oberlies, N.H., Kim, N.C., Brine, D.R., Collins, B.J., Handy, R.W., Sparacino, C.M., Wani, M.C. and Wall, M.E. (2004) Analysis of herbal teas made from the leaves of comfrey (*Symphytum officinale*): Reduction of N-oxides results in order of magnitude increases in the measurable concentration of pyrrolizidine alkaloids. *Public Health Nutrition*, **7**(7), 919–924.

Panter, K.E. and James, L.F. (1990) Natural plant toxicants in milk: A review. *Journal of Animal Sciences*, **68**(3), 892–904.

Parker, C.E., Verma, S., Tomer, K.B., Reed, R.L. and Buhler, D.R. (1990) Determination of *Senecio* alkaloids by thermospray liquid chromatography/mass spectrometry. *Biomedical and Environmental Mass Spectrometry*, **19**(1), 1–12.

Parvais, O., Vanderstricht, B., Vanhaelenfastre, R. and Vanhaelen, M. (1994) TLC detection of pyrrolizidine alkaloids in oil extracted from the seeds of *Borago officinalis*. *Journal of Planar Chromatography – Modern TLC*, **7**(1), 80–82.

Peterson, J.E. and Culvenor, C.C.J. (1983) Hepatotoxic pyrrolizidine alkaloids, in *Handbook of Natural Toxins: Plant and Fungal Toxins* (eds. R.F. Keeler and A.T. Tu). Marcel Dekker, New York, pp. 637–671.

Pieters, L.A. and Vlietinck, A.J. (1985) Quantitative proton Fourier transform nuclear magnetic resonance spectroscopic analysis of mixtures of pyrrolizidine alkaloids from *Senecio vulgaris*. *Fresenius' Zeitschrift fuer Analytische Chemie*, **321**, 355–358.

Pieters, L.A. and Vlietinck, A.J. (1986) Comparison of high-performance liquid chromatography with ¹H nuclear magnetic resonance spectroscopy for the quantitative analysis of pyrrolizidine alkaloids from *Senecio vulgaris*. *Journal of Liquid Chromatography*, **9**(4), 745–755.

Pieters, L.A. and Vlietinck, A.J. (1987) Quantitative analysis of pyrrolizidine alkaloid mixtures from *Senecio vulgaris* by carbon-13 nuclear magnetic resonance spectroscopy. *Magnetic Resonance in Chemistry*, **25**(11), 8–10.

Pieters, L.A.C., Vanzoelen, A.M., Vrieling, K. and Vlietinck, A.J. (1989a) Determination of the pyrrolizidine alkaloids from *Senecio jacobaea* by H-1 and C-13 NMR-spectroscopy. *Magnetic Resonance in Chemistry*, **27**(8), 754–759.

Pieters, L.A., Hartmann, T., Janssens, J. and Vlietinck, A.J. (1989b) Comparison of capillary gas-chromatography with ¹H and ¹³C nuclear magnetic resonance spectroscopy for the quantitation of pyrrolizidine alkaloids from *Senecio vernalis*. *Journal of Chromatography*, **462**, 387–391.

Prakash, A.S., Pereira, T.N., Reilly, P.E.B. and Seawright, A.A. (1999) Pyrrolizidine alkaloids in human diet. *Mutation Research – Genetic Toxicology and Environmental Mutagenesis*, **443**(1), 53–67.

Ridker, P.M., Ohkuma, S., McDermott, W.V., Trey, C. and Huxtable, R.J. (1985) Hepatic veno-occlusive disease associated with the consumption of pyrrolizidine containing dietary supplements. *Gastroenterology*, **88**, 1050–1054.

Ridker, P.M. and McDermott, W.V. (1989) Comfrey herb tea and veno-occlusive disease. *Lancet*, **1**(8639), 657–658.

Rizk, A.-F.M. (1991a) *Poisonous Plant Contamination of Edible Plants*. CRC Press, Boca Raton, FL.

Rizk, A.-F.M. (1991b) *Naturally Occurring Pyrrolizidine Alkaloids*. CRC Press, Boca Raton, FL.

Roeder, E. (1990) Carbon-13 NMR spectroscopy of pyrrolizidine alkaloids. *Phytochemistry*, **29**, 11–29.

Roeder, E. (1995) Medicinal plants in Europe containing pyrrolizidine alkaloids. *Pharmazie*, **50**(2), 83–98.

Roeder, E. (1999) Analysis of pyrrolizidine alkaloids. *Current Organic Chemistry*, **3**(6), 557–576.

Roeder, E. (2000) Medicinal plants in China containing pyrrolizidine alkaloids. *Pharmazie*, **55**(10), 711–726.

Roeder, E. and Pflueger, T. (1995) Analysis of pyrrolizidine alkaloids: A competitive enzyme-linked immunoassay (ELISA) for the quantitative determination of some toxic pyrrolizidine alkaloids. *Natural Toxins*, **3**, 305–309.

Roitman, J.N. (1983) Ingestion of pyrrolizidine alkaloids: A health hazard of global proportions, in *Xenobiotics in Foods and Feed* (eds. J.W. Finley and D.E. Schwass). American Chemical Society, Washington, DC, pp. 345–378.

Roseman, D.M., Wu, X. and Kurth, M.J. (1996) Enzyme-linked immunosorbent assay detection of pyrrolizidine alkaloids: Immunogens based on quaternary pyrrolizidinium salts. *Bioconjugate Chemistry*, **7**(2), 187–195.

Schaneberg, B.T., Molyneux, R.J. and Khan, I.A. (2004) Evaporative light scattering detection of pyrrolizidine alkaloids. *Phytochemical Analysis*, **15**(1), 36–39.

Schoch, T.K., Gardner, D.R. and Stegelmeier, B.L. (2000) GC/MS/MS detection of pyrrolic metabolites in animals poisoned with the pyrrolizidine alkaloid riddelliine. *Journal of Natural Toxins*, **9**(2), 197–206.

Schoental, R. (1968) Toxicology and carcinogenic action of pyrrolizidine alkaloids. *Cancer Research*, **28**(11), 2237–2246.

Segall, H.J., Wilson, D.W., Lamé, M.W., Morin, D. and Winter, C.K. (1991) Metabolism of pyrrolizidine alkaloids, in *Handbook of Natural Toxins*, Vol. 6 (eds. R.F. Keeler and A.T. Tu). Marcel Dekker, New York, pp. 3–26.

Sherma, J. (2003) High-performance liquid chromatography/mass spectrometry analysis of botanical medicines and dietary supplements: A review. *Journal of AOAC International*, **86**(5), 873–881.

Smith, L.W. and Culvenor, C.C.J. (1981) Plant sources of hepatotoxic pyrrolizidine alkaloids. *Journal of Natural Products*, **44**(2), 129–152.

Sperl, W., Stuppner, H., Gassner, I., Judmaier, W., Dietze, O. and Vogel, W. (1995) Reversible hepatic veno-occlusive disease in an infant after consumption of pyrrolizidine-containing herbal tea. *European Journal of Pediatrics*, **154**(2), 112–116.

Stegelmeier, B.L., Edgar, J.A., Colegate, S.M., Gardner, D.R., Schoch, T.K., Coulombe, R.A. and Molyneux, R.J. (1999) Pyrrolizidine alkaloid plants, metabolism and toxicity. *Journal of Natural Toxins*, **8**(1), 95–116.

Stelljes, M.E., Kelley, R.B., Molyneux, RJ. and Seiber, J.N. (1991) GC-MS determination of pyrrolizidine alkaloids in four *Senecio* species. *Journal of Natural Products*, **54**(3), 759–773.

Stewart, M.J. and Steenkamp, V. (2001) Pyrrolizidine poisoning: A neglected area in human toxicology. *Therapeutic Drug Monitoring*, **23**(6), 698–708.

Tandon, B.N., Tandon, H.D., Tandon, R.K., Narndranathan, M. and Joshi, Y.K. (1976) An epidemic of veno-occlusive disease of liver in Central India. *Lancet*, **2**, 271–272.

Than, K.A., Stevens, V., Knill, A., Gallagher, R., Gaul, K.L., Edgar, J.A. and Colegate, S.M. (2005) Plant-associated toxins in animal feed: Screening and confirmation assay development. *Animal Feed Science and Technology*, **121**(1–2), 5–21.

van Dam, N.M., Witte, L., Theuring, C. and Hartmann, T. (1994) Distribution, biosynthesis and turnover of pyrrolizidine alkaloids in *Cynoglossum officinale*. *Phytochemistry*, **39**(2), 287–292.

Weston, C.F., Cooper, B.T., Davies, J.D. and Levine, D.F. (1981) Veno-occlusive disease of the liver secondary to ingestion of comfrey. *British Medical Journal*, **44**, 129–144.

Winter, C.K., Segall, H.J. and Jones, A.D. (1988) Determination of pyrrolizidine alkaloid metabolites from mouse liver microsomes using tandem mass spectrometry and gas chromatography/mass spectrometry. *Biomedical and Environmental Mass Spectrometry*, **15**(5), 265–273.

Winter, C.K. and Segall, H.J. (1989) Metabolism of pyrrolizidine alkaloids, in *Toxicants of Plant Origin* (ed. P.R. Cheeke). CRC Press, Boca Raton, FL, pp. 1–22.

Witte, L., Rubiolo, P., Bicchi, C. and Hartmann, T. (1993) Comparative analysis of pyrrolizidine alkaloids from natural sources by gas chromatography–mass spectrometry. *Phytochemistry*, **32**(1), 187–196.

World Health Organisation (1988) International Programme on Chemical Safety. *Environmental Health Criteria 80: Pyrrolizidine Alkaloids*. WHO, Geneva.

Wretensjo, I. and Karlberg, B. (2003) Pyrrolizidine alkaloid content in crude and processed borage oil from different processing stages. *Journal of the American Oil Chemists Society*, **80**(10), 963–970.

Wuilloud, J.C.A., Gratz, S.R., Gamble, B.M. and Wolnik, K.A. (2004) Simultaneous analysis of hepatotoxic pyrrolizidine alkaloids and N-oxides in comfrey root by LC-ion trap mass spectrometry. *Analyst*, **129**(2), 150–156.

Yang, J.C., Yan, J., Doerge, D.R., Chan, P.C. and Fu, P.P. (2001) Metabolic activation of the tumorigenic pyrrolizidine alkaloid riddelliine, leading to DNA adduct formation in vivo. *Chemical Research in Toxicology*, **14**(1), 101–109.

Zhao, X.L, Chan, M.Y. and Ogle, C.W. (1989) The identification of pyrrolizidine alkaloid-containing plants – A study on 20 herbs of the compositae family. *American Journal of Chinese Medicine*, **17**(1–2), 71–78.

Zuendorf, I., Wiedenfeld, H., Roeder, E. and Dingermann, T. (1998) Generation and characterization of monoclonal antibodies against the pyrrolizidine alkaloid retrorsine. *Planta Medica*, **64**, 259–263.

3   Glucosinolates

Ruud Verkerk and Matthijs Dekker

Summary

Glucosinolates are amino acid-derived secondary plant metabolites found exclusively in cruciferous plants. The majority of cultivated plants that contain glucosinolates belong to the family of *Brassicaceae* such as Brussels sprouts, cabbage, broccoli and cauliflower. These are the major source of glucosinolates in the human diet of which about 120 different glucosinolates have been characterized. Glucosinolates and their breakdown products are of particular interest because of their nutritive and antinutritional properties, their potential adverse effects on health, their anticarcinogenic properties and finally the characteristic flavour and odour they give to many vegetables. This chapter reviews the biosynthesis, nature and occurrence of glucosinolates, their stability in biological systems, analysis and biological effects including toxicity.

3.1   INTRODUCTION

Glucosinolates are amino acid-derived secondary plant metabolites found exclusively in cruciferous plants. These sulfur-containing glycosides occur at highest concentrations in the families *Resedaceae,Capparaceae* and *Brassicaceae.*

The majority of cultivated plants that contain glucosinolates belong to the family of *Brassicaceae*. Mustard seed, used as a seasoning, is derived from *B. nigra, B. juncea* (L.) Coss and *B. hirta* species. Vegetable crops include cabbage, cauliflower, broccoli, Brussels sprouts and turnip of the *B. oleracea* L., *B. rapa* L., *B. campestris* L. and *B. napus* L. species. Kale of the *B. oleracea* species is used for forage, pasture and silage. *Brassica* vegetables such as Brussels sprouts, cabbage, broccoli and cauliflower are the major source of glucosinolates in the human diet. They are frequently consumed by humans from Western and Eastern cultures (McNaughton and Marks, 2003). In the Netherlands, the average consumption of these vegetables is more than 36 g *Brassica* per person per day (Godeschalk, 1987). The typical flavour of brassica vegetables is largely due to glucosinolate-derived volatiles.

Glucosinolates and their breakdown products are of particular interest in food research because of their nutritive and antinutritional properties, the adverse effects of some glucosinolates on health, their anticarcinogenic properties and finally because they are responsible for the characteristic flavour and odour of many vegetables (Mithen *et al.*, 2000). The versatility of these compounds is also demonstrated by the fact that glucosinolates are quite toxic to some insects and therefore could be included as one of many natural pesticides. However, a small number of insects, such as the cabbage aphids, use glucosinolates to locate their favourite plants as feed and to find a suitable environment to deposit their eggs (Barker *et al.*, 2006). Furthermore, glucosinolates show antifungal and antibacterial properties (Fahey *et al.*, 2001).

There are currently about 120 different glucosinolates characterized, of which only a limited number have been investigated thoroughly. A considerable amount of data on levels of total and individual glucosinolates is now available. The levels of total glucosinolates in plants may depend on variety, cultivation conditions, climate and agronomic practice, while the levels in a particular plant vary between the parts of the plant. Generally the same glucosinolates occur in a particular sub-species regardless of genetic origin, and in most species only between one and four glucosinolates are found in relatively high concentrations.

Glucosinolates themselves are chemically stable and biologically inactive while they remain separated within sub-cellular compartments throughout the plant. However, tissue damage caused by pests, harvesting, food processing or chewing initiates contact with the endogenous enzyme myrosinase which leads to hydrolysis releasing a broad range of biologically active products such as isothiocyanates (ITCs), organic cyanides, oxazolidinethiones and ionic thiocyanate on enzymatic degradation by myrosinase in the presence of water. The anticarcinogenic mechanisms by which these compounds may act include the induction of detoxification enzymes and the inhibition of the activation of promutagens/procarcinogens (Wattenberg, 1992; Dragsted *et al.*, 1993; Jongen, 1996; Mithen *et al.*, 2000).

Glucosinolate breakdown products exert a variety of toxic and antinutritional effects in higher animals amongst which the adverse effects on thyroid metabolism are the most thoroughly studied (Tripathi and Mishra, 2007). Tiedink *et al.* (1990, 1991) investigated the role of indole compounds and glucosinolates in the formation of *N*-nitroso compounds in vegetables. These studies revealed that the indole compounds present in *Brassica* vegetables can be nitrosated and thereby become mutagenic. However, the nitrosated products are stable only in the presence of large amounts of free nitrite.

3.2 NATURE AND OCCURRENCE

The chemistry and occurrence of glucosinolates and their breakdown products have been reviewed extensively by Fahey *et al.* (2001).

Several glucosinolates are isolated in the pure state. The first crystalline glucosinolate was isolated from the seed of white mustard in 1830 and since then the elucidation of their structures and chemistry has continued (Gildemeister and Hofmann, 1927). The common structure of glucosinolates is shown in Fig. 3.1. The side chain determines the chemical and biological nature of glucosinolates. They are considered to be (*Z*)-*cis*-*N*-hydroximinosulfate esters possessing a side chain R and a sulfur-linked D-glucopyranose moiety. Natural glucosinolates contain exclusively a β-D-glucopyranosyl linkage (Blanc-Muesser *et al.*, 1990).

The side chain of the glucosinolates is variable and is the basis for their structural heterogenity and for the biological activity of the enzymatic and chemical breakdown products.

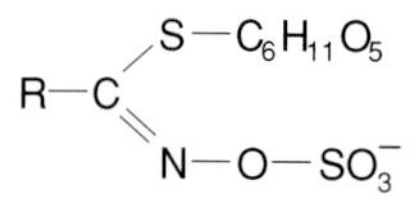

$$R-C \underset{N-O-SO_3^-}{\overset{S-C_6H_{11}O_5}{\big<}}$$

Fig. 3.1 General structure of glucosinolates.

Table 3.1 Glucosinolates commonly found in *Brassica* vegetables.

Trivial name	Chemical name (side chain R)
Aliphatic glucosinolates	
Glucoiberin	3-Methylsulfinylpropyl
Progoitrin	2-Hydroxy-3-butenyl
Sinigrin	2-Propenyl
Gluconapoleiferin	2-Hydroxy-4-pentenyl
Glucoraphanin	4-Methylsulfinylbutyl
Glucoalyssin	5-Methylsulfinylpentyl
Glucocapparin	Methyl
Glucobrassicanapin	4-Pentenyl
Glucocheirolin	3-Methylsulfonylpropyl
Glucoiberverin	3-Methylthiopropyl
Gluconapin	3-Butenyl
Indole glucosinolates	
4-Hydroxyglucobrassicin	4-Hydroxy-3-indolylmethyl
Glucobrassicin	3-Indolylmethyl
4-Methoxyglucobrassicin	4-Methoxy-3-indolylmethyl
Neoglucobrassicin	1-Methoxy-3-indolylmethyl
Aromatic glucosinolates	
Glucosinalbin	*p*-Hydroxybenzyl
Glucotropaeolin	Benzyl
Gluconasturtiin	2-Phenethyl

Table 3.1 gives an overview of the most commonly found glucosinolates in *Brassica* vegetables, however an extensive list of 120 different glucosinolates identified in higher plants has been described by Fahey *et al.* (2001).

Glucosinolates are prevalent in about 16 botanical families of the order *Capparales*, such as the *Capparaceae*, *Brassicaceae*, *Caricaceae* and *Resedaceae* (Fahey *et al.*, 2001). For the human diet, representatives of the *Brassicaceae* are of particular importance as vegetables (e.g. cabbage, Brussels sprouts, broccoli, cauliflower), root vegetables (e.g. radish, turnip and swede), leaf vegetables (e.g. rocket salad) and seasonings and relishes (e.g. mustard, wasabi) and sources of oil (Holst and Williamson, 2004). They occur in all parts of the plants, but in different profiles and concentrations. Usually, a single plant species contains up to four different glucosinolates in significant amounts while, as many as 15 different glucosinolates can be found in the same plant. The highest concentrations are usually found in the seeds, except for indol-3-ylmethyl and *N*-methoxyindol-3-ylmethyl glucosinolates, which are rarely found in seeds (Tookey *et al.*, 1980). Several reviews have presented and discussed the variation in glucosinolate composition and profiles of various representatives of the *Brassicaceae*. Occurrence and concentrations of glucosinolates vary according to difference in species and varieties, tissue type, physiological age, environmental conditions (agronomic practices, climatic and ecophysiological conditions), presence of pest infestation (Rosa *et al.*, 1997; Fahey *et al.*, 2001; Holst and Williamson, 2004; Schreiner, 2005).

3.3   BIOSYNTHESIS

The pathway of glucosinolate biosynthesis has been studied since the 1960s and the identity of many intermediates, enzymes and genes involved is now known. The biosynthesis of glucosinolates was recently reviewed extensively by Halkier and Gershenzon (2006). Knowledge of biosynthetic pathways of glucosinolates has increased as research advanced from traditional in vivo feeding studies and biochemical characterization of the enzymatic activities in plant extracts to identification and characterization of the biosynthetic genes encoding the involved enzymes. Especially the studies of glucosinolates in the model plant *Arabidopsis* facilitated the progress.

Kjaer and Conti (1954) suggested that amino acids may be natural precursors of the aglycone moiety of glucosinolates based on the similarities between the carbon skeletons of some amino acids and the glucosinolates. This hypothesis was confirmed by studies of the different biosynthetic stages. Most of these studies involved the administration of variously labelled compounds (^{3}H, ^{14}C, ^{15}N or ^{35}S) to plants and the assessment of their relative efficiencies as precursors on the basis of the extent of incorporation of isotope into the glucosinolate. The classification of glucosinolates as shown in Table 3.1 depends on the amino acid from which they are derived; aliphatic glucosinolates derived from alanine, leucine, isoleucine, methionine or valine; aromatic glucosinolates derived from phenylalanine or tyrosine; and indole glucosinolates are derived from tryptophane (Sørensen, 1990).

The biosynthesis of glucosinolates from amino acids can be divided into three separate steps. The first step is the chain elongation of aliphatic and aromatic amino acids by inserting methylene groups into their side chains. Second, the metabolic modification of the amino acids (or chain-extended derivatives of amino acids) takes place via an aldoxime intermediate. The same modifications also occur in the biosynthetic route of cyanogenic glycosides. However, the co-occurrence of glucosinolates and cyanogenic glycosides in the same plant is very rare (an example is *Carica papaya*). The biosynthesis of the cyanogenic glycosides has been elucidated in more detail by Halkier and Lindberg-Møller (1991) and by Koch *et al.* (1992). Third, following the formation of the aldoxime, the glucosinolate is formed by various secondary transformations such as S-insertion, glucosylation and sulfation. Further modification of the side chain can occur in the formed glucosinolate by, for example, oxidation and/or elimination reactions. The different steps in the synthesis are discussed below in more detail.

3.3.1   Amino acid modification

The glucosinolates can be divided into two groups by origin: those derived from common amino acids and those derived from modified amino acids. The modification of common amino acids is mainly in the form of side chain elongation. A general route for this was proposed by Kjaer (1976). Various enzymes are involved in these steps (Halkier and Gershenzon, 2006). The parent amino acid is deaminated to form the corresponding 2-oxo acid. Next is a three-step cycle of (i) condensation with acetyl-CoA, (ii) isomerization and (iii) oxidation–decarboxylation to yield a 2-oxo acid with one more methylene group than the starting compound. The resulting chain-extended 2-oxo acid can undergo additional chain-elongation cycles, each adding one further methylene group, or, following transamination, can enter the glucosinolate core biosynthetic pathway. Up to nine elongation cycles are known to occur in plants.

3.3.2   Conversion of amino acids

The glucosinolate core pathway converts the amino acid to an *S*-alkylthiohydroximate via two consecutive reactions that are catalysed by structurally specific cytochrome P450s,

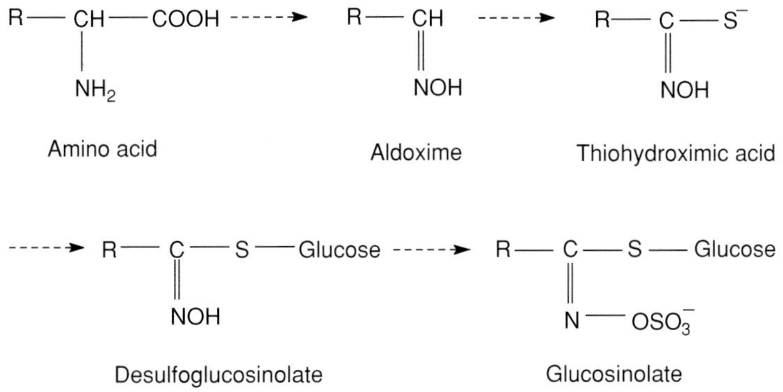

R—CH—COOH -----▸ R—CH -----▸ R—C—S⁻
$\quad\quad$ | $\quad\quad\quad\quad\quad\quad$ ‖ $\quad\quad\quad\quad\quad\quad$ ‖
$\quad\quad$ NH₂ $\quad\quad\quad\quad\quad\quad$ NOH $\quad\quad\quad\quad\quad\quad$ NOH

$\quad\quad$ Amino acid $\quad\quad\quad\quad\quad$ Aldoxime $\quad\quad\quad$ Thiohydroximic acid

-----▸ R—C—S—Glucose -----▸ R—C—S—Glucose
$\quad\quad\quad\quad\quad$ ‖ $\quad\quad\quad\quad\quad\quad\quad\quad\quad\quad$ ‖
$\quad\quad\quad\quad\quad$ NOH $\quad\quad\quad\quad\quad\quad\quad\quad\quad$ N—OSO₃⁻

$\quad\quad\quad$ Desulfoglucosinolate $\quad\quad\quad\quad\quad\quad$ Glucosinolate

Fig. 3.2 The simplified biosynthesis of the glucosinolate core structure.

encoded by the *CYP79* and *CYP83* gene families. *C-S* lyase activity results in the formation of thiohydroximates that are converted to desulfo-glucosinolates by a non-specific *S*-glucosyltransferase. The final glucosinolate is produced by sulfation by one of three structurally specific sulfotransferases. Subsequently, various secondary side-chain modifications can occur, including oxidation, hydroxylation, alkenylation, acylation or esterification (Halkier and Gershenzon, 2006).

3.3.3 Secondary transformations

The formed parent glucosinolate is subject to a wide range of further modifications of the R group. The R group of glucosinolates derived from methionine (and chain-elongated homologs) is especially subject to further modifications, such as the stepwise oxidation of the sulfur atom in the methylthioalkyl side chain leading to methylsulfinylalkyl and methylsulfonylalkyl moieties. Methylsulfinylalkyl side chains can be further modified by oxidative cleavage to produce alkenyl or hydroalkenyl chains. These reactions are of biological as well as biochemical interest because they influence the direction of glucosinolate hydrolysis and the resulting activity of the hydrolysis products (Halkier and Gershenzon, 2006).

The simplified biosynthetic pathway for glucosinolates is shown in Fig. 3.2 (based on Halkier and Gershenzon, 2006).

3.4 HYDROLYSIS

As mentioned before, hydrolysis products rather than intact glucosinolates are responsible for the various biological effects. Most glucosinolates are chemically stable and to lesser extent thermal stable. Therefore, hydrolysis is mainly enzymatically driven by the endogenous enzyme myrosinase. Upon consumption of intact glucosinolates without the presence of active myrosinase (e.g. cooked vegetables) glucosinolates can also be hydrolysed by ß-glucosidases from the intestinal flora (Shapiro *et al.*, 1998, 2006).

3.4.1 Myrosinase

Myrosinase (thioglucoside glucohydrolase EC 3.2.3.1) is the trivial name for the enzyme (or group of enzymes) responsible for the hydrolysis of glucosinolates. All plants that contain glucosinolates also contain myrosinase. Myrosinase is widely distributed, occurring in my-rosin cells of seeds, leaves, stems and roots of glucosinolate-containing plants and the activity appears to be higher in the young tissues of the plant (Bones and Rossiter, 1996).

Myrosinases have generally been well characterized by various approaches such as an-alytical gel electrophoresis, immunohistochemical techniques, light microscopy and elec-tron microscopy (Buchwaldt *et al.*, 1986; Thangstad *et al.*, 1990). The various studies have demonstrated the presence of several myrosinase isoenzymes (MacGibbon and Allison, 1970; Buchwaldt *et al.*, 1986). Different patterns were found depending on whether the extracts were made from the leaf, stem, root or seed. Little is known about the substrate specificity of myrosinase isoenzymes. There are two myrosinases isolated by James and Rossiter (1991) that degrade different glucosinolates at different rates. However, both isoenzymes show high-est activity against aliphatic glucosinolates and least activity against indole glucosinolates. Members of a given class of glucosinolates are degraded at approximately the same rate in vitro. It is also possible that the specificity is affected by associated factors like epithiospec-ifier protein, myrosinase-binding protein or other myrosinase-associated proteins or compo-nents. Myrosinase activity varies by plant species, organ and stage of development (Bones, 1990), but activity is also affected by seasonal conditions and climatic factors (Charron *et al.*, 2005).

Ascorbic acid has been shown to modulate myrosinase activity in some species; it inhibits at high concentrations and activates at low levels. Activation appears to be the result of a conformational change in the protein structure, leading to an enhanced reaction rate when the effector-binding sites are occupied (Ohtsuru and Hata, 1973).

The complexity of the glucosinolate/myrosinase system indicates an important role in cru-ciferous plants (Bones and Rossiter, 1996). The glucosinolate/myrosinase system may have several functions in the plant: (i) plant defence against fungal diseases and pest infestation; (ii) sulfur and nitrogen metabolism; and (iii) growth regulation.

Plant breeding strategies over past decades have concentrated on reducing the glucosinolate content of rapeseed to improve the acceptability of rapeseed meal and meet the increasingly stringent requirements of the rapeseed processing industry. One approach to reducing the un-desired breakdown products of glucosinolates would be to change the amount of myrosinase available for hydrolysis of the glucosinolates.

3.4.2 Hydrolysis products

Hydrolysis products of glucosinolates contribute significantly to the typical flavour of *Bras-sica* vegetables. Myrosinase catalyse the hydrolysis of glucosinolates by splitting off the glucose. The unstable aglucone (thiohydroxymate-O-sulfonate) then eliminates sulfate by a Lossen rearrangement (Fig. 3.3). The structure of the resulting products depends on a variety of factors. Whether ITCs or nitriles are formed depends on the specific glucosinolates, the part of the plant where they are located, the treatment of plant material before the hydrolysis of glucosinolates and conditions during hydrolysis, especially pH. Isothiocyanates are usually produced at pH 5–7, while nitriles are the major degradation products under acidic conditions. Most hydrolysis products are stable except for glucosinolates possessing a ß-hydroxylated side chain; ß-hydroxy-ITCs are unstable and spontaneously cyclise to oxazolidine-2-thiones (e.g. goitrin). Indole glucosinolates such as glucobrassicin form unstable ITCs and undergo further hydrolysis to give 3-indolemethanol, 3-indoleacetonitrile and 3,3′ diindolylmethane and subsequently condenses into dimers, trimers or tetramers (Holst and Williamson, 2004).

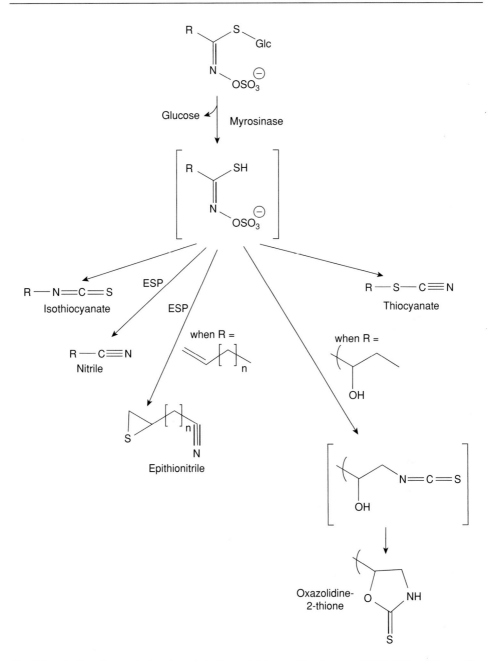

Fig. 3.3 Outline of glucosinolate hydrolysis (from Halkier and Gershenzon, 2006). ESP, epithiospecifier protein; R, variable side chain. Reprinted with permission, from the *Annual Review of Plant Biology*, Volume 57, copyright 2006, by Annual Reviews www.annualreviews.org

Hanley *et al.* (1990) isolated an indole isothiocyanate from neoglucobrassicin degradation under specific experimental conditions. Epithioalkanes are produced from the hydrolysis of alkenyl glucosinolates when myrosinase co-occurs with a labile protein known as epithiospec-ifier protein (ESP), a small protein that is not ubiquitous in the *Cruciferae* (Matusheski *et al.*, 2006).

3.5   ANALYTICAL METHODS

Understanding the involvement of glucosinolates and their biological active breakdown products in the many different scientific fields requires the development of rapid, reliable and reproducible methods for their identification and quantification. However, rarely does a method provide rapidity, simplicity, sensitivity and accuracy to discriminate among all compounds of interest. Especially since there occur a large amount of different glucosinolates in *Brassicas* and the fact that each glucosinolate can produce different breakdown products makes their analysis very complicated.

Analytical methods can be divided into those for total glucosinolates, individual glucosinolates and the breakdown products. Over the past four decades, increased knowledge of the diversity of the glucosinolates, their enzymatically released products and factors influencing their release have led to a multiplicity of analytical methods. Griffiths *et al.* (1998) and Kiddle *et al.* (2001) have presented an extensive overview of the wide variety of analytical methods of glucosinolates in plant tissue.

Since glucosinolates coexist with myrosinase in the plant, processes like grinding or cutting of fresh tissue in the presence of water will initiate a rapid hydrolysis of these compounds. For analysis of intact glucosinolates, inhibition of myrosinase activity is therefore essential. Before disruption of the material, samples should be completely dried by freeze-drying or frozen in liquid nitrogen. The use of aqueous methanol for extraction, in combination with high temperatures, also inhibits myrosinase (Heaney and Fenwick, 1993).

3.5.1   Total glucosinolates

Glucosinolates yield equimolar amounts of glucose on hydrolysis with myrosinase. This is true for almost all glucosinolates and spectroscopic methods for total glucosinolates based on the measurement of enzymatically released glucose or sulfate are relatively rapid and simple to apply (Schnug, 1987; Heaney *et al.*, 1988). Van Doorn *et al.* (1998) described a more specific determination of the glucosinolates sinigrin and progoitrin by raising antisera to these glucosinolates.

Measuring total glucosinolate levels has received less attention because of the great progress in analytical methodology for individual glucosinolates that leads to much more useful information. However, the high costs and labour input required for obtaining glucosinolate data is a serious hinder to generating large sets of samples, which is usually necessary in screening breeding programmes. Therefore, the use of fast analytical techniques such as near-infrared spectroscopy (NIRS) result in many advantages since analysis can be carried out with a considerable saving of time and at relative low costs. This technique has mostly applied for glucosinolate content in seeds of *Brassica* vegetables where these compounds are present in significant higher concentrations than those usually found in leaves (Velasco and Becker, 1998; Font *et al.*, 2004). Font *et al.* (2005) have described the quantification of glucosinolates in leaves of *Brassica* napa by NIRS. They demonstrated the potential of NIRS for predicting the total glucosinolate content as well as the major individual glucosinolates found in *B. napus*.

3.5.2   Individual glucosinolates

For more detailed analysis, researchers have used either HPLC-MS analysis for intact glucosinolates or GC-MS analysis for the hydrolysis products. GC-MS can be very sensitive, but has some limitations as some glucosinolate side chains (e.g. indolyl or hydroxylated side chains) produce poorly volatile or unstable breakdown products.

Spinks *et al.* (1984) developed a reverse-phase HPLC method for quantitative analysis of desulfoglucosinolates which is, hitherto, most widely used. This method uses an on-column enzymatic desulfation treatment of plant extracts followed by HPLC detection of the resulting desulfoglucosinolates. For the analysis, desulfobenzyl or desulfosinigrin is used as internal standard and relative response factors are determined with purified standards. Validation of chromatographic profiles takes place by correspondence of glucosinolate retention times with standardized rapeseed extracts or other available standards. The intact glucosinolates can be analysed directly by HPLC, although this often gives a poor resolution. Desulfation results in better separation and cleaner samples and also makes the compounds easier to analyse by HPLC or mass spectrometric techniques. HPLC systems using a photodiode array (PDA) detector are very sensitive allowing detection of glucosinolate levels in the nano-molar range. Whilst spectral data of individual desulfoglucosinolates will allow initial confirmation of structural class, addition of MS detection increases the discriminatory power of the technique even further (Kiddle *et al.*, 2001).

Nowadays, mass spectrometry has become a powerful technique for quantification of bioactive compounds in complex biological matrices; however, most methods were qualitative. Griffiths *et al.* (1998) reported a liquid chromatography–atmospheric pressure chemical ionization–mass spectrometric (LC-APCI/MS) method for determining quantitatively desulfoglucosinolates.

Song *et al.* (2005) described the quantitative analysis of glucosinolates in vegetable extracts and blood plasma by negative ion electrospray LC-MS/MS.

Both LC-thermospray-MS and LC-APCI-MS have been employed for desulfoglucosinolates (Mellon *et al.*, 2002; Song *et al.*, 2005) and the latter has been found to give the best sensitivity, with at least an order of magnitude improvement over UV detection and the added advantage of structural confirmation from the mass spectrum. The sensitivity can be further improved by the use of synthetic per-deuterated desulfoglucosinolates as internal standards.

Tian *et al.* (2005) improved the LC-APCI/MS method by using liquid chromatography–electrospray ionization–tandem mass spectrometry (LC-ESI/MS/MS). This method has the advantage that the glucosinolates do not require desulfonation before the analysis, while the detection limits are 10-fold lower compared to conventional HPLC methods.

Matrix-assisted laser desorption/ionization time-of-flight mass spectrometry (MALDI-TOF MS) was introduced in 1987 and developed for use with nonvolatile and large biomolecules (Karas *et al.*, 1987). It has some advantages over other methods including speed of analysis, high sensitivity, good tolerance toward contaminants, and the ability to analyse complex mixtures. MALDI-TOF MS thus has potential for the analysis of plant metabolites, both in the plant itself and in foodstuffs, as a profile of the sample can be obtained in only a few minutes (Karas, 1996). Botting *et al.* (2002) applied MALDI-TOF for the characterization of a number of intact glucosinolates. The method was used for crude plant extracts to rapidly examine glucosinolate profiles. The method is much faster compared to LC-MS methods and required far less sample, but has the drawback of not being a quantitative analysis method.

One of the major problems in the analysis of glucosinolates is the paucity of high purity chromatographic standard glucosinolates available to researchers. Only a limited number of glucosinolates are commercially available. The glucosinolate 2-propenyl (sinigrin) is not a suitable internal standard because of the presence of this compound in most brassicaceous plants. On the other hand, benzylglucosinolate (glucotropaeolin) is commonly not present in *Brassicas* and has been used as an internal standard.

3.5.3 Breakdown products

The application of HPLC to the investigation of glucosinolate breakdown products has been limited due to the volatility of many compounds. Furthermore, thiocyanates and nitriles are not detectable spectrometrically.

Isothiocyanates and nitriles can be analysed by GC, HPLC with UV detection may be used for analysis of oxazolidinethiones and indoles. Quinsac *et al.* (1992) developed a method for analysing oxazolidinethiones in biological fluids with a high degree of selectivity. However, HPLC finds most use in the analysis of intact glucosinolates or desulfoglucosinolates. For identification and confirmation of structures both techniques can be coupled to mass spectrometry. Mass spectroscopy has proved to be an important tool in the identification and structural elucidation of glucosinolates and their breakdown products.

Zhang *et al.* (1992a) developed a spectroscopic quantitation of organic ITCs. Under mild conditions nearly all organic ITCs react quantitatively with an excess of vicinal dithiols to give rise to five-membered cyclic condensation products, with release of the corresponding free amines. The method can be used to measure 1 nmol or less of pure ITCs or ITCs in crude mixtures.

Song *et al.* (2005) reported an LC-MS/MS method for the analysis of ITCs and amine degradation products. Isothiocyanates were pre-derivatized and quantified by positive ion electrospray LC-MS/MS.

3.6 BIOLOGICAL EFFECTS

Since there is a large number of a different type of glucosinolates and different possible pathways for hydrolysis, a broad range of hydrolysis products can be found in various food sources. All these compounds show distinctive, concentration-dependent biological activities varying from acute toxicity to anticarcinogenic properties. The consumption of vegetables and fruits has always been seen as health promoting. The protective effect of *Brassica* vegetables against cancer has been suggested to be due in part to the relatively high content of glucosinolates. The presence of glucosinolates distinguishes them from other vegetables. Besides their beneficial health effects, some glucosinolates and breakdown products also show toxicological effects. The toxicity of glucosinolates, mainly determined by the aglycones, has been described in many studies.

3.6.1 Anticarcinogenicity

The most extensively studied mechanism for inhibition of carcinogenesis by glucosinolate-derived hydrolysis products are the modulation of the antioxidant potential, enhancement of detoxification mechanisms and the induction of apoptosis in undifferentiated cells (Mithen *et al.*, 2000).

Carcinogenesis is a multi-stage process in which at least three distinct phases can be recognized: initiation, promotion and progression. At each stage of the carcinogenic process there can be intervention. Wattenberg (1985) proposed a system of classification of dietary anticarcinogens based on the stage of carcinogenesis at which they act. Anticarcinogens

Table 3.2 Protection against chemical carcinogenesis in rat and mouse organs by a variety of isothiocyanates and glucosinolates.

Protective isothiocyanates
α-Naphthyl-NCS, β-naphthyl-NCS
Phenyl-$[CH_2]_n$-NCS, where $n = 0, 1, 2, 3, 4, 5, 6, 8, 10$
PhCH(Ph)CH$_2$-NCS, PhCH$_2$CH(Ph)-NCS
CH$_3$[CH]$_n$-NCS, where $n = 5, 11$
CH$_3$[CH$_2$]$_3$CH(CH$_3$)-NCS
Sulforaphane, CH$_3$S(O)[CH$_2$]$_4$-NCS
2-Acetylnorbornyl-NCS (three isomers)

Protective glucosinolates
Indolylmethyl glucosinolate (glucobrassicin)
Benzyl glucosinolate (glucotropaeolin)
4-Hydroxybenzyl glucosinolate (glucosinalbin)

Carcinogens employed
3′-Methyl-4′-dimethylaminoazobenzene
4-Dimethylaminoazobenzene
N-2-Fluorenylacetamide, acetylaminofluorene
7,12-Dimethylbenz[a]anthracene (DMBA)
Benzo[a]pyrene
Methylazoxymethanol acetate
N-Nitrosodiethylamine
4-(Methylnitroamino)-1-(3-pyridyl)-1-butanone (NNK)
N-Nitrosobenzylmethylamine (NBMA)
N-Butyl-N-(4-hydroxybutyl)nitrosamine

Tumour target organs
Rat: liver, lung, mammary gland, bladder, small intestine/colon, oesophagus
Mouse: lung, forestomach

Source: Talalay and Zhang (1996).

can then be divided into three major classes. The first consists of compounds that prevent the formation of carcinogens from precursor substances. The second class is called blocking agents. These have been found to be effective when given immediately before or during treatment with chemical carcinogens. The third class, called suppressing agents, is thought to act by preventing the progression of initiated cells to fully transformed tumour cells.

Isothiocyanates that arise in plants as a result of enzymatic cleavage of glucosinolates by the endogenous enzyme myrosinase are attracting increasing attention as chemical and dietary protectors against cancer. Their anticarcinogenic activities have been demonstrated in rodents (mice and rats) with a wide variety of chemical carcinogens (Table 3.2). The anticarcinogenic effects of ITCs can be explained by two different mechanisms. The first, a blocking effect, involves induction of Phase II enzymes, including quinone reductase in the small intestinal mucosa and liver (Zhang *et al.*, 1992b; Talalay and Zhang, 1996). These enzymes are involved in the detoxification in the body of foreign compounds (xenobiotics). Increased activity will therefore block exposure of target tissues to DNA damage (Fimognari and Hrelia, 2007). The second mechanism, a suppressing effect, involves suppression of tumour development via deletion of damaged cells from colonic mucosal crypts through the induction of programmed cell death (apoptosis). Smith *et al.* (1996) showed that dietary supplementation with the glucosinolate sinigrin, or its breakdown product allyl isothiocyanate, can protect against chemically-induced colorectal carcinogenesis by stimulation of apoptosis.

The most characterized isothiocyanate is sulforaphane because of its ability to simultaneously modulate multiple cellular targets involved in cancer development (Fimognari and Hrelia, 2007).

The evidence for anticarcinogenic effects of *Brassica* vegetables in humans is strongly supported by evidence obtained from epidemiological and human intervention studies as well as with experimental animals. Verhoeven *et al.* (1996) reviewed 7 cohort studies and 87 case–control studies and showed an inverse correlation between the consumption of individual *Brassica* vegetables and the risk of lung, stomach and second primary cancers. Broccoli consumption in particular shows a uniform protective effect, with no contrary evidence in any study. Consumption of *Brassica* vegetables, which might be expected to yield high levels of indoles and ITCs, was particularly strongly associated with a lower risk of colon cancer. Brennan *et al.* (2005) showed that the protective effects of a diet rich in cruciferous vegetables towards the occurrence of lung cancer is strongly dependent on the human genetic background (GSTM1 and GSTT1 background). Reduction of risks up to 72% was reported in this study. High intake of glucoraphanin can be obtained by the consumption of broccoli sprouts. The effect of this intake is studied in clinical trials now (Shapiro *et al.*, 2006).

3.6.2 Toxicity

Initially, glucosinolates were studied because of their potentially deleterious effects. A lot of attention was given to removal of glucosinolates from dietary sources and animal feed by breeding ('double zero' rapeseed) and processing of plant material.

In the field of animal production it is well-known that ingestion of substantial amounts of glucosinolates may result in a variety of toxic and antinutritional effects. The fodder and seed meals of genus *Brassica* such as crambe, kale, mustard, rape, cabbage and turnips are the main source of glucosinolates in animal diets. Major deleterious effects of glucosinolates ingestion in animals are reduced palatability, decreased growth and production. Nitriles are known to affect liver and kidney functions. The ITCs interfere with iodine availability, whereas 5-vinyl-2-oxazolidinethione (VOT) is responsible for the morphological and physiological changes of the thyroid. Other adverse effects of glucosinolate metabolites are goitrogenicity and mutagenicity. Deleterious effects of glucosinolates are greater in non-ruminant animals compared to ruminants. Also, in general, young animals appear to be more sensitive to glucosinolates than adults or older animals. Animal species are affected in different degrees of severity; pigs are more severely affected by dietary glucosinolates than rabbits, poultry and fish (Tripathi and Mishra, 2007).

The oil meal of *Brassica* origin is a good source of protein for animal feeding but its glucosinolate content limits its efficient utilization. Presently, very low-glucosinolate rapeseed varieties are available that contain less than 25 μmol g^{-1} of total glucosinolates. Furthermore, various processing techniques were applied to remove glucosinolates in order to minimize their deleterious effects on animals. Most of these methodologies, like microwave irradiation, micronization and extrusion, included hydrolysis or decomposition of glucosinolate before feeding (Tripathi and Mishra, 2007). Also, a reduction in glucosinolate content could be obtained by autoclaving rapeseed meal for 1.5 hours (Mansour *et al.*, 1993), treatment of meal with Cu^{2+} and the use of ammonia in conjugation with other processing (Keith and Bell, 1982). During seed processing most glucosinolate breakdown products are formed. The degree of degradation depends on seed properties and on processing conditions such as moisture level, pressure and temperature (Mawson *et al.*, 1993).

In principle, glucosinolate breakdown products can be capable of inducing goitrogenic effects in humans, but there is little or no epidemiological evidence that this is an important cause of human disease. Experimentally, consumption of 150 g of Brussels sprouts in the diets of adult volunteers had no effect on their levels of thyroid hormones (McMillan *et al.*, 1986).

The current evidence suggests that normal consumption of glucosinolate containing vegetables does not damage human health. In contrast, beneficial effects are prevailing. On the other hand, there is a need for further studies (e.g. human intervention studies) because the growing interest in the anticarcinogenic properties of glucosinolates and the possibility that this may lead to increased human exposure to these compounds raises important questions about the balance of adverse and beneficial effects of *Brassica* vegetables or derived products with enhanced levels.

3.7   TASTE VERSUS HEALTH

Many health-protective dietary phytonutrients found in vegetables are bitter, acrid or astringent and therefore aversive for consumers. As a result, the food industry routinely removes these compounds from plant foods through selective breeding and various debittering processes (Drewnowski and Gomez-Carneros, 2000). This poses an interesting dilemma for designing healthy foods (e.g. functional foods) because increasing the content of bitter phytonutrients for health may not always be accepted by consumers. Also, several of the glucosinolates play an important role in the characteristic flavour and odour of *Brassica* vegetables. It has been shown that the glucosinolates sinigrin and progoitrin are involved in the bitterness observed in Brussels sprouts (Fenwick *et al.*, 1983). Van Doorn *et al.* (1998) confirmed the role of sinigrin and progoitrin in taste preference by using taste trials with samples of Brussels sprouts. It appeared that consumers preferred Brussels sprouts with a low sinigrin and progoitrin content. In cabbage, sinigrin is an abundant glucosinolate that gives a pungent and bitter flavour. The stronger flavour in the heart of the cabbage is consistent with the presence of greater amounts of sinigrin found in cabbage heads. Low levels of 2-propenyl ITCs formed from sinigrin result in a flat and dull product (Rosa *et al.*, 1997). Pungency and bitterness caused by glucosinolate breakdown products play a role in the taste preference of consumers and are therefore important quality factors for *Brassica* vegetables and derived products.

It has been shown that the flavour and nutritional properties of *Brassica* vegetables could be improved by using molecular markers in selecting specific glucosinolate lines in breeding programmes (Campos-De Quiroz and Mithen, 1996). Developing vegetables less susceptible to diseases, less attractive to insects and with desirable agronomic storage and sensory characteristics by manipulating the glucosinolate levels resulted in crops with greater commercial value (Borek *et al.*, 1994; Brown and Morra, 1995).

In general, studies on phytonutrients and health rarely consider the taste of vegetables and other plant-based foods. Glucosinolates and ITCs are regarded as dietary protectors against cancer. However, the protection capacity, ascribed to inducing detoxifying enzyme, has been linked to concentrations of bitter glucosinolates such as sinigrin, progoitrin, glucobrassicin and glucoraphanin.

Various promising approaches are available to enhance levels of health-protective phytonutrients in *Brassica* vegetables and derived products, like engineering plant foods or processes in the production chain. However, when it comes to bitter phytonutrients, the demands of good taste and good health could be difficult to overcome or even be wholly incompatible.

Therefore, it is proposed to consider sensory factors and food preferences when studying phytonutrients and health (Drewnowski and Gomez-Carneros, 2000).

3.8   RESPONSES TO STRESS FACTORS

Glucosinolates and their breakdown products are considered to function as part of the plant's defence against insect attack and to act as phagostimulants (Chew, 1988).

There is now considerable information on the importance of glucosinolates in insect–plant interactions. However, less is known about the influence of biotic factors on glucosinolate metabolism in plants. It has been demonstrated that attack by aphids (Lammerink *et al.*, 1984), root flies (Birch *et al.*, 1990, 1992) and flea beetles (Koritsas *et al.*, 1991), changes both the total concentration of glucosinolates in different plant tissues and the relative proportions of aliphatic and aromatic compounds.

Other examples of stress-induced increases in levels of glucosinolates are mechanical wounding and infestation (Koritsas *et al.*, 1991), methyl jasmonate exposure (Doughty *et al.*, 1995), and grazing (Macfarlane Smith *et al.*, 1991) for intact plants or UV-irradiation (Monde *et al.*, 1991) and chopping (Verkerk *et al.*, 2001) for vegetables after harvesting. Apparently, besides a breakdown mechanism for glucosinolates, an induction mechanism of glucosinolate biosynthesis by stress factors is present in brassica vegetables.

3.9   EFFECTS OF PROCESSING

Many steps in the food supply chain, such as breeding, cultivation, storage, processing and preparation of vegetables, may have an impact on levels and thus intake of glucosinolates (Dekker *et al.*, 2000). *Brassica* vegetables are, prior to consumption, subjected to different ways of processing, culinary as well as industrial. Culinary treatments of *Brassica* vegetables such as chopping, cooking, steaming and microwaving have received more attention in past years and have been shown to affect the glucosinolate content considerably. Typically, post-harvest physical disruption of the plants such as chewing, chopping, blending, juicing, cooking, freezing/thawing, and high temperature leads to loss of cellular compartmentalization and subsequent mixing of glucosinolates and myrosinase to form ITCs. This influences the levels of glucosinolates, the extent of hydrolysis and the composition, flavour and aroma of the final products.

Culinary processing of *Brassica* vegetables has complex influences on the food matrix affecting the level of glucosinolates:

- Enzymatic hydrolysis by myrosinase
- Myrosinase inactivation
- Cell lysis and leaching of glucosinolates, breakdown products and myrosinase in cooking water
- Thermal degradation of glucosinolates and their breakdown products
- Increase of the chemical glucosinolate extractability
- Loss of enzymatic co-factors (e.g. ascorbic acid, iron)

Chopping of fresh plant tissues creates optimal conditions for myrosinase and a high degree of glucosinolate hydrolysis can be expected. In contrast to these expectations and reported findings, Verkerk *et al.* (2001) observed elevated levels of all indole and some aliphatic glu-cosinolates after chopping and prolonged exposure of *Brassica* vegetables to air of different kinds. In white cabbage a 15-fold increase of 4-methoxy and 1-methoxy-3-indolylmethyl glucosinolates was noted after 48 hours storage of chopped cabbage. Chopping and storage of broccoli vegetables resulted in a strong reduction of most glucosinolates, except for 4-hydroxy- and 4-methoxy-3-indolylmethyl glucosinolates, which increased 3.5- and 2-fold, respectively. It was hypothesized that chopping triggers a de novo synthesis of glucosino-lates, especially indolyl GS, by mimicking pest damage as defence mechanism in harvested *Brassica* vegetables.

Low-temperature storage processes such as freezing and refrigerating can alter the metabolism of glucosinolates. Freezing without previous inactivation of myrosinase results in an almost complete decomposition of glucosinolates after thawing (Quinsac *et al.*, 1994).

Ciska and Pathak (2004) have studied the fermentation process of white cabbage and further storage. During fermentation all glucosinolates were hydrolysed within 2 weeks resulting in 15 products of glucosinolate degradation such as thiocyanates, ITCs, cyanides, nitriles, indole compounds and others. Ascorbigen formed from glucobrassicin was found to be a dominating compound in fermented cabbage, reaching levels of about 14 μmol 100 g^{-1}. The content of breakdown products in fermented cabbage depends, besides on the content of the native glucosinolates in raw cabbage, on physicochemical properties such as volatility, stability and reactivity in an acidic environment and microbiological stability (Ciska and Pathak, 2004).

Boiling of *Brassica* vegetables in water reduces glucosinolate levels significantly, mostly by leaching into the cooking water and thermal degradation. The amount of losses depends on the sort of vegetable, cooking time, ratio vegetable/water and also on the type of glucosinolate (Rosa and Heaney, 1993; Ciska and Kozlowska, 2001; Vallejo *et al.*, 2002; Oerlemans *et al.*, 2006).

Interestingly, microwave preparation of *Brassica* vegetables resulted in high retention of glucosinolates (Verkerk and Dekker, 2004; Song and Thornalley, 2007). Moreover, Verkerk and Dekker (2004) observed an increase in glucosinolate levels for red cabbage associated with the applied energy input. They ascribed these findings to an increased extractability of glucosinolates by thermal treatment.

Steaming and stir-frying as a culinary treatment of vegetables seem to be milder processes since high retention of glucosinolates appeared to occur, mainly as a result of limited leaching and thermal degradation (Rungapamestry *et al.*, 2006; Song and Thornalley, 2007).

During industrial processing of *Brassica* vegetables (e.g. canning), the thermal treat-ment can affect glucosinolate levels considerably. Oerlemans *et al.* (2006) described thermal degradation of individual glucosinolates in red cabbage. Degradation of all the identified glu-cosinolates occurred when heated at temperatures above 100°C. The indole glucosinolates 4-hydroxy-glucobrassicin and 4-methoxyglucobrassicin appeared to be most susceptible to thermal degradation, even at temperatures below 100°C. Canning, the most severe heat treat-ment, will result in substantial thermal degradation (73%) of the total amount of glucosinolates (Table 3.3).

It is evident that processing can result in a strong decrease or increase of glucosinolates which could influence the properties, such as flavour and anticarcinogenicity, of brassica vegetables.

The variation in glucosinolate profile and content of consumed products due to variation in the entire vegetable supply chain (breeding, growing, storage, processing, and preparation) has been shown to be enormous. The level of glucosinolates varies over 100-fold in consumed

Table 3.3 Predicted effects of three different thermal treatments (blanching, cooking, canning) on the residual percentage of glucosinolates in red cabbage as a result of thermal degradation.

	Initial concentration (μmol/100 g FW)	Blanching (3 min, 95°C) (%)	Cooking (40 min, 100°C) (%)	Canning (40 min, 120°C) (%)
Glucoiberin	14.8	100	94	18
Progoitrin	23.8	100	93	38
Sinigrin	14.7	100	91	12
Glucoraphanin	48.2	100	90	15
Gluconapin	36.9	100	93	53
4-Hydroxyglucobrassicin	1.9	93	26	3
Glucobrassicin	8.8	99	72	1
4-Methoxyglucobrassicin	1.6	97	48	1
Total aliphatic gls	138.4	100	92	29
Total indole gls	12.34	98	62	2
Total gls	150.8	100	89	27

Reproduced from Oerlemans, K., Barrett, D.M., Bosch Suades, C., Verkerk, R. and Dekker, M. Thermal degradation of glucosinolates in red cabbage. *Food Chemistry*, **95**, 19–29. Copyright 2006 with permission from Elsevier.

vegetable products (Dekker and Verkerk, 2003). The effect of this variation on the health protecting effect for humans was simulated by a Monte Carlo approach by calibrating this protective effect to reported epidemiological studies (Fig. 3.4). These simulations predicted that increasing the average level of glucosinolates by 3-fold through optimization of the food

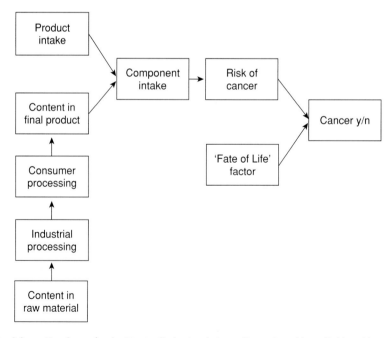

Fig. 3.4 Schematic scheme for the Monte Carlo simulations. (Reproduced from Dekker, M. and Verkerk, R. Dealing with variability in food production chains: A tool to enhance the sensitivity of epidemiological studies on phytochemicals. *European Journal of Nutrition*, **42**, 67–72. Copyright 2003 with kind permission from Springer Science and Business Media.)

supply chain could reduce the number of colon cancer cases by 45% (Dekker and Verkerk, 2003).

Conclusions

Glucosinolate research has been expanded from toxicological areas into the health-promoting areas. Much of this research nowadays is focused on the potential effects of glucosinolates, and their breakdown products, on biological processes associated with cellular damage and cancer prevention. Recent publications in the area of nutrigenomics show more insight into the effects of these compounds on gene expression related to health protection. Analysis of the effects of the supply chain on the delivery of glucosinolates to consumers shows enormous potential for optimization of the levels of these compounds in the final food products. The role and mechanisms of action of glucosinolates and their products in protecting plants against fungal and insect attack and the allelochemical effects on behaviour, health and growth of other species is another important area of research.

References

Barker, A.M., Molotsane, R., Müller, C., Schaffner, U. and Städler, E. (2006) Chemosensory and behavioural responses of the sawfly, *Athalia rosae*, to glucosinolates and isothiocyanates. *Chemoecology*, **16**, 209–218.

Birch, A.N.E., Griffiths, D.W. and MacFarlane-Smith, W.H. (1990) Change in forage and oilseed rape (*B. napus*) root glucosinolates in response to attack by turnip root fly (*Delia floralis*) larvae. *Journal of the Science of Food and Agriculture*, **51**, 309–320.

Birch, A.N.E., Griffiths, D.W., Hopkins, R.J., Macfarlane-Smith, W.H. and McKinlay, R.G. (1992) Glucosinolate responses of swede, kale, forage and oilseed rape to root damage by turnip root fly (*Delia floralis*) larvae. *Journal of the Science of Food and Agriculture*, **60**, 1–9.

Blanc-Muesser, M., Drigues, H., Joseph, B., Viaud, M.C. and Rollin, P. (1990) First synthesis of *alpha*-glucosinolates. *Tetrahedron Letters*, **31**, 3867–3868.

Bones, A.M. (1990) Distribution of ß-thioglucosidase activity in intact plants, cell and tissue cultures and regenerant plants of *Brassica napus* L. *Journal of Experimental Botany*, **227**, 737–744.

Bones, A.M. and Rossiter, J.T. (1996) The myrosinase-glucosinolate system, its organisation and biochemistry. *Physiologia Plantarum*, **97**, 194–208.

Borek, V., Morra, M.J., Brown, P.D. and McCaffrey, J.P. (1994) Allelochemicals produced during sinigrin decomposition in soil. *Journal of Agricultural Food Chemistry*, **42**, 1030–1034.

Botting, C.H., Davidson, N.E., Griffiths, D.W., Bennett, R.N. and Botting, N.P. (2002) Analysis of intact glucosinolates by MALDI-TOF mass spectrometry. *Journal of Agricultural Food Chemistry*, **50**, 983–988.

Brennan, P., Hsu, C., Moullan, N., Szeszenia-Dabrowska, N., Lissowska, J., Zaridze, D., Rudnai, P., Fabianova, E., Mates, D. and Bencko, V. (2005) Effect of cruciferous vegetables on lung cancer in patients stratified by genetic status: A mendelian randomisation approach. *The Lancet*, **366**, 1558–1560.

Brown, P.D. and Morra, M.J. (1995) Glucosinolate-containing plant tissues as bioherbicides. *Journal of Agricultural Food Chemistry*, **43**, 3070–3074.

Buchwaldt, L., Melchior Larsen, L., Ploger, A. and Sørensen, H. (1986) Fast polymer liquid chromatography isolation and characterization of plant myrosinase, *β*-thioglucoside glucohydrolase isoenzymes. *Journal of Chromatography*, **363**, 71–80.

Campos-De Quiroz, H. and Mithen, R. (1996) Molecular markers for low-glucosinolate alleles in oilseed rape (*Brassica napus* L.). *Molecular Breeding*, **2**(3), 277–281.

Charron, C.S., Saxton, A.M. and Sams, C.E. (2005) Relationship of climate and genotype to seasonal variation in the glucosinolate-myrosinase system. II: Myrosinase activity in ten cultivars of *Brassica oleracea* grown in fall and spring seasons. *Journal of the Science of Food and Agriculture*, **85**, 682–690.

Chew, F.S. (1988) Biological effects of glucosinolates, in *Biologically Active Natural Products for Potential Use in Agriculture* (ed. H.G. Cutler). American Chemical Society, Washington, DC, pp. 155–181.

Ciska, E. and Kozlowska, H. (2001) The effect of cooking on the glucosinolates content in white cabbage. *European Food Research and Technology*, **212**, 582–587.

Ciska, E. and Pathak, D.R. (2004) Glucosinolate derivatives in stored fermented cabbage. *Journal of Agricultural Food Chemistry*, **52**, 7938–7943.

Dekker, M., Verkerk, R. and Jongen, W.M.F. (2000) Predictive modelling of health aspects in the food production chain, a case study on glucosinolates in cabbage. *Trends in Food Science and Technology*, **11**(4), 174–181.

Dekker, M. and Verkerk, R. (2003) Dealing with variability in food production chains: A tool to enhance the sensitivity of epidemiological studies on phytochemicals. *European Journal of Nutrition*, **42**, 67–72.

Doughty, K.J., Kiddle, G.A., Pye, P.J., Wallsgrove, R.M. and Pickett, J.A. (1995) Selective induction of glucosinolates in oilseed rape leaves by methyl jasmonate. *Phytochemistry*, **38**(2), 347–350.

Dragsted, L.O., Strube, M. and Larsen, J.C. (1993) Cancer-protective factors in fruits and vegetables: Biochemical and biological background. *Pharmacology and Toxicology*, **72**(suppl 1), 116–135.

Drewnowski, A. and Gomez-Carneros, C. (2000) Bitter taste, phytonutrients, and the consumer: A review. *American Journal of Clinical Nutrition*, **72**, 1424–1435.

Fahey, J.W., Zalcmann, A.T. and Talalay, P. (2001) The chemical diversity and distribution of glucosinolates and isothiocyanates among plants. *Phytochemistry*, **56**, 5–51.

Fenwick, G.R., Griffiths, N.M. and Heaney, R.K. (1983) Bitterness in Brussels sprouts (*Brassica oleracea* L. var gemmifera): The role of glucosinolates and their breakdown products. *Journal of the Science of Food and Agriculture*, **34**, 73–80.

Fimognari, C. and Hrelia, P. (2007) Sulforaphane as a promising molecule for fighting cancer. *Mutation Research*, **635**, 90–104.

Font, R., Del Rio-Celestino, M., Fernández-Martinez, J.M. and De Haro-Bailón, A. (2004) Use of near-infrared spectroscopy for screening the individual and total glucosinolate content in Indian mustard seed (*Brassica juncea* L. Czern. & Coss.). *Journal of Agricultural Food Chemistry*, **52**, 3563–3569.

Font, R., Del Rio-Celestino, M., Cartea, E. and De Haro-Bailón, A. (2005) Quantification of glucosinolates in leaves of leave rape (*Brassica napus* ssp. Pabularia) by near-infrared spectroscopy. *Phytochemistry*, **66**, 175–185.

Gildemeister, E. and Hofmann, F. (1927) *Die Aetherische Ole*, 3rd edn, Vol. 1. Schimmel and Co, Leipzig, p. 145.

Godeschalk, F.E. (1987) Consumptie van voedingsmiddelen in Nederland in 1984 en 1985, in *Periodieke Rapportage*, Vol. 64. LEI, The Hague.

Griffiths, D.W., Birch, A.N.E. and Hillman, J.R. (1998) Anti-nutritional compounds in the Brassicaceae: Analysis, biochemistry, chemistry and dietary effects. *Journal of Horticultural Science and Biotechnology*, **73**, 1–18.

Halkier, B.A. and Lindberg-Møller, B. (1991) Involvement of cytochrome P-450 in the biosynthesis of dhurrin in *Sorghum bicolor* (L.) Moench. *Plant Physiology*, **96**, 10–17.

Halkier, B.A. and Gershenzon, J. (2006) Biology and biochemistry of glucosinolates. *Annual Review of Plant Biology*, **57**, 303–333.

Hanley, A.B., Parsley, K.R., Lewis, J.A. and Fenwick, G.R. (1990) Chemistry of indole glucosinolates: Intermediacy of indol-3-ylmethylisothiocyanates in the enzymic hydrolysis of indole glucosinolates. *Journal of the Chemical Society Perkin Transactions*, **1**, 2273–2276.

Heaney, R.K., Spinks, E.A. and Fenwick, G.R. (1988) Improved method for the determination of the total glucosinolate content of rapeseed by determination of enzymatically released glucose. *Analyst*, **113**, 1515–1518.

Heaney, R.K. and Fenwick, G.R. (1993) Methods for glucosinolate analysis, in *Methods in Plant Biochemistry*, Vol. 6 (ed. P.G. Waterman). Academic Press, London, pp. 531–550.

Holst, B. and Williamson, G. (2004) A critical review of the bioavailability of glucosinolates and related compounds. *Natural Product Reports*, **21**, 425–447.

James, D. and Rossiter, J.T. (1991) Development and characteristics of myrosinase in *Brassica napus* during early seedling growth. *Plant Physiology*, **82**, 163–170.

Jongen, W.M.F. (1996) Glucosinolates in brassica: Occurrence and significance as cancer-modulating agents. *Proceedings of the Nutrition Society*, **55**, 433–446.

Karas, M. (1996) Matrix-assisted laser desorption ionization MS; a progress report. *Biochemical Society Transactions*, **24**, 897–900.

Karas, M., Bachmann, D., Bahr, U. and Hillenkamp, F. (1987) Matrix-assisted ultraviolet laser desorption of non-volatile compounds. *International Journal of Mass Spectrometry and Ion Processes*, **78**, 53–68.

Keith, M.O. and Bell, J.M. (1982) Effects of ammoniation on the composition and nutritional quality of low glucosinolate rapeseed (canola) meal. *Canadian Journal of Animal Science*, **62**, 547–555.

Kiddle, G., Bennett, R.N., Botting, N.P., Davidson, N.E., Robertson, A.A.B. and Wallsgrove, R.M. (2001) High-performance liquid chromatographic separation of natural and synthetic desulphoglucosinolates and their chemical validation by UV, NMR and chemical ionisation–MS methods. *Phytochemical Analysis*, **12**, 226–242.

Kjaer, A. (1976) Glucosinolates in the cruciferae, in *The Biology and Chemistry of the Cruciferae* (eds. J.G. Vaughan, A.J. MacLeod and B.M.G. Jones). Academic Press, London, pp. 207–219.

Kjaer, A. and Conti, J. (1954) Isothiocyanates, VII: A convenient synthesis of erysoline. *Acta Chemica Scandinavica*, **8**, 295–298.

Koch, B., Nielson, V.S. and Halkier, B.A. (1992) Biosynthesis of cyanogenic glucosides in seedlings of cassava. *Archives of Biochemistry and Biophysics*, **292**, 141–150.

Koritsas, V.M., Lewis, J.A. and Fenwick, G.R. (1991) Glucosinolate responses of oilseed rape, mustard and kale to mechanical wounding and infestation by cabbage stem fleabeetle (*Psylliodes chrysocephala*). *Annual Applied Biology*, **118**, 209–221.

Lammerink, J., MacGibbon, D.B. and Wallace, A.R. (1984) Effect of cabbage aphid on total glucosinolate in the seed of oilseed rape. *New Zealand Journal of Agricultural Research*, **27**, 89–92.

Macfarlane Smith, W.H., Griffiths, D.W. and Boag, B. (1991) Overwintering variation in glucosinolate content of green tissue of rape (*Brassica napus*) in response to grazing by wild rabbit (*Oryctolagus cuniculus*). *Journal of the Science of Food and Agriculture*, **56**, 511–521.

MacGibbon, D.B. and Allison, R.M. (1970) A method for the separation and detection of plant glucosinolases (myrosinases). *Phytochemistry*, **9**, 541–544.

Mansour, E.H., Dworschák, E., Lugasi, A., Gaál, Ö., Barna, É. and Gergely, A. (1993) Effects of processing on the antinutritive factors and nutritive value of rapeseed products. *Food Chemistry*, 47, 247–252.

Matusheski, N.V., Swarup, R., Juvik, J.A., Mithen, R., Bennett, M. and Jeffery, E.H. (2006) Epithiospecifier protein from Broccoli (*Brassica oleracea* L. ssp. italica) inhibits formation of the anticancer agent sulforaphane. *Journal of Agricultural Food Chemistry*, **54**(6), 2069–2076.

Mawson, R., Heaney, R.K., Piskula, M. and Kozlowska, H. (1993) Rapeseed meal glucosinolates and their antinutritional effects. 1: Rapeseed production and chemistry of glucosinolates. *Die Nahrung*, **37**, 131–140.

McMillan, M., Spinks, E.A. and Fenwick, G.R. (1986) Preliminary observations on the effect of dietary Brussels sprouts on thyroid function. *Human Toxicology*, **5**, 15–19.

McNaughton, S.A. and Marks, G.C. (2003) Development of a food composition database for the estimation of dietary intakes of glucosinolates, the biologically active constituents of cruciferous vegetables. *British Journal of Nutrition*, **90**, 687–697.

Mellon, F.A., Bennett, R.N., Holst, B. and Williamson, G. (2002) Intact glucosinolate analysis in plant extracts by programmed cone voltage electrospray LC/MS: Performance and comparison with LC/MS/MS methods. *Analytical Biochemistry*, **306**, 83–91.

Mithen, R.F., Dekker, M., Verkerk, R., Rabot, S. and Johnson, I.T. (2000) The nutritional significance, biosynthesis and bioavailability of glucosinolates in human foods. *Journal of the Science of Food and Agriculture*, **80**, 967–984.

Monde, K., Takasugi, M., Lewis, J.A. and Fenwick, G.R. (1991) Time-course studies of phytoalexins and glucosinolates in UV irradiated turnip tissue. *Zeitschrift fur Naturforschung Section C Biosciences*, **46**(3–4), 189–193.

Oerlemans, K., Barrett, D.M., Bosch Suades, C., Verkerk, R. and Dekker M. (2006) Thermal degradation of glucosinolates in red cabbage. *Food Chemistry*, **95**, 19–29.

Ohtsuru, M. and Hata, T. (1973) General characteristics of the intracellular myrosinase from *A. niger.Agricultural and Biological Chemistry*, **37**, 2543–2548.

Quinsac, A., Ribaillier, D., Rollin, P. and Dreux, M. (1992) Analysis of 5-vinyl-1,3-oxazolidine-2-thione by liquid chromatography. *Journal of the American Official Analytical Chemists*, **75**, 529–536.

Quinsac, A., Charrier, A. and Ribaillier, D. (1994) Glucosinolates in etiolated sprouts of sea-kale (*Crambe maritima* L.). *Journal of the Science of Food and Agriculture*, **65**, 201–207.

Rosa, E.A.S. and Heaney, R.K. (1993) The effect of cooking and processing on the glucosinolate content: Studies on four varieties of Portuguese cabbage and hybrid white cabbage. *Journal of the Science of Food and Agriculture*, **62**, 259–265.

Rosa, E.A.S., Heaney, R.K., Fenwick, G.R. and Portas, C.A.M. (1997) Glucosinolates in crop plants. *Horticultural Reviews*, **19**, 99–215.

Rungapamestry, V., Duncan, A.J., Fuller, Z. and Ratcliffe, B. (2006) Changes in glucosinolate concentrations, myrosinase activity, and production of metabolites of glucosinolates in cabbage (*Brassica oleracea* var. capitata) cooked for different durations. *Journal of Agricultural Food Chemistry*, **54**, 7628–7634.

Schnug, E. (1987) Fine methode zur schnellen und einfachen bestimmung des gesamtglucosinolatgehaltes in grünmasse und samen von kruziferen durch die quantitative analyse enzymatisch freisetzharen sulfates. *Fat Science Technology*, **89**, 438–442.

Schreiner, M. (2005) Vegetable crop management strategies to increase the quantity of phytochemicals. *European Journal of Nutrition*, **44**(2), 85–94.

Shapiro, T.A., Fahey, J.W., Wade, K.L., Stephenson, K.K. and Talalay, P. (1998) Human metabolism and excretion of cancer chemoprotective glucosinolates and isothiocyanates of cruciferous vegetables. *Cancer Epidemiology Biomarkers and Prevention*, **7**(12), 1091–1100.

Shapiro, T.A., Fahey, J.W., Dinkova-Kostova, A.T., Holtzclaw, W.D., Stephenson, K.K., Wade, K.L., Lingxiang, Y. and Talalay, P. (2006) Safety, tolerance, and metabolism of broccoli sprout glucosinolates and isothiocyanates: A clinical phase I study. *Nutrition and Cancer*, **55**(1), 53–62.

Smith, T.K., Musk, S.R.R. and Johnson, I.T. (1996) Allyl isothiocyanate selectively kills undifferentiated HT 29 cells *in vitro* and suppresses aberrant crypt foci in the colonic mucosa of rats. *Biochemical Society Transactions*, **24**(3), s381.

Song, L., Morrison, J.J., Botting, N.P. and Thornalley, P.J. (2005) Analysis of glucosinolates, isothiocyanates, and amine degradation products in vegetable extracts and blood plasma by LC–MS/MS. *Analytical Biochemistry*, **347**, 234–243.

Song, L. and Thornalley, P.J. (2007) Effect of storage, processing and cooking on glucosinolate content of *Brassica* vegetables. *Food and Chemical Toxicology*, **45**, 216–224.

Spinks, E.A., Sones, K. and Fenwick, G.R. (1984) The quantitative analysis of glucosinolates in cruciferous vegetables, oilseeds and forage crops using high performance liquid chromatography. *Fette, Seifen, Anstrichmittel*, **86**(6), 228–231.

Sørensen, H. (1990) Glucosinolates: Structure properties function, in *Canola and Rapeseed* (ed. F. Shahdi). Van Nostrand, New York, pp. 149–172.

Talalay, P. and Zhang, Y. (1996) Chemoprotection against cancer by isothiocyanates and glucosinolates. *Biochemical Society Transactions*, **24**, 806–810.

Thangstad, O.P., Iversen, T.H., Sluphaug, G. and Bones, A. (1990) Immunocytochemical localization of myrosinase in *Brassica napus* L. *Planta*, **180**, 245–248.

Tian, Q.G., Rosselot, R.A. and Schwartz, S.J. (2005) Quantitative determination of intact glucosinolates in broccoli, broccoli sprouts, Brussels sprouts, and cauliflower by high-performance liquid chromatography-electrospray ionization-tandem mass spectrometry. *Analytical Biochemistry*, **343**, 93–99.

Tiedink, H.G.M., Hissink, A.M., Lodema, S.M., van Broekhoven, L.W. and Jongen, W.M.F. (1990) Several known indole compounds are not important precursors of direct mutagenic N-nitroso compounds in green cabbage. *Mutation Research*, **232**, 199–207.

Tiedink, H.G.M., Malingre, C.E., van Broekhoven, L.W., Jongen, W.M.F., Lewis, J. and Fenwick, G.R. (1991) The role of glucosinolates in the formation of N-nitroso compounds. *Journal of Agricultural Food Chemistry*, **39**, 922–926.

Tookey, H.L., VanEtten, C.H. and Daxenbichler, M.E. (1980) Glucosinolates, in *Toxic Constituents of Plant Foodstuffs* (ed. I.E. Liener). Academic Press, New York, pp. 103–142.

Tripathi, M.K. and Mishra, A.S. (2007) Glucosinolates in animal nutrition: A review. *Animal Feed Science and Technology*, **132**, 1–27.

Vallejo, F., Tomás-Barberán, F.A. and Garcia-Viguera, C. (2002) Glucosinolates and vitamin C content in edible parts of broccoli florets after domestic cooking. *European Food Research and Technology*, **215**, 310–316.

Van Doorn, H.E., Van Der Kruk, G.C., Van Holst, G.J., Raaijmakers-Ruijs, N.C.M.E., Postma, E., Groeneweg, B. and Jongen, W.M.F. (1998) The glucosinolates Sinigrin and Progoitrin are important determinants for taste preference and bitterness of Brussels sprouts. *Journal of the Science of Food and Agriculture*, **78**, 30–38.

Verkerk, R., Dekker, M. and Jongen, W.M.F. (2001) Post-harvest increase of indolyl glucosinolates as response to chopping and storage of *Brassica* vegetables. *Journal of the Science of Food and Agriculture*, **81**, 953–958.

Verkerk, R. and Dekker, M. (2004) Glucosinolates and myrosinase activity in red cabbage (*Brassica oleracea* L. Var. *Capitata* f. *rubra* DC.) after various microwave treatments. *Journal of Agricultural Food Chemistry*, **52**, 7318–7323.

Velasco, L. and Becker, H.C. (1998) Analysis of total glucosinolate content and individual glucosinolates in Brassica spp. by near-infrared reflectance spectroscopy. *Plant Breeding*, **117**, 97–102.

Verhoeven, D.T.H., Goldbohm, R.A., van Poppel, G., Verhagen, H. and van den Brandt, P.A. (1996) Epidemiological studies on Brassica vegetables and cancer risk. *Cancer Epidemiol Biomarkers and Prevention*, **5**, 733–751.

Wattenberg, L.W. (1985) Chemoprevention of cancer. *Cancer Research*, **45**, 1–8.

Wattenberg, L.W. (1992) Inhibition of carcinogenesis by minor dietary constituents. *Cancer Research*, **52**, 2085s–2091s.

Zhang, Y., Cho, C.-G., Posner, G.H. and Talalay, P. (1992a) Spectroscopic quantitation of organic isothiocyanates by cyclocondensation with vicinal dithiols. *Analytical Biochemistry*, **205**, 100–107.

Zhang, Y., Talalay, P., Cho, C.-G. and Posner, G.H. (1992b) A major inducer of anticarcinogenic protective enzymes from broccoli: Isolation and elucidation of structure. *Proceedings of the National Academy of Sciences of the USA*, **89**, 2399–2403.

4   Phycotoxins in Seafood

John W. Leftley and Fiona Hannah

Summary

Phycotoxins, produced by microscopic marine algae, may contaminate shellfish and fish, causing illness in humans, and are a world-wide problem. This chapter covers their origin, distribution, classification, aspects of their toxicology, methods of analysis and statutory regulation.

4.1   INTRODUCTION

Contamination of seafood with phycotoxins and the resultant effects on human health is a world-wide problem and research into all aspects of these compounds is being conducted in many countries. Consequently, there is a plethora of data and it is possible here to give only an overview of the principal phycotoxins. Of the literature cited in this chapter, the reader is referred especially to three comprehensive publications: Anderson *et al.* (2001), Hallegraeff *et al.* (2004) and van Egmond *et al.* (2004), the first two produced by the Intergovernmental Oceanographic Commission (IOC) of UNESCO (see Section 4.15) and the latter by the UN Food and Agriculture Organization, each of which amplify many of the topics covered below. In addition, detailed information on the chemistry and pharmacology of phycotoxins can be found in Botana (2007).

4.2   CAUSATIVE AND VECTOR ORGANISMS

The organisms that produce the phycotoxins described here are certain genera of unicellular algae, usually photosynthetic, which may be planktonic (free floating or swimming) or benthic, i.e. live on surfaces of, for example, plants and corals, or in or on marine sediments. With the exception of the diatoms that produce the toxins that cause amnesic shellfish poisoning (ASP), almost all the other toxic algae discussed here are dinoflagellates (Table 4.1). Detailed accounts of the taxonomy of the algae implicated in phycotoxin poisoning have been given by Steidinger (1993) and in part II of Hallegraeff *et al.* (2004). The phycotoxins can be regarded as secondary metabolites and their taxonomic distribution, biosynthesis and significance in the food chain have been reviewed by Shimizu (1996).

Seafood poisoning is often associated with the occurrence of algal 'blooms', where the microscopic algae may reach sufficient density to produce a visible discolouration of the water. These blooms, not necessarily toxic, are a natural phenomenon and occur when there is

Table 4.1 Phycotoxin poisoning: causative and vector organisms, clinical symptoms and treatment.

Paralytic shellfish poisoning (PSP)	Diarrhetic shellfish poisoning (DSP)	Amnesic shellfish poisoning (ASP)	Azaspiracid shellfish poisoning (AZP)	Neurotoxic shellfish poisoning (NSP)	Ciguatera fish poisoning (CFP)
Causative organisms					
Dinoflagellates Alexandrium catenella Alexandrium minutum Alexandrium tamarense	Dinoflagellates Dinophysis acuminata Dinophysis acuta Dinophysis fortii Dinophysis norvegica	Diatoms Amphora coffeaeformis Nitzschia navis-varingica Pseudo-nitzschia spp.[a]	Dinoflagellates Protoperidinium crassipes	Dinoflagellates Karenia brevis (= Gymnodinium breve Ptychodiscus brevis) (US) Karenia spp. (New Zealand)	Dinoflagellates Gambierdiscus toxicus ? Coolia spp. ? Ostreopsis spp. ? Prorocentrum spp.
Gymnodinium catenatum Pyrodinium bahamense	Prorocentrum lima			Raphidophytes[b] Chattonella antiqua C. marina Fibrocapsa japonica Heterosigma akashiwo	
Vector organisms[c]					
Shellfish and crustaceans, e.g. clams, cockles, gastropods, lobsters, mussels, oysters, scallops and whelks	Shellfish, e.g. clams, mussels and scallops	Shellfish and crustaceans, e.g. clams, crabs and mussels	Shellfish. Only mussels have so far caused human intoxication. Azaspiracids have also been detected in oysters, Manila clams, scallops and razor fish	Shellfish, e.g. clams, mussels, oysters and whelks. Humans may also be affected by direct contact with the toxic algae via sea spray or swimming	Carnivorous fishes, mainly barracudas, groupers, jacks, sea bass, snappers and surgeon fish which feed on herbivorous fish that inhabit coral reefs
Effect and symptoms					
Neurotoxic	Gastrointestinal disturbance	Neurotoxic	No neurotoxic symptoms have been reported	Neurotoxic	Complex, primarily neurotoxic

(continued)

Table 4.1 (continued)

	Paralytic shellfish poisoning (PSP)	Diarrhetic shellfish poisoning (DSP)	Amnesic shellfish poisoning (ASP)	Azaspiracid shellfish poisoning (AZP)	Neurotoxic shellfish poisoning (NSP)	Ciguatera fish poisoning (CFP)
Mild case	Within 30 min: tingling sensation or numbness around lips, gradually spreading to face and neck; prickly sensation in fingertips and toes; headache, dizziness, nausea, vomiting and diarrhoea	After 30 min to a few hours (seldom more than 12 h): diarrhoea, nausea, vomiting, abdominal pain	After 3–5 h: nausea, vomiting, diarrhoea, abdominal cramps	Similar to those for DSP – diarrhoea, nausea, vomiting, stomach cramps, chills – in the limited number of cases reported	After 3–6 h: chills, headache, diarrhoea, muscle weakness; muscle and joint pain; nausea and vomiting; irritation to eyes and nasal membranes where direct contact with algae	Symptoms develop within 12–24 h of eating fish. Gastrointestinal symptoms: diarrhoea, abdominal pain, nausea, vomiting
Extreme case	Muscular paralysis; pronounced respiratory difficulty; choking sensation; death, through respiratory paralysis may occur within 2–24 h after ingestion	Chronic exposure may promote tumour formation in the digestive system	Decreased reaction to deep pain; dizziness, hallucinations, confusion; short-term memory loss, sometimes permanent; seizure; damage to hippocampus in brain; coma; death in some extreme cases	None to date	Paraesthesia; altered perception of hot and cold; difficulty in breathing; double vision; trouble in talking and swallowing	Neurological symptoms: numbness and tingling of hands and feet; cold objects feel hot to touch; difficulty in balance; low heart rate and blood pressure; rashes; death in some extreme cases through respiratory failure
Average fatality rate	1–14%	0%	3%	0% to date	0%	<1%
Treatment	Patient has stomach pumped and is given artificial respiration. No lasting effects	Intravenous infusion of electrolytes. Recovery after 3 days irrespective of medical treatment	None at present other than life support systems if required	None at present. Activated charcoal recommended (Poisindex)	Support as required	No antidote or specific treatment is available; neurological symptoms may last for months and years; calcium and mannitol may help relieve symptoms

[a] A list of *Pseudo-nitzschia* spp. known to produce DA is given in Fehling et al. (2004) and see also van Egmond et al. (2004).
[b] Brevetoxins have been detected in these algae but they have not yet been implicated in NSP (Onoue and Nozawa, 1989).
[c] For a comprehensive list of affected species of shellfish, see van Egmond et al. (2004).
Updated from Lefley and Hannah (1998). See also Backer et al. (2004).

a particular combination of physical and chemical conditions that allows rapid growth of the organisms (Reynolds, 2006). Blooms of harmful algae are often confusingly referred to as 'red tides', although the actual colour of the water may be either red, brown or green (Anderson, 1994). The incidence of harmful algal blooms appears to be increasing all over the world (Hallegraeff, 2004), but it remains to be established whether this apparent increase is due to increased vigilance by the many countries that have monitoring programmes (see Section 4.13) or is a genuine phenomenon caused by as yet unknown factors, or is a combination of both.

Bivalve shellfish are the most common vectors of phycotoxins other than in ciguatera fish poisoning (CFP) (Table 4.1). This is because they are filter feeders and naturally ingest any phytoplankton or particles to which they may be attached and thereby concentrate any toxins in their tissues. The shellfish that have been most studied as regards accumulation of algal toxins are those consumed directly by humans, such as clams and mussels (Fernández *et al.*, 2004b). However, the importance of carnivorous gastropods and crustaceans as vectors should not be ignored. Shumway (1995) has provided comprehensive data on the occurrence of phycotoxins in these organisms.

4.3   CLASSIFICATION OF PHYCOTOXINS

The traditional way of grouping phycotoxins, which is still widely used, has been according to the poisoning syndromes that they engender in humans. It has become apparent, however, that this classification is inadequate – for example, regarding the classification of pecteno-toxins (PTXs) and yessotoxins (YTXs) with the diarrhetic toxins, as discussed in Section 4.10.2.2. The trend is to move to a classification of the phycotoxins based on their chemical taxonomy rather than the toxic symptoms they produce, as discussed by Quilliam (2004a) and see also Hess (2007). A Working Group of the Codex Alimentarius Commission of the UN FAO and WHO (Anon., 2006a; Toyofuku, 2006) has endorsed that view. Quilliam (2004a) suggests that this approach would permit seafood safety to be regulated in relation to permissible concentrations of specific toxins rather than on the result of specific assays, as is already the case with other environmental contaminants such as mycotoxins. It would allow a choice of appropriate validated analytical methods, ranging from rapid screening tests to mass spectrometry, as reviewed in Section 4.11.

In line with these suggestions, the toxins described below are categorized according to their structural affinities.

4.4   THE SAXITOXIN (STX) GROUP (PSP)

Paralytic shellfish poisoning (PSP) is caused by the ingestion of one or more of a group of basic, water-soluble, nitrogenous toxins originating from dinoflagellates belonging to the genera *Alexandrium*, *Gymnodinium* and *Pyrodinium* (Table 4.1). A comprehensive list of shellfish known to be affected is given in van Egmond *et al.* (2004).

PSP has been recognized for centuries but its cause was first understood only in the twentieth century (Kao, 1993). On a global scale PSP is the most common and widespread of the phycotoxin syndromes, as is the occurrence of PSP toxins in seafood (Fig. 4.1a). Up to 1993, there were about 2500 cases of human intoxication recorded throughout the world (Kao, 1993) and more cases are reported each year. The consequences of PSP can be

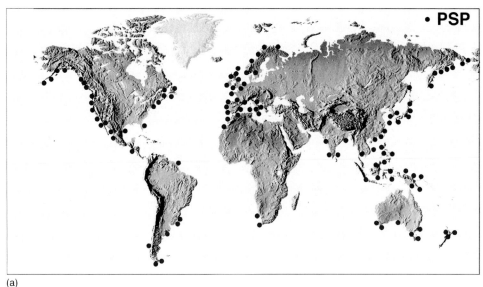

(a)

(b)

Fig. 4.1 The global distribution of the reported occurrence in seafood of the toxins causing (a) paralytic shellfish poisoning (PSP); (b) diarrhetic shellfish poisoning (DSP). Black dots ● indicate the location. After Anderson *et al.* (see Acknowledgements). See also maps in van Egmond *et al.* (2004).

very severe. For example, in Guatemala in 1987 about 186 people were admitted to hospital after ingesting PSP toxins, 26 of whom subsequently died. In Alaska 66 outbreaks of PSP occurred between 1973 and 1994, causing 143 human intoxications, including two fatalities. Detailed accounts of all aspects of PSP can be found in Kao (1993) and van Egmond *et al.* (2004).

Although dinoflagellates are undoubtedly the source of the PSP toxins, certain bacteria are also known to produce these compounds and the possible role of prokaryotes in PSP has

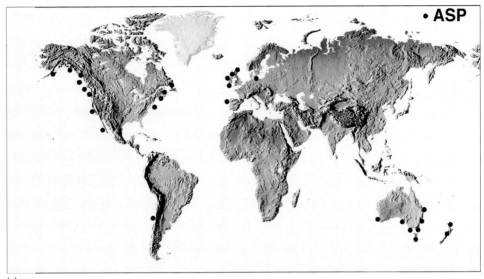

(c)

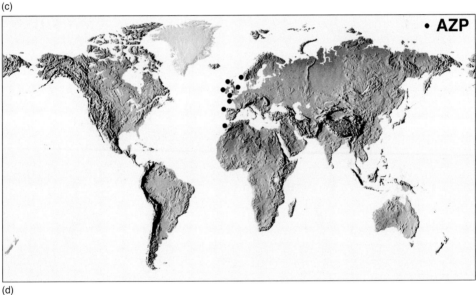

(d)

Fig. 4.1 The global distribution of the reported occurrence in seafood of the toxins causing (c) amnesic shellfish poisoning (ASP); (d) azaspiracid shellfish poisoning (AZP). Black dots ● indicate the location. After Anderson *et al.* (see Acknowledgements). See also maps in van Egmond *et al.* (2004).

come under scrutiny. Some photosynthetic cyanobacteria, for example, are a source of the PSP toxin saxitoxin as well as other unrelated toxins (see Section 4.10.4). It has been found that some marine bacteria produce saxitoxins, including bacteria associated with dinoflagellates that produce similar toxins (Shimizu *et al.*, 1996; Gallacher *et al.*, 1997; Uribe and Espejo, 2003; Azanza *et al.*, 2006). The exact relationship between these bacteria and dinoflagellates as regards toxin production remains to be clearly determined (Shimizu *et al.*, 1996; Kodama *et al.*, 2006).

(e)

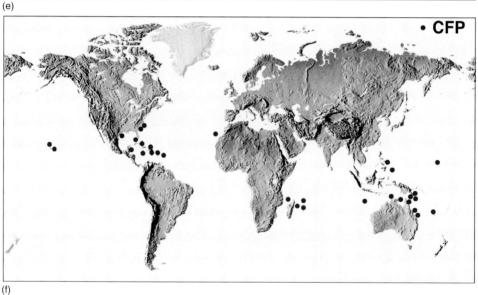

(f)

Fig. 4.1 The global distribution of the reported occurrence in seafood of the toxins causing (e) neuro-toxic shellfish poisoning (NSP); (f) ciguatera fish poisoning (CFP). Black dots • indicate the location. After Anderson *et al.* (see Acknowledgements). See also maps in van Egmond *et al.* (2004).

4.4.1   The toxins causing PSP: the saxitoxin family

More than 20 toxins in the saxitoxin group are known to exist at present, all of which are water-soluble. The basic molecule is a tetrahydropurine substituted at various positions, as shown in Fig. 4.2a. The specific toxicity of these compounds varies. For example, the most potent are the carbamate toxins (saxitoxin, neosaxitoxin and the gonyautoxins GTX-1 to GTX-4), which are 10–100 times more toxic than the N-sulfocarbamoyl derivatives, the B and C toxins (Cembella *et al.*, 1993; Wright, 1995). Relative toxicities of the saxitoxin family as determined by the mouse bioassay (see Section 4.11) are given in van Egmond *et al.* (2004).

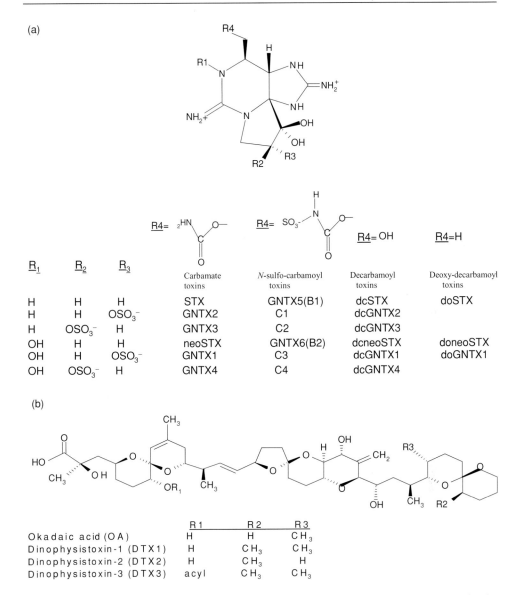

Fig. 4.2 (a) The chemical structures of the saxitoxin group (from van Egmond et al., 2004). (b) The chemical structures of the okadaic acid group (from van Egmond et al., 2004).

PSP toxins may form up to 0.2% of the wet weight of PSP-toxic dinoflagellates and the most predominant toxins are usually the sulfated derivatives (Laycock et al., 1994). However, some of these compounds may undergo degradation or biotransformation after being assimilated by shellfish and other organisms (Cembella et al., 1994; Oshima, 1995; Wright, 1995; Bricelj et al., 1996) and novel derivatives are reported from time to time (Negri et al., 2003; Llewellyn et al., 2004). Intoxicated shellfish found in cold or temperate waters most commonly contain the sulfated C toxins GTX-2 and -3 and also saxitoxin (Wright, 1995).

(c)

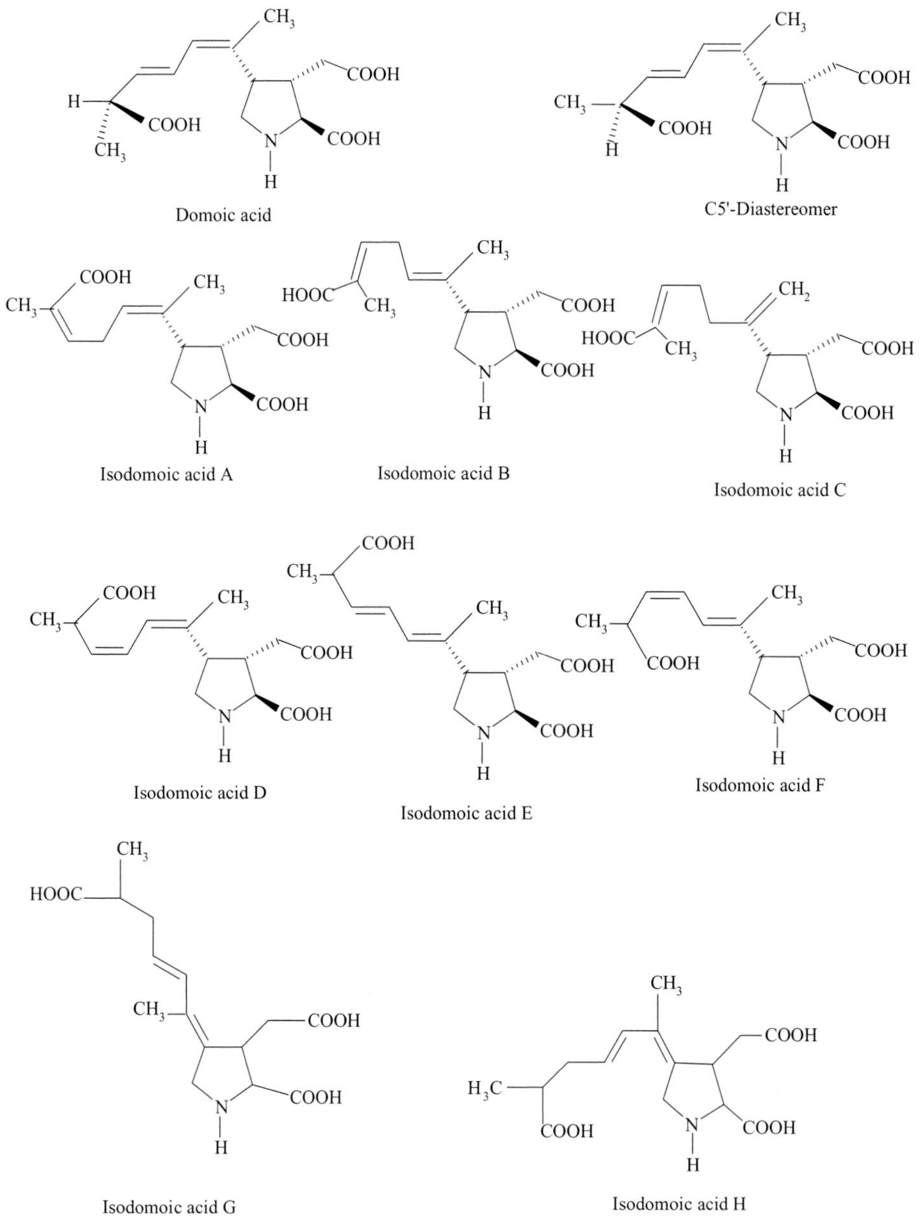

Fig. 4.2 (c) The chemical structures of the domoic acid group (from van Egmond et al., 2004).

4.4.2 Toxic effects

The symptoms of PSP poisoning in humans are very characteristic and range from a tingling sensation or numbness as a result of mild intoxication to death through respiratory paralysis in extreme cases (Table 4.1). At the biochemical level the effect of the saxitoxin family, especially saxitoxin itself, is well understood. They act by blocking very specifically both neuronal and muscular sodium channels, thereby affecting transmission of nerve impulses and

(d)

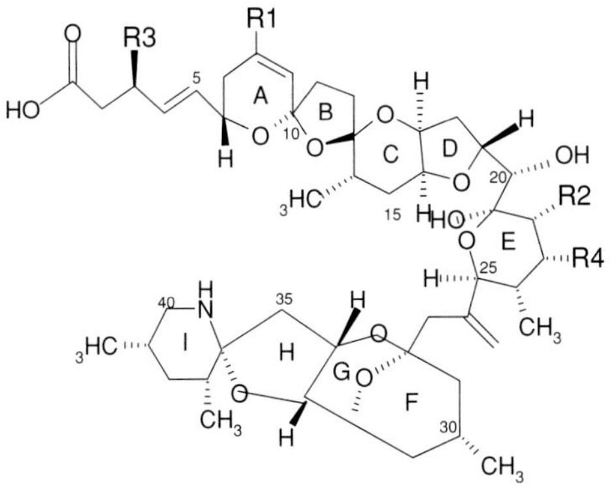

	R₁	R₂	R₃	R₄
azaspiracid-1 (AZA-1)	H	CH₃	H	H
azaspiracid-2 (AZA-2)	CH₃	CH₃	H	H
azaspiracid-3 (AZA-3)	H	H	H	H
azaspiracid-4 (AZA-4)	H	H	OH	H
azaspiracid-5 (AZA-5)	H	H	H	OH
azaspiracid-6 (AZA-6)	CH₃	H	H	H
azaspiracid-7 (AZA-7)	H	CH₃	OH	H
azaspiracid-8 (AZA-8)	H	CH₃	H	OH
azaspiracid-9 (AZA-9)	CH₃	H	OH	H
azaspiracid-10 (AZA-10)	CH₃	H	H	OH
azaspiracid-11 (AZA-11)	CH₃	CH₃	OH	H

Fig. 4.2 (d) The chemical structures of the azaspiracid group (figure adapted from van Egmond et al., 2004). The structures for AZ-6 to -11 are as suggested by James et al. (2004). See also Nicolaou et al. (2006).

muscle contraction (Table 4.2). Detailed accounts of the mechanism of action of saxitoxin and the related toxins can be found in Baden and Trainer (1993), Kao (1993) and Llewellyn (2006).

Studies on the toxicity and pharmacokinetics of saxitoxin on various laboratory animals have been reviewed by van Egmond *et al.* (2004). The acute toxicity of saxitoxin to mice (LD_{50}, µg kg^{-1} body weight [b.w.]) via various routes is given as: oral 260–263, intravenous 2.4–3.4 and intraperitoneal (i.p.) 9.0–11.6. By contrast, the LD_{50} (i.p.) for mice of sodium cyanide is 10 mg kg^{-1} b.w. (Kao, 1993).

4.5   THE OKADAIC ACID (OA) GROUP (DSP)

Three disparate groups of lipophilic polyether compounds originally made up the family of diarrhetic (or diarrhoeic) shellfish toxins: okadaic acid (OA) and the dinophysistoxins (DTXs), the pectenotoxins (PTXs) and yessotoxin (YTX) and its derivatives. The original collective term 'diarrhetic shellfish toxins' for the group is a misnomer since only OA and the DTXs and possibly a few of the PTXs are actually diarrhetic, whereas most of the PTXs and the YTX group have different physiological effects (Table 4.2 and see Section 4.10.2.2). There is a strong body of opinion that the PTXs and YTXs should be removed from the group (Aune et al., 2002; Anon., 2004, 2006a; Quilliam, 2004a) and they are therefore dealt with separately in Section 4.10.

(e)

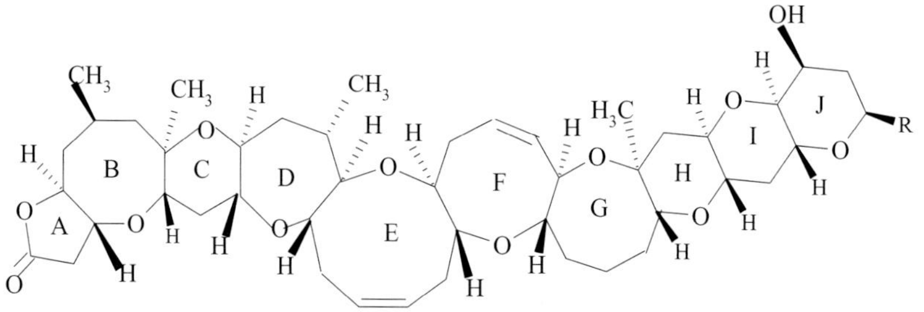

Type 1 (A) brevetoxins: PbTx-1,   R = CH₂C(=CH₂)CHO
                               PbTx-7,   R = CH₂C(=CH₂)CH₂OH
                               PbTx-10,   R = CH₂CH(CH₃)CH₂OH

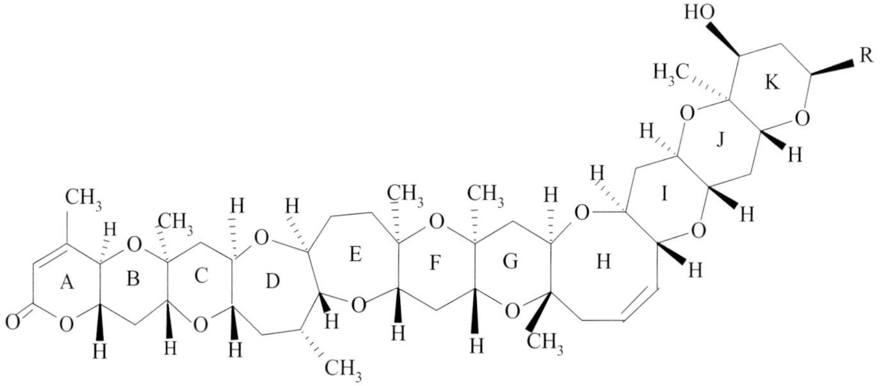

Type 2 (B) brevetoxins: PbTx-2   R = CH₂C(=CH₂)CHO
                               PbTx-2   R = CH₂C(=CH₂)COOH (oxidized)
                               PbTx-3   R = CH₂C(=CH₂)CH₂OH
                               PbTx-8   R = CH₂COCH₂Cl
                               PbTx-9   R = CH₂CH(CH₃)CH₂OH
                               PbTx-5   the K-ring acetate of PbTx-2
                               PbTx-6   the H-ring epoxide of PbTx-2

Fig. 4.2   (e) The chemical structures of the brevetoxin group (from van Egmond et al., 2004).

Diarrhetic shellfish poisoning (DSP) in humans is caused primarily by eating shellfish, such as mussels and clams contaminated with OA and/or DTXs, which originate mainly from certain species of the dinoflagellate genus *Dinophysis* (Table 4.1).

The global distribution of DSP toxins (OA and DTX toxins) in shellfish is shown in Fig. 4.1b and see van Egmond *et al.* (2004). This is the second most widespread intoxication of humans and shellfish caused by algae. As regards OA and the DTXs, the predominant toxin may vary according to geographic region. For example, OA is normally the major toxin in most of Europe whereas DTX-1 usually predominates in Japanese and Canadian shellfish,

(f)

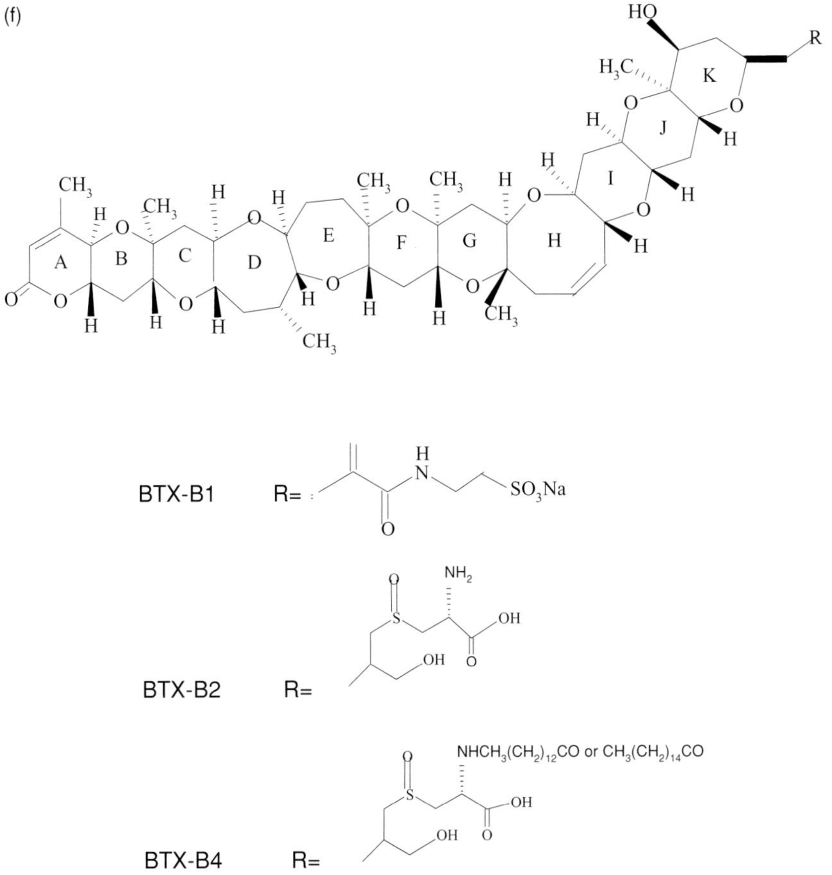

BTX-B1   R=

BTX-B2   R=

BTX-B4   R=

(g)

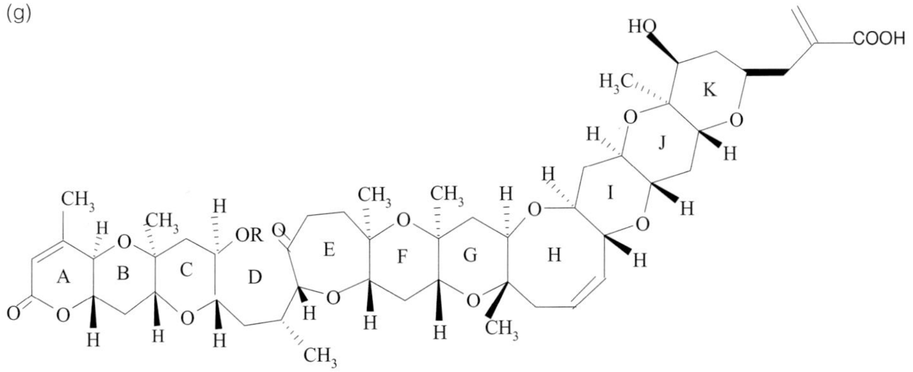

Fig. 4.2 (f) The chemical structures of the brevetoxin analogues BTX-B1, -B2 and -B4 isolated from contaminated shellfish (from van Egmond et al., 2004). (g) The chemical structure of the brevetoxin analogue BTX-B3 isolated from contaminated shellfish (from van Egmond et al., 2004).

(h)

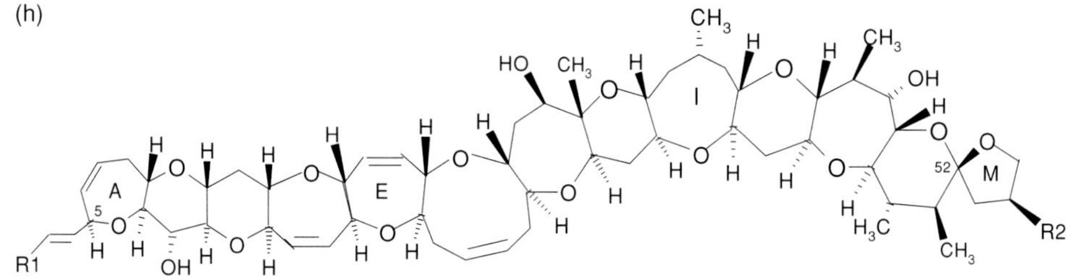

	R1	R2
P-CTX-1:	¹CH₂OHCHOH	OH
P-CTX-3 (P-CTX-2):	¹CH₂OHCHOH	H
P-CTX-4B (P-CTX-4A);	¹CH₂CH	H

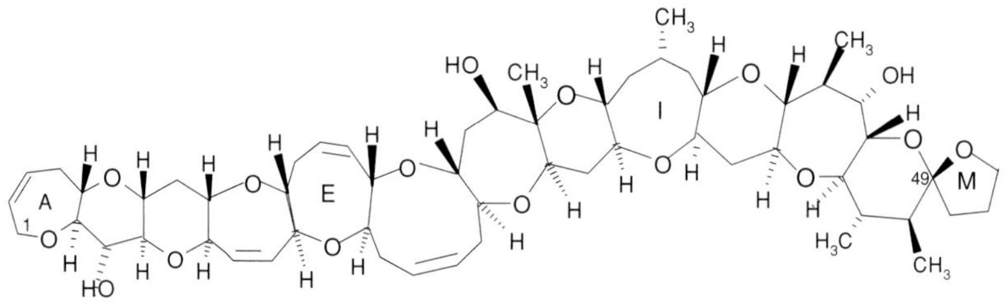

P-CTX-3C

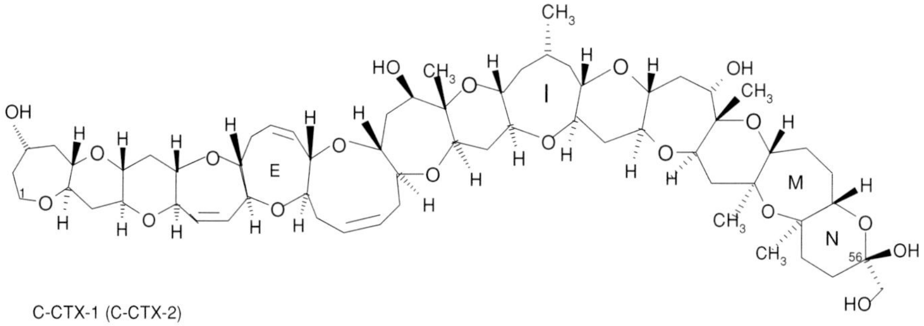

C-CTX-1 (C-CTX-2)

Fig. 4.2   (h) The chemical structures of Pacific (P) and Caribbean (C) ciguatoxins (CTXs). The energetically less favoured epimers, P-CTX-2 (52-epi P-CTX-3), P-CTX-4A (52-epi P-CTX-4B) and C-CTX-2 (56-epi C-CTX-1) are indicated in parentheses (from van Egmond et al., 2004). 2,3-Dihydroxy P-CTX-3C and 51-hydroxy P-CTX-3C have also been isolated from Pacific fish (Lewis, 2001).

and in Ireland DTX-2 predominates (Quilliam and Wright, 1995; Carmody et al., 1996) and has subsequently been found in France, Spain, Norway, Portugal and the UK (Hess, 2008). A wide-ranging review of DSP has been provided by van Egmond et al. (2004).

4.5.1   The toxins causing DSP: okadaic acid and the dinophysistoxins

Okadaic acid and dinophysistoxins (DTXs), the latter so named because they occur in the dinoflagellate genus *Dinophysis* (see Table 4.1), are acidic compounds (Fig. 4.2b (p. 59)

(i)

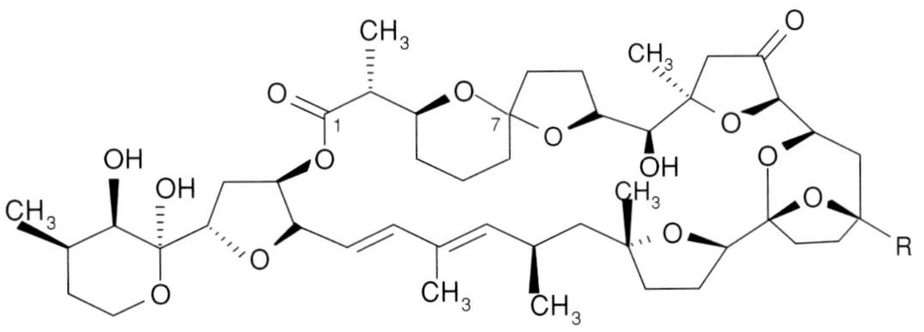

	R	C-7
pectenotoxin-1 (PTX1)	CH$_2$OH	R
pectenotoxin-2 (PTX2)	CH$_3$	R
pectenotoxin-3 (PTX3)	CHO	R
pectenotoxin-4 (PTX4)	CH$_2$OH	S
pectenotoxin-6 (PTX6)	COOH	R
pectenotoxin-7 (PTX7)	COOH	S

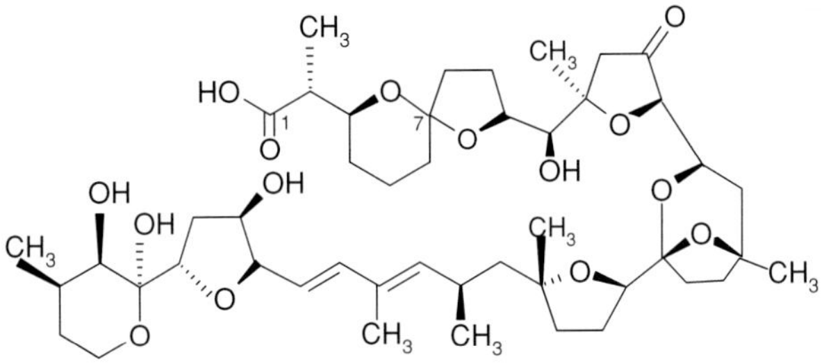

	C-7
pectenotoxin-2 seco acid (PTX2SA)	R
7-epi-PTX2SA	S

Fig. 4.2 (i) The chemical structures of some of the pectenotoxin group (from van Egmond *et al.*, 2004). See also Halim and Brimble (2006).

and see Quilliam, 2004a). The most important of this group of toxins as regards shellfish contamination are OA, DTX-1 and DTX-2, which may occur singly or together. The polyether backbone of OA, DTX-1 and DTX-2 may be acylated with a range of saturated and unsaturated fatty acids from C14 to C22 in the digestive glands of various shellfish to produce a mixture of compounds with diarrhetic activity, collectively known as 'DTX-3' (Quilliam and Wright, 1995; Wright, 1995; Quilliam, 2004a; van Egmond *et al.*, 2004). In addition to bivalves, there is evidence that fatty acid esters of OA can be produced in the digestive gland of Brown crabs (*Cancer pagurus*) (Torgersen *et al.*, 2005).

Two OA derivatives, DTX-4 and DTX-5a and -b, which are unusual in being water-soluble, have been isolated from the benthic dinoflagellates *Prorocentrum lima* and *P. maculatum* (Hu *et al.*, 1995b,c; Quilliam *et al.*, 1996). They possess an aliphatic side chain containing hydroxyl and sulfate groups and in the case of DTX-5 an amide moiety. Quilliam (2004a) suggests that DTX-4 and -5 have not been detected in shellfish because of the action of esterases.

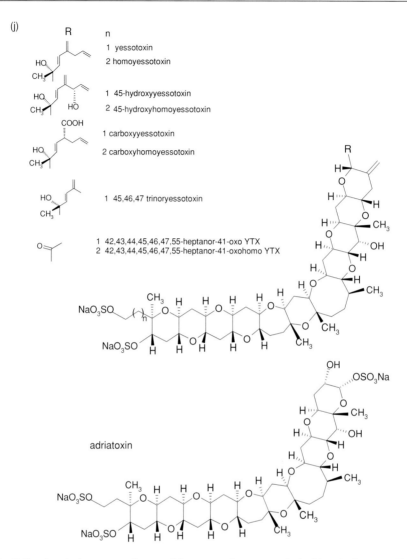

Fig. 4.2   (j) The chemical structures of some of the yessotoxin group and of adriatoxin (from van Egmond *et al.*, 2004). See also Bowden (2006) and Hess and Aasen (2007).

4.5.2 Toxic effects

The symptoms of DSP in humans are summarized in Table 4.1. Gastrointestinal disturbance is the main effect and the symptoms usually disappear within 3 days. Studies on the pathological effects of OA and DTXs as well as the PTXs and YTXs (see Section 4.10) have been confined mainly to suckling and adult mice and to rats – detailed accounts can be found in Terao *et al.* (1990a,b, 1993), Aune and Yndestad (1993) and in van Egmond *et al.* (2004). The acute toxicities (LD$_{50}$, µg kg^{-1} b.w.) to mice of OA, DTX-1 and DTX-3 are 200, 160 and 500 respectively (Fernández *et al.*, 2004a). Aune *et al.* (2007) demonstrated that DTX-2 has about 60% of the toxicity of OA when estimated by the mouse bioassay (i.p. injection) and an enzyme inhibition assay showed good correspondence.

Table 4.2   Sites of action and primary physiological effects of the major classes of phycotoxins subject to governmental regulation in Europe and elsewhere.

Toxin	Site of action	Primary physiological effects
Saxitoxin group	Site 1 on voltage-dependent sodium channels	Block sodium ion influx in neuronal and muscular sodium channel. Prevents propagation of the action potential essential for conduction of nerve impulses and contraction of muscles. Peripheral nervous system particularly affected in vertebrates
Okadaic acid group: OA, DTXs and derivatives	Catalytic subunit of protein phosphorylase phosphatases type 1 and 2A	Inhibit protein phosphorylase phosphatases Tumour promoters
Pectenotoxins	Not known	Hepatotoxic. Damage to cytoskeleton, e.g. F-actin depolymerization to cells *in vitro*
Yessotoxins	Not known. Possible interaction with calcium channels reported	Cardiotoxic
Azaspiracids	Not known	Affects liver, pancreas, thymus, spleen and gastrointestinal tract and T and B lymphocytes in rodents. AZA-1 has effects on cytoskeleton (F-actin). Effects on intracellular calcium in cells *in vitro*. See Tables 4.3 and 4.4
Domoic acid	Ionotropic class of glutamate receptors in central nervous system	Binds to glutamate receptors, i.e. competes with glutamate; receptor-induced depolarization and excitation
NSP toxins	Site 5 on voltage-dependent sodium channels	Induce channel-mediated sodium ion influx; depolarizes isolated muscle and nerve cells
Ciguatera toxins: *Ciguatoxin*	Site 5 on voltage-dependent sodium channels	Opening of sodium channels at resting potential and inability of open channels to be inactivated during subsequent depolarization
Maitotoxin	Calcium channels	Calcium ion influx, which may lead to cell death

Table updated from Leftley and Hannah (1998). See also Hampson and Manalo (1998), Burgess and Shaw (2001), van Egmond *et al.* (2004), Arias (2006), Bowden (2006) and Llewellyn (2006).

Intraperitoneal and oral doses of OA and DTXs cause marked changes in the small intestine, such as fluid accumulation and distension. Ultrastructural changes include degeneration of the intestinal absorptive epithelium. Terao *et al.* (1990a,b, 1993) also observed liver damage in rats and mice given i.p. and oral doses of OA and DTX-1. The effects of purified OA and DTX-1 on freshly prepared rat hepatocyte cells have been observed using light and electron microscopy (Aune *et al.*, 1991). OA and DTX-1 produced dose-dependent changes including blebbing of the cell surface and overall irregular shape.

Severe injuries to the intestinal mucosa of mice were observed within an hour of oral administration. OA injected into ligated loops of rat intestine produced progressive damage to the villi within 15 minutes and the degree of damage was dose-dependent. OA and DTX-1 induced liver damage in mice and rats after oral as well as i.p. administration. DTX-1 and DTX-3 also induced damage to the epithelium in the small intestine after both oral and i.p. dosing (van Egmond *et al.*, 2004).

A number of *in vitro* cytotoxicity studies have been carried out with various cell lines. Among the most significant findings were that protein and DNA synthesis in monkey kidney cells (Vero line) were both inhibited by OA in a concentration-dependent manner and it was concluded that the first and main site of action of OA was protein synthesis (van Egmond *et al.*, 2004). OA administered to human intestinal epithelial cells (T$_{84}$ line) increased the paracellular permeability and it was suggested that this might contribute to the diarrhetic effect of DSP toxins (Tripuraneni *et al.*, 1997).

The main concern about OA and the DTX toxins as regards human health is their role as potent tumour promoters. An effect of OA and DTX-1 on mouse skin was to induce ornithine carboxylase, an enzyme associated with various cancers that is known to be involved in signal transduction pathways involved in cell proliferation. OA has been observed to induce ornithine decarboxylase in rat stomach and also enhanced neoplastic changes in rat stomach after it had been pre-treated with a carcinogen (van Egmond *et al.*, 2004).

The possible effects of chronic exposure to these toxins in humans still have to be established (Aune and Yndestad, 1993; van Egmond *et al.*, 2004). An epidemiological study was carried out in France to investigate digestive cancer mortality in coastal regions related to consumption of seafood contaminated by DSP toxins. A very tenuous link was established between possible exposure to DSP toxins via seafood and some digestive cancers, mainly in men (Cordier *et al.*, 2000).

At a biochemical level OA and DTXs are powerful inhibitors of protein phosphatases PP1 and PP2 and 2A, which are important in regulating many important metabolic processes in eukaryotic cells, including cell proliferation (Table 4.2).

4.6   THE DOMOIC ACID (DA) GROUP (ASP)

Amnesic shellfish poisoning (ASP) was first recognized in 1987 on Prince Edward Island, on the east coast of Canada, when a serious incident occurred in which there were four human fatalities and over 100 cases of acute poisoning following the consumption of blue mussels (*Mytilus edulis*). ASP derives its name from the fact that one of the recognized symptoms of the poisoning is amnesia (Table 4.1). Several of the severely affected cases during the initial incident still suffered memory loss 5 years later (Todd, 1993). As amnesia is not always present, the syndrome is also sometimes referred to as domoic acid poisoning (DAP) (van Egmond *et al.*, 2004). The toxin responsible was identified as DA, a water-soluble, non-protein amino acid that was found to originate from the diatom *Pseudo-nitzschia multiseries* (syn. *Pseudo-nitzschia pungens* f. *multiseries*) (Ravn, 1995; Wright, 1995). This was the first recorded incidence of a diatom species being implicated in shellfish intoxication. Other diatoms are now known to produce DA and these are listed in Table 4.1.

Another, less serious, outbreak affecting humans occurred near Monterey Bay, California, in 1991. It was traced to razor clams. Prior to this incident the deaths of cormorants and pelicans in the area had been shown to be due to the consumption of anchovies that had accumulated DA after feeding on a bloom of diatoms composed of *Pseudo-nitzschia australis* (Wright, 1995). Deaths of sea lions off the Californian coast were similarly linked to a toxic bloom of this species (Scholin *et al.*, 2000).

Reports of DA in seafood products (predominantly shellfish and crabs) were initially confined to North America (Alaska, Bay of Fundy, British Columbia, California, Oregon, Prince Edward Island and Washington) while less significant concentrations were found in Australia, Europe, Japan and New Zealand (Hallegraeff, 2004). In 1999, a prolonged shellfishery closure was enforced in UK waters as a result of the presence of significant levels of DA (Gallacher *et al.*, 2001). High concentrations of DA are now a recurrent problem in UK and other European countries (van Egmond *et al.*, 2004). Significant amounts of DA have also been found in the benthic diatom *Nitzschia navis-varingica* in Korea, the Philippines and Japan (Kotaki *et al.*, 2000, 2004). The global occurrence of DA in seafood is shown in Fig. 4.1c.

Following the initial outbreaks in Canada and the US the regulatory authorities in those countries introduced monitoring and control measures and since then no serious cases of ASP have been reported. For reviews of DA and DA poisoning see Todd (1993) and van Egmond *et al.* (2004).

4.6.1   The toxins causing ASP (DAP): domoic acid and its isomers

Domoic acid (DA) is a water-soluble, acidic, non-protein amino acid. At least 10 isomers of DA, some of which are shown in Fig. 4.2c (p. 60), have been isolated (Quilliam, 2004b; van Egmond *et al.*, 2004) but only DA and its 5′-epi-diasteromer are thought to be of toxicological significance (Hess, 2008).

4.6.2   Toxic effects

The major symptoms of ASP are summarized in Table 4.1. These are gastrointestinal (abdominal cramps and diarrhoea) and neurological (crippling headaches and short-term memory loss) (Perl *et al.*, 1990).

The pathological effects of DA have been studied in rodents and cynomologous monkeys (summarized in Jeffery *et al.*, 2004 and van Egmond *et al.*, 2004). Intraperitoneal injection of mussel extract containing DA in mice produced symptoms such as scratching, tremors and seizures at both lethal and sublethal doses (Iverson *et al.*, 1989; Grimmelt *et al.*, 1990). The LD_{50} of DA was approximately 2.5–3.5 mg kg^{-1} b.w. Much higher doses of DA (35–60 mg kg^{-1} b.w.) were required to produce similar responses when orally administered. Necrosis of the hippocampal region in rat brain has been observed as well as a number of physiological effects on neurons (Baden and Trainer, 1993; Jeffery *et al.*, 2004). Similar neurotoxic responses were observed in monkeys to those seen in rodents and evidence of neuronal degeneration was also found in structures of the thalamus as well as the hippocampus (Schmued *et al.*, 1995).

At the biochemical level, DA acts competitively with glutamic acid as a neurotransmitter and has an antagonistic effect at several types of glutamate receptors, specifically the kainate and alpha-amino-5-methyl-3-hydroxyisoxazolone-4-propionate (AMPA) receptors where high specific affinity for DA relative to glutamate was recorded (Hampson and Manalo, 1998). For detailed reviews see Jeffery *et al.* (2004) and van Egmond *et al.* (2004).

4.7   THE AZASPIRACID (AZA) GROUP (AZP)

The family of toxins causing azaspiracid shellfish poisoning (AZP) are the most recently discovered of the marine phycotoxins in shellfish that pose a potentially serious hazard to human health. The toxins were isolated and characterized following a number of poisonings of humans in the Netherlands in 1995 due to consumption of mussels (*Mytilus edulis*) imported from the west coast of Ireland. Symptoms of diarrhoea and vomiting were similar in many respects to DSP but analysis of the shellfish tissue revealed insignificant amounts of DSP toxins (Furey *et al.*, 2003). The behaviour of mice used in the bioassay of the contaminated shellfish was different from that normally observed for DSP (van Egmond *et al.*, 2004).

Subsequently, a new family of toxins, the azspiracids (AZAs), was isolated from Irish mussels. Studies on natural phytoplankton gathered from the areas where contaminated mussels originated indicated that the source of the AZA is almost certainly the dinoflagellate *Protoperidinium crassipes*, previously thought to be inocuous. Phytoplankton from affected areas were collected and individual cells of various genera picked out and pooled for analysis. *Dinophysis* spp. contained the expected DSP toxins but AZAs were not detected (James *et al.*, 2003). Twiner *et al.* (2005) caution that laboratory culture studies are still needed to confirm this. The dinoflagellate is phagotrophic and the AZAs could originate from prey organisms. Other *Protoperidinium* spp. may prove to contain the toxins.

Following the incident in the Netherlands, poisonings occurred in Ireland in 1997, Italy and France in 1998 and in the UK in 2000 (James *et al.*, 2004; Anon., 2006b). In each case the cause was consumption of shellfish originating from the west coast of Ireland. Since then development of improved analytical techniques and rigorous monitoring for AZAs by the Irish authorities has ensured that there have been no further poisonings.

AZAs have so far been found in shellfish in NW Europe and sites on the Atlantic seaboard of France, Spain and most recently of Morocco (Taleb *et al.*, 2006) (Fig. 4.1d, p. 61).

In Ireland, the toxins have been detected in a range of shellfish: mussels (*Mytilus edulis*), oysters (*Crassostrea gigas*), scallops (*Pecten maximus*), cockles (*Cardium edule*), Manila clams (*Ruditapes phillipinarum*) (Furey *et al.*, 2003) and small amounts in razor fish (*Ensis siliqua*) (Hess *et al.*, 2001). The toxins have also been detected in mussels from SW Norway, Eastern England, Galicia in Spain and Morocco, and in scallops from Brittany (James *et al.*, 2004). AZAs were also identified in Norwegian Brown crabs (*Cancer pagurus*) in 2005 (Anon., 2006b). The AZAs may prove to be more widespread as they become included in national monitoring programmes.

4.7.1 The toxins causing AZP: the azaspiracids

In common with other lipophilic phycotoxins, the AZAs have an elongated fused polyether ring backbone. They also contain a nitrogen atom but differ from other nitrogenous phycotoxins such as gymnodimine and prorocentrolide (see Section 4.10.3) in having a cyclic amine instead of a cyclic imine group (Fig. 4.2d). Among other unique features the AZAs have several separate spirocyclic linkages (fusion of aliphatic rings where fusion occurs at a single atom) and a terminal carboxyl group (James *et al.*, 2004; van Egmond *et al.*, 2004; Nicolaou *et al.*, 2006). The unique structural characteristics of the AZAs are discussed in detail by James *et al.* (2004) who point out that, as a consequence of the carboxyl and amino terminal groupings in AZA-1, the molecule can exist as a zwitterion at physiological pH and this may facilitate its absorption and uptake.

The principal AZAs present in contaminated shellfish in Europe are the parent compound AZA-1 and its 8-methyl (AZA-2) and 22-desmethyl (AZA-3) analogues (Fig. 4.2d). AZA-1 is usually predominant with lesser amounts of AZA-2 and AZA-3 (James *et al.*, 2004) but the latter two are reported to be more toxic to mice than AZA-1, the lethal dose (i.p.) of AZA-1 being 200 μg kg^{-1} b.w. whilst that for AZA-2 and -3 being 110 and 140 μg kg^{-1} b.w. (Ofuji *et al.*, 1999) respectively. These three compounds are regarded as being a potential danger to human health.

Table 4.3 Effects of azaspiracids observed *in vivo*.

Study	Observed effects	Reference
Oral administration of purified AZA-1 in mice. Progressive changes observed over 24 h at doses of 500, 600 and 700 µg kg^{-1} body weight (b.w.)	Multiple organ damage: necrosis in the lamina propria of the small intestine and in lymphoid tissues such as thymus, spleen and Peyer's patches, T and B lymphocytes damaged, fatty changes observed in the liver	Ito *et al.* (2000)
Oral administration of purified AZA-1 in mice and chronic effects observed; 25 mice were administered AZA-1 twice at doses 300–450 µg kg^{-1} b.w. and recovery processes from severe injuries observed	Slow recoveries from injuries: erosion and shortened villi persisted in the stomach and small intestine for >3 months: oedema, bleeding, and infiltration of cells in the alveolar wall of the lung for 56 days; fatty changes in the liver for 20 days; necrosis of lymphocytes in the thymus and spleen for 10 days	Ito *et al.* (2002)
Oral administration of purified AZA-1 in mice and chronic effects observed: low doses of AZA equivalent to 50, 20, 5 and 1 µg kg^{-1} b.w. were administered twice a week up to 40 times to four groups of mice	Nine out of 10 mice at 50 µg kg^{-1} and 3 out of 10 at 20 µg kg^{-1} became so weak that they were sacrificed before completion of 40 injections. All these mice showed interstitial pneumonia and shortened small intestinal villi. Lung tumours were observed in 4 mice: 1 out of 10 at 50 µg kg^{-1} and 3 out of 10 at doses of 20 µg kg^{-1}. Tumours were not observed in 11 mice treated at lower doses and in 19 control mice. Hyperplasia of epithelial cells was observed in the stomach of 6 mice out of 10 at doses of 20 µg kg^{-1}	Ito *et al.* (2002)
Microinjection of AZA-1 into *Oryzias latipes* (Japanese medaka fish) embryos	Dose-dependent effect of purified AZA-1 observed. Doses of >40 pg AZA-1 per egg caused within 4 days reduced somatic growth and yolk absorption with delayed onset of pigmentation and blood circulation; heart rate lower *in ovo* and success of hatching decreased	Colman *et al.* (2005)

Eight other hydroxy analogues, which are thought to arise from metabolism in shellfish of the parent compounds, have been characterized (Fig. 4.2d). Two of these hydroxy analogues, AZA-4 and -5, were reported by Ofuji *et al.* (2001) to be less toxic than AZA-1 by a factor of 2–5 as estimated by the mouse bioassay (i.p.). Other AZA congeners have been discovered and are under investigation (Hess, 2008).

4.7.2 Toxic effects

Studies on mice have shown pathological changes to various organs that differ from those known to be caused by DSP toxins and also prolonged exposure of mice to the toxins caused widespread organ damage. Teratogenic effects in fish embryos have also been observed (Table 4.3).

Table 4.4 Effects of azaspiracids observed *in vitro*.

AZA congener	Cell type	Positive effects	Reference
AZA-1	Neuroblastoma cells	Decrease in F-actin concentration	Román et al. (2002)
AZA-1	Human lymphocytes	Increase in Ca^{2+} concentration and cAMP	Román et al. (2002)
AZA-1	Seven mammalian cell lines	Cytotoxicity (measured by cell viability) effects on all cells, which was both concentration and time dependent. Human T lymphocytes and rat epithelial pituitary cells particularly sensitive. In T lymphocytes arrangement of F-actin fibrils affected along with loss of pseudopodia	Twiner et al. (2005)
AZA-1	Murine spinal cord and frontal cortex neuronal networks	Inhibition of bioelectrical activity	Kulagina et al. (2006)
AZA-1	Two human epithelial cell lines: MCF-7 and CACO-2	Nanomolar concentrations reduced MCF-7 cell division and impaired cell-cell adhesion; affected the intracellular pool of the adhesion molecule E-cadherin causing accumulation of E-cahedrin fragments (EC_{50} 0.47 nM) – these effects found to be indistinguishable from those of yessotoxin in the same experimental system. Comparison of effects of AZA-1 in MCF-7 and Caco-2 cells found to be identical to those caused by YTX in same system	Ronzitti et al. (2007)
AZA-1	Primary murine neuronal cultures	AZA-1 increased cytosolic Ca^{2+} concentration and decreased cell viability. Effects of various ring fragments of AZA-1 were examined. The effect of AZA on cytosolic Ca^{2+} was found to depend on the presence of the ABCD or ABCDE ring domains but only the complete molecule induced neurotoxic effects	Vale et al. (2007)
AZA-1	Three human cell lines: colon carcinoma (CACO-2), lung carcinoma (NCI-H460) and neuroblastoma (BE(2)-M17)	AZA-1 induced rearrangement of stress fibres (actin filament bundles) and the loss of focal adhesion points in neuroblastoma and CACO-2 cells. The microtubule cytoskeleton was unaltered by AZA-1, but cell shape and internal morphology were affected. The toxin inhibited growth of lung carcinoma and neuroblastoma cells. Fifteen different fragments and/or stereomers of AZA-1 were tested for cytoskeletal rearrangement and cell growth inhibition. None had any activity, except for ABCD-epi-AZA-1, which was toxic	Vilariño et al. (2006)

AZA-1	Human neuroblastoma (BE(2)-M17)	Morphological and cytoskeletal changes were observed 24 h after addition of toxin, with no recovery after its removal. Two fragments and a stereomer of AZA-1 were used to analyse the structure–activity relationship. Only ABCD-epi-AZA-1 had a similar effect to AZA-1. AZA-1 appeared to bind irreversibly to its cellular target, needing moieties located in the ABCDE and FGHI rings of the molecule. AZA-1 induced activation of caspases (an early indicator of apoptosis), but these proteases were not the cause of cyto-skeleton disarrangement induced by AZA-1	Vilariño et al. (2007)
AZA-2 and AZA-3	Human lymphocytes	Effect on cytosolic Ca^{2+} concentration and increase in concentration of cAMP	Román et al. (2004)
AZA-4	Human T lymphocytes	Apparent inhibition of plasma membrane Ca^{2+} store-operated channels, an effect distinct from that of other AZA congeners previously observed	Alfonso et al. (2005)
AZA-1 to -5 and open ring synthetic analogues	Human lymphocytes	Modulation of cytosolic Ca^{2+} concentrations; effect on cytosolic pH by some congeners; synthetic analogues had little significant effect	Alfonso et al. (2006)

A paper by the Food Standards Agency of Ireland (Anon., 2001) points out shortcomings in some of the studies on mice: the numbers studied were small and a 'no observable effect limit' was not determined. Also, the doses used were high, equivalent to a human eating several kg of highly toxic shellfish at one sitting. The maximum permissible concentration of AZAs in shellfish tissue has been set by the European Commission at 160 μg kg^{-1} (see Table 4.7).

Comparatively little is known about the action of AZAs at the cellular and biochemical level. A number of *in vitro* studies have been carried out and are summarized in Table 4.4. Several of these studies have demonstrated an effect on F-actin, an important component of the cytoskeleton, and on intracellular calcium concentrations and on Ca^{2+} channels. However, work by Ronzitti *et al.* (2007) on the effect of AZA-1, OA and YTX (see Section 4.10.2) on F-actin and E-cadherin in two human epithelial cell lines, MCF-7 (breast cancer) and CACO-2 (colon cancer), showed that AZA-1 had little effect on F-actin. E-cadherin plays a role in cell–cell adhesion and abnormal regulation and functionality of the protein has been observed in human cancers. AZA-1 caused accumulation of E-cadherin fragments (the molecule lacking the intracellular domain, i.e. the part that interacts with the interior of the cell). The accumulation was time- and dose-dependent (EC$_{50}$ of 0.47 nM) and was not due to depolymerization of F-actin. The effects of AZA-1 and YTX on the two cell lines in this experimental system were identical, and apparently also in other *in vitro* systems reported in the literature, prompting the suggestion that AZA-1 and YTX may share molecular mechanisms in defined conditions.

James *et al.* (2004) have pointed out that AZAs may co-occur with DSP toxins and, since the latter are tumour promoters whilst AZAs appear to initiate tumours (Ito *et al.*, 2002), this has implications for human health and warrants further study.

4.8   THE BREVETOXIN (BTX) GROUP (NSP)

Neurotoxic shellfish poisoning (NSP), sometimes also referred to as neurological shellfish poisoning (Cembella *et al.*, 2004), is caused by ingestion of a group of lipophilic polyether toxins, the brevetoxins (PbTXs – originally so abbreviated because of their origin in the alga *Ptychodiscus brevis*), produced by the naked dinoflagellate *Karenia brevis* (synonyms *Gymnodinium breve* and *Ptychodiscus brevis*). As well as being present in shellfish, the brevetoxins may also affect humans directly through contact with these algae causing asthma-like symptoms when inhaled as an aerosol from sea spray (Table 4.1). Dense blooms of this alga have caused large-scale fish kills as well as mortalities in birds, marine mammal and other marine species (Tester *et al.*, 2000; Lansberg, 2002).

Prior to 1993, the incidence of NSP was restricted to the Gulf of Mexico, the east coast of Florida, and occasionally the coast of North Carolina (van Egmond *et al.*, 2004). An outbreak of NSP occurred in New Zealand in 1993 and affected 180 people. The causative organism was similar but not identical to *K. brevis* (Jasperse, 1993; Hallegraeff, 2004) and several *Karenia* species have been implicated here (Haywood *et al.*, 2004). The global distribution of NSP toxins is shown in Fig. 4.1d.

4.8.1   The toxins causing NSP: the brevetoxins

The brevetoxins (BTXs) comprise a group of lipophilic polyethers that are similar in some ways to ciguatoxins (see Figs. 4.2e, 4.2f and 4.2h, pp. 62–64). The BTXs are grouped as type A or B, based on the backbone structures of BTX-1 and BTX-2 respectively (Fig. 4.2e). Three Type-A BTXs and six Type-B have so far described (Baden *et al.*, 2005). In addition, BTX analogues have been isolated from contaminated shellfish during a NSP incident in New Zealand (details in van Egmond *et al.*, 2004) and the chemical structures of these, BTX-B1, -B2, -B3 and -B4, are shown in Figs. 4.2f and 4.2g. Various shellfish appear to exhibit different metabolism of the BTXs, with BTX-B1 only present in cockles (*Austrovenus stutchburyi*) while -B2, -B3 and -B4 have been found in green shell mussels (*Perna canaliculus*) and a study of Eastern oysters (*Crassostrea virginica*) from the Gulf of Mexico revealed further BTX analogues (Wang *et al.*, 2004).

BTXs are also found in four species of algae other than dinoflagellates: *Chattonella antiqua*, *C. marina*, *Fibrocapsa japonica* and *Hetrosigma akashiwo*, all of which belong to the class Raphidophyceae. Although these algae have caused fish deaths they have not yet been implicated in human poisonings (van Egmond *et al.*, 2004).

4.8.2 Toxic effects

Some of the symptoms of NSP are similar to those seen in mild cases of PSP and usually resolve within a few days (Table 4.1). The symptoms include nausea, cramps, paraesthesias of lips, face and extremities, weakness and difficulty in movement, paralysis, seizures and coma.

The physiological action of the brevetoxins is well understood and is similar to that of the ciguatoxins. They act specifically on site-5 voltage-dependent sodium channels and induce influx of sodium ions into affected cells (Table 4.2). For example, PbTx-3 inhibits the inactivation of Na^+ channels and prolongs the mean open time of these channels (Jeglitsch *et al.*, 1998). Detailed descriptions of the pharmacological action of the BTXs can be found in Baden and Trainer (1993), Premazzi and Volterra (1993) and van Egmond *et al.* (2004).

Brevenal, a fused-ring polyether aldehyde, and its dimethyl acetate derivative have been isolated from *Karenia* spp. cells. These compounds appear to act in an antagonistic way to BTXs, possibly moderating their effect (Bourdelais *et al.*, 2004, 2005).

4.9 THE CIGUATERA TOXIN (CTX) GROUP (CFP)

Ciguatera fish poisoning (CFP) in humans is caused by ingestion of a complex group of polyether toxins, the ciguatera toxins (CTXs), which accumulate in the flesh of fish, living on and around tropical coral reefs. The name ciguatera is derived from the Spanish word *cigua*, the common name given to small marine snails reputed to cause indigestion and the name became transposed to the intoxication arising from eating reef fish (van Egmond *et al.*, 2004; Yasumoto, 2005). The occurrence of CFP toxins in fish in a specific area is often sporadic; a given species of fish caught in one area may have a high level of toxicity but the same species nearby may be relatively free of toxin (Lehane and Lewis, 2000). A list of fish species associated with ciguatera is given in van Egmond *et al.* (2004).

CFP is concentrated mainly in tropical and subtropical areas, especially islands, in a band between latitudes 35°N and 35°S (see Fig. 4.1f), and is a serious public health problem in those countries, where fish constitute the main source of dietary protein. With the advent of rapid air transport, CFP may now occur in countries outside those latitudes to which contaminated fish are exported, or symptoms may develop in tourists after they have returned home. Often misdiagnosed and under-reported, over 25,000 people are affected by CFP each year but it is rarely fatal. For detailed accounts of CFP, see Yasumoto and Satake (1996), Yasumoto (1998), Lewis (2001), Lehane and Lewis (2000) and Nicholson and Lewis (2006).

4.9.1 The toxins causing CFP

4.9.1.1 The ciguatoxins

These are lipid-soluble polyether compounds consisting of 13–14 rings fused by ether linkages into a rigid, ladder-like structure. The origin of the toxins is the dinoflagellate *Gambierdiscus toxicus* (Lewis and Holmes, 1993). These dinoflagellates produce less polar and less potent CTXs such as P-CTX-4B (formerly known as gambiertoxins or GT-4B) that are transformed into more polar CTXs in the liver of fish by oxidative metabolism and spiroisomerization (Nicholson and Lewis, 2006). Recently, other *Gambierdiscus* spp. have been proposed as causative agents of ciguatera but their toxicity remains to be investigated (Nicholson and Lewis, 2006). CTXs undergo various biotransformations (mainly oxidation) as they pass up the food chain through various fish. This results in numerous CTX congeners, up to 23 in the Pacific region (Yasumoto, 2005) and the polar (more oxidized) forms have been found to be the more toxic of these (Lehane and Lewis, 2000).

Three structurally distinct CTX families occur in the Pacific Ocean, the Caribbean region and in the Indian Ocean (Hamilton *et al.*, 2002) and the letters P-, C- and I- are used as a prefix to indicate the origin of the CTX homologue. The structures of several Pacific and Caribbean CTXs are shown in Fig. 4.2h.

The main Pacific CTXs, P-CTX-1, P-CTX-2 and P-CTX-3, are present in fish in variable amounts (Lehane and Lewis, 2000). Variation in toxin structure is seen mainly at both ends of the toxin molecules and occurs as a result of oxidation (Naoki *et al.*, 2001; Yasumoto *et al.*, 2001). C-CTX-1 is less polar than the equivalent P-CTX-1. The latter is the major toxin in carnivorous fish and if found in quantities $>0.1\,\mu g\,kg^{-1}$ fish it is considered to pose a health risk (de Fouw *et al.*, 2001).

The regional differences, together with the transformations taking place within the food chain, may help to explain the wide range of clinical symptoms that has been observed with this intoxication (see Bagnis, 1993 and also Lehane and Lewis, 2000).

Detailed information on the isolation and structure of CTXs can be found in Legrand *et al.* (1992), Lewis *et al.* (1991, 1994), Yasumoto and Satake (1996) and Yasumoto (2005).

4.9.1.2 The maitotoxins

Maitotoxin (MTX) is a water-soluble compound that is often associated with CTXs in the viscera of herbivorous fish. It is one of the most potent marine toxins known, with an LD_{50} in mice of 0.17 $\mu g\,kg^{-1}$ b.w. when administered i.p. (Baden and Trainer, 1993). Murata *et al.* (1994) isolated MTX from cultures of the dinoflagellate *G. toxicus* and determined its structure and partial stereochemistry. It is a large and complex bisulfated molecule (MW 3422 Da) containing 32 ether rings, not all of which are contiguous. The complete structure and stereochemistry of MTX have been determined by Nonomura *et al.* (1996) and Zheng *et al.* (1996). Relatives of MTX, MTX-2 and -3, have been identified by mass spectrometry (Lewis *et al.*, 1994). MTXs are thought to play only a minor role in ciguatera poisoning because of the low concentrations found in herbivorous fish and their low oral potency (Lewis, 2001).

4.9.2 Toxic effects of CTXs

The symptoms produced by CFP toxins are complex and the effects may be long lasting (Table 4.1). They fall into the main categories of gastrointestinal, neurological, cardiovascular and diffuse pain. At the ultrastructural level, CTXs cause microscopic oedema of neural tissue (Bagnis, 1993) and CTX-1B causes nodal swelling in frog myelinated nerve fibres (Benoit *et al.*, 1996). Light and electron microscopy showed that i.p. or oral administration of P-CTX-1 or P-CTX-4C to male mice also causes swelling and focal necrosis on cardiac myocytes, the adrenal medulla and autonomic nerves (Terao *et al.*, 1991).

CTXs act by binding to, and causing persistent activation of, the voltage-gated sodium channels (Arias, 2006). This results in an influx of sodium ions, cell depolarization and the appearance of spontaneous action potentials in excitable cells. As a consequence of increased sodium permeability, the plasma membrane is unable to maintain the internal environment and volume control of the cell. This results in alteration of bioenergetic mechanisms, cell and mitochondrial swelling and bleb formation on cell surfaces (van Egmond *et al.*, 2004).

The high toxicity of MTX is explained by the fact that it activates both voltage-sensitive and receptor-operated calcium channels in the plasma membrane of cells, resulting in calcium influx and ultimately cell death due to calcium overload. The biochemical and physiological effects of CTXs are summarized in Table 4.2. Detailed reviews can be found in Baden and Trainer (1993), Yasumoto and Satake (1996), van Egmond *et al.* (2004) and Nicholson and Lewis (2006).

4.10 OTHER PHYCOTOXINS

All the toxins described so far are those that are of major concern to the regulatory authorities and the pectenotoxins and yessotoxins, described below, are also currently included with these. In addition, there are also a number of other phycotoxins that are found in seafood that are not presently subject to regulation, although some are under review.

4.10.1 The pectenotoxin group

The pectenotoxins are a group of neutral macrolide polyether lactones so named because they were first isolated from the Japanese scallop *Patinopecten yessoensis*. Halim and Brimble (2006) have reviewed the chemistry of the molecules. There are at least 14 members of the PTX family presently known, two of which, PTX-5 and -10, remain to be fully characterized. PTX-4 and PTX-7 are stereomers of PTX-1 and -6 respectively. PTX-8 and -9 appear to be chemical artefacts, not having been identified in any natural products.

PTX-2 appears to be the parent compound as it has only been found around the world in various species of the dinoflagellate genus *Dinophysis*: *D. acuta* (New Zealand, Norway, Spain and Ireland), *D. acuminata* (Norway and New Zealand), *D. caudata* (Spain), *D. fortii* (Japan and Italy), and *D. rotundata* (Norway) and never in shellfish (Wilkins *et al.*, 2006). PTX-12, which occurs as two diasteromers, has been isolated from *Dinophysis* spp., mussels (*Mytilus edulis*) and cockles (*Cerastoderma edule*) in Norway (Miles *et al.*, 2004b). PTX-11 (34S hydroxypectenotoxin-2), PTX-13 and -14 have all been isolated from *Dinophysis acuta* in New Zealand (Miles *at al.*, 2006; Suzuki *et al.*, 2006).

It is thought that PTX-1, -3 and -6 are derived from PTX-2 by oxidative enzymatic reactions in the hepatopancreas of shellfish. Shellfish also hydrolyse PTX-2 to the less toxic open-ring compound PTX-2 seco acid, which epimerizes non-enzymatically to 7-*epi*-PTX-2 seco acid. The seco acids of PTX-12 have been detected in Norwegian shellfish (Miles *et al.*, 2004b). Three series of fatty acid esters of PTX-2 seco acid and 7-*epi*-PTX-2 seco acid have also been detected in Irish blue mussels (*Mytilus edulis*) (Wilkins *et al.*, 2006). The structures of some of the PTX family, including the PTX-2 seco acids are shown in Fig. 4.2i (p. 65); see also Halim and Brimble (2006).

4.10.1.1 *Toxic effects*

Pectenotoxins have not been unequivocally associated with human intoxication. In two incidents in Australia PTX-2 seco acid and 7-epi-PTX-2 seco acid were detected in shellfish that caused poisonings with symptoms similar to those of DSP (Table 4.1) (Burgess and Shaw, 2001) but the cause was subsequently shown to be due to OA esters also present (Anon., 2004).

In mice the lethal dose ($\mu g\ kg^{-1}$ b.w.) by i.p. injection of various PTXs is reported to be: PTX-1 250, PTX-2 230–260, PTX-3 350, PTX-4 770 and PTX-6 500 (van Egmond *et al.*, 2004). In contrast, mice given oral doses of PTX-2 and PTX-2 seco acid up to 5000 $\mu g\ kg^{-1}$ b.w. showed no signs of toxicity although by the i.p. route the LD_{50} for PTX-2 was 219 $\mu g\ kg^{-1}$ (Miles *et al.*, 2004a).

Zhou *et al.* (1994) studied the effect of PTX-1 on chick embryo liver cells and found that it caused actin accumulation at the cell peripheries and cytoplasm, due to disruption of actin microfilaments within myofibrils, and microtubules were reduced in number and lost their radial arrangement. The damage was reversible if the cells were exposed to a concentration <0.5 $\mu g\ mL^{-1}$ PTX-1 for less than 4 hours. Salmon and rat hepatocytes exposed to 2 μM PTX-1 underwent apoptosis, with changes (chromatin hypercondensation and cell shrinkage) observed within an hour. The salmon cells were more sensitive (Fladmark *et al.*, 1998).

PTX-2 was found to have significant cytotoxic effects on a number of human cancer cell lines the most marked effect being on lung (A-549), colon (HT-29) and breast (MCF-7). A number of other lines (breast, melanoma, CNS, colon, lung, renal and ovarian) showed variable sensitivity with a 100-fold range in LC_{50} (Jung *et al.*, 1995).

PTX-6 caused depolymerization of F-actin in neuroblastoma cells, which was time- and dose-dependent, with a half-maximal effect at ca. 700 nM at 24 hours (Leira *et al.*, 2002). Other cytotoxic effects have been discussed in Burgess and Shaw (2001) and van Egmond *et al.* (2004).

4.10.2 The yessotoxin group

Yessotoxin, so called because it was originally isolated from *Patinopecten yessoensis*, resembles the BTXs and CTXs in that it has contiguously fused ether rings (Fig. 4.2j, p. 66) but differs in that it has a longer backbone of 47 carbon atoms, a terminal side chain of nine carbons, two sulfate esters and lacks carbonyl groups (Murata *et al.*, 1987; Takahashi *et al.*, 1996). Twenty-two structural analogues have so far been isolated variously from the dinoflagellates *Protoceratium reticulatum* and *Lingulodinium polyedrum* and from shellfish, including the congener adriatoxin, which is thought to be a shellfish metabolite (Bowden, 2006). A glycoyessotoxin that contains a β-arabinofuranose moiety has been isolated from *P. reticulatum* (Paz *et al.*, 2006). Some of the YTX family are shown in Fig. 4.2j; see also Bowden (2006) and Hess and Aasen (2007).

YTXs have been found in shellfish in Japan (Murata *et al.*, 1987; Satake *et al.*, 1996), the Adriatic (Ciminiello *et al.*, 1997) and also in Norway (Aasen *et al.*, 2005b) where YTX has also been found to occur simultaneously in mussels with DTX-1 (Lee *et al.*, 1988) and with OA esters (Torgersen *et al.*, 2005).

4.10.2.1 Toxic effects

Terao *et al.* (1990a,b) studied the effect of intraperitoneal injection of YTX and desulfated YTX (dsYTX) on mice. YTX was found to affect the heart whereas the dsYTX targeted the liver and pancreas. At a dose of >300 µg kg^{-1} b.w. YTX affected the heart muscle cells within 3 hours of administration. After 24 hours dsYTX had not affected the heart but caused marked fatty degeneration and intracellular necrosis in the liver and pancreas. Also, the concentration in the liver of triglycerides was 60-fold and phospholipids double those of mice treated with YTX. At this dose level mice died after 3 hours with YTX and after 24 hours with dsYTX. Oral administration of 500 µg kg^{-1} b.w. of YTX had no discernible effect whereas mice given dsYTX showed fatty degeneration of the liver (van Egmond *et al.*, 2004).

Aune *et al.* (2002) compared intraperitoneal and oral toxicity of YTX to mice. By intraperitoneal injection YTX was highly toxic (lethal dose in the range 100–214 µg kg^{-1} b.w.) whereas oral doses equivalent to 10 mg kg^{-1} b.w. were not lethal. At that dose swelling of heart muscle cells was observed and the minimum oral dose of YTX producing observable effects was 2.5 mg kg^{-1} b.w.

In human lymphocytes, YTX was observed to produce a calcium influx through certain channels sensitive to specific inhibitors and this Ca^{2+} entry was not affected by OA. YTX also produced an inhibition of capacitative Ca^{2+} entry produced by emptying of internal calcium stores. The inhibitory effect of YTX was dependent on the time of its addition. It was suggested that YTX may interact with calcium channels in a way similar to BTXs and MTX (de la Rosa *et al.*, 2001).

YTX has a powerful neurotoxic effect on cultured rat cerebellar neurons. The concentration of YTX that produced a 50% decrease in neuronal survival at 48 hours was ca. 20 nM (Pérez-Gómez *et al.*, 2006). Other toxicological studies have been reviewed by van Egmond *et al.* (2004) and Bowden (2006).

4.10.2.2 Diarrhetic effects of pectenotoxins and yessotoxins

The PTXs and YTxs were originally included with the OA group of toxins because they are often found associated with them and thought to cause DSP. However, the evidence that PTXs have a diarrhetic effect is contradictory. In bioassays carried out with purified PTX-2 or PTX-2 SA neither was overtly toxic to mice by the oral route at doses up to 5000 µg kg^{-1} b.w. and no diarrhoea was seen to be caused by either compound (Miles *et al.*, 2004a). Similarly, no signs of toxicity were recorded in mice following an oral dose of purified PTX-11 equivalent to 5000 µg kg^{-1} b.w. No diarrhoea was observed in animals administered with PTX-11 either orally or by i.p. injection (Suzuki *et al.*, 2006).

Contrary evidence has been presented by Ito (2006) who observed the diarrhetic effects of OA and PTX-2 delivered orally in mice and rats. In mice, both toxins at sublethal doses produced tissue damage in the small intestine after 60 minutes, characterized by oedema with OA and vacuolation of epithelial cells with PTX-2. Maximum fluid accumulation occurred after 90 minutes, after which there was rapid recovery. The minimum adverse effect doses were estimated to be 75 μg kg^{-1} b.w. for OA and 400 μg kg^{-1} b.w. for PTX-2. When administered singly, OA at 50 μg kg^{-1} b.w. and PTX2 at 300 μg kg^{-1} b.w. did not result in fluid accumulation but when co-administered they appeared to have a synergistic effect. Rats showed a greater tolerance to the toxins. Both OA and PTX-2 diluted in saline caused intestinal fluid accumulation at 400 μg kg^{-1} b.w. and above. The minimum effective dose for OA was lowered to 200 μg kg^{-1} when triolein oil was used as carrier and that for PTX-2 lowered to 300 μg kg^{-1} b.w. when 2% lecithin-water was the carrier.

YTXs have not been shown to be diarrhetic. When YTX was given orally by gavage to four-day old suckling mice at doses of 0.1, 0.2 and 0.4 μg per mouse as a 1% suspension in Tween® 60 solution, no intestinal fluid accumulation was observed after 4 hours. When equivalent doses of OA or DTX-1 were given they all produced fluid accumulation (Ogino et al., 1997). Aune et al. (2002) did not observe diarrhoea in mice given YTX both i.p. and orally, even at an oral dose of 10 mg kg^{-1} b.w. See also van Egmond et al. (2004) and Bowden (2006).

Another significant difference between the DSP toxins and PTXs and YTXs is their effect on protein phosphatases. Luu et al. (1993) found PTX-1 and PTX-6 to be inactive against protein phosphatases PP1 and PP2A and YTX does not inhibit PP2A (Bowden, 2006), whereas the DSP toxins are potent inhibitors of protein phosphatases.

The occurrence of PTXs and YTXs in shellfish is still subject to regulation within the European Union (EU) (see Table 4.7) and in other countries, but there are moves to deregulate these toxins as discussed in Section 4.13.

4.10.3   The cyclic imine group

4.10.3.1   *The spirolides*

These toxins were first detected and identified in shellfish in Canada in 1995 after mouse bioassays indicated the presence of possibly unknown toxins. These proved to be a novel group of macrocyclic imines with a spiro-linked tricyclic ether ring system and an unusual seven-membered spiro-linked cyclic iminium moiety and were named spirolides (SPXs) (Hu et al., 1995a, 1996a, 2001). The structures of some of the group are shown in Quilliam (2004a). Spirolides A to D are bioactive whereas E and F, which have open cyclic imine rings and are found only in shellfish, are not (Fernández et al., 2004a).

The source of the SPXs was identified as the dinoflagellate *Alexandrium ostenfeldii* (Cembella et al., 2000). Studies on the biosynthesis of 13-desmethyl spirolide C in this dinoflagellate using stable isotope labelled precursors have shown that most of the C atoms in this congener are derived from polyketides and that glycine is incorporated intact into the cyclic imine moiety (MacKinnon et al., 2006a).

SPXs have been found to be widespread in W European waters. In Norway, they have been detected in natural phytoplankton dominated by *A. ostenfeldii* and also in mussels (Aasen et al., 2005a). The main spirolide was shown to be 20-methyl spirolide G. Subsequently, a complex mixture of fatty acid esters of 20-methyl spirolide G were detected in Norwegian mussels (Aasen et al., 2006). These esters are the products of toxin metabolism by the shellfish in the same way that they produce fatty acid acyl esters of DSP toxins. Cultures of two clonal isolates of *A. ostenfeldii* from Danish waters were found to contain 13-desmethyl spirolide C and two new SPXs, 13,19-didesmethylspirolide C and spirolide G. The latter was the first spirolide to be isolated that contained a 5:6:6-trispiroketal ring system and it may prove to have different toxicological properties compared to other SPXs (MacKinnon et al., 2006b).

Small amounts of 13-desmethyl spirolide C were detected in mussels and other shellfish in NW Spain (González et al., 2006). In Italy, 13-desmethyl spirolide C has been identified in a cultured isolate of *A. ostenfeldii* from the N Adriatic (Ciminiello et al., 2006) and in France, spirolide C was detected in shellfish from the Atlantic coast of S Brittany (Amzil et al., cited in Hess, 2008).

4.10.3.2   Toxic effects

Pathological studies of some SPXs on rats and mice following i.p. injection have been undertaken involving histological examination of the internal organs and brain and some biomarkers for neural injury visualized by immuno-histochemistry and transcriptional analysis. The effect of the toxins appeared to be species-specific and confined to the brain. In mice the brain stem and hippocampus appeared to be major targets for the toxins but not in rats. Transcriptional analysis in rat brain showed major changes in biomarkers for neuronal injury, including those related to converting stimuli into intracellular changes in neurons, whereas there were no similar changes in mice (Gill et al., 2003).

Studies with mice using a mixture of SPXs of predominantly 13-desmethyl spirolide C showed the lethal dose given i.p. to be 40 µg kg^{-1} b.w. and the lethal oral dose to be about 25 times greater (Fernández et al., 2004a). These compounds have not been implicated in any human intoxication (Fernández et al., 2004a) and there is currently no regulatory limit for cyclic imines in shellfish for human consumption but the situation is under review (Hess, 2008).

4.10.3.3   The gymnodimines

Gymnodimine was first isolated from oysters in New Zealand and the dinoflagellate *Gymnodinium* sp. (Seki et al., 1995, 1996; Mackenzie et al., 1996). The molecule is composed of a 16-membered carbocyclic ring, a butenolide ring and a cyclic imine. B and C analogues have been isolated (for structures see Quilliam, 2004a). Although gymnodimine was at one time subject to regulation in New Zealand (Fernández et al., 2004a), it has now been deregulated (Hess, 2008). The LD$_{50}$ of gymnodimine for mice is reported to be 96 µg kg^{-1} b.w. (i.p.) and by oral administration 755 µg kg^{-1} b.w. Mice showed no signs of toxicity when it was administered in food that provided a dose equivalent to ca. 7.5 mg kg^{-1} b.w. (Munday et al., 2004). Consequently, gymnodimine is not now considered a hazard to human health (Munday et al., 2004), rather it is considered to be an interference in the mouse bioassay for other lipophilic toxins (Hess, 2008).

4.10.3.4   The prorocentrolides

The toxic macrocycle lactone, prorocentrolide (PC), was isolated from the benthic dinoflagellate *Prorocentrum lima* (Torigoe et al., 1988) where it co-occurs with OA and DTX-1. The molecule includes a C27 macrolide and a hexahydroisoquinoline in its unique structure. PC was found to be lethal to mice at a dose (i.p.) of 400 µg kg^{-1} b.w. A sulfated analogue of PC, prorocentrolide-B, was isolated from a tropical strain of *Prorocentrum maculosum* (Hu et al., 1996b). It killed mice by i.p. injection but the critical amount was not reported. Below the critical dose the mice recovered completely.

Spiro-prorocentrimine (SPC), has been isolated from an isolate of *Prorocentrum* sp., strain PM08, from Taiwan. It is a spiro-linked cyclic imine with an *ortho*, *para*-disubstituted

3'-cyclohexene in addition to its macrolide skeleton (Lu *et al.*, 2001). Mouse toxicity assays of SPC indicated an LD_{99} (i.p.) of 2.5 mg kg^{-1} b.w. These workers also isolated the PC analogues, 4-hydroxyprorocentrolide and 14-O-acetyl-4-hydroxyprorocentrolide from a Taiwanese strain of *P lima*, PL01.

None of the prorocentrolides has yet been implicated in human intoxication.

4.10.3.5 The pinnatoxins

The pinnatoxins consist of a 20-membered ring with 5,6-bicyclo, 6,7-azaspiro and 6,5,6-triketal moieties in their structure and four congeners, A, B, C and D have been described (Uemura *et al.*, 1995; Chou *et al.*, 1996; Kuramoto *et al.*, 2004; Quilliam, 2004a). The toxins originate from the bivalve *Pinna* spp., such as *P. muricata*, which occurs in tropical and temperate shallow zones of the Pacific and Indian oceans. The adductor muscle of this shellfish is eaten in China and Japan, where it is reported to cause food poisoning. Because extracts from the digestive glands of a number of *Pinna* spp. produce the same acute symptoms in mice as does the muscle tissue from the shellfish it has been suggested that *Pinna* spp. become toxic as the result of feeding on organisms such as toxic dinoflagellates. The toxin from *P. attenuata* may possibly be a Ca^{2+} channel activator (Kuramoto *et al.*, 2004).

4.10.4 The cyanobacterial toxins

Various cyanobacteria (blue-green algae) produce a number of potent toxins. One major group comprises the microcystins and the nodularins, which are peptides, and cylindrospermopsin, a cyclic guanidine alkaloid, all of which are hepatotoxic. A second major group is made up of neurotoxins including the alkaloids anatoxin-a and homoanatoxin-a, a guanidinium phosphate ester, anatoxin-a(s), that inhibits the enzyme acetylcholinesterase, and also some of the saxitoxin family, such as saxitoxin, neosaxitoxin, C toxins and gonyautoxins (Carmichael and Falconer, 1993; Negri and Jones, 1995).

Blooms of toxic cyanobacteria are predominantly a freshwater phenomenon, although they may also occur in estuarine and marine environments (Long and Carmichael, 2004). Such blooms have caused illness and death in livestock. The effects of cyanobacterial toxins on human health through potable water supplies and recreational use of affected fresh water are well documented (Carmichael and Falconer, 1993; Codd *et al.*, 1999, 2005; Chorus *et al.*, 2000; Kuiper-Goodman *et al.*, 1999).

The potential for intake of cyanobacterial toxins with food exists but there are few clinical reports of such poisoning. Measurement of cyanobacterial toxins in foodstuffs is recommended, however, particularly in crustaceans and shellfish harvested from waters with high cyanobacterial concentrations (Falconer, 1993). Vasconcelos (1995) found in an experimental study that mussels (*Mytilus galloprovincialis*) could accumulate microcystin when exposed to the toxic estuarine cyanobacterium *Microcystis aeruginosa*. Microcystins have been found to accumulate in freshwater and marine mussels (Eriksson *et al.*, 1989; Chen *et al.*, 1993). Falconer *et al.* (1992) found cyanobacterial toxins present in edible mussels (*Mytilus edulis*) harvested in an estuary containing toxic *Nodularia* species.

Severe food poisoning episodes have been recorded after ingestion of the edible red seaweed *Gracilaria* sp. contaminated with a toxin similar to aplysiatoxin, a contact irritant, which was thought to have been produced by the cyanobacterium *Lyngbya majuscula* growing on the surface of the red seaweed (Nagai *et al.*, 1996; Yasumoto, 2005). Toxins from *Lyngbya* were implicated in fatalities following consumption of marine turtle meat in Madagascar (Yasumoto, 2005).

There is evidence that chronic exposure to some cyanobacterial toxins may promote tumours (Falconer, 1991; Falconer and Humpage, 1996; Kuiper-Goodman *et al.*, 1999) and this has increased awareness of these compounds.

4.10.5 Miscellaneous phycotoxins

There are many other toxic compounds that are known to occur in marine microalgae, for example palytoxin, which is normally associated with fish poisonings in the tropics. It is a huge and complex molecule (MW ca. 3300 Da) and is amongst the most potent non-peptide toxins known, being about 10 times more potent than tetrodotoxin. It acts by binding to Na^+/K^+-ATPase and opening the ion channel. It has been reported to be associated with algae causing nuisance blooms in the Mediterranean and has been detected by bioassay in shellfish in Greece, although it is not thought to be a danger to human health (Hess, 2008). The primary source is thought to be members of the dinoflagellate genus *Ostreopsis*. The palytoxin analogue, ostreocin, was isolated from *O. siamensis* (Usami *et al.*, 1995). Also, the palytoxin analogues, mascarenotoxin-A and -B, have been isolated from blooms of *Ostreopsis mascarenensis* in the Indian Ocean and it is suggested that they could be a cause of palytoxin poisoning (Lenoir *et al.*, 2004).

Prorocentin, a novel polyketide, has been isolated from an OA-producing strain of *P. lima* by Lu *et al.* (2005). It is reported to be a C35 polyketide chain with four methyl groups, has an all-trans triene moiety, an epoxide, a furan ring, and trans-fused/spiro-linked tricyclic ether rings. It is suggested that it may share part of its biosynthetic pathway with OA. The compound showed inhibitory activity against human malignant melanoma and human colon adenocarcinoma cell lines. The biosynthesis of the polyketide toxins has been reviewed by Shimizu (2003).

The amphidinolides are family of toxic 26- and 27-membered toxic macrocycles that originate from the dinoflagellate genus *Amphidinium*. Amphidinol W, for example, is reported to exhibit cytotoxicity to murine lymphoma cells *in vitro* with an IC_{50} of 3.9 μg mL^{-1} (Kobayashi *et al.*, 2003 and see Kobayashi and Tsuda, 2004).

Gymnocins A and B, which are cytotoxic polyethers, occur in *Karenia* (*Gymnodinium*) *mikimotoi* (Satake *et al.*, 2002, 2005. Cooliatoxin, found in *Coolia montis* (Holmes *et al.*, 1995) may in fact be dsYTX as its molecular weight was reported to correspond to a monosulfated analogue of YTX (Bowden, 2006).

Although most of these toxins have not been implicated in human poisonings they may possibly be present in the food chain and some of them, such as the gymnocins, are ichthytoxic. The structure and occurrence of some of these toxins have been reviewed by Yasumoto (1990), Premazzi and Volterra (1993), Shimizu (1996, 2003) and Kobayashi and Tsuda (2004).

4.11 DETECTION OF PHYCOTOXINS IN SEAFOOD AND ALGAE

Only a general overview of the methodology of detection (summarized in Table 4.5) is provided here. In-depth coverage of all the topics touched upon can be conveniently found in a single volume (Hallegraeff *et al.*, 2004), which gives both theoretical and practical details.

Table 4.5   Summary of assays and analytical methods used to detect phycotoxins.

Toxin group	In vivo assays[a]	In vitro assays[b]	Physico-chemical analyses[c]
Saxitoxins[R]	Mouse ✓ Fly Locust (1)	Cultured mouse neuroblastoma cells (2) Immunoassay[R] Receptor binding	HPLC ✓ (3) Capillary electrophoresis (4) Liquid chromatography-mass spectrometry (LC-MS)
Okadaic acid, DTXs	Mouse (four variants) Rat	Cultured mouse neuroblastoma cells (5) Enzyme inhibition (protein phosphatases PP1 and 2A) Immunoassay[R]	HPLC LC-MS (6) (7) UPLC-MS (8) Capillary electrophoresis
Pectenotoxins	Mouse		LC-MS (6) (7) UPLC-MS (8)
Yessotoxins	Mouse	MCF cells with slot-blot procedure (9) Immunoassay[R] (10)	HPLC LC-MS (6) (7) UPLC-MS (8) Capillary electrophoresis (11)
Domoic acid	Mouse (has been used but insensitive)	Immunoassay (ELISA)[R] ✓* Immunobiosensor (12) Receptor binding	HPLC ✓ (13) LC-MS (7) Capillary electrophoresis (14)
Azaspiracids	Mouse	Antibodies available, could be used for method development (15)	LC-MS (6) (7) UPLC-MS (8)
Brevetoxins	Mouse	Cultured mouse neuroblastoma cells Immunoassay Receptor binding	LC-MS Capillary electrophoresis
Ciguatoxins *Ciguatoxin*	Mouse	Immunoassay[R]*(16) Receptor binding	HPLC Ionspray-MS
Maitotoxin	Mouse		HPLC Ionspray-MS

[a] See Fernández et al. (2004a); [b] see Cembella et al. (2004), Lewis (2004); [c] see Quilliam (2004a,b), Luckas et al. (2004), Lewis (2004).
✓ = AOAC International Official Method of Analysis (OM); [R] = Test kits commercially available; * = see text Section 4.11.4.
Details of all the above methods, including many practical protocols, can be found in one reference manual (Hallegraeff et al., 2004) in which all the above articles [a,b,c] appear. Numbers in parentheses indicate *additional* references describing methodology: (1) Cook et al. (2006); (2) Okumura et al. (2005); (3) Lawrence et al. (2005); (4) Wu et al. (2006); (5) Leira et al. (2003); (6) Stobo et al. (2005); (7) McNabb et al. (2005); (8) Fux et al. (2007)§; (9) Pierotti et al. (2007); (10) Briggs et al. (2004); (11) de la Iglesia et al. (2007); (12) Traynor et al. (2006); (13) Lopez-Rivera et al. (2005); (14) Kvasnicka et al. (2006); (15) Forsyth et al. (2006); (16) Campora et al. (2006). See also van Egmond et al. (2004).
§ This method has also been used for analysis of cyclic imines (see text Section 4.10.3).

Another useful source of information regarding current developments is the annual report on marine and freshwater toxins by the Committee on Natural Toxins and Food Allergens of the Association of Official Analytical Chemists International, e.g. Hungerford (2005, 2006).

4.11.1   Assays and analyses

Sullivan (1993) has drawn a useful distinction between assays and analyses for toxins. An assay is a method that produces a single response from the collective responses of all the individual components. Thus, it indicates overall toxicity but does not identify which toxins are present and their relative abundance. An example is the mouse bioassay discussed below. An analysis is a method that attempts to differentiate the various specific components present in a mixture and to quantify them individually, for example the separation and quantification of PSP and DSP toxins by high performance liquid chromatography.

4.11.2   Mammalian bioassays

The mouse bioassay is presently the mainstay for detecting phycotoxins and is recognized as the standard reference method for PSP, DSP, NSP and ciguatera toxins in most countries that carry out monitoring for these toxins. The basic procedure involves i.p. injection of an extract of the sample containing the toxin and observing the symptoms. There are recommended strains and weights of mice, and the protocol varies depending on the toxins being assayed (van Egmond *et al.*, 1992, 2004; Sullivan, 1993; Lewis, 2004; Fernández *et al.*, 2004a). Four variants of the mouse bioassay for DSP arc in use and a rat bioassay is also employed for the detection of these toxins (Fernández *et al.*, 2004a).

Article 5 of a European Commission Decision dated 15 March 2002 (see Section 4.13 and Table 4.7), laying down rules relating to maximum permitted levels of certain biotoxins and methods of analysis for marine bivalve molluscs and other seafood states: 'When the results of the analyses performed demonstrate discrepancies between the different methods, the mouse bioassay should be considered as the reference method'.

4.11.2.1   Disadvantages of the mouse bioassay

The mouse bioassay has a number of disadvantages, which have been summarized by Sullivan (1993):

(i) *Presence of multiple toxins*: Samples often contain more than one member of a family of toxins, each of which may have a different degree of toxicity, as is normally the case with PSP, DSP, NSP and ciguatera toxins. The bioassay therefore provides only a measure of the *total* toxicity, which is expressed relative to the effects of a reference toxin. This is adequate for toxicity monitoring but can provide no indication of the relative amounts of individual toxins in the sample. An additional problem that has been recognized is the presence of 'side toxins', i.e. toxins that have been extracted that may cause symptoms in mice, or are detected by other bioassays, but which are not related to the toxin(s) actually being assayed (Luu *et al.*, 1993; Zhao *et al.*, 1993; Amzil *et al.*, 1996; Mackenzie *et al.*, 1996).

(ii) *Inherent variability*: This can exceed ±20% of the true value compared to chromato-graphic techniques, which have a variability of <10%.

(iii) *Route of administration of the toxin*: Intraperitoneal injection bypasses the gastro-intestinal tract, the normal route of exposure to phycotoxins in humans, and its functions (absorption, metabolism and distribution), which may be important in decreasing or increasing the toxicity of the food.

(iv) *Public opposition*: There is growing public opposition, including militant action, to the use of animals for experimental purposes.

A strong scientific case against the use of the mouse bioassay for monitoring phycotox-ins, particularly in mussels, has been presented by the German Federal Institute for Risk

Assessment (Bundesinstitut für Risikobewertung [BfR]) and the German National Reference Laboratory for the Control of Marine Biotoxins (Anon., 2005a,b).

It has been recognized internationally for some time that there is an urgent need to replace mouse bioassays for phycotoxins with other methods. What are required for monitoring seafoods for public health reasons are rapid, specific and relatively inexpensive assays that can be used to screen large numbers of samples, ideally 'at the dock-side', and can be undertaken by relatively unskilled personnel (see Mackintosh *et al.*, 2002). Where such assays indicate the presence of toxins above the regulatory limit they can be validated by accurate and precise reference methods. Much has been done in recent years towards achieving these goals, but relatively few methods have been developed to a stage where they have gained international acceptance and also been designated as an Official Method (OM) by the Association of Official Analytical Chemists International (AOAC).

4.11.3 Instrumental (physico-chemical) analysis

Instrumental analyses for phycotoxins have been developed using such methods as HPLC (Table 4.5). There are standard HPLC analyses for PSP and DSP toxins using sensitive fluorescence detection and for DA using UV detection. AOAC International has to date validated two HPLC methods – for PSP toxins, OM 2005.6 (Lawrence *et al.*, 2005; Hungerford, 2006) and DA, OM 1991.26 (cited in Quilliam, 2004b). An advantage of fluorescence detection used with HPLC is that it can be more sensitive than mass spectrometry.

The method of choice for detailed analysis of phycotoxins is liquid chromatography-mass spectrometry (LC-MS) as evidenced by many of the papers cited in this chapter. Quilliam (2004a) concluded that, for lipophilic toxins, LC-MS is the most comprehensive and efficient technique, meeting all the needs of both monitoring and research laboratories. He lists many points in favour, including universal detection capability, high sensitivity, high selectivity and specificity, precision and accuracy, wide linear range of response, ability to separate complex mixtures and to deal with structural diversity, and minimal sample preparation.

LC-MS methods have been developed for screening for a range of lipophilic toxins simultaneously, for example OA, DTX-1 and -2, YTX, homoYTX, 45-hydroxy-YTX, 45-hydroxyhomo-YTX, PTX-1 and -2 and azaspiracid-1, -2 and -3 (Stobo *et al.*, 2005). A method using liquid chromatography with tandem mass spectrometry (LC/MS/MS) has been developed that is highly sensitive and specific for hydrophilic DA group toxins as well as DSP toxins and a wide range of other lipophilic algal toxins and metabolites found in shellfish (McNabb *et al.*, 2005). Hess (2008) has briefly reviewed current developments in LC-MS technology, the main one being the use of tandem mass spectrometry used in association with UPLC (ultra-performance liquid chromatography) that operates at very high pressure and facilitates sample resolution and speed of throughput (see Fux *et al.*, 2007).

There are a number of problems associated with instrumental analyses, for example the high initial capital cost of the equipment, especially for MS. Lengthy clean-up procedures may be required, especially for lipophilic toxins, and this limits the rate of throughput. For fluorescence detection by HPLC, e.g. PSP toxins, the appropriate fluorescent derivatives must be prepared and some of these may be unstable (see for example Lawrence *et al.*, 2005). Details of analytical protocols and discussions of associated problems can be found in Luckas *et al.* (2004), Quilliam (2004a,b), Lewis (2004) and Lawrence *et al.* (2005).

There is a shortage of pure analytical standards for some phycotoxins (Quilliam, 2004a), although this problem is being addressed by the international community (Hess, 2008). The problems of obtaining fit-for-purpose shellfish reference materials for internal and external quality control in the analysis of phycotoxins have been discussed by Hess *et al.* (2006).

4.11.4   In vitro assays

4.11.4.1   Cytotoxicity assays

Cytotoxicity assays using tissue culture microplate techniques have been developed for PSP, NSP and CFP toxins and for the DSP toxins; details can be found in Cembella *et al.* (2004) and see Table 4.5.

One problem with cytotoxicity assays is that they can respond to non-specific inhibitors. Tissue extracts are 'dirty' and may contain substances that affect the assay. For example, zinc in some mussel extracts can be toxic to mice and is presumably also toxic to cultured cells (Sullivan, 1993). Manger *et al.* (1995) observed non-specific toxic effects with crab viscera extracts but the problem was overcome by dilution. Another problem is that these assays may respond to known and unknown 'side toxins' also present in the sample, such as some of those described in Section 4.10.

4.11.4.2   Receptor binding assays

The PSP, NSP, ASP and CFP toxins all share a common property in that they bind very specifically to certain receptor sites (Table 4.2). A number of binding assays for these toxins have been developed that exploit this property. They use synaptosome preparations from animal brains and usually measure competition for the specific binding sites between a radioactive substrate and the toxin(s) in the sample (Table 4.5). The objection to using animal preparations may be overcome by using cloned receptor proteins (Goldin *et al.*, 1986; Cembella *et al.*, 2004). Binding assays are sensitive, specific and represent an integrated measure of the potencies of a mixture of a family of toxins such as PSP. Good correlations have been obtained between the standard mouse bioassay and binding assays for PSP toxins and the NSP toxin, BTX (Cembella *et al.*, 2004).

4.11.4.3   Immunoassays

Cembella *et al.* (2004) have discussed some of the problems encountered in developing immunoassays, such as the lack of pure toxins for conjugation and the low antigenicity of some of the lower molecular weight toxin molecules. Also, antibodies are usually prepared from a single purified toxin, such as saxitoxin, whereas natural samples may contain several closely related derivatives and the antibody may show limited cross-reactivity to these compounds. A novel approach to this problem has been the use of laboratory synthesized hapten-containing rings F-I of AZA to generate antibodies that crossreact with AZAs via their common C28-C40 domain. The antibodies had a similar affinity for AZA-2, -3, and -6 as they did for AZA-1 and they may prove suitable for analysis of AZAs in shellfish samples (Forsyth *et al.*, 2006). A similar approach has been used to generate antibodies to ciguatoxin CTX-1 (Lewis, 2004) and has been suggested for gymnodimines (Kong *et al.*, 2005).

Both monoclonal and polyclonal antibodies have been produced to various phycotoxins and these have been visualized using techniques such as enzyme-linked immunosorbent assays (ELISA) (Cembella *et al.*, 2004) and plasmon resonance (Traynor *et al.*, 2006). A number of immunoassays have been developed for some of the PSP, DSP and NSP toxins and ciguatoxin (Table 4.5) and some of these are available commercially. An integrated ELISA screening system has been used for monitoring DA, BTX, OA and STX group toxins found in New Zealand shellfish (Garthwaite *et al.*, 2001). An ELISA kit for DA (ASP) manufactured by Biosense® (http://www.biosense.com) has been validated by AOAC International as a First Action Official Method, number 2006.02.

Rapid qualitative tests suitable for various phycotoxins have been developed suitable for 'dockside' use, for example Cigua-Check® (http://cigua.oceanit.com) for CTXs (Lewis, 2004) and MIST-Alert® (http://www.jellett.ca) for STX group (PSP) toxins (see Mackintosh *et al.*, 2002; Cembella *et al.*, 2004; Inami *et al.*, 2004). The MIST-Alert® assay for STXs has been approved by the US Food and Drugs Administration.

In general, immunoassays for phycotoxins are many times more sensitive than instrumental methods. Provided problems of low cross-reactivity and false positives can be overcome, this type of assay can be useful both for initial rapid screening of suspect samples and for quantitative laboratory use.

4.11.4.4   Enzyme inhibition assays for OA group toxins

Enzyme inhibition assays have been developed for OA and DTX-1 based on the fact that they are potent inhibitors of protein phosphatases PP1 and PP2A. A radioisotope assay involving ^{32}P and PP1 and PP2A was developed by Holmes (1991). It involves HPLC prior to the assay to resolve phosphatase inhibitors into discrete fractions, since the assay responds to all phosphatase inhibitors. The method can quantify OA and DTX-1. Luu *et al.* (1993) subsequently applied Holmes' assay to identify DSP toxins in cultures of the dinoflagellate *Prorocentrum lima* in natural populations of phytoplankton and in mussels obtained both from Holland and from Canada. The most interesting result was that at least six phosphatase inhibitors distinct from known DSP toxins were found in mussel extracts. These did not include PTXs and YTX, which were found to be inactive against PP1 and 2A.

A protein phosphatase assay for OA and DTX-1 in shellfish extracts using PP2A and ^{32}P was described by Honkanen *et al.* (1996a,b) who highlighted the problems involved in using ^{32}P and PP2A. A non-radioactive colorimetric assay using PP1 has been described that can detect OA and DTX-1 with a limit of about 1.5 ng total toxin per gram of mussel hepatopancreas (Nunez and Scoging, 1997). A PP2A assay and two HPLC methods were compared to the mouse bioassay by Ramstad *et al.* (2001) and it was concluded that all three methods would be suitable for replacing the mouse bioassay. These enzyme assays can also be used to detect microcystins from cyanobacteria (Cembella *et al.*, 2004).

The main problem, then, with protein phosphatase assays for OA and DTX-1, is that they respond to all phosphatase inhibitors in crude extracts. Two other disadvantages are the use of ^{32}P, which can be avoided by using a colorimetric end-point but with lower sensitivity – although a more sensitive fluorometric method has been developed (Mountfort *et al.*, 1999, 2001) – and the relative scarcity and expense of the enzymes.

4.12   DEPURATION OF PHYCOTOXINS

4.12.1   Natural depuration

If shellfish contaminated with phycotoxins are left for sufficient time they will depurate (detoxify) themselves naturally once the causative algae are no longer present in the water.

Some of the extensive literature on this subject has been reviewed by Fernández *et al.* (2004b) and by van Egmond *et al.* (2004), the latter covering all the major toxins responsible for PSP, DSP, ASP and NSP.

In order to speed up depuration, the affected shellfish may be harvested and moved to a site where the water is free of toxic algae. However, this is not always a satisfactory solution as the rates of depuration vary greatly between species and on the type of toxin they carry, as well as physical factors. Fernández *et al.* (2004b) have provided a comprehensive compilation of the approximate retention times taken for a number of phycotoxins to fall below either statutory limits or levels of detection in a wide range of shellfish. These times range from 2 days to more than 2 years. In general, shellfish depurate PSP toxins relatively slowly and various methods of enhancing depuration have been tried with limited success, including variation in temperature or salinity, chlorination, ozonation and electrical stimulation (Fernández *et al.*, 2004b).

4.12.2 Studies on cooking as a method of depuration

Both domestic and commercial cooking have been studied as ways of reducing toxicity (Table 4.6), primarily for bivalve shellfish contaminated with PSP toxins, and have been shown to be largely ineffective. Most of the seafood toxins causing human intoxication are heat stable.

The main effect of boiling or pre-cooking with steam (e.g. domestic cooking and some commercial canning), seems to be to leach out and dilute the water-soluble PSP toxins in the tissues. Since a good proportion of the cooking liquor is discarded this lowers the total concentration of toxins in the shellfish (Quayle, 1969; Prakash *et al.*, 1971; Berenguer *et al.*, 1993; Vieites *et al.*, 1999). Cooking by retorting (110°C for 80 min or 122°C for 22 min) of PSP-contaminated clams was reported to reduce significantly the concentration of toxins in tissues, for example from >1800 µg STX-eq. 100 g^{-1} tissue to <9 µg STX-eq. 100 g^{-1} in the digestive gland (Noguchi *et al.*, 1980a,b).

Following various studies on cooking of the Mediterranean cockle (*Acanthocardia tuberculatum*) the European Commission issued a Decision (EC) 96/77 (see note at the end of reference section) applying to Spain which permits heating as a means of partially detoxifying cockles contaminated with PSP. Harvesting of *A. tuberculatum* when PSP levels in the edible parts are >80 µg STX-eq. 100 g^{-1} tissue but <300 µg STX-eq. 100 g^{-1}. If analysis of the heated product shows that the amount of STX is below the EU statutory limit of 800 µg kg^{-1} then the cockles may be marketed and sold for human consumption (Anderson *et al.*, 2001; van Egmond *et al.*, 2004). Fernández *et al.* (2004b) emphasize that the effectiveness of canning as a means of reducing PSP toxin levels below the statutory limit depends on the initial toxicity and so must be used with caution.

Ohta *et al.* (1992) investigated the effect of extrusion processing on various tissues of contaminated clams mixed with defatted soya protein. Extrusion was carried out at 130°C and 170°C and at elevated pressures. This considerably reduced the toxin content of the mixture, by up to 82% at 130°C and up to 97% at 170°C. The reduction in moisture content of the mixture after heating was only 1–2% so the decrease in toxicity did not appear to be due to leaching. HPLC analysis of the uncooked mixture showed it contained gonyautoxins-1, -2, -3 and -4 and saxitoxin, but after processing at either of the two temperatures it contained only saxitoxin, which has a higher specific toxicity. Nagashima *et al.* (1991) examined the thermal stability of PSP toxins under various conditions. Scallop homogenates showed little residual toxicity after heating to 120°C for 2 hours, whereas the toxicity of oyster homogenates doubled when heated to 100°C for up to 2 hours and then declined linearly on further heating.

Table 4.6 The effects of cooking and freezing on the phycotoxin content of some shellfish.

Toxin group	Organism	Process	Effect on toxin content in tissues	References
Saxitoxins (STXs)	American lobster (*Homarus americanus*)	Cooking	Reduction	Lawrence et al. (1994), Watson-Wright et al. (1991)
	Butter clams (*Saxidomus giganteus*)	Cooking	Reduction	Quayle (1969)
		Freezing	Minimal effect	Quayle (1969)
	Mediterranean cockle (*Acanthocardia tuberosum*)	Cooking	Reduction	Berenguer et al. (1993)
	Mussels (*Mytilus edulis*)	Cooking	Reduction	Prakash et al. (1971)
	(*Mytilus galloprovincialis*)	Cooking	Reduction	Vieites et al. (1999)
	Soft clams (*Mya arenaria*)	Cooking	Reduction	Prakash et al. (1971)
	Scallops (*Patinopecten yessoensis*)	Cooking	Reduction	Noguchi et al. (1980a,b), Ohta et al. (1992)
		Freezing	Migration to other tissues	Noguchi et al. (1984)
Okadaic acid (OA)	Mussels (*Mytilus* sp.)	Cooking	No reduction	Vernoux et al. (1994)
Domoic acid (DA)	Crabs (*Cancer magister*)	Cooking	Reduction	Hatfield et al. (1995)
		Chilling and freezing	Migration to other tissues	Hatfield et al. (1995)
	Mussels (*Mytilus edulis*)	Cooking	Migration from hepatopancreas to other tissues in whole mussels after steaming or autoclaving	McCarron and Hess (2006)
	Scallops (*Pecten maximus*)	Cooking	Leaching of up to 30% of toxin into packing media	Leira et al. (1998)
		Freezing	Migration to other tissues	Leira et al. (1998)
Azaspiracids (AZAs)	Mussels (*Mytilus edulis*)	Cooking	Concentrated in tissue due to loss of water from matrix after steaming	Hess et al. (2005)

The increase in toxicity (as measured by the mouse bioassay) was attributed to chemical conversion of the toxins to more toxic derivatives.

The DSP toxins contained in mussel tissue are stable to cooking at >95°C for 15 minutes, as estimated by the mouse bioassay (Vernoux *et al.*, 1994). Comparison of raw and cooked meat showed that heating did not change the concentration of toxins or their partitioning between the digestive gland (the most toxic organ) and the rest of the meat.

DA in the viscera of Dungeness crabs (*Cancer magister*) cooked by normal commercial processes in salt or fresh water could be reduced by up to 67% (Hatfield *et al.*, 1995). After cooking some DA was detectable in the body or leg meat of raw crabs, although it was not originally present in these tissues. The majority of the toxin was extracted and diluted into the cooking liquor.

Changes in DA concentrations in scallops (*Pecten maximus*) were followed during the normal commercial canning processes of pickled scallops and scallops in brine following frozen storage. No significant destruction of DA was observed during the canning process but the transfer of DA to packing media exceeded 30% of the total toxin content in the canned product (Leira *et al.*, 1998).

The effect of heat treatment on the DA content of soft tissues of mussels (*Mytilus edulis*) was investigated by McCarron and Hess (2006). DA concentrations in whole flesh, hepatopancreas and remaining tissue were measured in fresh, steamed and autoclaved mussel flesh. Relative decreases in DA and tissue fluid following heat treatments of whole flesh were similar. Concentrations of DA measured pre- and post-treatment were approximately equal. The concentration of DA in the hepatopancreas decreased but increased in the tissue remainder indicating that heat treatment caused some organ disruption of the mussels. It was concluded that heating either by steaming or autoclaving at 121°C are not suitable methods to reduce DA in mussels during commercial processing.

The effect of steaming as a sample pre-treatment of mussels (*Mytilus edulis*) containing azaspiracids (AZAs) was investigated by Hess *et al.* (2005). AZAs were found to be concentrated indirectly through the loss of water or juice from the matrix. The cooked shellfish tissues had a concentration of AZAs twice that of the uncooked shellfish, both for whole flesh and for digestive gland tissue.

Both CFP toxins in fish tissue and NSP toxins in shellfish are stable to cooking and boiling (Premazzi and Volterra, 1993; Anderson *et al.*, 2001; van Egmond *et al.*, 2004). Salting, drying, smoking or marinating do not reduce ciguatoxin in contaminated tissue (van Egmond *et al.*, 2004).

Fernández *et al.* (2004b) observed that, with the possible exception of the cooking method described by Berenguer *et al.* (1993), '... there have been no useful methods devised for effectively reducing phycotoxins in contaminated shellfish. All methods tested to date have been either unsafe, economically unfeasible or yielded products unacceptable in appearance and taste'.

4.12.3   The effects of freezing and chilling

There are a few reports on the effects of freezing and chilling on the mobility of water-soluble toxins in tissues. Quayle (1969) reported an experiment where 32 lots of butter clams (*Saxidomus giganteus*) (mean toxicity 74 µg STX-eq. 100 g^{-1} tissue, range 34–256 µg STX-eq.) were frozen for 3 months. At the end of that time mouse bioassays showed virtually no change in toxicity (mean 79 µg STX-eq. 100 g^{-1}, range 33–342 µg STX-eq.).

Noguchi *et al.* (1984) examined the distribution of PSP toxicity (measured by the mouse bioassay) in highly toxic specimens of the clam *Patinopecten yessoensis* and the effect of freezing. In unfrozen clams containing very high concentrations of toxins (117 000–198 000 μg STX-eq. 100 g^{-1} digestive gland) the concentration in the adductor muscle was 54–234 μg STX-eq. 100 g^{-1}. In less toxic unfrozen clams containing 18 000–52 200 μg STX-eq. 100 g^{-1} digestive gland the adductor muscle, washed or unwashed, was not detectably toxic. These less toxic specimens were kept frozen (temperature not specified) for 5 months; when thawed by heating to 100°C for 5 minutes or by standing at 17–21°C for 4 hours, the toxicity of the non-washed adductor muscle was <36 μg STX-eq. 100 g^{-1} but rose to between 54 and 162 μg STX-eq. 100 g^{-1} when allowed to thaw slowly at 5°C for 24 hours. When the whole clams were subjected to three freeze-thaw cycles the toxicity of the adductor muscle rose to 972 μg STX-eq. 100 g^{-1}, which indicated migration of toxins between tissues. Shumway (unpublished, cited in Shumway and Cembella, 1993) obtained similar results.

Studies have been carried out on whole cooked Dungeness crab (*Cancer magister*) where DA was confined mainly to the viscera; small amounts were also detectable in leg and body meat. After storage of cooked crabs for 90 days at –23°C slightly increased concentrations of DA were found in the leg and body meat. In cooked crabs held at 1°C for 1–6 days there was evidence of diffusion of toxin from the viscera to the body meat (Hatfield *et al.*, 1995).

Leira *et al.* (1998) monitored changes in DA levels of toxic in whole scallops (*Pecten maximus*) after 60 and 180 days of frozen storage. After 180 days, the concentration of DA decreased by 43%, mostly in the hepatopancreas, due to diffusion into other tissues.

4.13 MONITORING AND REGULATION

Given that natural depuration and cooking are largely ineffective, the first line of defence against phycotoxins endangering human health is the monitoring of toxin levels in seafoods and the implementation of legislation to prevent harvesting and marketing if toxin concentrations exceed statutory limits.

Monitoring programmes exist on local, regional and national scales in countries where phycotoxins are a problem and these have been reviewed by Fernández *et al.* (2004b). A valuable reference work is that by Anderson *et al.* (2001), which provides details for over 20 countries of national management plans, monitoring strategies, toxin tolerance limits, cost-benefit analyses, responsible authorities and many more useful facts.

In addition to causing public health problems, toxic phytoplankton can have a significant effect on the use of coastal waters for mariculture, commercial fisheries and recreation (Premazzi and Volterra, 1993; Anderson *et al.*, 2001). Given the need to minimize economic losses and the potential risk to the public caused by these outbreaks, two monitoring strategies have been used: screening of natural phytoplankton populations for the presence of harmful species and the regular testing of seafoods.

4.13.1 Phytoplankton monitoring

National strategies and methodologies adopted to monitor harmful algae have been reviewed in detail by Anderson *et al.* (2001), van Egmond *et al.* (2004) and in part III of Hallegraeff *et al.* (2004). By itself, phytoplankton monitoring does not provide sufficient protection to public health but it is often used as an early warning system on which to base intensive sampling programmes of toxicity testing of shellfish (or finfish) tissue in an area where toxic species have been detected.

4.13.2 Monitoring of shellfish tissues for toxicity

When shellfish resources are well developed, intensive monitoring has proven cost-effective and has permitted flexible management so as to optimize the areas of shellfish open to harvesting and minimize economic loss. Tolerance limits for toxins in seafood have been established above which harvesting or marketing is prohibited and this approach is discussed in Fernández *et al.* (2004b), who also list 'action levels' for phycotoxins in seafood used in various countries. Table 4.7 illustrates the approach taken by the European Union which has laid down detailed regulations concerning not only the permitted concentrations in seafood but also the methods of bioassay and analysis. There is still disagreement in some countries regarding the prescribed bioassays (Anon., 2005a,b). The EU itself is committed to replacing animal bioassays: 'Biological methods shall be replaced by alternative detection methods as soon as reference materials for detecting the toxins . . . are readily available, the methods have been validated and this Chapter has been amended accordingly'. (Chapter V of Section VI of Annex III to Regulation (EC) No. 853/2004 – see note in reference section).

4.13.3 Risk analysis

It is never going to be possible to eliminate the possibility of phycotoxins being present in seafood, hence, for example, the EU regulations to minimize the risk to public health and the trend is towards risk analysis based on sound scientific data. An example of rigorous risk analysis is that for azaspiracids carried out by the Food Standards Agency for Ireland (Anon., 2006b).

Lupin (2004) has defined risk analysis as being composed of three elements: risk assessment, risk management and risk communication. van Egmond *et al.* (2004) concluded that risk assessment of phycotoxins has not been carried out in a clear-cut fashion and has become merged to some extent with risk management, thereby complicating the process. They made a number of recommendations, including the suggestion that formal risk assessment should be carried out by international bodies, such as the World Health Organization, and this has subsequently begun to come about.

A working group (WG) of the Codex Committee on Fish and Fishery Products (CCFFP) of the FAO/WHO Codex Alimentarius Commission (Anon., 2006a) was convened to consider the recommendations of an international expert working group (Anon., 2004) and how these might be incorporated into the 'Draft Standard for Live and Raw Molluscs and the section of the Code on Live and (Raw) Bivalve Molluscs' of the Codex. Among the guiding principles set for the CCFFP WG were (i) The WG should recommend marine biotoxin levels in a manner consistent with the approach taken for setting levels for other naturally occurring toxicants in Codex Standard (compare this with comments made in Section 4.3) and (ii) that 'Marine biotoxin standards should not be set where there is a lack of evidence of harm to humans, either from human clinical data, epidemiological studies or animal voluntary feeding studies' with a footnote that 'Before regulating, where only intraperitoneal studies exist, these must be complemented by oral studies. Among these, voluntary feeding should take priority over gavage'.

One of the recommendations of the CCFFP WG was that PTXs and YTXs should be deregulated (Table 4.7), based on the marked difference in toxicity when administered orally

Table 4.7 Permissible concentrations of toxins in tissue of bivalve molluscs, echinoderms, tunicates and marine gastropods and methods of bioassay or analysis as prescribed by European Union Legislation in 2006.

Toxin group	Syndrome	Prescribed method of assay or analysis (1); see also Table 4.5	Parts of shellfish soft parts to be analysed	Maximum concentrations permitted by EU legislation	Maximum concentrations recommended by FAO/WHO CAC (3)	Relevant EC legal instruments (see reference section for www URLs)
Hydrophilic						
Saxitoxins (STXs)	PSP	Bioassay: AOAC mouse bioassay (MBA). This may be carried out in association, if necessary, with another method for detecting STX and any of its analogues for which standards are available. If the results are challenged, the reference method must be the biological method	Entire body or any part edible separately	800 µg kg^{-1}	800 µg kg^{-1}	Directive 91/492/EEC Decision 2002/225/EC (5) Regulation (EC) No. 853/2004 (6) Regulation (EC) No. 2074/2005 (7)
Domoic acid (DA)	ASP	HPLC method or any other recognized method, e.g. ELISA. If the results are challenged, the reference method must be HPLC	Entire body or any part edible separately	20 mg kg^{-1}	20 mg kg^{-1}	Directive 97/61/EC (4) Decision 2002/225/EC (5) Regulation (EC) No. 853/2004 (6) Regulation (EC) No. 2074/2005 (7)
Lipophilic						
Okadaic acid (OA) and Dinophysistoxins (DTXs)	DSP	Bioassay: Three MBAs and one rat bioassay are prescribed. Choice of MBA depends on the particular lipophilic toxins and the solvents used for extraction. Alternative methods may be used separately or in combination (HPLC with fluorimetric detection, LC-MS, ELISA, enzyme inhibition (PPA) assay) provided that they are not less effective than the biological methods and that their implementation provides an equivalent level of public health protection (2)	Whole body or hepatopancreas. Specified weights of these to be used in some MBAs	160 µg kg^{-1} of OA equivalents*	160 µg kg^{-1} of OA equivalents	Decision 2002/225/EC (5) Regulation (EC) No. 853/2004 (6) Regulation (EC) No. 2074/2005 (7)

Azaspiracids (AZAs)	AZP	Bioassays as for OA/DTX. Alternative methods (above) must be able to detect AZA-1, -2 and -3	Entire body to be used when single MBA with acetone extraction used	160 µg kg^{-1} of AZA equivalents	160 µg kg^{-1} – to be reviewed as more data become available	Decision 2002/225/EC (5) Regulation (EC) No. 853/2004 (6) Regulation (EC) No. 2074/2005 (7)
Pectenotoxins (PTXs)	DSP?	Bioassays as for OA and DTX. Alternative methods (above) must be able to detect PTX-1 and -2	As for OA/DTX	Included with OA equivalents (above*) in total	Deregulation recommended – see Section 4.13.3 in text	Decision 2002/225/EC (5) Regulation (EC) No. 853/2004 (6) Regulation (EC) No. 2074/2005 (7)
Yessotoxins (YTXs)		Bioassays as for OA and DTX. Alternative methods (above) must be able to detect YTX, 45 OH YTX, homo YTX and 45 OH homo YTX	As for OA/DTX	1.0 mg kg^{-1} YTX equivalents	Deregulation recommended – see Section 4.13.3 in text	Decision 2002/225/EC (5) Regulation (EC) No. 853/2004 (6) Regulation (EC) No. 2074/2005 (7)

Notes: (1) EC Regulation (EC) No. 2074/2005, Annex III. (2) It is stipulated: (a) If new analogues of public health significance are discovered, they should be included in the analysis. (b) Standards must be available before chemical analysis is possible. (c) Total toxicity shall be calculated using conversion factors based on the toxicity data available for each toxin. (d) The performance characteristics of these methods shall be defined after validation following an internationally agreed protocol. (3) Codex Alimentarius Commission of the UN Food & Agriculture Organisation and the World Health Organisation, Joint Food Standards Programme, Codex Committee on Fish and Fisheries Products, Working Group – see Anon. (2006a). (4) Amends Annex to Directive 91/492/EEC to include domoic acid. (5) Lays down detailed rules for implementation of Directive 91/492EEC as regards maximum levels of specified biotoxins and methods of analysis. (6) Lays down specific hygiene rules. (7) Amends Reg. 853/2004: Annex 3 details required testing methods – see (1).

as compared to i.p. (see Sections 4.10.1 and 4.10.2) and the application of risk factors, with the proviso that should data become available, the toxicological effects of PTX and YTX to humans would be reassessed.

A full account of the background to the WG and its detailed recommendations can be found in Anon. (2006a). In addition, a useful overview of the work of the FAO, WHO and IOC to provide scientific advice on marine biotoxins has been provided by Toyofuku (2006).

4.14   FUTURE PROSPECTS

The phycotoxins represent a particularly problematical group of compounds as regards detection and quantitation in seafood. The original algal toxins may undergo chemical or biological transformations in the vector organisms and the lack of pure reference material for many of these variants makes quantitation difficult, although this problem is being addressed (Hess, 2008). The lack of sufficient quantities of pure toxins also hinders toxicological studies. In this respect, synthetic organic chemists may be of some assistance. The complex stereochemistry of many of the larger phycotoxins has long been viewed as a challenge by chemists (see Halim and Brimble, 2006) and they have achieved some remarkable results, for example, the total synthesis of AZA-1 (Nicolau et al., 2006), ciguatoxin CTX3C (Inoue et al., 2004) and of palytoxin carboxylic acid (Kishi, 1989) as well as of fragments of some toxins (see references in Section 4.11.4.3).

When the molecular structure of some of the lipophilic phycotoxins have been unequivocally determined then there may be a better understanding of the relationship between their structure and their interactions at cellular and biochemical level, for example toxicity related to the imine functionality in cyclic imines (Kong et al., 2005), similarities between AZA and YTX in their action at cellular level (Ronzitti et al., 2007) and observations that the open ring seco acids of the PTXs are less toxic than their parent compounds (Halim and Brimble, 2006; Ares et al., 2007).

Pharmacologically or chemically dissimilar phycotoxins may occur together, for example DTX-1 and YTX have been found together in mussels (Lee et al., 1988) and the co-existence of OA, PTXs, YTXs and AZAs in Norwegian mussels has been observed (Yasumoto and Aune, unpublished data, cited in van Egmond et al., 2004). PSP and DSP toxins have been found simultaneously in shellfish (Gago-Martinez et al., 1996). Little is known about chronic exposure to such mixtures of phycotoxins, for example where possibly carcinogenic AZAs may occur with tumour promoters such as OA, as discussed by James et al. (2004) and further research on this topic has been recommended (van Egmond et al., 2004).

4.15   A NOTE ON THE IOC HARMFUL ALGAL BLOOM
PROGRAMME

The Intergovernmental Oceanographic Commission (IOC) of UNESCO has long been aware of the detrimental effects that harmful algal blooms (HAB) can have on national economies and aims to support research to mitigate these effects on fisheries, aquaculture, human health, recreation areas and ecosystems (Bernal, 2004). To this end, the IOC maintains a Science and Communication Centre on Harmful Algae, which serves as an important international channel of communication for information on many aspects of HAB. The Centre publishes a regular international newsletter, *Harmful Algae News*, a best-selling manual on HAB (Hallegraeff *et al.*, 2004) and other literature and maintains various information databases, including an international directory of experts. Further information can be obtained from the IOC website http://www.ioc.unesco.org/hab/.

Acknowledgements

The authors thank Dr Donald Anderson and Ms Judy Kleindinst of the Woods Hole Oceanographic Institution, Massachusetts, USA, for supplying the maps in Figs. 4.1a–4.1f and both Ir Hans van Egmond of the National Institute of Public Health, Bilthoven, the Netherlands and the United Nations Food and Agriculture Organization for permission to reproduce Figs. 4.2a–4.2j. Thanks are also due to Dr Philipp Hess of the Marine Institute, Oranmore, Eire, for helpful comments.

References

Aasen, J., MacKinnon, S.L., LeBlanc, P., Walter, J.A., Hovgaard, P., Aune, T. and Quilliam, M.A. (2005a) Detection and identification of spirolides in Norwegian shellfish and plankton. *Chemical Research in Toxicology*, **18**(3), 509–515.

Aasen, J., Samdal, I.A., Miles, C.O., Dahl, E., Briggs, L.R. and Aune, T. (2005b) Yessotoxins in Norwegian blue mussels (*Mytilus edulis*): Uptake from *Protoceratium reticulatum*, metabolism and depuration. *Toxicon*, **45**(3), 265–272.

Aasen, J.A., Hardstaff, W., Aune, T. and Quilliam, M.A. (2006) Discovery of fatty acid ester metabolites of spirolide toxins in mussels from Norway using liquid chromatography/tandem mass spectrometry. *Rapid Communications in Mass Spectrometry*, **20**(10), 1531–1537.

Alfonso, A., Román, Y., Vieytes, M.R., Ofuji, K., Satake, M., Yasumoto, T. and Botana, L.M. (2005) Azaspiracid-4 inhibits Ca^{2+} entry by stored operated channels in human T lymphocytes. *Biochemical Pharmacology*, **69**(11), 1627–1636.

Alfonso, A., Vieytes, M.R., Ofuji, K., Satake, M., Nicolaou, K.C., Frederick, M.O. and Botana, L.M. (2006) Azaspiracids modulate intracellular pH levels in human lymphocytes. *Biochemical and Biophysical Research Communications*, **346**(3), 1091–1099.

Amzil, Z., Marcaillou-Le Baut, C. and Bohec, M. (1996) Unexplained toxicity in molluscs gathered during phytoplankton monitoring, in *Harmful and Toxic Algal Blooms*, *Proceedings of the Seventh International Conference on Toxic Phytoplankton*, 1995 (eds. T. Yasumoto, Y. Oshima and Y. Fukuyo). Intergovernmental Oceanographic Commission of UNESCO, Paris, pp. 543–546.

Anderson, D.M. (1994) Red tides. *Scientific American*, **271**(2), 52–58.

Anderson, D.M., Andersen, P., Bricelj, V.M., Cullen, J.J. and Rensel, J.E. (2001) *Monitoring and Management Strategies for Harmful Algal Blooms in Coastal Waters*, APEC #201-MR-01.1, Asia Pacific Economic Program, Singapore, and Intergovernmental Oceanographic Commission Technical Series No. 59, Paris, 268 pp. Available from: http://www.whoi.edu/cms/files/Monitoring_Management_Report_24193.pdf

Anon. (2001) FSAI (2001) *Risk assessment of azaspiracids (AZAs) in shellfish*. Food Safety Authority of Ireland, 29 pp.

Anon. (2004) Report of the Joint FAO/IOC/WHO *ad hoc* Expert Consultation on Biotoxins in Bivalve Molluscs, Oslo, Norway, 26–30 September 2004. Full report 40 pp; summary 8 pp. Available from: ftp://ftp.fao.org/es/esn/food/biotoxin_report_en.pdf (full) *and* http://unesdoc.unesco.org/images/0013/001394/139421e.pdf (summary)

Anon. (2005a) Analysis of marine biotoxins: Approaches to validation and regulatory acceptance of alternative methods to the mouse bioassay as method of reference, Position Paper nr. 013/2005 of the BfR of 7 April 2005, 42 pp. Available from: http://www.bfr.bund.de/cm/245/position_paper_analysis_of_marine_biotoxins.pdf

Anon. (2005b) Mouse bioassay not suitable as a reference method for the regular analysis of algae toxins in mussels. BfR Expert Opinion No. 032/2005 of 26 May 2005, 7 pp. Available from: http://www.bfr.bund.de/cm/245/mouse_bioassay_not_suitable_as_a_reference_method_for_the_regular_analysis_of_algae_toxins_in_mussels.pdf

Anon. (2006a) Joint FAO/WHO Food Standards Programme, Codex Committee on Fish and Fishery Products, Twenty-eighth Session, Beijing, China, 18–22 September 2006, Agenda Item 6 CX/FFP 06/28/6-Add.1, Proposed Draft Standard for Live and Raw Bivalve Molluscs: Report of the Working Group meeting to assess the advice from the Joint FAO/WHO/IOC *Ad Hoc* Expert Consultation on Biotoxins in Bivalve Molluscs, 16 pp. Available from: ftp://ftp.fao.org/Codex/ccffp28/fp2806ae.pdf

Anon. (2006b) Risk assessment of azaspiracids (AZAs) in shellfish, August 2006. A report of the Scientific Committee of the Food Safety Authority of Ireland (FSAI), 39 pp. Available from: http://www.fsai.ie/publications/other/AZAs_risk_assess_aug06.pdf

Arias, H.R. (2006) Marine toxins targeting ion channels. *Marine Drugs*, **4**, 37–69.

Ares, I.R., Louzao, M.C., Espina, B., Vieytes, M.R., Miles, C.O., Yasumoto, T. and Botana, L.M. (2007) Lactone ring of pectenotoxins: A key factor for their activity on cytoskeletal dynamics. *Cellular Physiology and Biochemistry*, **19**(5–6), 283–292.

Aune, T., Yasumoto, T. and Engeland, E. (1991) Light and scanning electron microscopic studies on effects of marine algal toxins toward freshly prepared hepatocytes. *Journal of Toxicology and Environmental Health*, **34**, 1–9.

Aune, T. and Yndestad, M. (1993) Diarrhetic shellfish poisoning, in *Algal Toxins in Seafood and Drinking Water* (ed. I.R. Falconer). Academic Press, London, pp. 87–104.

Aune, T., Sørby, R., Yasumoto, T., Ramstad, H. and Landsverk, T. (2002) Comparison of oral and intraperitoneal toxicity of yessotoxin towards mice. *Toxicon*, **40**(1), 77–82.

Aune, T., Larsen, S., Aasen, J.A.B., Rehmann, N., Satake, M. and Hess, P. (2007) Relative toxicity of dinophysistoxin-2 (DTX-2) compared with okadaic acid, based on acute intraperitoneal toxicity in mice. *Toxicon*, **49**(1), 1–7.

Azanza, M.P.V., Azanza, R.V., Vargas, V.M.D. and Hedreyda, C.T. (2006) Bacterial endosymbionts of *Pyrodinium bahamense* var. *compressum*. *Microbial Ecology*, **52**(4), 756–764.

Backer, L.C., Fleming, L.E., Rowan, A.D. and Baden, D.G. (2004) Epidemiology, public health and human diseases associated with harmful marine algae, in *Manual on Harmful Marine Algae* (eds. G.M. Hallegraeff, D.M. Anderson and A.D. Cembella). UNESCO, Paris, pp. 723–749.

Baden, D.G. and Trainer, V.J. (1993) Mode of action of toxins of seafood poisoning, in *Algal Toxins in Seafood and Drinking Water* (ed. I.R. Falconer). Academic Press, London, pp. 49–74.

Baden, D.G., Bourdelais, A.J., Jacocks, H., Michelliza, S. and Naar, J. (2005) Natural and derivative brevetoxins: Historical background, multiplicity, effects. *Environmental Health Perspectives*, **113**(5), 621–625.

Bagnis, R. (1993) Ciguatera fish poisoning, in *Algal Toxins in Seafood and Drinking Water* (ed. I.R. Falconer). Academic Press, London, pp. 105–115.

Benoit, E., Juzans, P., Legrand, A.M. and Molgo, J. (1996) Nodal swelling produced by ciguatoxin-induced selective activation of sodium channels in myelinated nerve-fibers. *Neuroscience*, **71**(4), 1121–1131.

Berenguer, J.A., Gonzalez, L., Jimenez, I., Legarda, T.M., Olmedo, J.B. and Burdaspal, P.A. (1993) The effect of commercial processing on the paralytic shellfish poison (PSP) content of naturally contaminated *Acanthocardia tuberculatum* L. *Food Additives and Contaminants*, **10**(2), 217–230.

Bernal, P. (2004) Preface to *Manual on Harmful Marine Algae* (eds. G.M. Hallegraeff, D.M. Anderson and A.D. Cembella). UNESCO, Paris, pp. 5–6.

Botana, L. (ed.) (2007) *Chemistry and Pharmacology of Marine Toxins*. Blackwell Publishing, Ames, USA, 368 pp.

Bourdelais, A.J., Campbell, S., Jacocks, H., Naar, J., Wright, J.L.C., Carsi, J. and Baden, D.G. (2004) Brevenal is a natural inhibitor of brevetoxin action in sodium channel receptor binding assays. *Cellular and Molecular Neurobiology*, **24**(4), 553–563.

Bourdelais, A.J., Jacocks, H.M., Wright, J.L.C., Bigwarfe, P.M. and Baden, D.G. (2005) A new polyether ladder compound produced by the dinoflagellate *Karenia brevis*. *Journal of Natural Products*, **68**(1), 2–6.

Bowden, B.F. (2006) Yessotoxins – polycyclic ethers from dinoflagellates: Relationships to diarrhetic shellfish toxins. *Toxin Reviews*, **25**, 137–157.

Bricelj, V.M., Cembella, A.D., Laby, D., Shumway, S.E. and Cucci, T.L. (1996) Comparative physiological and behavioral responses to PSP toxins in two bivalve molluscs, the softshell clam, *Mya arenaria*, and surfclam, *Spisula solidissima*, in *Harmful and Toxic Algal Blooms, Proceedings of the Seventh International Conference on Toxic Phytoplankton*, 1995 (eds. T. Yasumoto, Y. Oshima and Y. Fukuyo). Intergovernmental Oceanographic Commission of UNESCO, Paris, pp. 405–408.

Briggs, L.R., Miles, C.O., Fitzgerald, J.M., Ross, K.M., Garthwaite, I. and Towers, N.R. (2004) Enzyme-linked immunosorbent assay for the detection of yessotoxin and its analogues. *Journal of Agricultural and Food Chemistry*, **52**(19), 5836–5842.

Burgess, V. and Shaw, G. (2001) Pectenotoxins – an issue for public health: A review of their comparative toxicology and metabolism. *Environment International*, **27**, 275–283.

Campora, C.E., Hokama, Y. and Ebesu, J.S.M. (2006) Comparative analysis of purified Pacific and Caribbean ciguatoxin congeners and related toxins using a modified ELISA technique. *Journal of Clinical Laboratory Analysis*, **20**(3), 121–125.

Carmichael, W.W. and Falconer, I.R. (1993) Diseases related to freshwater blue-green algal toxins, and control measures, in *Algal Toxins in Seafood and Drinking Water* (ed. I.R. Falconer). Academic Press, London, pp. 187–209.

Carmody, E.P., James, K.J. and Kelly, S.S. (1996) Dinophysistoxin-2: The predominant diarrhoetic shellfish toxin in Ireland. *Toxicon*, **34**(3), 351–359.

Cembella, A.D., Shumway, S.E. and Lewis, N.I. (1993) Anatomical distribution and spatio-temporal variation in paralytic shellfish toxin composition in two bivalve species from the Gulf of Maine. *Journal of Shellfish Research*, **12**(2), 389–403.

Cembella, A.D., Shumway, S.E. and Laroque, R. (1994) Sequestering and putative biotransformation of paralytic shellfish toxins by the sea scallop *Placopecten magellanicus*: Seasonal and spatial scales in natural populations. *Journal of Experimental Marine Biology and Ecology*, **180**, 1–22.

Cembella, A.D., Lewis, N.I. and Quilliam, M.A. (2000) The marine dinoflagellate *Alexandrium ostenfeldii* (Dinophyceae) as the causative organism of spirolide shellfish toxins. *Phycologia*, **67**(1), 67–74.

Cembella, A.D, Doucette, G.J. and Garthwaite, I. (2004) *In vitro* assays for phycotoxins, in *Manual on Harmful Marine Algae* (eds. G.M. Hallegraeff, D.M. Anderson and A.D. Cembella). UNESCO, Paris, pp. 297–345.

Chen, D.Z.X., Boland, M.P., Smillie, M.A., Klix, H., Ptak, C., Andersen, R.J. and Holmes, C.F.B. (1993) Identification of protein phosphatase inhibitors of the microcystin class in the marine environment. *Toxicon*, **31**(11), 1407–1414.

Chorus, I., Falconer, I.R., Salas, H.J. and Bartram, J. (2000) Health risks caused by freshwater cyanobacteria in recreational waters. *Journal of Toxicology and Environmental Health, Part B*, **3**(4), 323–347.

Chou, T., Kamo, O. and Uemura, D. (1996) Pinnatoxin-A, a potent shellfish poison from *Pinna muricata*. *Tetrahedron Letters*, **37**(23), 4023–4026.

Ciminiello, P., Fattorusso, E., Forino, M., Magno, S., Poletti, R., Satake, M., Vivani, R. and Yasumoto, T. (1997) Yessotoxin in mussels of the northern Adriatic Sea. *Toxicon*, **35**(2), 177–183.

Ciminiello, P., Dell'Aversano, C., Fattorusso, E., Magno, S., Tartaglione L., Cangini, M., Pompei, M., Guerrini, F., Boni, L. and Pistocchi, R. (2006) Toxin profile of *Alexandrium ostenfeldii* (Dinophyceae) from the Northern Adriatic Sea revealed by liquid chromatography–mass spectrometry. *Toxicon*, **47**(5), 597–604.

Codd, G.A., Bell, S.G., Kaya, K., Ward, C.J., Beattie, K.A. and Metcalf, J.S. (1999) Cyanobacterial toxins, exposure routes and human health. *European Journal of Phycology*, **34**(4), 405–415.

Codd, G.A., Morrison, L.F. and Metcalf, J.S. (2005) Cyanobacterial toxins: Risk management for health protection. *Toxicology and Applied Pharmacology*, **203**(3), 264–272.

Cook, A.C., Morris, S., Reese, R.A. and Irving, S.N. (2006) Assessment of fitness for purpose of an insect bioassay using the desert locust (*Schistocerca gregaria* L.) for the detection of paralytic shellfish toxins in shellfish flesh. *Toxicon*, **48**(6), 662–671.

Colman, J.R., Twiner, M.J., Hess, P., McMahon, T., Satake, M., Yasumoto, T., Doucette, G.J. and Ramsdell, J.S. (2005) Teratogenic effects of azaspiracid-1 identified by microinjection of Japanese medaka (*Oryzias latipes*) embryos. *Toxicon*, **45**(7), 881–890.

Cordier, S., Monfort, C., Miossec, L., Richardson, S. and Belin, C. (2000) Ecological analysis of digestive cancer mortality related to contamination by diarrhetic shellfish poisoning toxins along the coasts of France. *Environmental Research Section A*, **84**, 145–150.

de Fouw, J.C., van Egmond, H.P. and Speijers, G.J.A. (2001) Ciguatera fish poisoning: A review. RIVM Report No. 388802021, National Institute for Public Health and the Environment, 3720BA, Bilthoven, The Netherlands. Available from: http://www.rivm.nl/bibliotheek/rapporten/388802021.pdf

de la Iglesia, P., Gago-Martinez, A. and Yasumoto, T. (2007) Advanced studies for the application of high-performance capillary electrophoresis for the analysis of yessotoxin and 45-hydroxyyessotoxin. *Journal of Chromatography A*, **1156**(1–2), 160–166.

de la Rosa, L.A., Alfonso, A., Vilariño, N., Vieytes, M.R. and Botana, L.M. (2001) Modulation of cytosolic calcium levels of human lymphocytes by yessotoxin, a novel marine phycotoxin. *Biochemical Pharmacology*, **61**(7), 827–833.

Eriksson, J.E., Meriluto, J.A.O. and Lindholm, T. (1989) Accumulation of a peptide toxin from the cyanobacterium *Oscillatoria agardhii* in the freshwater mussel *Anodonta cygnea*. *Hydrobiologia*, **183**, 211–216.

Falconer, I.R. (1991) Tumour promotion and liver injury caused by oral consumption of cyanobacteria. *Environmental Toxicology and Water Quality*, **6**, 177–184.

Falconer, I.R. (1993) Measurement of toxins from blue-green algae in water and foodstuffs, in *Algal Toxins in Seafood and Drinking Water* (ed. I.R. Falconer). Academic Press, London, pp. 165–175.

Falconer, I.R., Choice, A. and Hosja, W. (1992) Toxicity of the edible mussel (*Mytilus edulis*) growing naturally in an estuary during a water bloom of the blue-green alga *Nodularia spumigena*. *Environmental Toxicology and Water Quality*, **7**, 119–123.

Falconer, I.R. and Humpage, A.R. (1996) Tumour promotion by cyanobacterial toxins. *Phycologia*, **35**(suppl. 6), 74–79.

Fehling, J., Green, D.H., Davidson, K., Bolch, C.J. and Bates, S.S. (2004) Domoic acid production by *Pseudo-nitzschia seriata* (Bacillariophyceae) in Scottish waters. *Journal of Phycology*, **40**(4), 622–630.

Fernández, M.L., Richard, J.D.A. and Cembella, A. (2004a) *In vivo* assays for phycotoxins, in *Manual on Harmful Marine Algae* (eds. G.M. Hallegraeff, D.M. Anderson and A.D. Cembella). UNESCO, Paris, pp. 347–380.

Fernández, M.L., Shumway, S. and Blanco, J. (2004b) Management of shellfish resources, in *Manual on Harmful Marine Algae* (eds. G.M. Hallegraeff, D.M. Anderson and A.D. Cembella). UNESCO, Paris, pp. 657–692.

Fladmark, K.E., Serres, M.H., Larsen, N.L., Yasumoto, T., Aune, T. and Døskeland, S.O. (1998) Sensitive detection of apoptogenic toxins in suspension cultures of rat and salmon hepatocytes. *Toxicon*, **36**(8), 1101–1114.

Forsyth, C.J., Xu, J., Nguyen, S.T., Ingunn, A., Samdal, I.A., Briggs, L.R., Rundberget, T., Sandvik, M. and Miles, C.O. (2006) Antibodies with broad specificity to azaspiracids by use of synthetic haptens. *Journal of the American Chemical Society*, **128**(47), 15114–15116.

Furey, A., Moroney, C., Braña Magdalena, A., Saez, M.J.F., Lehane, M. and James, K.J. (2003) Geographical, temporal and species variation of the polyether toxins, azaspiracids, in shellfish. *Environmental Science and Technology*, **37**, 3078–3084.

Fux, E., McMillan, D., Bire R. and Hess, P. (2007) Development of an ultra-performance liquid chromatography–mass spectrometry method for the detection of lipophilic marine toxins. *Journal of Chromatography A*, **1157**(1–2), 273–280.

Gago-Martinez, A., Rodriguez-Vazquez, J.A., Thibault, P. and Quilliam, M.A. (1996) Simultaneous occurrence of diarrhetic and paralytic shellfish poisoning toxins in Spanish mussels in 1993. *Natural Toxins*, **4**(2), 72–79.

Gallacher, S., Flynn, K.J., Franco, J.M., Brueggemann, E.E. and Hines, H.B. (1997) Evidence for production of paralytic shellfish toxins by bacteria associated with *Alexandrium* spp. (Dinophyta) in culture. *Applied and Environmental Microbiology*, **63**(1), 239–245.

Gallacher, S., Howard, F.G., Hess, P., Macdonald, E.M., Kelly, M.C., Bates, L.A., Brown, N., MacKenzie, M., Gillibrand, P.A. and Turrell, W.R. (2001) The occurrence of amnesic shellfish poisons in shellfish from Scottish waters, in *Harmful Algal Blooms 2000* (eds. G.M. Hallegraeff, S.I. Blackburn, C.J. Bolch and R.J. Lewis). Intergovernmental Oceanographic Commission of UNESCO, Paris, pp. 30–33.

Garthwaite, I., Ross, K.M., Miles, C.O., Briggs, L.R., Towers, N.R., Borrell, T. and Busby, B. (2001) Integrated enzyme-linked immunosorbent assay screening system for amnesic, neurotoxic, diarrhetic, and paralytic shellfish poisoning toxins found in New Zealand. *Journal of AOAC International*, **84**(5), 1643–1648.

Gill, S., Murphy, M., Clausen, J., Richard, D., Quilliam, M., MacKinnon, S., LeBlanc, P., Mueller, R. and Pulido, O. (2003) Neural injury biomarkers of novel shellfish toxins, spirolides: A pilot study using immunochemical and transcriptional analysis. *Neurotoxicology*, **24**(4–5), 593–604.

Goldin, A.L., Snutch, T., Lubbert, H., Dowsett, A., Marshall, J., Auld, V., Downey, W., Fritz, L.C., Lester, H.A., Dunn, R., Catterall, W.A. and Davidson, N. (1986) Messenger RNA coding for only the alpha subunit of the rat brain Na channel is sufficient for expression of functional channels in *Xenopus* oocytes. *Proceedings of the National Academy of Sciences USA (Neurobiology)*, **83**, 7503–7507.

González, A.V., Rodríguez-Velasco, M.L., Ben-Gigirey, B. and Botana, L.M. (2006) First evidence of spirolides in Spanish shellfish. *Toxicon*, **48**(8), 1068–1074.

Grimmelt, B., Nijjar, M.S., Brown, J., Macnair, N., Wagner, S., Johnson, G.R. and Amend, J.F. (1990) Relationship between domoic acid levels in the blue mussel (*Mytilus edulis*) and toxicity in mice. *Toxicon*, **28**(5), 501–508.

Halim, R. and Brimble, M.A. (2006) Synthetic studies towards the pectenotoxins: A review. *Organic and Biomolecular Chemistry*, **4**, 4048–4058.

Hallegraeff, G.M. (2004) Harmful algal blooms: A global overview, in *Manual on Harmful Marine Algae* (eds. G.M. Hallegraeff, D.M. Anderson and A.D. Cembella). UNESCO, Paris, pp. 25–49.

Hallegraeff, G.M., Anderson, D.M. and Cembella, A.D. (eds.) (2004) *Manual on Harmful Marine Algae*. UNESCO, Paris, 793 pp.

Hamilton, B., Hurbungs, M., Jones, A. and Lewis, R.J. (2002) Multiple ciguatoxins present in Indian Ocean reef fish. *Toxicon*, **40**(9),1347–1353.

Hampson, D.R. and Manalo, J.L. (1998) The activation of glutamate receptors by kainic acid and domoic acid. *Natural Toxins*, **6**(3–4), 153–158.

Hatfield, C.L., Gauglitz, E.J., Barnett, H.J., Lund, J.A.K., Wekell, J.C. and Eklund, M. (1995) The fate of domoic acid in Dungeness crab (*Cancer magister*) as a function of processing. *Journal of Shellfish Research*, **14**(2), 359–363.

Haywood, A.J., Steidinger, K.A., Truby, E.W., Bergquist, P.R., Bergquist, P.L., Adamson, J. and Mackenzie, L. (2004) Comparative morphology and molecular phylogenetic analysis of three new species of the genus *Karenia* (Dinophyceae) from New Zealand. *Journal of Phycology*, **40**(1), 165–179.

Hess, P. (2008) What's new in Toxins? in *Proceedings of the Twelfth International Conference on Harmful Algae*, Copenhagen, 4–8 September 2006 (eds. Ø. Moestrup, G. Doucette, H. Enevoldsen, A. Godhe, G. Hallegraeff, J. Lewis, B. Luckas, N. Lundholm, K. Rengefors, K. Sellner, K. Steidinger and P. Tester). International Society for the Study of Harmful Algae (ISSHA) and the Intergovernmental Oceanographic Commission (IOC) of UNESCO.

Hess, P., McMahon, T., Slattery, D., Swords, D., Dowling, G., McCarron, M., Clarke, D., Devilly, L., Gibbons, W., Silke, J. and O'Cinneide, M. (2001) *Biotoxin Chemical Monitoring in Ireland 2001, Proceedings of the Second Irish Marine Science Biotoxin Workshop*, pp. 8–18. Available from: http://www.marine.ie/NR/rdonlyres/200434CF-F390-4C71-B96E-8FC84060BF31/0/2nd_proceedings2001.pdf

Hess, P., Nguyen, L., Aasen, J., Keogh, M., Kilcoyne, J., McCarron, P. and Aune, T. (2005) Tissue distribution, effects of cooking and parameters affecting the extraction of azaspiracids from mussels, *Mytilus edulis*, prior to analysis by liquid chromatography coupled to mass spectrometry. *Toxicon*, **46**(1), 62–71.

Hess, P., McCarron, P. and Quilliam, M.A. (2006) Fit-for-purpose shellfish reference materials for internal and external quality control in the analysis of phycotoxins. *Analytical and Bioanalytical Chemistry*, **387**(7), 2463–2474.

Hess, P. and Aasen, J.B. (2007) Chemistry, origins and distribution of yessotoxin and its analogues, in *Chemistry and Pharmacology of Marine Toxins* (ed. L. Botana). Blackwell Publishers, Ames, USA, pp. 187–202.

Holmes, C.F.B. (1991) Liquid chromatography-linked protein phosphatase bioassay: A highly sensitive marine bioscreen for okadaic acid and related diarrhetic shellfish toxins. *Toxicon*, **29**(4–5), 469–477.

Holmes, M.J., Lewis, R.J., Jones, A. and Hoy, A.W.W. (1995) Cooliatoxin, the first toxin from *Coolia montis* (Dinophyceae). *Natural Toxins*, **3**(5), 355–362.

Honkanen, R.E., Mowdy, D.E. and Dickey, R.W. (1996a) Detection of DSP-toxins, okadaic acid and dinophysis toxin-1 in shellfish by the serine/threonine protein phosphatase assay. *Journal of AOAC International*, **79**(6), 1336–1343.

Honkanen, R.E., Stapleton, J.D., Bryan, D.E. and Abercrombie, J. (1996b) Development of a protein phosphatase-based assay for the detection of phosphatase inhibitors in crude whole cell and animal extracts. *Toxicon*, **34**(11–12), 1385–1392.

Hu, T.M., Curtis, J.M., Oshima, Y., Quilliam, M.A., Walter, J.A., Watson-Wright, W.M. and Wright, J.L.C. (1995a) Spirolide-B and spirolide-D, two novel macrocycles isolated from the digestive glands of shellfish. *Journal of the Chemical Society, Chemical Communications*, **20**, 2159–2161.

Hu, T.M., Curtis, J.M., Walter, J.A., McLachlan, J.L. and Wright, J.L.C. (1995b) Two new water-soluble DSP toxin derivatives from the dinoflagellate *Prorocentrum maculosum* – possible storage and excretion products of the dinoflagellate. *Tetrahedron Letters*, **36**(51), 9273–9276.

Hu, T.M., Curtis, J.M., Walter, J.A. and Wright, J.L.C. (1995c) Identification of DTX-4, a new water-soluble phosphatase inhibitor from the toxic dinoflagellate *Prorocentrum lima*. *Journal of the Chemical Society, Chemical Communications*, **5**, 597–599.

Hu, T.M., Curtis, J.M., Walter, J.A. and Wright, J.L.C. (1996a) Characterization of biologically inactive spirolides E and F: Identification of the spirolide pharmacophore. *Tetrahedron Letters*, **37**(43), 761–764.

Hu, T.M., De Freitas, A.S.W., Curtis, J.M., Oshima, Y., Walter, J.A. and Wright, J.L.C. (1996b) Isolation and structure of prorocentrolide-B, a fast-acting toxin from *Prorocentrum maculosum*. *Journal of Natural Products*, **59**(11), 1010–1014.

Hu, T., Burton, I.W., Cembella, A.D., Curtis, J.M., Quilliam, M.A., Walter, J.A. and Wright, J.L. (2001) Characterization of spirolides A, C, and 13-desmethyl C, new marine toxins isolated from toxic plankton and contaminated shellfish. *Journal of Natural Products*, **64**(3), 308–312.

Hungerford, J.M. (2005) Committee on Natural Toxins and Food Allergens, Marine and Freshwater Toxins (report). *Journal of AOAC International*, **88**(1), 299–313.

Hungerford, J.M. (2006) Committee on Natural Toxins and Food Allergens, Marine and Freshwater Toxins (report). *Journal of AOAC International*, **89**(1), 248–269.

Inami, G.B., Crandall, C., Csuti, D., Oshiro, M. and Brenden, R.A. (2004) Feasibility of reduction in use of the mouse bioassay: Presence/absence screening for saxitoxin in frozen acidified mussel and oyster extracts from the coast of California with *in vitro* methods. *Journal of AOAC International*, **87**(5), 1133–1142.

Inoue, M., Miyazaki, K., Uehara, H., Maruyama, M. and Hirama, M. (2004) First- and second-generation total synthesis of ciguatoxin CTX3C. *Proceedings of the National Academy of Sciences USA*, **101**(33), 12013–12018.

Ito, E. (2006) Verification of diarrhetic activities of PTX-2 and okadaic acid *in vivo*. Poster PO.08-01 presented at the *12th International Conference on Harmful Algae*, Copenhagen, Denmark, 4–8 September 2006, International Society for the Study of Harmful Algae. Available from: http://www.bi.ku.dk/hab/programme.asp

Ito, E., Satake, M., Ofuji, K., Kurita, N., McMahon, T., James, K. and Yasumoto, T. (2000) Multiple organ damage caused by a new toxin azaspiracid, isolated from mussels produced in Ireland. *Toxicon*, **38**(7), 917–930.

Ito, E., Satake, M., Ofuji, K., Higashi, M., Harigaya, K., McMahon, T. and Yasumoto, T. (2002) Chronic effects in mice caused by oral administration of sublethal doses of azaspiracid, a new marine toxin isolated from mussels. *Toxicon*, **40**(2), 193–203.

Iverson, F., Truelove, J., Nera, E., Tryphonas, L., Campbell, J. and Lok, E. (1989) Domoic acid poisoning and mussel-associated intoxication: Preliminary investigations into the response of mice and rats. *Food and Chemical Toxicology*, **27**, 377–384.

James, K.J., Moroney, C., Roden, C., Satake, M., Yasumoto, T., Lehane, M. and Furey, A. (2003) Ubiquitous 'benign' alga emerges as the cause of shellfish contamination responsible for the human toxic syndrome, azaspiracid poisoning. *Toxicon*, **41**(2), 145–151.

James, K.J., Saez, M.J.F., Furey, A. and Lehane, M. (2004) Azaspiracid poisoning, the food-borne illness associated with shellfish consumption. *Food Additives and Contaminants*, **21**(9), 879–892.

Jasperse, J.A. (ed.) (1993) *Marine Toxins and New Zealand Shellfish, Proceedings of a Workshop on Research Issues*, June 1993, The Royal Society of New Zealand, Miscellaneous Series, No. 24, 68 pp.

Jeffery, B., Barlow, T., Moizer, K., Paul, S. and Boyle, C. (2004) Amnesic shellfish poison. *Food and Chemical Toxicology*, **42**(4), 545–557.

Jeglitsch, G., Rein, K., Baden, D.G. and Adams, D.J. (1998) Brevetoxin-3 (PbTx-3) and its derivatives modulate single tetrodotoxin-sensitive sodium channels in rat sensory neurons. *Journal of Pharmacology and Experimental Therapeutics*, **284**, 516–525.

Jung, J.H., Sim, J.S. and Lee, C.-O. (1995) Cytotoxic compounds from a two-sponge association. *Journal of Natural Products*, **58**(11), 1722–1726.

Kao, C.Y. (1993) Paralytic shellfish poisoning, in *Algal Toxins in Seafood and Drinking Water* (ed. I.R. Falconer). Academic Press, London, pp. 75–86.

Kishi, Y. (1989) Natural products synthesis: Palytoxin. *Pure and Applied Chemistry*, **61**(3), 313–324.

Kobayashi, J., Shimbo, K., Kubota, T. and Tsuda, M. (2003) Bioactive macrolides and polyketides from marine dinoflagellates. *Pure and Applied Chemistry*, **75**(2–3), 337–342.

Kobayashi, J. and Tsuda, M. (2004) Amphidinolides, bioactive macrolides from symbiotic marine dinoflagellates. *Natural Product Reports*, **21**, 77–93.

Kodama, M., Doucette, G.J. and Green, D.H. (2006) Relationships between bacteria and harmful algae, in *Ecology of Harmful Algae. Vol. 189: Ecological Studies* (eds. E. Granéli and J.T. Turner). Springer-Verlag, Berlin, pp. 243–255.

Kong, K., Moussa, Z. and Romo, D. (2005) Studies toward a marine toxin immunogen: Enantioselective synthesis of the spirocyclic imine of (−)-gymnodimine. *Organic Letters*, **7**(23), 5127–5130.

Kotaki, Y., Koike, K., Yoshida, M., Thuoc, C.V., Huyen, N.T.M., Hoi, N.C., Fukuyo, Y. and Kodama, M. (2000) Domoic acid production in *Nitzschia* sp. (Bacillariophyceae) isolated from a shrimp-culture pond in Do Son, Vietnam. *Journal of Phycology*, **36**(6), 1057–1060.

Kotaki, Y., Lundholm, N., Onodera, H., Kobayashi, K., Bajarias, F.F.A., Furio, E.F., Iwataki, M., Fukuyo, Y. and Kodama, M. (2004) Wide distribution of *Nitzschia navis-varingica*, a new domoic acid-producing benthic diatom found in Vietnam. *Fisheries Science*, **70**(1), 28–32.

Kuiper-Goodman, T., Falconer, I.R. and Fitzgerald, J. (1999) Human health aspects, in *Toxic Cyanobacteria in Water* (eds. I. Chorus and J. Bartram). E. and F.N. Spon, London, pp. 113–153.

Kulagina, N.V., Twiner, M.J., Hess, P., McMahon, T., Satake, M., Yasumoto, T., Ramsdell, J.S., Doucette, G.J., Ma, W. and O'Shaughnessy, T.J. (2006) Azaspiracid-1 inhibits bioelectrical activity of spinal cord neuronal networks. *Toxicon*, **47**(7), 766–773.

Kuramoto, M., Arimoto, H. and Uemura, D. (2004) Bioactive alkaloids from the sea: A review. *Marine Drugs*, **1**, 39–54.

Kvasnicka, F., Sevcik, R. and Voldrich, M. (2006) Determination of domoic acid by on-line coupled capillary isotachophoresis with capillary zone electrophoresis. *Journal of Chromatography A*, **1113**(1–2), 255–258.

Lansberg, J.H. (2002) The effects of harmful algal blooms on aquatic organisms. *Reviews in Fisheries Science*, **10**(2), 113–390.

Lawrence, J.F., Maher, M. and Watson-Wright, W. (1994) Effect of cooking on the concentration of toxins associated with paralytic shellfish poison in lobster hepatopancreas. *Toxicon*, **32**(1), 57–64.

Lawrence, J.F., Niedzwiadek, B. and Menard, C. (2005) Quantitative determination of paralytic shellfish poisoning toxins in shellfish using prechromatographic oxidation and liquid chromatography with fluorescence detection: Collaborative study. *Journal of AOAC International*, **88**(6), 1714–1732.

Laycock, M.V., Thibault, P., Ayer, S.W. and Walter, J.A. (1994) Isolation and purification procedures for the preparation of paralytic shellfish poisoning toxin standards. *Natural Toxins*, **2**(4), 175–183.

Lee, J.-S., Tangen, K., Dahl, E., Hovgaard, P. and Yasumoto, T. (1988) Diarrhetic shellfish toxins in Norwegian mussels. *Nippon Suisan Gakkaishi*, **54**(11), 1953–1957.

Leftley, J.W. and Hannah, F. (1998) Phycotoxins in seafood, in *Natural Toxicants in Foods* (ed. D.H. Watson). Sheffield Academic Press, Sheffield, pp. 182–224.

Legrand, A.-M., Fukui, M., Cruchet, P., Ishibashi, Y. and Yasumoto, T. (1992) Characterization of ciguatoxins from different fish species and wild *Gambierdiscus toxicus*, in *Proceedings of the Third International Conference on Ciguatera Fish Poisoning*, Puerto Rico (ed. T.R. Tosteson). Polyscience Publications, Quebec, pp. 25–32.

Lehane, L. and Lewis, R.J. (2000) Ciguatera: Recent advances but the risk remains. *International Journal of Food Microbiology*, **61**(2–3), 91–125.

Leira, F.J., Vieites, J.M., Botana, L.M. and Vyeites, L.M. (1998) Domoic acid levels of naturally contaminated scallops as affected by canning. *Journal of Food Science*, **63**(6), 1081–1083.

Leira, F., Cabadoa, A.G., Vieytes, M.R., Román, Y., Alfonso, A., Botana, L.M., Yasumoto, T., Malaguti, C. and Rossini, G.P. (2002) Characterization of F-actin depolymerization as a major toxic event induced by pectenotoxin-6 in neuroblastoma cells. *Biochemical Pharmacology*, **63**(11), 1979–1988.

Leira, F., Alvarez, C., Cabado, A.G., Vieites, J.M., Vieytes, M.R. and Botana, L.M. (2003) Development of an F actin-based live-cell fluorimetric microplate assay for diarrhetic shellfish toxins. *Analytical Biochemistry*, **317**(2), 129–135.

Lenoir, S., Ten-Hage, L., Turquet, J., Quod, J.P., Bernard, C. and Hennion, M.-C. (2004) First evidence of palytoxin analogues from an *Ostreopsis mascarenensis* (Dinophyceae) benthic bloom in southwestern Indian Ocean. *Journal of Phycology*, **40**(6), 1042–1051.

Lewis, R.J. (2001) The changing face of ciguatera. *Toxicon*, **39**(1), 97–106.

Lewis, R.J. (2004) Detection of toxins associated with ciguatera fish poisoning, in *Manual on Harmful Marine Algae* (eds. G.M. Hallegraeff, D.M. Anderson and A.D. Cembella). UNESCO, Paris, pp. 267–277.

Lewis, R.J., Sellin, M., Poli, M.A., Norton, R.S., MacLeod, J.K. and Sheil, M.M. (1991) Purification and characterization of ciguatoxins from moray eel (*Lycodontis javanicus*, Muraenidae). *Toxicon*, **29**(9), 1115–1127.

Lewis, R.J. and Holmes, M.J. (1993) Origin and transfer of toxins involved in ciguatera. *Comparative Biochemistry and Physiology C – Pharmacology, Toxicology and Endocrinology*, **106**(3), 615–628.

Lewis, R.J., Holmes, M.J., Alewood, P.F. and Jones, A. (1994) Ionspray mass spectrometry of ciguatoxin-1, maitotoxin-2 and -3, and related marine polyether toxins. *Natural Toxins*, **2**(2), 56–63.

Llewellyn, L.E. (2006) Saxitoxin, a toxic marine natural product that targets a multitude of receptors. *Natural Product Reports*, **23**, 200–222.

Llewellyn, L.E., Negri, A. and Quilliam, M.A. (2004) High affinity for the rat brain sodium channel of newly discovered hydroxybenzoate saxitoxin analogues from the dinoflagellate *Gymnodinium catenatum*. *Toxicon*, **43**(1), 101–104.

Long, B.M. and Carmichael, W.W. (2004) Marine cyanobacterial toxins, in *Manual on Harmful Marine Algae* (eds. G.M. Hallegraeff, D.M. Anderson and A.D. Cembella). UNESCO, Paris, pp. 279–296.

Lopez-Rivera, A., Suarez-Isla, B.A., Eilers, P.P., Beaudry, C.G., Hall, S., Fernandez-Amandi, M., Furey, A. and James, K.J. (2005) Improved high-performance liquid chromatographic method for the determination of domoic acid and analogues in shellfish: Effect of pH. *Analytical and Bioanalytical Chemistry*, **381**(8), 1540–1545.

Lu, C.-K, Leeb, G.-S., Huang, R. and Chou, H.-N. (2001) Spiro-prorocentrimine, a novel macrocyclic lactone from a benthic *Prorocentrum* sp. of Taiwan. *Tetrahedron Letters*, **42**(9), 1713–1716.

Lu, C.-K., Chou, H.N., Lee, C.K. and Lee, T.H. (2005) Prorocentin, a new polyketide from the marine dinoflagellate *Prorocentrum lima*. *Organic Letters*, **7**(18), 3893–3896.

Luckas, B., Hummert, C. and Oshima, Y. (2004) Analytical methods for paralytic shellfish poisons, in *Manual on Harmful Marine Algae* (eds. G.M. Hallegraeff, D.M. Anderson and A.D. Cembella). UNESCO, Paris, pp. 191–209.

Lupin, H.M. (2004) Introduction to the analysis of regulations of interest to public health safety of molluscs and mollusc products. Available from: http://www.globefish.org/index.php?id=2073

Luu, H.A., Chen, D.Z.X., Magoon, J., Worms, J., Smith, J. and Holmes, C.F.B. (1993) Quantification of diarrhetic shellfish toxins and identification of novel protein phosphatase inhibitors in marine phytoplankton and mussels. *Toxicon*, **31**(1), 75–83.

Mackenzie, L., Haywood, A., Adamson, J., Truman, P., Till, D., Seki, T., Satake, M. and Yasumoto, T. (1996) Gymnodimine contamination of shellfish in New Zealand, in *Harmful and Toxic Algal Blooms, Proceedings of the Seventh International Conference on Toxic Phytoplankton*, 1995 (eds. T. Yasumoto, Y. Oshima and Y. Fukuyo). Intergovernmental Oceanographic Commission of UNESCO, Paris, pp. 97–100.

MacKinnon, S.L., Cembella, A.D., Burton, I.W., Lewis, N., LeBlanc, P. and Walter, J.A. (2006a) Biosynthesis of 13-desmethyl spirolide C by the dinoflagellate *Alexandrium ostenfeldii*. *Journal of Organic Chemistry*, **71**(23), 8724–8731.

MacKinnon, S.L., Walter, J.A., Quilliam, M.A., Cembella, A.D., LeBlanc, P., Burton, I.W., Hardstaff, W.R. and Lewis, N.I. (2006b) Spirolides isolated from Danish strains of the toxigenic dinoflagellate *Alexandrium ostenfeldii*. *Journal of Natural Products*, **69**(7), 983–987.

Mackintosh, F.H., Gallacher, S., Shanks, A.M. and Smith, E.A. (2002) Assessment of MIST Alert™, a commercial qualitative assay for detection of paralytic shellfish poisoning toxins in bivalve molluscs. *Journal of AOAC International*, **85**(3), 632–641.

Manger, R.L., Leja, L.S., Lee, S.Y., Hungerford, J.M., Hokama, Y., Dickey, R.W., Granade, H.R., Lewis, R., Yasumoto, T. and Wekell, M.M. (1995) Detection of sodium channel toxins: Directed cytotoxicity assay of purified ciguatoxins, brevetoxins, saxitoxins, and seafood extracts. *Journal of AOAC International*, **78**(2), 521–527.

McCarron, P. and Hess, P. (2006) Tissue distribution and effects of heat treatments on the content of domoic acid in blue mussels, *Mytilus edulis*. *Toxicon*, **47**(4), 473–479.

McNabb, P., Selwood, A.I., Holland, P.T., Aasen, J., Aune, T., Eaglesham, G., Hess, P., Igarishi, M., Quilliam, M., Slattery, D., van de Riet, J., van Egmond, H., van den Top, H. and Yasumoto, T. (2005) Multiresidue method for determination of algal toxins in shellfish: Single-laboratory validation and interlaboratory study. *Journal of AOAC International*, **88**(3), 761–772.

Miles, C.O., Wilkins, A.L., Munday, R., Dines, M.H., Hawkes, A.D., Briggs, L.R., Sandvik, M., Jensen, D.J., Cooney, J.M., Holland, P.T., Quilliam, M.A., MacKenzie, A.L., Beuzenberg, V. and Towers, N.R. (2004a) Isolation of pectenotoxin-2 from *Dinophysis acuta* and its conversion to pectentotoxin-2 seco acid, and preliminary assessment of their acute toxicities. *Toxicon*, **43**(1), 1–9.

Miles, C.O., Wilkins, A.L., Samdal, I.A., Sandvik, M., Petersen. D., Quilliam, M.A., Naustvoll, L.J., Rundberget, T., Torgersen, T., Hovgaard, P., Jensen, D.J. and Cooney, J.M. (2004b) Novel pectenotoxin, PTX-12, in *Dinophysis* spp. and shellfish from Norway. *Chemical Research in Toxicology*, **17**, 1423–1433.

Miles, C.O., Wilkins, A.L., Hawkes, A.D., Jensen, D.J., Selwood, A.I, Beuzenberg, V., MacKenzie, A.L., Cooney, J.M. and Holland, P.T. (2006) Isolation of pectenotoxins -13 and -14 from *Dinophysis acuta* in New Zealand. *Toxicon*, **48**(2), 152–159.

Mountfort, D.O., Kennedy, G., Garthwaite, I., Quilliam, M. and Hannah, D.J. (1999) Evaluation of the fluorometric protein phosphatase inhibition assay in the determination of okadaic acid in mussels. *Toxicon*, **37**(6), 909–922.

Mountfort, D.O., Suzuki, T. and Truman, P. (2001) Protein phosphatase inhibition assay adapted for determination of total DSP in contaminated mussels. *Toxicon*, **39**(2–3), 383–390.

Munday, R., Towers, N.R., Mackenzie, L., Beuzenberg, V., Holland, P.T. and Miles, C.O. (2004) Acute toxicity of gymnodimine to mice. *Toxicon*, **44**(2), 173–178.

Murata, M., Kumagai, M., Lee, J.-S. and Yasumoto, T. (1987) Isolation and structure of yessotoxin, a novel polyether compound implicated in diarrhetic shellfish poisoning. *Tetrahedron Letters*, **28**, 5869–5872.

Murata, M., Naoki, H., Matsunaga, S., Satake, M. and Yasumoto, T. (1994) Structure and partial stereochemical assignments for maitotoxin, the most toxic and largest natural non-biopolymer. *Journal of the American Chemical Society*, **116**(16), 7098–7107.

Nagai, H., Yasumoto, T. and Hokama, Y. (1996) Aplysiatoxin and debromoaplysiatoxin as the causative agents of a red alga *Gracilaria coronopifolia* poisoning in Hawaii. *Toxicon*, **34**(7), 753–761.

Nagashima, Y., Noguchi, T., Tanaka, M. and Hashimoto, K. (1991) Thermal degradation of paralytic shellfish poison. *Journal of Food Science*, **56**(6), 1572–1575.

Naoki, H., Fujita, T., Cruchet, P., Legrand, A.-M., Igarashi, T. and Yasumoto, T. (2001) Structural determination of new ciguatoxin congeners by tandem mass spectrometry, in *Mycotoxins and Phycotoxins in Perspective at the Turn of the Millennium, Proceedings of the Xth International IUPAC Symposium on Mycotoxins and Phycotoxins* (May 2000, Guarujá, Brazil) (eds. W.J. de Koe, R.A. Samson, H.P. van Egmond, J. Gilbert and M. Sabino). Ponsen and Looyen, Wageningen, the Netherlands, pp. 475–482.

Negri, A.P. and Jones, G.J. (1995) Bioaccumulation of paralytic shellfish poisoning (PSP) toxins from the cyanobacterium *Anabaena circinalis* by the freshwater mussel *Alathyria condola*. *Toxicon*, **33**(5), 667–678.

Negri, A., Stirling, D., Quilliam, M., Blackburn, S., Bolch, C., Burton, I., Eaglesham, G., Thomas, K., Walter, J. and Willis, R. (2003) Three novel hydroxybenzoate saxitoxin analogues isolated from the dinoflagellate *Gymnodinium catenatum*. *Chemical Research in Toxicology*, **16**(8), 1029–1033.

Nicholson, G.M. and Lewis, R.J. (2006) Ciguatoxins: Cyclic polyether modulators of voltage-gated ion channel function. *Marine Drugs*, **4**, 82–118.

Nicolaou, K.C., Koftis, T.V., Vyskocil, S., Petrovic, G., Tang, W., Frederick, M.O., Chen, D.Y.-K., Li, Y., Ling, T. and Yamada, Y.M.A. (2006) Total synthesis and structural elucidation of azaspiracid-1: Final assignment and total synthesis of the correct structure of azaspiracid-1. *Journal of the American Chemical Society*, **128**(9), 2859–2872.

Noguchi, T., Ueda, Y., Onoue, Y., Kono, M., Koyama, K., Hashimoto, K., Seno, Y. and Mishima, S. (1980a) Reduction in toxicity of PSP infested scallops during canning process. *Bulletin of the Japanese Society of Scientific Fisheries*, **46**(10), 1273–1277.

Noguchi, T., Ueda, Y., Onoue, Y., Kono, M., Koyama, K., Hashimoto, K., Takeuchi, T., Seno, Y. and Mishima, S. (1980b) Reduction in toxicity of highly PSP-infested scallops during canning process and storage. *Bulletin of the Japanese Society of Scientific Fisheries*, **46**(11), 1339–1344.

Noguchi, T., Nagashima, Y., Maruyama, J., Kamimura, S. and Hashimoto, K. (1984) Toxicity of the adductor muscle of markedly-infested scallop *Patinopecten yessoensis*. *Bulletin of the Japanese Society of Scientific Fisheries*, **50**(3), 517–520.

Nonomura, T., Sasaki, M., Matsumori, N., Murata, M., Tachibana, K. and Yasumoto, T. (1996) The complete structure of maitotoxin–2. Configuration of the C135-C142 side-chain and absolute configuration of the entire molecule. *Angewandte Chemie* (International Edition in English), **35**(15), 1675–1678.

Nunez, P.E. and Scoging, A.C. (1997) Comparison of a protein phosphatase inhibition assay, HPLC assay and enzyme-linked immunosorbent assay with the mouse bioassay for the detection of diarrhetic shellfish poisoning toxins in European shellfish. *International Journal of Food Microbiology*, **36**(1), 39–48.

Ofuji, K., Satake, M., McMahon, T., Silke, J., James, K.J., Naoki, H., Oshima, Y. and Yasumoto, T. (1999) Two analogs of azaspiracid isolated from mussels, *Mytilus edulis*, involved in human intoxication in Ireland. *Natural Toxins*, **7**(3), 99–102.

Ofuji, K., Satake, M., McMahon, T., James, K.J., Haoki, H., Oshima, Y. and Yasumoto, T. (2001) Structures of azaspiracid analogs, azaspiracid-4 and azaspiracid-5, causative toxins of azaspiracid poisoning in Europe. *Bioscience, Biotechnology and Biochemistry*, **65**(3), 740–742.

Ogino, H., Kumagai, M. and Yasumoto, T. (1997) Toxicologic evaluation of yessotoxin. *Natural Toxins*, **5**(6), 255–259.

Ohta, T., Nomata, H., Takeda, T. and Kaneko, H. (1992) Studies on shellfish poison of scallop *Patinopecten yessoensis*–1. Reduction in toxicity of PSP infested scallops during extrusion processing. *Scientific Reports of the Hokkaido Fisheries Experimental Station*, **38**, 23–30.

Okumura, M., Tsuzuki, H. and Tomita, B.-I. (2005) A rapid detection method for paralytic shellfish poisoning toxins by cell bioassay. *Toxicon*, **46**(1), 93–98.

Onoue, Y. and Nozawa, K. (1989) Separation of toxin from harmful red tides occurring along the coast of Kagoshima Prefecture, in *Red Tides: Biology, Science and Toxicology* (eds. T. Okaichi, D.M. Anderson and T. Nemoto). Elsevier Science, New York, pp. 361–374.

Oshima, Y. (1995) Chemical and enzymatic transformations of paralytic shellfish toxins in marine organisms, in *Harmful Marine Algal Blooms, Proceedings of the Sixth International Conference on Toxic Marine Phytoplankton*, 1993 (eds. P. Lassus, G. Arzul, E. Erard-Le Denn, P. Gentien and C. Marcaillou-Le Baut). Lavoisier, Paris, pp. 475–480.

Paz, B., Riobó, P., Souto, M.L., Gil, L.V., Norte, M., Fernández, J.J. and Franco, J.M. (2006) Detection and identification of glycoyessotoxin A in a culture of the dinoflagellate *Protoceratium reticulatum*. *Toxicon*, **48**(6), 611–619.

Pérez-Gómez, A., Ferrero-Gutierrez, A., Novelli, A., Franco, J.M., Paz, B. and Fernández-Sánchez, M.T. (2006) Potent neurotoxic action of the shellfish biotoxin yessotoxin on cultured cerebellar neurons. *Toxicological Sciences*, **90**(1), 168–177.

Perl, T.M., Bedard, L., Kosatsky, T., Hockin, J.C., Todd, E.C. and Remis, R.S. (1990) An outbreak of toxic encephalopathy caused by eating mussels contaminated with domoic acid. *New England Journal of Medicine*, **322**, 1775–1780.

Pierotti, S., Albano, C., Milandri, A., Callegari, F., Poletti, R. and Rossinni, G.P. (2007) A slot blot procedure for the measurement of yessotoxins by a functional assay. *Toxicon*, **49**(1), 36–45.

Prakash, A., Medcof, J.C. and Tennant, A.D. (1971) Paralytic shellfish poisoning in eastern Canada. *Bulletin of the Fisheries Research Board of Canada*, **177**, 1–87.

Premazzi, G. and Volterra, L. (1993) *Microphyte Toxins: A Manual for Toxin Detection, Environmental Monitoring and Therapies to Counteract Intoxications*, EUR 14854. Office for Official Publications of the European Communities, Luxembourg.

Quayle, D.B. (1969) Paralytic shellfish poisoning in British Columbia. *Bulletin of the Fisheries Research Board of Canada*, **168**, 1–69.

Quilliam, M.A. (2004a) Chemical methods for lipophilic shellfish toxins, in *Manual on Harmful Marine Algae* (eds. G.M. Hallegraeff, D.M. Anderson and A.D. Cembella). UNESCO, Paris, pp. 211–245.

Quilliam, M.A. (2004b) Chemical methods for domoic acid, the amnesic shellfish poisoning (ASP) toxin, in *Manual on Harmful Marine Algae* (eds. G.M. Hallegraeff, D.M. Anderson and A.D. Cembella). UNESCO, Paris, pp. 247–265.

Quilliam, M.A. and Wright, J.L.C. (1995) Methods for diarrhetic shellfish poisons, in *Manual on Harmful Marine Microalgae* (eds. G.M. Hallegraeff, D.M. Anderson and A.D. Cembella). Intergovernmental Oceanographic Commission Manuals and Guides No. 33. UNESCO, Paris, pp. 95–111.

Quilliam, M.A., Hardstaff, W.R., Ishida, N., McLachlan, J.L., Reeves, A.R., Ross, N.W. and Windust, A.J. (1996) Production of diarrhetic shellfish poisoning (DSP) toxins by *Prorocentrum lima* in culture and development of analytical methods, in *Harmful and Toxic Algal Blooms, Proceedings of the Seventh International Conference on Toxic Phytoplankton*, 1995 (eds. T. Yasumoto, Y. Oshima and Y. Fukuo). Intergovernmental Oceanographic Commission of UNESCO, Paris, pp. 289–292.

Ramstad, H., Larsen, S. and Aune, T. (2001) The repeatability of two HPLC methods and the PP2A assay in the quantification of diarrhetic toxins in blue mussels (*Mytilus edulis*). *Toxicon*, **39**(4), 515–522.

Ravn, H. (1995) *Amnesic Shellfish Poisoning (ASP)*, HAB Publication Series, Vol. 1, Intergovernmental Oceanographic Commission Manuals and Guides No. 31, Paris, 19 pp.

Reynolds, C.S. (2006) *The Ecology of Phytoplankton*. Cambridge University Press, Cambridge, 535 pp.

Román, Y., Alfonso, A., Louzao, M.C., de la Rosa, L.A., Leira, F., Vieites, J.M., Vieytes, M.R., Ofuji, K., Satake, M., Yasumoto, T. and Botana, L.M. (2002) Azaspiracid-1, a potent, nonapoptotic new phycotoxin with several cell targets. *Cellular Signalling*, **14**(8), 703–716.

Román, Y., Alfonso, A., Vieytes, M.R., Ofuji, K., Satake, M., Yasumoto, T. and Botana, L.M. (2004) Effects of azaspiracids -2 and -3 on intracellular cAMP, [Ca^{2+}] and pH. *Chemical Research in Toxicology*, **17**(10), 1338–1349.

Ronzitti, G., Hess, P., Rehmann, N. and Rossini, G.P. (2007) Azaspiracid-1 alters the E-cadherin pool in epithelial cells. *Toxicological Sciences*, **95**(2), 427–435.

Satake, M., Terasawa, K., Kadowaki, Y. and Yasumoto, T. (1996) Relative configuration of yessotoxin and isolation of two new analogs from toxic scallops. *Tetrahedron Letters*, **37**(33), 5955–5958.

Satake, M., Shoji, M., Oshima, Y., Naoki, H., Fujita, T. and Yasumoto, T. (2002) Gymnocin-A, a cytotoxic polyether from the notorious red tide dinoflagellate, *Gymnodinium mikimotoi*. *Tetrahedron Letters*, **43**(33), 5829–5832.

Satake, M., Tanaka, Y., Ishikura, Y., Oshima, Y., Naoki, H. and Yasumoto, T. (2005) Gymnocin-B with the largest contiguous polyether rings from the red tide dinoflagellate, *Karenia* (formerly *Gymnodinium*) *mikimotoi*. *Tetrahedron Letters*, **46**(20), 3537–3540.

Schmued, L.C., Scallett, A.C. and Slikker, W. (1995) Domoic acid induced neuronal degeneration in the primate forebrain revealed by degeneration specific histochemistry. *Brain Research*, **695**(1), 64–70.

Scholin, C.A., Gulland, F., Douchette, G.J., Benson, S., Busman, M., Chavez, F.P., Cordaro, J., DeLong, R., De Vogelaere, A., Harvey, J., Haulena, M., Lefebvre, K., Lipscomb, T., Loscutoff, S., Lowenstine, L.J., Marin, R.I., Miller, P.E., McLellan, W.A., Moeller, P.D.R., Powell, C.L., Rowles, T., Silvagni, P., Silver, M., Spraker, T., Trainer, V. and Van Dolah, F.M. (2000) Mortality of sea lions along the central California coast linked to a toxic diatom bloom. *Nature*, **403**, 80–84.

Seki, T., Satake, M., Mackenzie, L., Kaspar, H.F. and Yasumoto, T. (1995) Gymnodimine, a new marine toxin of unprecedented structure isolated from New Zealand oysters and the dinoflagellate *Gymnodinium* sp. *Tetrahedron Letters*, **36**(39), 7093–7096.

Seki, T., Satake, M., Mackenzie, L., Kaspar, H.F. and Yasumoto, T. (1996) Gymnodimine, a novel toxic imine isolated from the Foveaux Strait oysters and *Gymnodinium* sp., in *Harmful and Toxic Algal Blooms*, *Proceedings of the Seventh International Conference on Toxic Phytoplankton*, 1995 (eds. T. Yasumoto, Y. Oshima and Y. Fukuyo). Intergovernmental Oceanographic Commission of UNESCO, Paris, pp. 495–498.

Shimizu, Y. (1996) Microalgal metabolites: A new perspective. *Annual Review of Microbiology*, **50**, 431–465.

Shimizu, Y. (2003) Microalgal metabolites. *Current Opinion in Microbiology*, **6**(3), 236–243.

Shimizu, Y., Giorgio, C., Koerting-Walker, C. and Ogata, T. (1996) Nonconformity of bacterial production of paralytic shellfish poisons – neosaxitoxin production by a bacterium strain from *Alexandrium tamarense* Ipswich strain and its significance, in *Harmful and Toxic Algal Blooms*, *Proceedings of the Seventh International Conference on Toxic Phytoplankton*, 1995 (eds. T. Yasumoto, Y. Oshima and Y. Fukuyo). Intergovernmental Oceanographic Commission of UNESCO, Paris, pp. 359–362.

Shumway, S.E. (1995) Phycotoxin-related shellfish poisoning: Bivalve molluscs are not the only vectors. *Reviews in Fisheries Science*, **3**(1), 1–31.

Shumway, S.E. and Cembella, A.D. (1993) The impact of toxic algae on scallop culture and fisheries. *Reviews in Fisheries Science*, **1**(2), 121–150.

Steidinger, K.A. (1993) Some taxonomic and biologic aspects of toxic dinoflagellates, in *Algal Toxins in Seafood and Drinking Water* (ed. I.R. Falconer). Academic Press, London, pp. 1–28.

Stobo, L.A., Lacaze, J.P., Scott, A.C., Gallacher, S., Smith, E.A. and Quilliam, M.A. (2005) Liquid chromatography with mass spectrometry – detection of lipophilic shellfish toxins. *Journal of AOAC International*, **88**(5), 1371–1382.

Sullivan, J.J. (1993) Methods of analysis for algal toxins: Dinoflagellate and diatom toxins, in *Algal Toxins in Seafood and Drinking Water* (ed. I.R. Falconer). Academic Press, London, pp. 29–48.

Suzuki, T., Walter, J.A., LeBlanc, P., MacKinnon, S., Miles, C.O., Wilkins, A.L., Munday, R., Beuzenberg, V., MacKenzie, A.L., Jensen, D.J., Cooney, J.M. and Quilliam, M.A. (2006) Identification of pectenotoxin-11 as 34S-hydroxypectenotoxin-2, a new pectenotoxin analogue in the toxic dinoflagellate *Dinophysis acuta* from New Zealand. *Chemical Research in Toxicology*, **19**(2), 310–318.

Takahashi, H., Kusumi, T., Kan, Y., Satake, M. and Yasumoto, T. (1996) Determination of the absolute configuration of yessotoxin, a polyether compound implicated in diarrhetic shellfish poisoning, by NMR spectroscopic method using a chiral anisotropic reagent, methoxy-(2-naphthyl)acetic acid. *Tetrahedron Letters*, **37**(39), 7087–7090.

Taleb, H., Vale, P., Amanhir, R.M., Benhadouch, A. and Sagou, R. (2006) First detection of azaspiracid outside European coastal waters. Poster PO.09-06 presented at the *12th International Conference on Harmful Algae*, Copenhagen, Denmark, 4–8 September 2006, International Society for the Study of Harmful Algae. Available from: http://www.bi.ku.dk/hab/programme.asp

Terao, K., Ito, E., Oarada, M., Murata, M. and Yasumoto, T. (1990a) Histopathological studies on experimental marine toxin poisoning – 5. The effects in mice of yessotoxin isolated from *Patinopecten yessoensis* and of a desulfated derivative. *Toxicon*, **28**(9), 1095–1104.

Terao, K., Ito, E., Yasumoto, T. and Yamaguchi, K. (1990b) Enterotoxic, hepatotoxic and immunotoxic effects of dinoflagellate toxins on mice, in *Toxic Marine Phytoplankton*, *Proceedings of the Fourth International Conference on Toxic Marine Phytoplankton*, 1989 (eds. E. Granéli, B. Sundström, L. Edler and D.M. Anderson). Elsevier, New York, pp. 418–423.

Terao, K. Ito, E., Oarada, M., Ishibashi, Y., Legrand, A.-M and Yasumoto, T. (1991) Light and electron microscopic studies of pathological changes induced in mice by ciguatoxin poisoning. *Toxicon*, **29**(6), 633–643.

Terao, K., Ito, E., Ohkusu, M. and Yasumoto, T. (1993) A comparative study of the effects of DSP-toxins on mice and rats, in *Toxic Phytoplankton Blooms in the Sea, Proceedings of the Fifth International Conference on Toxic Marine Phytoplankton*, 1991 (eds. T.J. Smayda and Y. Shimizu). Elsevier, Amsterdam, pp. 581–586.

Tester, P.A., Turner, J.T. and Shea, D. (2000) Vertical transport of toxins from the dinoflagellate *Gymnodinium breve* through copepods to fish. *Journal of Plankton Research*, **22**(1), 47–62.

Todd, E.C.W. (1993) Domoic acid and amnesic shellfish poisoning – a review. *Journal of Food Protection*, **56**, 69–83.

Torgersen, T., Aasen, J. and Aune, T. (2005) Diarrhetic shellfish poisoning by okadaic acid esters from Brown crabs (*Cancer pagurus*) in Norway. *Toxicon*, **46**(5), 572–578.

Torigoe, K., Murata, M., Yasumoto, T. and Iwashita, T. (1988) Prorocentrolide, a toxic nitrogenous macrocycle from a marine dinoflagellate, *Prorocentrum lima*. *Journal of the American Chemical Society*, **110**, 7876–7877.

Toyofuku, H. (2006) Joint FAO/WHO/IOC activities to provide scientific advice on marine biotoxins (research report). *Marine Pollution Bulletin*, **52**(12), 1735–1745.

Traynor, I.M., Plumpton, L., Fodey, T.L., Higgins, C. and Elliott, C.T. (2006) Immunobiosensor detection of domoic acid as a screening test in bivalve molluscs: Comparison with liquid chromatography-based analysis. *Journal of AOAC International*, **89**(3), 868–872.

Tripuraneni, J., Koutsouris, A., Pestic, L., De Lanerolle, P. and Hecht, G. (1997) The toxin of diarrhetic shellfish poisoning, okadaic acid, increases intestinal epithelial paracellular permeability. *Gastroenterology*, **112**(1), 100–108.

Twiner, M.J., Hess P., Dechraoui, M.-Y.B., McMahon, T., Samons, M.S., Satake M., Yasumoto T., Ramsdell, J.S. and Doucette, G.J. (2005) Cytotoxic and cytoskeletal effects of azaspiracid-1 on mammalian cell lines. *Toxicon*, **45**(7), 891–900.

Uemura, D., Chou, T., Haino, T., Nagatsu, A., Fukuzawa, S., Zheng, S.Z. and Chen, H. (1995) Pinnatoxin A: A toxic amphoteric macrocycle from the Okinawan bivalve *Pinna muricata*. *Journal of the American Chemical Society*, **117**, 1155–1156.

Uribe, P. and Espejo, R.T. (2003) Effect of associated bacteria on the growth and toxicity of *Alexandrium catenella*. *Applied and Environmental Microbiology*, **69**(1), 659–662.

Usami, M., Satake, M., Ishida, S., Inoue, A., Kan, Y. and Yasumoto, T. (1995) Palytoxin analogs from the dinoflagellate *Ostreopsis siamensis*. *Journal of the American Chemical Society*, **117**(19), 5389–5390.

Vale, C., Nicolaou, K.C., Frederick, M.O., Gomez-Limia, B., Alfonso, A., Vieytes, M.R. and Botana, L.M. (2007) Effects of azaspiracid-1, a potent cytotoxic agent, on primary neuronal cultures: A structure–activity relationship study. *Journal of Medicinal Chemistry*, **50**(2), 356–363.

van Egmond, H.P., Speijers, G.J.A. and van den Top, G.J.A. (1992) Current situation on worldwide regulation for marine phycotoxins. *Journal of Natural Toxins*, **1**(1), 67–85.

van Egmond, H.P., van Apeldoorn, M.E. and Speijers, G.J.A. (2004) *Marine Biotoxins, Food and Nutrition Paper 80*. Food and Agriculture Organization of the United Nations, Rome, 278 pp. Available from: http://www.fao.org/docrep/007/y5486e/y5486e00.htm

Vasconcelos, V.M. (1995) Uptake and depuration of the heptapeptide toxin microcystin-LR in *Mytilus galloprovincialis*. *Aquatic Toxicology*, **32**, 227–237.

Vernoux, J.P., Bansard, S., Simon, J.F., Nwal-Amang, D., Le-Baut, C., Gleizes, E., Fremy, J.M. and Lasne, M.C. (1994) Cooked mussels contaminated by *Dinophysis* sp.: A source of okadaic acid. *Natural Toxins*, **2**(4), 184–188.

Vieites, J.M., Botana, L.M., Vieytes, M.R. and Leira, F.J. (1999) Canning process that diminishes paralytic shellfish poison in naturally contaminated mussels (*Mytilus galloprovincialis*). *Journal of Food Protection*, **62**(5), 515–519.

Vilariño, N., Nicolaou, K.C., Frederick, M.O., Cagide, E., Ares, I.R., Louzao, M.C., Vieytes, M.R. and Botana, LM. (2006) Cell growth inhibition and actin cytoskeleton disorganization induced by azaspiracid-1 structure–activity studies. *Chemical Research in Toxicology*, **19**(11), 1459–1466.

Vilariño, N., Nicolaou, K.C., Frederick, M.O., Vieytes, M.R. and Botana, L.M. (2007) Irreversible cytoskeletal disarrangement is independent of caspase activation during *in vitro* azaspiracid toxicity in human neuroblastoma cells. *Biochemical Pharmacology*, **74**(2), 327–335.

Wang, Z., Plakas, S.M., El Said, K.R., Jester, E.L.E., Granade, H.R. and Dickey, R.W. (2004) LC/MS analysis of brevetoxin metabolites in the Eastern oyster (*Crassostrea virginica*). *Toxicon*, **43**(4), 455–465.

Watson-Wright, W., Gillis, M., Smyth, C., Trueman, S., McGuire, A., Moore, W., McLachlan, D. and Sims, G. (1991) Monitoring of PSP in hepatopancreas of lobster from Atlantic Canada, in *Proceedings of the*

Second Canadian Workshop on Harmful Marine Algae, Canadian Technical Reports on Fisheries and Aquatic Science No. 1799, pp. 27–28 (microfiche).

Wilkins, A.L., Rehmann, N., Torgersen, T., Rundberget, T., Keogh, M., Petersen, D., Hess, P., Rise, F. and Miles, C.O. (2006) Identification of fatty acid esters of pectenotoxin-2 seco acid in blue mussels (*Mytilus edulis*) from Ireland. *Journal of Agricultural and Food Chemistry*, **54**(15), 5672–5678.

Wright, J.L.C. (1995) Dealing with seafood toxins: Present approaches and future options. *Food Research International*, **28**(4), 347–358.

Wu, Y., Ho, A.Y.T., Qian, P.-Y., Leung, K.S.-Y., Cai, Z. and Lin, J.-M. (2006) Determination of paralytic shell-fish toxins in dinoflagellate *Alexandrium tamarense* by using isotachophoresis/capillary electrophoresis. *Journal of Separation Science*, **29**, 399–404.

Yasumoto, T. (1990) Marine microorganisms toxins – An overview, in *Toxic Marine Phytoplankton, Proceedings of the Fourth International Conference on Toxic Marine Phytoplankton*, 1989 (eds. E. Granéli, B. Sundström, L. Edler and D.M. Anderson). Elsevier, New York, pp. 3–8.

Yasumoto, T. (1998) Fish poisonings due to toxins of microalgal origin in the Pacific. *Toxicon*, **36**(11), 1515–1518.

Yasumoto, T. (2005) Chemistry, etiology, and food chain dynamics of marine toxins, *Proceedings of the Japan Academy, Series B Physical and Biological Sciences*, **81**(2), 43–51.

Yasumoto, T. and Satake, M. (1996) Chemistry, etiology and determination methods of ciguatera toxins. *Journal of Toxicology – Toxin Reviews*, **15**(2), 91–107.

Yasumoto, T., Igarashi, T. and Satake, M. (2001) Chemistry of phycotoxins – Structural elucidation, in *Mycotoxins and Phycotoxins in Perspective at the Turn of the Millennium, Proceedings of the Xth International IUPAC Symposium on Mycotoxins and Phycotoxins* (May 2000, Guarujá, Brazil) (eds. W.J. De Koe, R.A. Samson, H.P. van Egmond, J. Gilbert and M. Sabino). Wageningen, the Netherlands, Ponsen & Looyen, pp. 465–474.

Zhao, J., Lembeye, J., Cenci, G., Wall, B. and Yasumoto, T. (1993) The determination of okadaic acid and dinophysistoxin-1 in mussels from Chile, Italy and Ireland, in *Toxic Phytoplankton Blooms in the Sea, Proceedings of the Fifth International Conference on Toxic Marine Phytoplankton*, 1991 (eds. T.J. Smayda and Y. Shimizu). Elsevier, Amsterdam, pp. 587–592.

Zheng, W.J., Demattei, J.A., Wu, J.P., Duan, J.J.W., Cook, L.R., Oinuma, H. and Kishi, Y. (1996) Complete relative stereochemistry of maitotoxin. *Journal of the American Chemical Society*, **118**(34), 7946–7968.

Zhou, Z.-H., Komiyama, M., Terao, K. and Shimada, Y. (1994) Effects of pectenotoxin-1 on liver cells *in vitro*. *Natural Toxins*, **2**(3), 132–135.

Note: How to access the EU legislation cited in Table 4.7 and in Section 4.13.2*

Go to http://eur-lex.europa.eu./ Access the 'welcome' page

Select: 'simple search' > 'search by document number' > 'CELEX NUMBER'

Enter the relevant CELEX number given in parentheses below:

Directive 91/492 [31991L0492]

Directive 97/61 [31997L0061]

Decision 96/77 [31996D0077]*

Decision 2002/225 [32002D0225]

Regulation 853/2004 [02004R0853-20060101]

Regulation 2074/2005 [32005R2074]

5 Mushroom Toxins

Jana Hajslová and Vera Schulzova

Summary

Of almost 10 000 known mushroom species, some are much sought after for human consumption, several are cultivated, but approximately 250 species can cause unpleasant effects when consumed and a few of them may be responsible for fatal poisoning. In this chapter, the incidence and levels of occurrence of these biologically active compounds (amatoxins, phallotoxins, virotoxins, orellanine, muscarine, ibotenic acid, muscimol, psilocybin, psilocin, coprine, gyromitrin, agaratine and other phenylhydrazines) in poisonous and cultivated mushrooms is reviewed. The available analytical methods are summarized and data is reviewed on exposure assessment of these compounds from mushrooms in the diet.

5.1 INTRODUCTION

Human poisonings due to natural toxins is in many cases associated with consumption of raw or cooked fruiting bodies of mushrooms representing various species of higher fungi. Currently, there are almost 10 000 mushroom species known, most of them are innocuous, with some that are sought for their delicious flavour and several of them that are even cultivated. However, there are also approximately 250 species that will cause physical discomfort when consumed; moreover, few of them (often called toadstools) may be responsible for fatal poisoning. It should be noted, there is no simple rule or test available that would help non-specialist to distinguish whether the mushroom is edible or poisonous. Typically, collectors (so called mycophagists) do not have enough expertise in wild mushroom identification and, therefore, they collect only those species which they know from their own experience and are able to recognize on the basis of distinct morphological features. Unfortunately, poisonings due to confusion of edible species with toxic ones are reported every year since some individuals take unnecessary risk eating species with which they are not familiar. Outbreaks may occur after ingestion of fresh (raw) toxic species or home processed, e.g. stir-fried, cooked and/or canned mushrooms. Risk of multiple outbreaks exists when processed (preserved or frozen) toadstools are carried to another location and consumed at other time.

Mushroom poisonings are generally acute and are manifested by various symptoms. Their severity and overall prognosis depend on which particular species and what amount (dose of toxin) was consumed. Typically, poisonous species contain one or more toxic compounds which are unique to few other ones. Therefore, cases of mushroom poisonings generally do not resemble each other unless they are caused by the same, or very closely related mushroom species. Unfortunately, compared to existing relatively comprehensive knowledge

Table 5.1   Overview of major mushroom toxins.

Type of toxicity	Toxin groups according to structure	Typical toxic symptoms	Genus of higher fungi producing toxin
Protoplasmic poison	Amatoxins, phallotoxins, virotoxins	Nausea, vomiting, diarrhoea abdominal pain, kidney and liver failure, hypoglycaemia (for phallotoxins and virotoxins this applies only when applied intravenously)	*Amanita*
	Hydrazines	Nausea, vomiting, diarrhoea, delirium, seizures, coma	*Gyromitra*
	Orellanine	Nausea, vomiting, diarrhoea, polydipsia, polyuria, headache, myalgias	*Cortinarius*
Neurotoxins	Muscarine	Salivation, urination, tearing, diarrhoea, sweating, vomiting, urination, respiratory failure	*Amanita, Inocybe, Clitocybe*
	Ibotenic acid, muscimol	Drowsiness, dizziness followed by hyperactivity, hallucinations, seizures, twitching	*Amanita*
	Psilocybin, psilocin	Euphoria, hallucinations or fear, rage and violence	*Psilocybe*
Gastrointestinal irritants	Mostly unidentified (poorly described) toxins	Nausea, vomiting, abdominal cramping, diarrhoea (may be bloody)	*Agaricus, Amanita, Chlorophyllum, Omphalotus*
Disulfiram-like toxins	Coprine	Symptoms short-lived acute toxic syndromes such as flushing, sweating, headache, nausea, vomiting, rapid heart rate, shortness of breath and chest pain appear only when alcohol is consumed within 72 h after consumption	*Coprinus*

on most of natural toxins occurring in common food crops, the chemistry of many of the mushroom toxins (namely those which are not deadly toxic) has not been fully explored yet. The lack of comprehensive data makes positive identification of processed mushrooms cases of intoxication either very difficult or almost impossible. With regard to these limitations, more than on the basis of structure, mushroom toxins are often classified with regard to their toxic effects. Using this criterion, listed in Table 5.1 and discussed in within this chapter, there are characterized four major categories of mushroom toxins.

It should be noted, that both qualitative and quantitative analytical characterization of mushroom toxins is a very demanding task, which is both complicated and time-consuming, regardless of whether the fungal fruiting body or biological material such stomach contents, serum and/or urine, are the subject of examination (the latter samples are collected in case of poisoning suspicion). The exact chemical natures of most of the toxins that produce the milder symptoms are unknown. Chromatographic techniques are most common in mushroom toxins analysis. Thin layer chromatography (TLC) was used in most of older studies, while more up-to-date methods utilize gas chromatography (GC) and/or high performance chromatography (HPLC). Although conventional detectors may enable reliable analysis of some target analytes, the more progress in analysis of both parent compounds and their toxic breakdown products has been achieved by introduction of mass spectrometry (MS).

In the paragraphs below, the main toxins occurring both in mushrooms known as poisonous and those representing 'edible' species are characterized. The current knowledge on occurrence routes, methods of analysis, incidence and levels of occurrence as well as exposure assessment is presented. Besides general review articles (Faulstich and Wieland, 1992, 2005; Hajslová, 1995; Spoerke and Rumack, 1994; Wieland and Faulstich, 1983) and other scientific papers concerned with the most important mushroom toxins, also some official websites (e.g. http://www.cfsan.fda.gov/~mow/chap40.html; http://calpoison.org/public/mushrooms.html; http://www.maps.org/research/psilo/azproto.html) were consulted.

5.2 POISONOUS MUSHROOMS

The incidence of accidental ingestion of poisonous mushrooms largely varies over the world depending on local tradition, life-style, nutritional factors and, obviously, on climatic and environmental conditions that are related to their occurrence.

5.2.1 Amatoxins, phallotoxins and virotoxins

5.2.1.1 Occurrence route or mechanism of formation

Amatoxins, extremely (deadly) poisonous protoplasmic toxins of some *Amanita* species, consist of at least nine similar bicyclic octapeptides shown in Fig. 5.1: α-amanitin, β-amanitin,

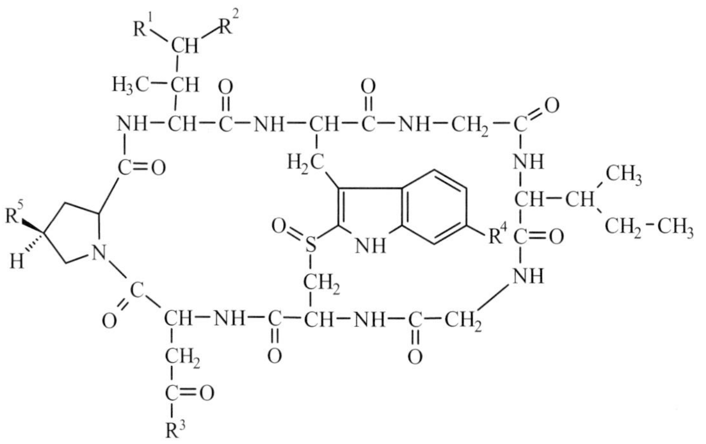

α-amanitin, $R^1 = CH_2OH$, $R^2 = OH$, $R^3 = NH_2$, $R^4 = OH$, $R^5 = OH$
β-amanitin, $R^1 = CH_2OH$, $R^2 = OH$, $R^3 = OH$, $R^4 = OH$, $R^5 = OH$
γ-amanitin, $R^1 = CH_3$, $R^2 = OH$, $R^3 = NH_2$, $R^4 = OH$, $R^5 = OH$
ε-amanitin, $R^1 = CH_3$, $R^2 = OH$, $R^3 = OH$, $R^4 = OH$, $R^5 = OH$
amanin, $R^1 = CH_2OH$, $R^2 = OH$, $R^3 = OH$, $R^4 = H$, $R^5 = OH$
amaninamid, $R^1 = CH_2OH$, $R^2 = OH$, $R^3 = NH_2$, $R^4 = H$, $R^5 = OH$
amanullin, $R^1 = CH_3$, $R^2 = H$, $R^3 = NH_2$, $R^4 = OH$, $R^5 = OH$
amanullic acid, $R^1 = CH_3$, $R^2 = H$, $R^3 = OH$, $R^4 = OH$, $R^5 = OH$
proamanullin, $R^1 = CH_3$, $R^2 = H$, $R^3 = NH_2$, $R^4 = OH$, $R^5 = H$

Fig. 5.1 Structures of the amatoxins.

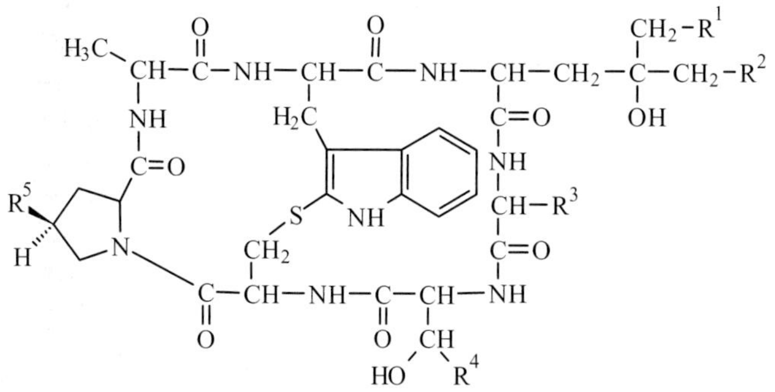

phalloidin, R^1 = OH, R^2 = H, R^3 = CH$_3$, R^4 = CH$_3$, R^5 = OH
phalloin, R^1 = H, R^2 = H, R^3 = CH$_3$, R^4 = CH$_3$, R^5 = OH
prophalloin, R^1 = H, R^2 = H, R^3 = CH$_3$, R^4 = CH$_3$, R^5 = H
phallisin, R^1 = OH, R^2 = OH, R^3 = CH$_3$, R^4 = CH$_3$, R^5 = OH
phallacin, R^1 = H, R^2 = II, R^3 = CH(CH$_3$)$_2$, R^4 = COOH, R^5 = OH
phallacidin, R^1 = OH, R^2 = H, R^3 = CH(CH$_3$)$_2$, R^4 = COOH, R^5 = OH
phallasacin, R^1 = OH, R^2 = OH, R^3 = CH(CH$_3$)$_2$, R^4 = COOH, R^5 = OH

Fig. 5.2 Structures of the phallotoxins.

γ-amanitin, ε-amanitin, amanullinic acid, amanin, amaninamid, amanullin and proamanullin. Amatoxins occur often together with hepatotoxic phallotoxins, another group of bicyclic peptides with seven amino acids in the ring (see Fig. 5.2). They are represented mainly by phalloidin, other six actin-binding compounds identified until now are prophalloin, phalloin, phallisin, phallicidin, phallacin and phallisacin.

Virotoxins, represent the third, minor group of cyclic peptides similar to phallotoxins. This group consists of six monocyclic heptapeptides. The structures of viroidin, desoxyviroidin, [Ala]viroidin, [Ala]deoxyoviroidin, viroisin and desoxoviroisin are shown in Fig. 5.3 (Vetter, 1998).

5.2.1.2   Methods of analysis

Since botanical identification of the fungus that was eaten is in many poisoning cases impossible, diagnosis and its management is now based on various laboratory methods typically radioimunoassays (RIA) that were developed for examination of amatoxins (mainly α- and β-anmanitin) in serum, urine or gastrointestinal fluids. The test kits use either 3H tracer or 125 I tracer. The sensitivity of commercial RIA assays is around 0.1 ng mL^{-1} for serum and 0.25 ng mL^{-1} for urine (Gonmori and Yoshioka, 2003).

As an alternative, sensitive HPLC methods employing conventional detectors are available (Gonmori and Yoshioka, 2003). Also, the possibility to use of capillary zone electrophoresis (CZE) with photodiode array detection has been demonstrated (Brüggemann et al., 1996). Nowadays, combined liquid chromatography mass spectrometry (LC/MS) is a prominent technique for the comprehensive analysis of amatoxins (and many other mushroom toxins), both in Amanita fruiting bodies and biological fluids (Drummer, 1999; Maurer et al., 1998, 2000; Zhang et al., 2005). It should be noted, however, that LC/MS availability for routine use is still limited, actually, most clinical laboratories do not use even the older RIA technique,

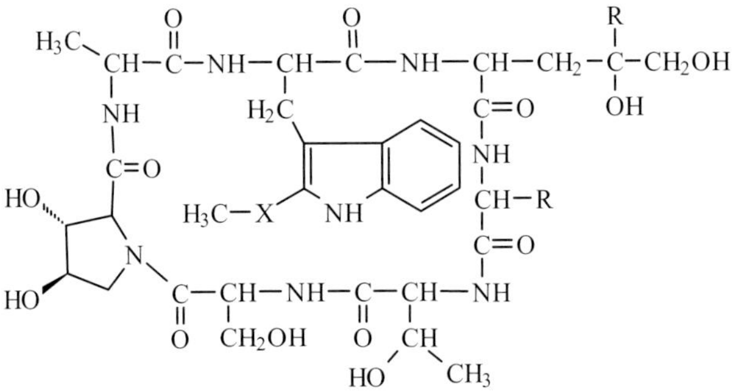

viroidin, X = SO$_2$, R^1 = CH$_3$, R^2 = CH(CH$_3$)$_2$
deoxyviroidin, X = SO, R^1 = CH$_3$, R^2 = CH(CH$_3$)$_2$
[Ala^1]viroidin, X = SO$_2$, R^1 = CH$_3$, R^2 = CH$_3$
[Ala^1]deoxyviroidin, X = SO, R^1 = CH$_3$, R^2 = CH$_3$
viroisin, X = SO$_2$, R^1 = CH$_2$OH, R^2 = CH(CH$_3$)$_2$
deoxyviroisin, X = SO, R^1 = CH$_2$OH, R^2 = CH(CH$_3$)$_2$

Fig. 5.3 Structures of the virotoxins.

hence the diagnosis based entirely on symptomology and recent dietary history is still common is many places.

5.2.1.3 Incidence and levels of occurrence

Amanita phalloides, the most poisonous known mushroom (so-called green Death Cap), is commonly found all over Europe, North America and some other areas with mild climates. It may contain up to 80 mg kg^{-1} of fresh tissue α-amanitin and 50 mg kg^{-1} β-amanitin. These compounds account in most cases for more than 90% of total amatoxins. In addition to them, phalloidin is present, its content may approach 100 mg kg^{-1}. Phalloidin is also found in the edible (and sought after) Blusher (*Amanita rubescens*) (Litten, 1975).

Occasionally, a closely related sub-species *Amanita phalloides* var. *verna* (in the older papers misclassified as separate species, *A. verna*), can be found. Its cup instead of olive green is pale or even white and the content of amatoxins may be lower compared to green cup. Another white species, *A. virosa* (the Destroying Angel), differs from the above poisonous mushrooms in its toxin pattern: α-amanitin may be completely replaced by amaninamide. The overall concentration of amatoxins in *A. virosa* varies in the range 40–200 mg kg^{-1} fresh tissue (Faulstich, 2005). In this species, also virotoxins are found exclusively. There are several other fungi genera, such as *Galerina* and *Lepiota*, containing amatoxins, nevertheless, their content is several times lower as compared to previous highly poisonous representatives of *Amanitus* genus. The distribution of toxins in various tissues of *Amanita* species was studied by several groups. In an older study (Enjalbert *et al.*, 1993) *Amanita phalloides* mushrooms representing three specimens at two carpophore development stages were examined. Substantial differences in the tissue toxin content were found. The ring displayed a very high level of toxins, whereas the bulb had the lowest toxin content. Compositional differences in relation to the nature of the tissue were also noted. The highest amatoxin content was found in the ring, gills and cap, whereas the bulb and volva were the richest in phallotoxins.

Furthermore, variability in the toxin composition was observed, it is assumed that the differences in the distribution of individual toxins in the tissues might be related to the carpophore developmental stage. Generally, young fruit body contains lower, and the well-developed fungus higher concentrations of toxins, but their concentrations in certain species are variable, even among mushrooms collected in the same region (Vetter, 1998). The follow-up study (Enjalbert *et al.*, 1999) conducted on 25 *Amanita phalloides* carpophores collected from three sites in France that differed in their geological and pedological characteristics confirmed the above conclusions.

5.2.1.4 Exposure assessment

Often the cause of poisoning by these deadly *Amanitas* is, akin to case of other mushrooms, i.e. their mistaking for edible species. In this respect white-coloured species are the most dangerous since they are erroneously considered to be edible *Macrolepiota* or *Agaricus*. It should be emphasized that amatoxins peptides are not destroyed by cooking and can be kept for years if dried, they will only decompose slowly when exposed to ultraviolet (UV) light for several months. Regarding detoxification by organisms, unfortunately, no protease known would cleave the peptide bonds in the cyclic peptide. In other words, the consumer's exposure correlates with amount of fresh mushrooms consumed and their content of toxins. The LOD_{50} of major amatoxins in *Amanita phalloides*, α-amanitin and β-amanitin, was in experiments with mice estimated in the range 0.3–0.6 mg kg^{-1} of body weight. Given the estimated lethal dose for humans (0.1 mg kg^{-1} body weight), a full-grown mushroom (25 g) will be sufficient to kill a human (Faulstich, 2005).

The major toxic mechanism associated with amatoxins poisoning is the inhibition of RNA polymerase II, a vital enzyme in cell metabolism. In this way, mRNA synthesis is affected (amatoxins inhibit protein biosynthesis at the transcriptional level), leading to cell death. The liver is the principal organ affected, as it is the organ which is first encountered after absorption in the gastrointestinal symptom, though other organs, especially the kidneys, are susceptible.

The course of amanitin intoxication has three chronological phases. A latent phase takes approximately 6–24 hours and rarely exceeds 48 hours. A gastrointestinal phase lasts typically 2–3 days and is associated with abdominal pain, vomiting and diarrhoea causing dehydration, hypovolaemia, electrolyte and acid–base disorders. The third phase, so called hepatic, begins 36–48 hours after ingestion. Hepatitis becomes clinically evident with the onset of jaundice on the 3rd–4th day after ingestion; hepatic coma, bleeding and anuria may occur in intoxicated patients. When liver damage is reversible, patients usually make a slow and steady recovery. In fatal cases, death occurs within 6–16 days. In spite of improved knowledge of amatoxins poisoning, fatalities are still relatively high (according to some estimates up to 30%) since there is no specific therapy available (Faulstich and Wieland, 1992; Karlson-Stiber and Persson, 2003). In total, more than 90% of all fatal cases of mushroom poisoning in Europe are due to *Amanita phalloides*.

Compared to amatoxins, phallotoxins are highly toxic to liver cells (Wieland and Govindan, 1974), their intoxication mechanism is believed to be due to the specific binding of the toxin to F-actin, which subsequently inhibits the depolymerization of F-actin into G-actin. However, since their absorption from the gastrointestinal tract is not significant, they do not seem to play a major role in human toxicity. Virotoxins are closely related to phallotoxins with respect to their structure and toxicity, but they do not exert any acute toxicity after ingestion in humans (Karlson-Stiber and Persson, 2003).

5.2.2 Orellanine

5.2.2.1 Occurrence route or mechanism of formation

Orellanine is a nefrotoxin exclusively occurring in mushrooms representing the *Cortinarius* genus. The chemical composition of orellanine remained unknown until mid of the 1970s when it was identified as a hydroxylated bipyridyl-N, N'-dioxide (Oubrahim *et al.*, 1998). In the most stable form of orellanine, the nitrogen atoms are positively charged. It is supposed, that reactive semi-quinone, see Fig. 5.4, is produced in cells by a peroxidase reaction. The semi-quinone is a radical that probably causes intracellular depletion of glutathione and ascorbate as the toxic event. Orellanine is in *Cortinarius* mushrooms typically accompanied by the corresponding monooxide, orellinine, which is less toxic. When irradiated by UV light, non-toxic stable orelline is formed through the loss of N-oxides (Ruedl *et al.*, 1989).

5.2.2.2 Methods of analysis

Several analytical methods were developed both for the analysis of mushrooms and biological samples, such as plasma and renal tissue. In the later case, TLC can be used for separation prior to proof of the presence of toxins under UV light. Orellanine is visible as navy blue, orellinine as dark blue, and orelline as light blue fluorescent spots on silica TLC plates (Horn *et al.*, 1997). Quantitative analysis of orellanine in plasma samples, or in (rat) urine samples, was performed by two-dimensional TLC on cellulose employing spectrophotometric evaluation of orelline produced by UV-induced decomposition of orellanine. HPLC was used for the analysis of orellanine in mushroom extracts (Faulstich, 2005). Very simple and quick methods for the identification of orellanine in mushroom isolates based in TLC and electrophoresis was developed by Oubrahim *et al.* (1997).

5.2.2.3 Incidence and levels of occurrence

The orellanine levels and its tissue distribution were studied in *Cortinarius orellanus* and *Cortinarius rubellus* species (Koller *et al.*, 2002). The analysis of caps showed the content of toxin to be (expressed on dry weight basis) 9400 mg kg^{-1} and 7800 mg kg^{-1} in stems 4800 mg kg^{-1} and 4200 mg kg^{-1}, and in spores 3100 mg kg^{-1} and 900 mg kg^{-1}, respectively. In mycorrhiza roots from *C. rubellus*, the orellanine content was 0.03%. In another study (Faulstich, 2005), the amount of orellanine was determined as ca. 14 000 mg kg^{-1} dry weight in *C. orellanus*, and 9000 mg kg^{-1} dry weight in *C. speciocissimus*.

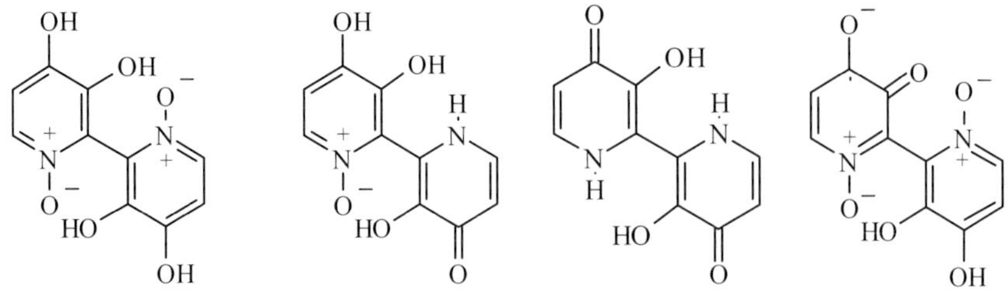

Fig. 5.4 Structures of orellanine and its degradation products orellinine, orelline and radical semiquinone of orellanine.

5.2.2.4   *Exposure assessment*

Bipyridines with positively charged nitrogen atoms were already known to be poisonous before the structure of orellanine was elucidated. At the molecular level, orellanine was shown to be an inhibitor of alkaline phosphatase (Ruedl *et al.*, 1989). Symptoms of orellanine poisoning are typical of renal damage, developing over several days, in some cases up to 2 weeks after the mushroom meal (Danel *et al.*, 2001). During the latent period, mild gastrointestinal disorders occur, which may be overlooked. Accordingly, patients present themselves at hospital only at the stage when renal failure has developed. In this respect, orellanin is particularly insidious. Lethal doses of orellanine are known for the mouse only, corresponding to 15–20 mg kg^{-1} body weight for intraperitoneal and 33–90 mg kg^{-1} body weight for oral administration (Faulstich, 2005).

5.2.3   Muscarine

5.2.3.1   *Occurrence route or mechanism of formation*

L-(+)-Muscarine, [(4R)-4-hydroxy-5-methyl-oxolan-2-yl]methyl-trimethyl-azanium (Fig. 5.5) is a toxic chiral quaternary amine occurring in mainly mushrooms of *Clitocybe* and *Inocybe* genera. Traces of three other stereoisomers (theoretically may exist 8 isomeric substances), epimuskarin, epiallomuskatin and allomuskarin (Fig. 5.5) were also detected in these toxic fungi, nevertheless, they possess only low biological activity.

5.2.3.2   *Methods of analysis*

Muscarine can be detected by high performance thin layer chromatography (HPTLC). Using Dragendorff reagent, muscarine appears as an orange spot on the plate (Stijve, 1981).

Recently, Wai-cheung *et al.* (2007) developed a new multi-detection method using HPLC separation of mushroom toxins employing an HILIC column with amide-based stationary phase that enables hydrophilic interactions and tandem mass spectrometry (MS-MS) with electrospray ionization (ESI+) for high sensitive detection. This LC/MS-MS method was successfully employed to simultaneously separate several polar mushroom toxins, including amanitins and phallotoxins.

5.2.3.3   *Incidence and levels of occurrence*

Muscarine is present in all the parts of respective most poisonous genera,*Inocybe* and *Clitocybe*. In a specimen like *I. patouillardi*, *I. fastigiata*, *I. geophylla* and *C. dealbata*, it accounts for 1000–3000 mg kg^{-1} of dry weight. While *Inocybe* mushrooms are mycorrhizal on conifers or broad-leafed trees, *Clitocybe* mushrooms are saprophytic and grow on forest litter or grassland humus. Both these genera occur commonly in summer and autumn and have a worldwide distribution.

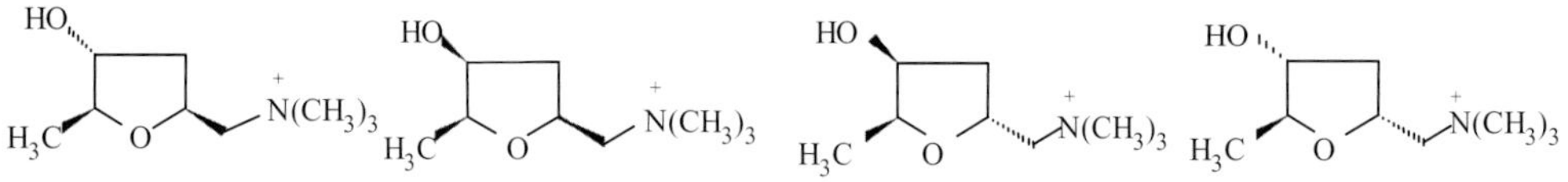

Fig. 5.5   Structures of muscarine, epimuscarine, epiallomuscarine and allomuscarine.

Small amounts of muscarine, at a maximum of 90 mg kg^{-1}, are also present in *A. muscaria*. Although this species got its name due to the presence of muscarine (this toxin was for the first time isolated and identified), the main toxic principle in it is muscimol and its precursor ibotenic acid (Faulstich, 2005). Harmless levels of muscarine, typically not exceeding 20 mg kg^{-1}, occur in several other genera, for instance *Amanita, Boletus, Hygrocybe, Lactarius, Mycena* and *Russula*.

5.2.3.4 Exposure assessment

Due to certain structural similarity, muscarine mimics the action of the neurotransmitter acetylcholine at metabotropic receptors that are also known under the name *muscarinic* acetylcholine receptors. It should be noted that muscarine is not destroyed by heating during cooking (Lambert *et al.*, 2000). So when a mushroom meal with a high muscarine content is ingested, poisoning characterized by profuse sweating, perspiration, salivation, urination, gastric upset, emesis and lachrymation occurs, and vomiting may occur. Symptoms onset is soon, typically within 15–30 minutes after ingestion of the mushroom. With large doses, these symptoms may be followed by abdominal pain, severe nausea, diarrhoea, blurred vision and laboured breathing. Intoxication generally subsides within 2 hours. The species such as *Amanta muscaria* containing less muscarine are only occasionally responsible for cholinomimetic signs. Generally, fatalities due to muscarine are rare, and are mostly limited to victims with pre-existing health problems.

5.2.4 Ibotenic acid and muscimol

5.2.4.1 Occurrence route or mechanism of formation

Ibotenic acid and muscimol are neurotoxins occurring in some species of *Amanita* genus, namely *Amanita muscaria* so called Fly Agaricus and *Amanita pantherina*, Panthercap.Ibotenic acid, α-amino-2,3-dihydro-2-oxo-5-oxazoleacetic acid, is isoxazole from which active species, muscimol, 5-(aminomethyl)-3(2H)-isoxazolone, is originated through decarboxylation (Fig. 5.6).

5.2.4.2 Methods of analysis

Chromatographic methods are most commonly used for the analysis of *A. muscaria* toxins. When using TLC and/or paper chromatography, ninhydrin is used for their detection, and muscinol develops as a yellow spot after heating (Faulstich, 2005). Alternatively, a GC/MS method can be employed (Tsujikawa, 2006). Recently, an LC/MS-MS method employing electrospray ionization (ESI) has been used for identification and analysis of these toxins in selected *Amanita* mushrooms. Also, LC/UV method after pre-column derivatization with dansyl chloride has been described (Tsujikawa, 2007).

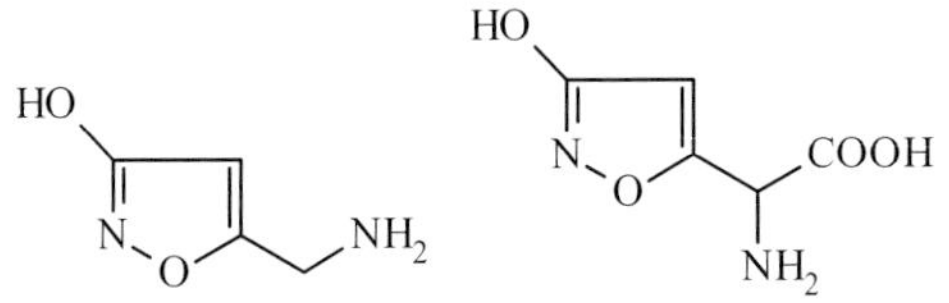

Fig. 5.6 Structures of muscimol and ibotenic acid.

5.2.4.3 Incidence and levels of occurrence

Species containing ibotenic acid and muscinol are widely distributed throughout the planet. *Amanita muscaria* but also *Amanita pantherina* and *Amanita gemmata*, grow in summer and autumn under coniferous and deciduous trees, from the lowland up to the sub-alpine zone. They occur practically all over the temperate and sub-tropical zones of all continents (Seeger and Stijve, 1978).

It is thought that, in *A. muscaria*, the layer just below the skin of the cap contains the highest amount of muscimol, and is therefore the most psychoactive portion. Toxic doses of ibotenic acid (30–60 mg) and muscimol (ca. 6 mg) can be found in single specimens of *A. muscaria*. Commonly, two to four mushrooms of *A. muscaria* are ingested to produce mind-altering effects (Faulstich, 2005).

Recently, levels of muscimol and ibotenic acid in *Amanita* mushrooms naturally grown and circulated in the drug market have been described (Tsujikawa *et al.*, 2007). The mean levels of muscimol determined in *A. muscaria* cap were 156 mg kg^{-1} (ranging from 46 to 1203 mg kg^{-1}) and in stem 178 mg kg^{-1} (82–292 mg kg^{-1}). For ibotenic acid, the mean content in cap was 863 mg kg^{-1} (182–1839 mg kg^{-1}) and 1125 mg kg^{-1} in stem (627–1998 mg kg^{-1}).

5.2.4.4 Exposure assessment

The consumption of *A. muscaria* has been associated with various tribal and religious ceremonies. Ibotenic acid and muscimol contained in it are potent neurotoxins mimicking the natural transmitters glutamic acid and aspartic acid on neurons in the central nervous system (CNS) with specialized receptors for amino acids. The latter compound is a selective γ-aminobutyric acid (GABA)$_A$ agonist, and acts on the CNS in a way similar to diazepam (Michelot and Melendez-Howell, 2003). Commonly, mind-altering effects are experienced after eating two to four mushrooms. When ibotenic acid is ingested, a small portion is decarboxylated into more biologically active muscimol. To enhance the conversion process, some individuals have deliberately roasted these mushrooms prior to eating.

Ibotenic acid produces psychedelic effects in human beings at doses in the range of 50–100 mg. Peak intoxication is reached approximately 2–3 hours after oral ingestion, consisting of one or all of the following; visual distortions/hallucinations, loss of equilibrium, muscle twitching (commonly mislabelled as convulsions), and altered sensory perception. These effects generally last for 6–8 hours, varying with dose and have been shown to lack 'structured' hallucinations in most cases, and the effects are frequently compared to a lucid dream state. The psychoactive dose of muscimol is 15–20 mg.

Patients may appear to be intoxicated, presenting nausea, vomiting and diarrhoea. They have colour hallucinations, slow pulse, hypotension, irritability and lack of coordination. Children may develop fever and seizures. Fatalities rarely occur in adults, but in children, accidental consumption of large quantities of these mushrooms may cause convulsions, coma and other neurologic problems for up to 12 hours.

5.2.5 Psilocybin, psilocin

5.2.5.1 Occurrence route or mechanism of formation

Psilocybin, 4-phosphoryloxy-N,N-dimethyl-tryptamine, is a psychedelic pro-drug occurring in *Psilocybe* and several other mushroom genera together with a closely related psilocin,

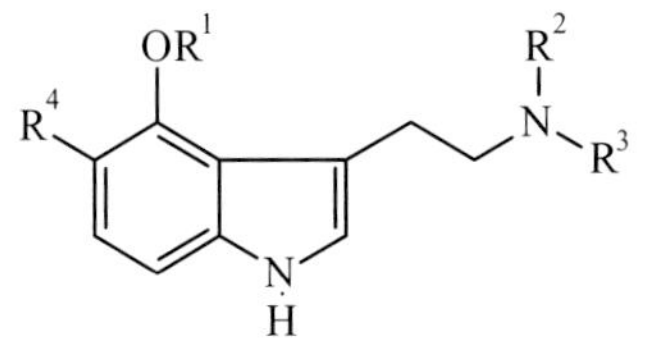

Psilocybin, $R^1 = PO_3H_2$, $R^2 = CH_3$, $R^3 = CH_3$, $R^4 = H$
Psilocin, $R^1 = H$, $R^2 = CH_3$, $R^3 = CH_3$, $R^4 = H$

Fig. 5.7   Structures of psilocybin and psilocin.

4-hydroxy-N,N-dimethyl-tryptamine (see Fig. 5.7). In the body, psylocibin is enzymatically dephosphorylated by phosphatases yielding pharmacologically active psilocin. This reaction takes place also under strongly acidic conditions. Generally, psilocin is relatively unstable in solution due to its phenolic group on the benzene ring at the 4-position in the indole ring.

5.2.5.2   Methods of analysis

For qualitative analysis of psilocybin and psilocin, TLC can be used. The visualization of toxin spots is obtained either by N,N-dimethyl-aminobenzaldehyde (Ehlich reagent) or N,N-dimethyl aminocinnamic aldehyde. For accurate GC/MS or HPLC employing fluorescence detector (FLD) or MS detection with electrospray ionization can be used (Bogusz, 2000; Faulstich, 2005; Saito *et al.*, 2004, 2005). As an alternative, capillary-zone electrophoresis (CZE) and flow injection analysis (FIA) were reported as methods of choice for the determination of these toxins in mushroom extracts (Anastos *et al.*, 2005). Several methods such as HPLC with electrochemical detection (ECD) or DNA-based tests were applied for examination of biological fluids human plasma and urine for the presence of pisloscybin and psilocin (the latter compound is excreted as glucuronide) (Faulstich, 2005).

5.2.5.3   Incidence and levels of occurrence

Mushrooms containing these hallucinogenic indole derivatives were known already in the 16th century in the Mayan culture of ancient Mexico. Species from genera *Psilocybe, Conocybe, Panaeolus* and *Gymnopilus* may contain psilocybin in amounts as high as 2000–16 000 mg kg^{-1} dry weight, and a second psychoactive component, psilocin, in amounts ranging from 0 to 10 000 mg kg^{-1} dry weight (Faulstich, 2005); the average is 5000 and 5000 mg kg^{-1}, respectively. The quantity of toxins in the mushrooms can vary widely, depending on species, phase of growth, mushroom size and drying conditions (both psilocybin and psilocin are temperature-sensitive). Anastos *et al.* (2005) found only small content of psylocin and psylocibin in dried mushrooms *Psilocybe subaeruginosa* and *Hypholoma auranitiaca*, also Saito *et al.* (2005) reported levels of psylocibin in *Psylocybe* mushroom lower than 20 mg kg^{-1} dried samples. In dry form, the mushrooms are available in the black market in various countries (Faulstich, 2005).

5.2.5.4 Exposure assessment

Psilocybin mushrooms are commonly called 'magic mushrooms', or more simply 'shrooms', they are seldom mistaken for food fungi by innocent hunters of wild mushrooms. For recreational and entheogenic use, most common are *Psilocybe semilanceata* (liberty cap), and *Psilocybe cubensis* (golden tops). When ingested, psilocybin is absorbed through the lining of the mouth and stomach, then it is metabolized mostly in the liver where it becomes psilocin (it is broken down by the enzyme monoamine oxidase). The toxin affects CNS with the production of a syndrome, often described as pleasant, somewhat similar to alcohol intoxication, sometimes accompanied by hallucinations. Its intensity and duration depends on many factors including dosage, individual physiology (metabolism), setting etc. Typically, following dosage 10–50 mg psilocybin corresponding to approximately 1–5 g dried mushroom or 10–50 g wet mushrooms effect begins in 10–60 minutes and lasts 2–6 hours.

Poisonings by these mushrooms are rarely fatal in adults. The most severe cases of psilocybin poisoning occur in small children, where large doses may cause the hallucinations accompanied by fever, convulsions, coma and death.

It should be noted that psilocybin and psilocin are listed as Schedule I drugs under the United Nations 1971 Convention on Psychotropic Substances (Schedule I drugs are illicit drugs that are claimed to have no known therapeutic benefit). Parties to the treaty are required to restrict use of the drug to medical and scientific research under strictly controlled conditions.

5.2.6 Coprine

5.2.6.1 Occurrence route or mechanism of formation

Coprine, N5-1-hydroxycyclopropyl-L-glutamine is a poison occurring in the popular Inky Cap mushroom (*Coprinus atramentarius*) and several other species. This unusual amino acid, consisting of cyclopropanone and glutamine, is metabolized to 1-aminocyclopropanol and 1,1-cyclopropandiol (cyclopropanone hydrate). The latter compound reacts with essential SH-group of acetaldehyde dehydrogenase forming respective thio-hemiketal. Biologically active components exhibit similar metabolic effects as disulfiram (see Fig. 5.8).

5.2.6.2 Methods of analysis

No method of analysis for the determination of coprine in mushrooms has been published in the available literature, only analysis employing HPLC with electro-chemical detection for its and other disulfiram-like compounds in rat brain in vitro has been described (Nilsson and Tottmat, 1987).

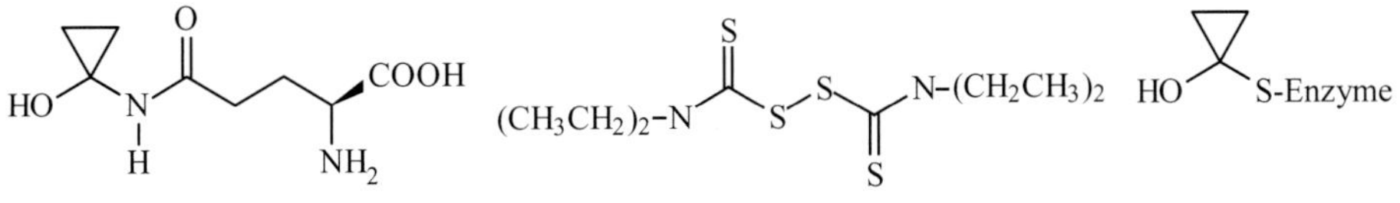

Fig. 5.8 Structures of coprine, disulfiram and thio-hemiketal.

5.2.6.3 *Incidence and levels of occurrence*

Coprinus atramentarius, or inky cap, has a delicious flavour when young. Like other inky caps, for instance *C. insignis*, *C. quadrifidus*, *C. variegates* and *C. atramentarius* grows in tufts. It is commonly associated with buried wood and is found in grassland, meadows, disturbed ground, and open terrain in autumn. It appears in spring, summer and autumn across the Northern Hemisphere, but has also been found in Australia. Traces of coprine were also found in other mushrooms like *Clytocibe claviceps*, *Pholiota squarrosa* and/or *Boletus luridus*.

5.2.6.4 *Exposure assessment*

Although generally considered edible, when consumed with alcohol, *Coprinopsis atramentaria* and other related species are toxic. Coprine hydrolysis products originating in the human body interfere with the breakdown of alcohol at the intermediate stage, since they block acetaldehyde dehydrogenase. Consumption of alcoholic beverages within 72 hours after eating these mushrooms will cause headache, nausea and vomiting, flushing and cardiovascular disturbances that last for 2–3 hours. Although very unpleasant, the syndrome has not been associated with any fatalities. The symptoms are described as similar to those observed after application of disulfiram (trade names for disulfiram in different countries are Antabuse or Antabus), that is a drug used to support the treatment of chronic alcoholism by producing an acute sensitivity to alcohol. Coprine and disulfiram both caused an increase in the acetaldehyde/ethanol ratio, coprine being more potent than disulfiram.

5.3 EDIBLE MUSHROOMS

Their special, delicious flavour has made some mushrooms essential ingredients in a variety of meals throughout the ages. The demand for edible mushrooms is tremendous nowadays and is still growing. The market needs are saturated predominantly by commercially cultivated species – their total annual production is estimated to be well over 1.2 million tons, of which *Agaricus bisporus* forms more than half.

The popularity of mushrooms as dietary constituents is, in addition to their attractive organoleptic properties, based on their low energy content hence suitability for 'low calorie' diets. Without any exception, the content of lipids is very low and, at the same time, they are a good source of dietary fibre and minerals. On the other hand, many wild and even some cultivated edible mushrooms contain toxic constituents (Benjamin, 1995), fortunately, as shown below, the content can be largely reduced by appropriate processing.

Characterized below, hydrazine derivatives gyromitrin and agaritin (the first occurring in certain False Morels, *Gyromitra esculenta*, and the latter one in many *Agaricus* species), are the most probable mushroom toxins to which humans can be exposed via the diet. It should be noted, that no regulations for toxins occurring in edible mushrooms exist, the existing toxicological data do not allow setting acceptable daily intake (ADI) values.

5.3.1 Gyromitrin

5.3.1.1 *Occurrence route or mechanism of formation*

Gyromitrin, acetaldehyde-*N*-methyl-*N*-formylhydrazone (AMFH), see Fig. 5.9, is a hydrazine derivative that was for the first time recognized as the main natural toxin (protoplasmic

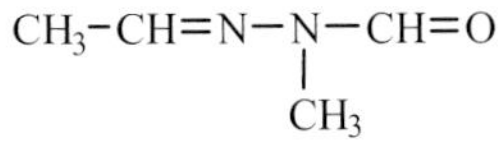

$$CH_3-CH=N-\underset{\underset{\displaystyle CH_3}{|}}{N}-CH=O$$

Fig. 5.9 Structure of gyromitrin.

poison) of False Morel, *Gyromitra esculenta*. This fairly unstable compound is easily oxidized by oxygen when exposed to ambient air, even at room temperature. Under cooking conditions, hydrolysis of gyromitrin occurs yielding *N*-methyl-*N*-formylhydrazine (MFH), which is more stable than its precursor compound (Pyysalo and Niskanen, 1977; Coulet, 1982). *N*-methylhydrazine (monomethylhydrazine, MMH) that originates from MFH (Fig. 5.10) by its further decomposition, represents the real toxic principle responsible for human poisoning by some species of *Gyromitra* genus (Berger and Guss, 2005a,b). It was estimated that 25–30% of gyromitrin can be biotransformed via MFH to MMH in vivo. The highly reactive *N*-nitroso-*N*-methylformamide (NMFA), shown in Fig. 5.10, originates through this process catalysed by liver microsomal monooxygenases. It should be noted, that in addition to gyromitrin, eight other homologous *N*-methyl-*N*-formylhydrazones were identified in *Gyromitra esculenta*. Acetaldehyde is replaced in their molecule by the following higher aldehydes: propanal, n-butanal, 3-methylbutanal, n-pentanal, n-hexanal, n-octanal, *trans*-2-octenal, and cis-2-octenal. The weight ratio of gyromitrin to the total content of these hydrazones was found to be approximately 88:12 (Pyysalo and Niskanen, 1977).

5.3.1.2 Methods of analysis

The analysis of gyromitrin and its breakdown products is a rather complicated task. Currently, no methods validated by interlaboratory studies are available. Enforcing effective

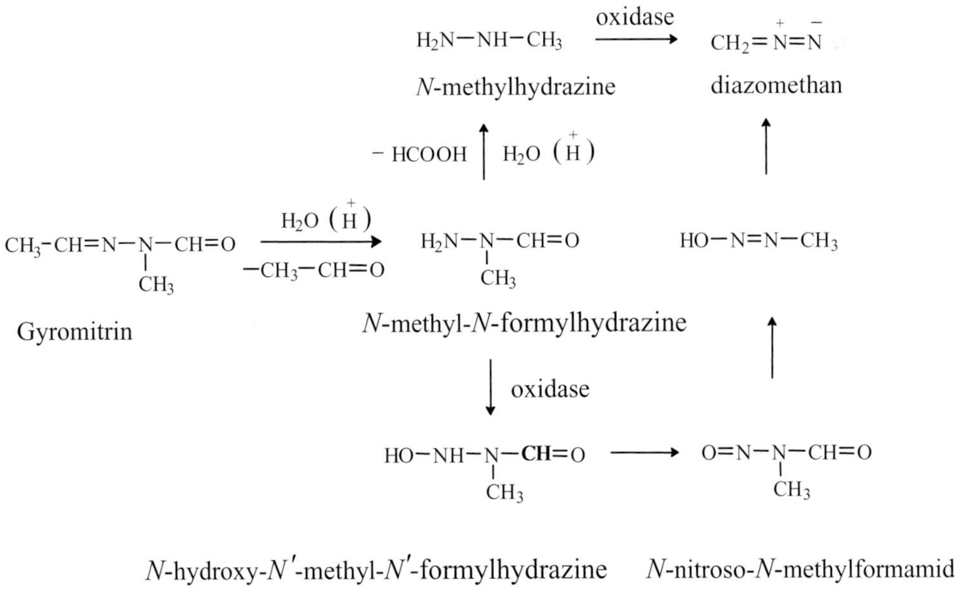

$$H_2N-NH-CH_3 \xrightarrow{\text{oxidase}} \overset{+}{CH_2}=N=\overset{-}{N}$$

N-methylhydrazine   diazomethan

$$- HCOOH \uparrow H_2O\ (\overset{+}{H})$$

$$CH_3-CH=N-\underset{\underset{\displaystyle CH_3}{|}}{N}-CH=O \xrightarrow[-CH_3-CH=O]{H_2O\ (\overset{+}{H})} H_2N-\underset{\underset{\displaystyle CH_3}{|}}{N}-CH=O \qquad HO-N=N-CH_3$$

Gyromitrin   *N*-methyl-*N*-formylhydrazine

$$\downarrow \text{oxidase}$$

$$HO-NH-\underset{\underset{\displaystyle CH_3}{|}}{N}-CH=O \longrightarrow O=N-\underset{\underset{\displaystyle CH_3}{|}}{N}-CH=O$$

N-hydroxy-*N'*-methyl-*N'*-formylhydrazine   *N*-nitroso-*N*-methylformamid

Fig. 5.10 Degradation and metabolism of gyromitrin.

control of these substances in the False Morels is thus not easy. Early methods used for the determination of gyromitrin were based on a titration of an alcohol extract of the False Morel with potassium iodate, exploiting the ability of gyromitrin and its hydrolysis products to be oxidized. Obviously these approaches were fairly unspecific giving only an indication of total hydrazine derivatives, moreover, the sensitivity was low while labour – demands high (TemaNord, 1995). Another applicable method is based on electrochemical oxidation of methylhydrazine (Slanina *et al.*, 1993). In some studies (TemaNord, 1995), the released MMH from bound to an unknown high-molecular component in which it is supposed to occur, was achieved by heating a water suspension of fruiting bodies in a sealed glass tube at 120°C for several hours. Gas chromatography and TLC were then used for its identification.

Recently, Arshadi *et al.* (2006) developed a simple analytical technique employing GC/MS for the rapid determination of low levels (0.3 mg kg^{-1}) of MMH. In this way the content of gyromitrin (and its homologues) in air-dried False Morel could be monitored. Prior to the determinative step, all MMH obtained by conversion of gyromitrin is derivatized using pentafluorobenzoyl chloride (to form the stable derivative tris-pentafluorobenzoyl methylhydrazine, tris-PFB-MH).

5.3.1.3 Incidence and levels of occurrence

The fruiting bodies of False Morel (*Gyromitra esculenta* and *G. gigas*) (also known by a variety of common names such as 'red mushroom', the 'beefsteak mushroom' or the 'lorchel') appear mostly in spring on coniferous, occasionally also on deciduous sandy woodland. These mushrooms are somewhat similar in appearance to the 'true' morel (*Morchella* sp.). Morels are very popular for mushroom hunters and gourmets in some countries, namely, in northern and eastern Europe because of their delicious taste (interestingly, these mushrooms have not yet been successfully cultivated on a large scale). According to some studies (Hajslová, 1995), the content of gyromitrin, representing the most abundant toxic hydrazone in freshly picked False Morels, can be as high as 3000 mg kg^{-1}. The levels of MFH could reach 500 mg kg^{-1}; the concentration of MMH typically ranges from 40 to 350 mg kg^{-1} (Hajslová, 1995). As indicated above, it is supposed that the greater part of gyromitrin and MFH in False Morel is chemically bound (probably as glycosides) to higher molecular weight molecules. Generally, the toxin content in False Morels may largely vary with growth conditions, such as altitude and temperature (List and Luft, 1967).

5.3.1.4 Exposure assessment

Morels containing toxic gyromitrin are considered to be edible mushrooms in some countries, especially in northern and eastern Europe. Some *Gyromitra* species contain gyromitrin, whereas only traces or none toxin are present in others. Unfortunately, toxic and non-toxic species are often mixed-up, and specific cases of mistaken mushroom identity appear frequently. In this respect, the early False Morel *Gyromitra esculenta* is a dangerous fungus since it is easily confused with the true morel *Morchella esculenta*. Several poisonings have occurred after consumption of fresh or cooked *Gyromitra*. Alternatively, the reason of poisoning may be 'contamination' of commercially available 'morels' with *G. esculenta* (Toth, 2000).

For exposure assessment, changes occurring during common culinary practices have to be considered, for example gyromitrin is rather unstable under processing. However, regarding, drying, this process does not lead to significant detoxification as most of the precursors that can be hydrolysed to MMH are eliminated. For instance, Larsson and Eriksson (1989) reported that drying in the open air at room temperature for 3 months reduced the original MMH levels in the range 30–71%, so relatively high amounts of this toxin were still left in the dried mushrooms (410–610 mg kg^{-1} in a particular case). In some commercial dried False Morels, levels as high as 1000–3000 mg kg^{-1} of this hydrazine were reported (Larsson and Eriksson, 1989). On the other hand, a significant decrease of hydrazines can be achieved by boiling. On average, only 10–15% of MMH remained in False Morels after the boiling experiment. The concentrations of MMH in drained canned mushrooms varied from 6 to 65 mg kg^{-1}, which corresponds to 3–30 mg kg^{-1} in fresh tissue (Larsson and Eriksson, 1989). However, one should be aware, that at the same time when elimination of gyromitrin takes place, intoxication of humans may also occur due to inhalation of the fumes emitted during cooking in an uncovered pot.

Hydrazines released from gyromitrin are cytotoxic, but far less so than amatoxins and/or orellanine (Karlson-Stiber and Persson, 2003). However, FMH and MMH were reported to be carcinogenic, possibly due to methylation of guanine moieties in DNA. The lethal dose for hydrazines for humans has been estimated as 20–50 mg kg^{-1} body weight, less for children, i.e., 10–30 mg kg^{-1} body weight (Bergman and Hellenas, 1992).

Acute intoxication by False Morels somehow resembles *Amanita* poisoning, but is less severe. There is generally a latent period of 6–10 hours (or even more) after ingestion during which no symptoms are evident, then there is a sudden onset of abdominal discomfort (a feeling of fullness), severe headache, vomiting and sometimes diarrhoea occurs. The toxin affects primarily the liver, but there are additional disturbances to blood cells and the central nervous system. However, life-threatening poisonings, or even fatalities, are nowadays rare. Considering historical data, the mortality rate based on historical data is relatively low (2–4%).

5.3.2 Agaritine and other phenylhydrazines

5.3.2.1 Occurrence route or mechanism of formation

Within the research concerned with the metabolism of 2-oxo acids in *Basidiomycota*, one of two large phyla, a glutamin-containing compound was isolated from press-juice of *Agaricus bisporus* mushrooms as early as the 1960s (Levenberg, 1961, 1964). This was the first report of a natural phenylhydrazine derivative occurring in relatively high amounts in common *Agaricus* mushrooms, was identified as β-*N*-(γ-L(+)-glutamyl)-4-hydroxymethyl phenylhydrazine (Fig. 5.11) and was given the trivial name agaritine. In follow-up studies, several other nitrogen–nitrogen bond containing substances – related phenylhydrazine

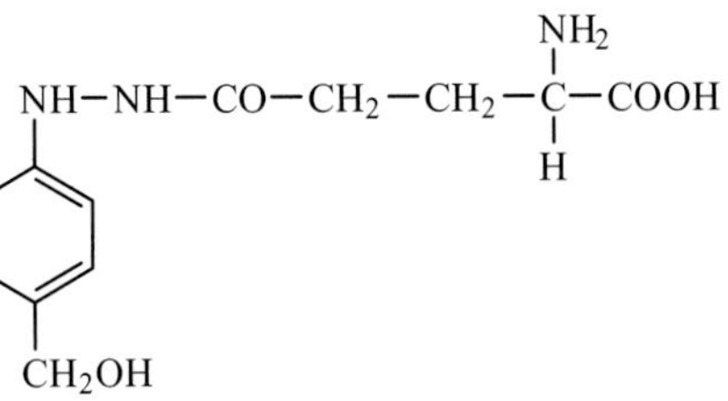

$$NH-NH-CO-CH_2-CH_2-\underset{\underset{H}{|}}{\overset{\overset{NH_2}{|}}{C}}-COOH$$

CH$_2$OH

Fig. 5.11 Structure of agaritine.

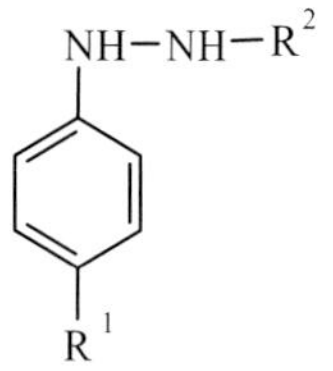

4-(karboxy)phenylhydrazine, R^1 = COOH, R^2 = H
4-(hydroxymethyl)phenylhydrazine, R^1 = CH$_2$OH, R^2 = H
β-N-[γ-L(+)-glutamyl]-4-(carboxy)phenylhydrazine, R^1 = COOH, R^2 = CH$_2$CH$_2$CH(NH$_2$)COOH
β-N-[γ-L(+)-glutamyl]-4-(formyl)phenylhydrazine, R^1 = CH=O, R^2 = CH$_2$CH$_2$CH(NH$_2$)COOH

Fig. 5.12 Structures of precursors and degradation products of agaritine.

derivatives (Fig. 5.12), and also 4-(hydroxymethyl)benzenediazonium ion (Fig. 5.13) were found in *Agaricus* mushrooms (Levenberg, 1961; Ross *et al.*, 1982; Chauhan *et al.*, 1984, 1985), the latter one only in the basal stalk. Of these substances, agaritine is most prevalent, usually occurring at average levels between 200 and 500 mg kg^{-1}fresh weight; 4-(carboxy)phenylhydrazine, β-N-(γ-L(+)-glutamyl)-4-(carboxy)phenylhydrazine and the 4-(hydroxymethyl)benzenediazonium ion (Figs. 5.12 and 5.13) were found in much smaller amounts (Andersson and Gry, 2004). The first two compounds were postulated as possible biosynthetic precursors of agaritine, whereas the 4-(hydroxymethyl)benzenediazonium ion together with 4-(hydroxymethyl)-phenylhydrazine are breakdown products. Hydrolysis of agaritine, which is supposed to be catalysed in mushroom tissue by γ-glutamyl transferase (EC 2.3.2.1), results in releasing L-glutamic acid and 4-(hydroxymethyl)phenylhydrazine. This intermediate, however, has never been detected in *Agaricus* mushrooms, obviously due to its high instability; In the same way as similar hydrazines, it is easily oxidized (Hajslová, 1995), yielding the respective 4-(hydroxymethyl) benezendiazonium ion in this particular case. In addition to enzymatic formation of this catabolite, a direct formation pathway of diazonium ion from agaritine is proposed by some authors (Ross *et al.*, 1982). Most probably, oxidative transformation of the desglutamyl moiety of its precursor takes place, either on the intact molecule or after cleavage of glutamate residue. An interesting hypothesis was postulated by Stijve *et al.* (1986). Trying to explain why agaritine is produced by *Agaricus* species, he concluded that the above catabolites concomitantly produced in vivo help to inhibit competitive fungi that may attack these mushrooms. Besides increasing production with time during mushroom aging (what makes up for its increasing vulnerability), the proposed fungistatic role of agaritine is supported by several other finding such as seldom, if ever, occurrence of mould on *Agaricus* mushrooms body or significantly higher content of this

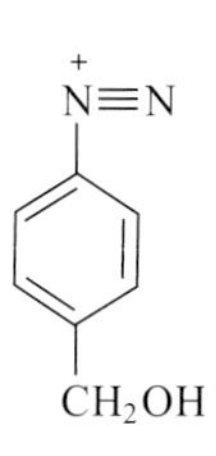

Fig. 5.13 Structure of 4-(hydroxymethyl)benzendiazonium ion.

toxin in wild growing species and strains as compared to cultivated ones that are growing in a protected environment (Anderson *et al.*, 2006).

5.3.2.2 Methods of analysis

The oldest methods employed for quantification of agaritine in mushrooms were time and labour demanding. In his pioneering study, Levenberg (1961, 1964) used an enzymatic hydrolysis for releasing 4-(hydroxymethyl)phenylhydrazine from the parent toxin. After addition of glyoxylic acids, the hydrazone originating from this agaritine breakdown product was determined spectrophotometrically. Gravimetric determination of agaritine, involving a multi-step purification procedure was described by Kelly *et al.* (1962). The first procedure employing HPLC for the determination of agaritine in fresh and processed *Agaricus* mushrooms was reported by Liu *et al.* (1982), although only high concentrations of target analyte could be reliably determined. Some improvement in the performance characteristics of an HPLC/UV method was achieved by Speroni and Bellman (1982). Schulzova *et al.* (2002) validated a reversed phase HPLC method employing diode-array detector (DAD) for the analysis of agaritine extracted with methanol, both from fresh and processed mushrooms. A detection limit of 0.2 mg kg^{-1} in fresh mushroom was reported. Similar parameters were obtained by an HPLC/FLD method developed by Nagaoka *et al.* (2006). Agaritine and its three derivatives were determined in crude extract obtained from *Agaricus* mushroom and products thereof after their conversion to the corresponding fluorescent products with 3,4-dihydro-6,7-dimethoxy-4-methyl-3-oxoquinoxaline-2-carbonyl chloride (DMEQ-COCl). Highly sensitive and accurate analytical procedure (limit detection as low as 0.003 mg kg^{-1}) employing liquid chromatography coupled with electrospray ionization (ESI) tandem mass spectrometry LC/MS-MS was implemented by Kondo *et al.* (2006a,b). Similar approach, i.e. tandem mass spectrometry (LC/MS-MS, ESI in the negative mode) was employed not only for analysis of several edible and processed mushroom species, but also for examination of agaritine-administered mouse plasma (Kondo *et al.*, 2006a,b). LC/MS-MS was also shown applicable for identification and quantification of agaritine in spores of *Agaricus bisporus* (Janak *et al.*, 2006).

5.3.2.3 Incidence and levels of occurrence

The presence of agaritine seems to be limited to the *Agaricus genus*. Reported levels of this natural toxin in fresh mushrooms vary considerably; nevertheless, completely agaritine-free easily cultivatable mushrooms are not known. Liu *et al.* (1982) found levels of agaritine in *Agaricus* mushrooms to range from 330 to 1730 mg kg^{-1} in 14 lots of button mushrooms from 10 different commercial growers in Pennsylvania. According to other studies (Liu *et al.*, 1982), the agaritine content ranged from 440 to 720 mg kg^{-1} in two different mushroom batches (Ross *et al.*, 1982), from 160 to 650 mg kg^{-1} in four varieties of white *Agaricus* mushrooms and from 240 to 650 mg kg^{-1} in brown ones, from 80 to 250 mg kg^{-1} in two different strains (Stijve *et al.*, 1986). In general, higher levels of agaritine can be expected in young still-closed mushrooms. Enzymatic breakdown of agaritine yielding 4-(hydroxymethyl)hydrazine is considered to be responsible for a lower content of parent compound in older fruiting bodies. An extensive study by Speroni *et al.* (1983) concerned with the influence of growing conditions showed, that all the production steps – composting as well as spawning and cropping – influence the final levels of agaritine in cultivated mushrooms. Agaritine levels were found to be lower in mushrooms grown on natural composts (containing mostly horse manure) than in a case of blended or synthetic substrates.

Limited data are available on levels of related hydrazines. Concentrations 10–11 mg kg^{-1} of 4-(carboxy)phenylhydrazine, 16–42 mg kg^{-1} of β-N-(γ-L(+)-glutamyl)-4-(carboxy)phenylhydrazine and 0.6–4 mg kg^{-1} of 4-(hydroxymethyl)benzenediazonium ion were reported to be present in *A. bisporus* (Andersson and Gry, 2004).

5.3.2.4 Exposure assessment

The occurrence of a phenylhydrazine derivative in such a common delicacy as *Agaricus* mushrooms has been of great health concern since hydrazines are an established class of chemical carcinogens. Life-long feeding of Swiss albino mice with fresh, dry-baked and/or freeze-dried *A. bisporus* fungi resulted in tumour development in various tissues of the experimental animals (Toth and Sornson, 1984; Toth, 1986; Toth *et al.*, 1997, 1998). On this account, agaritine, the dominating hydrazine in these mushrooms, was considered as the most likely reason for the carcinogenicity (it is believed to be metabolized to the reactive, 4-hydroxymethylbenzenediazonium ion via 4-(hydroxymethyl)phenylhydrazine) (Fischer *et al.*, 1984). However, experiments in which agaritine was administered to mice orally as a pure compound in drinking water or by gavage failed to elicit a carcinogenic response (Toth *et al.*, 1981), while other mushroom hydrazines induced tumours (Toth *et al.*, 1982; Toth and Erickson, 1986; McManus *et al.*, 1987). The observation that agaritine was degraded in aqueous solution in the presence of oxygen (Hajslová *et al.*, 2002) may contribute to the apparent discrepancy in results of toxicological studies. Generally, it seems, that direct acting mutagenicity of mushrooms is not related to the concentration of agaritine (Toth, 1995). The metabolism and bioactivation of agaritine and other mushroom hydrazines in whole mushroom homogenate and by mushroom tyrosinase was investigated to get more knowledge in this area (Walton *et al.*, 1998, 2001). Also, more recent results of short-term tests in animals (Andersson and Gry, 2004) indicated the need of more studies explaining the relationship between mutagenicity and hydrazine derivatives occurring in *Agaricus* mushrooms.

Since *Agaricus* mushrooms are consumed raw only exceptionally, the changes occurring during their handling and processing have to be taken into consideration for the estimation of consumer's dietary intake. As documented in one of conducted market studies, large differences exist between unprocessed and heat processed *Agaricus* mushrooms (Andersson and Gry, 2004). While the content of agaritine in various fresh *Agaricus bisporus* mushrooms was not below 200 mg kg^{-1} in any of examined sample, its levels in the 35 different trademarks of canned mushroom products obtained in retail market, were by one order of magnitude lower. In average, whole canned mushrooms (25 samples) contained 14.9 mg agaritine per kg product, mean content in cut mushrooms (10 various products) was 18.1 mg kg^{-1}. Agaritine levels in brine were generally slightly lower than those determined in solid parts of the mushroom. On a portion basis, somewhat higher amounts of agaritine may be found in some other food products (mushroom soup and pasta sauce) containing *A. bisporus* (Andersson *et al.*, 1999). In another extensive study, a reduction of agaritine levels by 2–47% was observed in mushrooms stored at 4°C and 12°C for 7 days. The decrease of agaritine content after 14 days ranged from 36 to 76%. Significant changes in agaritine levels were observed in frozen (−25°C, 30 days) and then thawed mushrooms, the losses being about 74% (Ross *et al.*, 1982). On the other hand, relatively high levels of agaritine (1000–4600 mg kg^{-1}) were reported in dry powdered and sliced mushrooms which are often used as a seasoning for soups and gravies. Blanching in boiling water led to leaching of approximately one half of the agaritine into the blanch-water, blanching and boiling for 5 minutes reduced the original content of agaritine by 57 and 75% in brown and white strains, respectively (Fischer *et al.*, 1984).

Table 5.2 The influence of storage and household processing on the agaritine content of *Agaricus* mushrooms.

Process	Conditions	Time	Amount of agaritine remaining in the mushroom[a]
Storage			
Drying	25°C	24 h	82%
Drying	50°C	7.5 h	76%
Drying	40–60°C	7 h	81%
Freezing without thawing	−18°C	7 days	75%
Freezing with thawing	−18°C	7 days	52%
Freezing without thawing	−18°C	30 days	41%
Freezing with thawing	−18°C	30 days	23%
Household processing			
Cooking	Boiling water	5 min	44%
Cooking	Boiling water	60 min	12%
Dry baking	200°C	10 min	77%
Deep-frying	150°C	10 min	50%
Deep-frying	170°C	5 min	52%
Frying	150°C	10 min	43%
Microwave heating	1000 W, 2450 MHz[b]	1 min	35%

[a] 100% = agaritine content in fresh mushrooms before processing.
[b] 20 grams sliced mushrooms.
Reproduced from Schulzova, V., Hajslova, J., Peroutka, R., Gry, J. and Andersson, H.C. Influence of storage and household processing on the agaritine content of the cultivated Agaricus mushroom. *Food Additives and Contaminants*, 19(9), 853–863. Copyright 2002 with permission from Taylor & Francis Ltd, http://www.tandf.co.uk/journals (http://www.informaworld.com).

As illustrated in Table 5.2, depending on the cooking procedure, household-processing of cultivated mushrooms generally reduces the agaritine content, the degree of reduction being dependant on the length and conditions of storage and was usually in the range 20–75% (for more details see Table 5.2). No reduction in agaritine content was observed during freeze-drying (Schulzova *et al.*, 2002).

In spite of apparent elimination of agaritine through the above household/industrial practices, one should be aware, that it is not known to what extent agaritine and other phenylhydrazines occurring in the cultivated mushroom are degraded into other biologically active compounds during these processes. On this account the risk assessment based on the currently available data is hardly possible (Andersson and Gry, 2004).

To evaluate the human health risk related to the consumption of Agaricus mushrooms, studies carried out in accordance with modem test guidelines, together with information on the concentration of hydrazine derivatives in mushrooms used for experiments as well as epidemiological data are needed

References

Anastos, N., Barnett, N.W., Lewis, S.W., Gathergood, N., Scammells, P.J. and Sims, D.N. (2005) Determination of psilocin and psilocybin using flow injection analysis with acidic potassium permanganate and tris(2,2'-bipyridyl) ruthenium(II) chemiluminiscence determination respectively. *Talanta*, **67**, 354–359.
Anderson, Ch., Borovicka, J. and Schulzova, V. (2006) Occurrence of agaritine in various species of *Agaricus*. *Svensk Mykologisk Tidskrift*, **27**(2), 29–39.

Andersson, H.C., Hajslova, J., Schulzova, V., Panovska, Z., Hajkova, L. and Gry, J. (1999) Agaritine content in processed foods containing the cultivated mushroom (*Agaricus bisporus*) on the Nordic and Czech market. *Food Additives and Contaminants*, **16**(10), 439–446.

Andersson, H.C. and Gry, J. (2004) *Phenylhydrazines in the Cultivated Mushroom (Agaricus bisporus) – Occurrence, Biological Properties, Risk Assessment and Recommendations*. TemaNord 2004:558. Nordic Council of Ministers, Copenhagen.

Arshadi, M., Nilsson, C. and Magnusson, B. (2006) Gas chromatography–mass spectrometry determination of the pentafluorbenzoyl derivative of methylhydrazine in False Morel (*Gyromitra esculenta*) as a monitor for the content of the toxin gyromitrin. *Journal of Chromatography A*, **1125**, 229–233.

Benjamin, D.R. (1995) *Mushrooms: Poisons and Panaceas – A Handbook for Naturalists, Mycologists and Physicians*. WH Freeman and Company, New York.

Berger, K.J. and Guss, D.A. (2005a) Mycotoxins revisited: Part I. *Journal of Emergency Medicine*, **28**(1), 53–62.

Berger, K.J. and Guss, D.A. (2005b) Mycotoxins revisited: Part II. *Journal of Emergency Medicine*, **28**(2), 175–183.

Bergman, K. and Hellenas, K.E. (1992) Methylation of rat DNA by the mushroom poison gyrometin and its metabolite monomethylhydrazine. *Cancer Letters*, **61**, 165–170.

Bogusz, M.J. (2000) Liquid chromatography–mass spectrometry as a routine method in forensic sciences: A proof of maturity. *Journal of Chromatography B*, **748**, 3–19.

Brüggemann, O., Meder, M. and Freitag, R. (1996) Analysis of amatoxins α-amanitin ang β-amanitin in toadstool axtrects and body fluids by capillary zone electrophoresis with photodiode array detection. *Journal of Cromatography A*, **744**, 167–176.

Chauhan, Y., Nagel, D., Issenberg, P. and Toth, B. (1984) Identification of P-hydrazinobenzoic acid in the commercial mushroom *Agaricus bisporus*. *Journal of Agricultural Food Chemistry*, **32**, 1067–1069.

Chauhan, Y., Nagel, D., Gross, M., Cerny, R. and Toth, B. (1985) Isolation of N2-[γ-L-(+)-glutamyl]-4-carboxyphenylhydrazine in the cultivated mushroom *Agaricus bisporus*. *Journal of Agricultural Food Chemistry*, **33**, 817–820.

Coulet, M. (1982) Poisoning by *Gyromita*: A possible mechanism. *Medical Hypotheses*, **8**, 325.

Danel, V.C., Saviuc, P.F. and Garon, D. (2001) Main features of *Cortinarius* spp. poisoning: A literature review. *Toxicon*, **39**, 1053–1060.

Drummer, O.H. (1999) Chromatographic screening techniques in systematic toxicological analysis. *Journal of Chromatography B*, **733**, 27–45.

Enjalbert, F., Gallion, C., Jehl, F. and Monteil, H. (1993) Toxin content, phallotoxin and amatoxin composition of *Amanita phalloides* tissues. *Toxicon*, **6**, 803–807.

Enjalbert, F., Cassanas, G., Salhi, S.L., Guinchard, Ch. and Chaumont, J.P. (1999) Comptes Rendus de l'Académie des Sciences, Series III. *Sciences de la Vie*, **322**(10), 855–886.

Faulstich, H. (2005) Mushroom toxins, in *Toxins in Foods* (eds. W.M. Dabrowski and Z.E. Sikorski). CRC Press, Boca Raton, FL pp. 65–83.

Faulstich, H. and Wieland, T. (1992) Mushroom Poisons, in *Food Poisoning* (ed. A.T. Tu). Marcel Dekker, New York.

Fischer, B., Lüthy, J. and Schlatter, C. (1984) Gehaltstimmung von Agaritin im Zuchtchampignon (*Agaricus bisporus*) mittels Hochleistungsflüssigchromatographie (HPLC). *Zeitschrift fuer Lebensmitteluntersuchung und Forschung*, **179**, 218–223.

Gonmori, K. and Yoshioka, N. (2003) The examination of mushroom poisonings at Akita University. *Legal Medicine*, **5**, 83–86.

Hajslová, J. (1995) Mushrooms Toxins, in *Natural Toxic Compounds of Food* (ed. J. Davídek). CRC Press, Boca Raton, FL, Part J, pp. 137–143.

Hajslová, J., Hajkova, L., Schulzova V., Frandsen, H., Gry, J. and Andersson, H.C. (2002) Stability of agaritine – a natural toxicant of *Agaricus* mushrooms. *Food Additives and Contaminants*, **19**(11), 1028–1033.

Horn, S., Horina, J.H., Krejs, G.J., Holzer, H. and Ratschek, M. (1997) Endstage renal failure from mushroom poisoning with *Cortinarius orellanus*: Report of four CASE and review of literature. *American Journal of Kidney Diseases*, **30**, 282–286.

Janak, K., Størmer, C.F. and Koller, G.E.B. (2006) The content of agaritine spores from Agaricus bisporus. *Food Chemistry*, **99**, 521–524.

Karlson-Stiber, C. and Persson, H. (2003) Cytotoxic fungi – an overview. Review. *Toxicon*, **42**(4), 339–349.

Kelly, R.B., Daniels, E.G. and Hinman, J.W. (1962) Agaritine: Isolation, degradation and synthesis. *Journal of Organic Chemistry*, **27**, 3229–3231.

Koller, G.E.B., Høiland, K., Janak, K. and Størmer, F.C. (2002) The presence of orellanine in spores and basidiocarp from Cortinarius orellanus and Cortinarius rubellus. *Mycologia*, 94(5), 752–756.

Kondo, K., Watanabe, A., Iwanaga, Y., Abe, I., Tanaka, H., Nagaoka, M.H, Akiyama, H. and Maitani, T. (2006a) Analysis of agaritine in mushrooms and in agaritine – administered mice using liquid chromatography–tandem mass spectrometry. *Journal of Chromatography B*, **834**, 55–61.

Kondo, K., Watanabe, A., Iwanaga, Y., Abe, I., Tanaka, H., Nagaoka, M.H., Akiyama, H. and Maitani, T. (2006b) Determination of genotic phenylhydrazine agaritine in mushrooms using liquid chromatography–electrospray ionization tandem mass spectrometry. *Food Additives and Contaminants*, **23**(11), 1179–1186.

Lambert, H., Zitoli, J.L., Pierrot, M. and Manel, J. (2000) Intoxications par les Agaricus mushrooms: Syndromes mineurs. *Encycl Méd Chir Toxicologie – Pathologie professionnelle*, **16077B10**, 10.

Larsson, B.K. and Eriksson, A.T. (1989) The analysis and occurrence of hydrazine toxins in fresh and processed false morels, *Gyromitra esculenta*. *Zeitschrift für Lebensmitteluntersuchung und Forschung*, **189**, 438–442.

Levenberg, B. (1961) Structure and enzymatic cleavage of agaritine, a phenylhydrazine of L-glutamic acid isolated from *Agaricaceae*. *Journal of American Chemical Society*, **83**, 503–504.

Levenberg, B. (1964) Isolation and structure of agaritine, a γ-glutamyl-substituted arylhydrazine derivative from *Agaricacaea*. *The Journal of Biological Chemistry*, **239**, 2267–2273.

List, P.H. and Luft, P. (1967) Gyromitrin, das Gift der Frühjahrslorchel *Gyromitra (Helvella) esculenta* FR. *Tetrahedrom Letters*, **20**, 1893.

Litten, W. (1975) The most poisonous mushrooms. *Scientific American*, **232**(3), 90–101.

Liu, J.W., Beelman, R.B., Lineback, D.R. and Speroni, J.J. (1982) Agaritine content of fresh and processed mushrooms (Agaricus bisporus). *Journal of Food Science*, **47**, 1542–1548.

Maurer, H.H. (1998) Liquid-chromatography–mass spectrometry in forensic clinical toxicology. *Journal of Chromatography B*, **713**, 3–25.

Maurer, H.H., Schmidtt, CH.J., Webser, A.A. and Kraemer, T. (2000) Validated electrospray liquid chromatographic–mass spectrometric assay for the determination of the mushroom toxins α-amanitin and β-amanitin in urine after immunoaffinity extraction. *Journal of Chromatography B*, **748**, 125–135.

McManus, B.M., Toth, B. and Patil, K.D. (1987) Aortic rupture and aortic smooth muscle tumors in mice. Induction by *p*-hydrozinobenzoic acid hydrochloride of the cultivated mushroom. *Agaricus bisporus*. *Laboratory Investigation*, **57**, 78–85.

Michelot, D. and Melendez-Howell, L.M. (2003) Amanita muscaria: Chemistry, biology, toxicology and ethnomycology. *Mycological Research*, **107**, 146–231.

Nagaoka, M.H., Nagaoka, H., Kondo, K., Akiyama, H. and Maitani, T. (2006) Measurement of a genotoxic hydrazine, agaritine, and its derivatives by HPLC with fluorescence derivatization in the agaricus mushroom and its products. *Chemical and Pharmaceutical Bulletin*, **54**(6), 922–924.

Nilsson, G.E. and Tottmat, O. (1987) Effects of biogenic aldehydes and aldehyde dehydrogenase inhibitors on rat brain tryptophan hydroxylase activity in vitro. *Brain Research*, **409**(2), 374–379.

Oubrahim, H., Richard, J.M., Esnault-Cantin, D., Murandi-Seigle, F. and Trecourt, F. (1997) Novel methods for identification and quantification of the mushroom neprhrotoxin orellanine. Thin-layer chromatography and electrophoresis screening of mushrooms with electron spin resonance determination of the toxin. *Journal of Chromatography A*, **758**, 145–157.

Oubrahim, H., Richard, J.M. and Esnault-Cantin, D. (1998) Peroxidase-mediated oxidation, a possible pathway for activation of the fungal nepohrotoxin orellanine and related compounds. ESR and spin-trapping studies. *Free Radical Research*, **28**, 497–505.

Pyysalo, H. and Niskanen, A. (1977) On the occurrence of *N*-methyl-*N*-formyl-hydrazones in fresh and processed false motel, *Gyromitra esculenta*. *Journal of Agricultural Food Chemistry*, **25**(3), 644–647.

Ross, A.E., Nagel, A.E. and Toth, D.L. (1982) Occurrence, stability and decomposition of β-A-(γ-L(+)-glutamyl)-4-hydroxymethylphenylhydrazine (agaritine) from the mushroom *Agaricus bisporus*. *Food Chemistry and Toxicology*, **20**, 903–907.

Ruedl, C., Gstrauntaler, G. and Moser, M. (1989) Differential inhibitory activity of the fungal toxin orellanine and alkaline phosphatase isoenzymes. *Biochimica et Biophysica Acta*, **991**, 280–283.

Saito, K., Toyóoka, T., Fukushima, T., Kato, M., Shirota, O. and Goda, Y. (2004) Determination of psilocin in magic mushrooms and rat plasma by liquid chromatography with fluorimetry and electrospray ionization mass spectrometry. *Analytica Chimica Acta*, **527**, 149–156.

Saito, K., Toyóoka, T., Kato, M., Fukushima, T., Shirota, O. and Goda, Y. (2005) Determination of psilocybin in hallucinogenic mushrooms by reverse-phase liquid chromatography with fluorescence detection. *Talanta*, **66**, 562–568.

Schulzova, V., Hajslova, J., Peroutka, R., Gry, J. and Andersson, H.C. (2002) Influence of storage and household processing on the agaritine content of the cultivated Agaricus mushroom. *Food Additives and Contaminants*, **19**(9), 853–863.

Seeger, R. and Stijve, T. (1978) *Amanita muscaria, Amanita pantherina* and others, in *Amanita Toxins and Poisoning* (eds. H. Faulstich, B. Kommerell and T. Wieland). Verlag Gerhard Witzstrock, Baden-Baden, pp. 3–16.

Slanina, P., Cekan, E., Halen, B., Bergman, K. and Samuelson, R. (1993) Toxicological studies of the false morel (*Gyromitra esculenta*): Embryotoxicity of monomethylhydrazine in the rat. *Food Additives and Contaminants*, **10**(4), 391–398.

Speroni, J.J. and Bellman, R.B. (1982) High performance liquid chromatographic determination of agaritine in cultivated mushrooms. *Journal of Food Science*, **47**, 1539–1541.

Speroni, J.J., Bellman, R.B. and Schisler, L.S. (1983) Factors influencing the agaritine content in cultivated mushrooms, *Agaricus bisporus*. *Journal of Food Protection*, **46**, 506–509.

Spoerke, D.G. and Rumack, B.H. (eds.) (1994) *Handbook of Mushroom Poisoning, Diagnosis and Treatment*. CRC Press, Boca Raton, FL.

Stiber, C.K. and Persson, H. (2003) Cytotoxic fungi – an overview. Review. *Toxicon*, **42**, 339–349.

Stijve, T. (1981) High performance thin-layer chromatographic determination of the toxic principle of some poisonous mushrooms. *Mitteilungen Gebiete Lebensmittelunters Hygiene*, **72**, 2432–2436.

Stijve, T., Fumeaux, R. and Philippossian, G. (1986) Agaritine, a p-hydroxymethylphenylhydrazine derivative in cultivated mushrooms (*Agaricus bisporus*), and in some of its wild-growing relatives. *Deutsche Lebensm. Rdsch.*, **82**, 243–248.

TemaNord (1995) *Hydrazones in the False Morel*. The Nordic Council of Ministers, Copenhagen, 561 pp.

Toth, B. (1986) Carcinogenicity by N2-[γ-L-(+)-glutamyl]-4-carboxyphenylhydrazine of *Agaricus bisporus* in mice. *Anticancer Research*, **6**, 917–920.

Toth, B. (1995) Mushroom toxins and cancer (Review). *International Journal of Oncology*, **6**, 137–145.

Toth, B. (2000) A review of the natural occurrence, synthetic production and use of carcinogenic hydrazines and related chemicals. *In Vivo*, **14**, 299–320.

Toth, B., Raha, C.R., Wallcave, L. and Nagel, D. (1981) Attempted tumor induction with agaritine in mice. *Anticancer Research*, **1**, 255–258.

Toth, B., Nagel, D. and Ross, A. (1982) Gastric tumorigenesis by a single dose of 4-(hydroxymethyl)benzenediazonium ion of *Agaricus bisporus*. *British Journal of Cancer*, **46**, 17–422.

Toth, B. and Sornson, H. (1984) Lack of carcinogenicity of agaritine by subcutaneous administration in mice. *Mycopathologia*, **85**, 75–79.

Toth, B. and Erickson, J. (1986) Cancer induction in mice by feeding of the uncooked cultivated mushroom of commerce *Agaricus bisporus*. *Cancer Research*, **46**, 4007–4011.

Toth, B., Erickson, J., Gannett, P.M. and Patil, K. (1997) Carcinogenesis by the cultivated baked *Agaricus bisporus* mushroom in mice. *Oncology Reports*, **4**, 931–936.

Toth, B., Gannett, P., Visek, W.J. and Patil, K. (1998) Carcinogenesis studies with the lyophilized mushroom *Agaricus bisporus* in mice. *In Vivo*, **12**, 239–244.

Tsujikawa, K., Mohri, H., Kuwayama, K., Miyaguchi, H., Iwata, Y., Gohda, A., Fukushima, S., Inoue, H. and Kishi, T. (2006) Analysis of hallucinogenic constituents in Amanita mushrooms circulated in Japan. *Forensic Science International*, **164**(2–3), 172–178.

Tsujikawa, K., Kuwayama, K., Miyaguchi, H., Kanamori, T., Iwata, Y., Inoue, Y.T. and Kishi, T. (2007) Determination of muscimol and ibotenic acid in Amanita mushrooms by high-performance liquid chromatography and liquid chromatography–tandem mass spectrometry. *Journal of Chromatography B*, **852**(1–2), 430–435.

Vetter, J. (1998) Toxins of *Amanita phalloides*. Review article. *Toxicon*, **36**(1), 13–24.

Wai-cheung, Ch., Sau-ching, T. and Sai-tim, S. (2007) Separation of polar mushroom toxins by mixed-mode hydrophilic and ionic interaction liquid chromatography–electrospray ionization–mass spectrometry. *Journal of Chromatographic Science*, **45**(2), 104–111.

Walton, K., Walker, R. and Ionnides, C. (1998) Effect of baking and freeze-drying on the direct and indirect Mutagenicity of extracts from the edible mushroom *Agaricus bisporus*. *Food and Chemical Toxicology*, **36**, 315–320.

Walton, K., Coombs, M.M., Walker, R. and Ionnides, C. (2001) The metabolism and bioactivation of agaritine and other mushroom hydrazines by whole mushroom homogenate and mushroom tyrosinase. *Toxicology*, **161**, 165–177.

Wieland, T. and Faulstich, H. (1983) Peptide toxins from Amanita, in *Handbook of Natural Toxins* (eds. R.F. Keeler and A.T. Tu). Marcel Dekker, New York, Part VI, pp. 585–636.

Wieland, T. and Govindan, V.M. (1974) Phallotoxins bind to actins. *FEBS Letters*, **46**(1), 351–353.

Zhang, P., Chen, Z.H., Hu, J.S., Wei, B.Y., Zhang, Z.G. and Hu, W.Q. (2005) Production and characterization of Amanitin toxins from a pure culture of *Amanita exitialis*. *FEMS Microbiology Letters*, **252**, 223–228.

6   Mycotoxins

Keith A. Scudamore

Summary

The name mycotoxin combines the Greek word for fungus 'mykes' and the Latin word 'toxicum' meaning poison. Moulds can be of great benefit to man as nutritious foods or as the source of antibiotics and other useful chemicals. The term 'mycotoxin' is usually reserved for the toxic chemical products formed by a few fungal species that readily colonise crops in the field or after harvest and thus pose a potential threat to human and animal health through the ingestion of food products prepared from these commodities.

Any crop growing in the field or that is stored for more than a few days is at risk from mould growth and potential mycotoxin formation. Mycotoxins occur worldwide and affect important food sources such as cereals, nuts, dried fruit, coffee, cocoa, spices, oil seeds, dried legumes and fruit. Once formed mycotoxins are mostly stable and persistent, thus the best method of control is prevention. They may also be found in beer and wine resulting from contaminated barley, other cereals and grapes used in their production. Mycotoxins also enter the human food chain via meat or other animal products such as eggs, milk and cheese as the result of livestock ingesting contaminated animal feed.

Mycotoxins cause a diverse range of toxic effects and are of concern for the long-term health of the human population even when present in low amounts, as some of the most common mycotoxins are carcinogenic, genotoxic, or may target the kidney, liver or immune system.

The risk that such mycotoxins pose to man is under continuous review and this has resulted in statutory or guideline maximum permissible limits for some mycotoxins. Sampling and analysis represents a demanding challenge for the analyst. Failure to achieve a satisfactory performance in either area may lead to unacceptable consignments being accepted or satisfactory loads being unnecessarily rejected.

6.1   INTRODUCTION

6.1.1   Mycotoxins and their study

Mycotoxins are toxic secondary metabolites produced by certain species of fungi. They have a range of diverse chemical and physical properties and toxicological effects on man and animals. While many hundreds of such metabolites have been identified, only 20–30 have been shown to be naturally occurring contaminants of human or animal food (Watson, 1985). Only one group of major importance, the fumonisins, have been identified in the past

Table 6.1 Physical properties of some mycotoxins.

Mycotoxin	Molecular formula	Molecular weight	Melting point	UV absorption (nm)
Ochratoxin A	$C_{20}H_{18}ClNO_6$	403	169	36 800 (213), 6400 (332) ethanol
Deoxynivalenol	$C_{15}H_{20}O_6$	296	131–135	Maximum at 218 ethanol
T-2 toxin	$C_{24}H_{34}O_9$	466	150–151	Maximum at 187 cyclohexane
Diacetoxyscirpenol	$C_{19}H_{26}O_7$	366	162–164	None
Zearalenone	$C_{18}H_{22}O_5$	318	164	29 700 (236), 13 910 (274), 6020, (316) ethanol
Fumonisin B$_1$	$C_{34}H_{59}NO_{15}$	721	Powder	Low
Fumonisin B$_2$	$C_{34}H_{59}NO_{14}$	705	Powder	Low
Patulin	$C_7H_6O_4$	154	111	14 600 (275) ethanol
Citrinin	$C_{13}H_{14}O_5$	250	179	22 280 (222), 8279 (253), 4710 (319) ethanol
Cyclopiazonic acid	$C_{20}H_{20}N_2O_3$	336	246	20 400 (282) methanol
Sterigmatocystin	$C_{18}H_{12}O_6$	324	246	15 200 (325) benzene
Moniliformin	C_4HO_3Na	120		
Tenuazonic acid	$C_{10}H_{15}NO_3$	197	Oil	5000 (218), 12 500 (277) acid methanol; 11 500 (240), 14 500 (280) methanol
Altenuene	$C_{15}H_{16}O_6$	292	190–191	30 000 (240), 10 000 (278), 6600 (319) ethanol
Alternariol	$C_{14}H_{10}O_5$	258	350	38 000 (258) ethanol
Alternariol monomethyl ether	$C_{15}H_{12}O_5$	272	267	

20 years (Gelderblom *et al.*, 1988). The most important mycotoxins found in food are listed in Table 6.2 together with the principal fungal species responsible for their production.

This chapter discusses the key mycotoxins in turn with particular emphasis on their occurrence and methods for their determination. Physical and chemical properties of the most commonly encountered mycotoxins are given in Tables 6.1 and 6.2. The main toxicological properties are considered but the reader should consult other works for their detailed toxicology. In a toxicological study the administration to an animal of a pure crystalline mycotoxin often leads to results at variance with those achieved when similar quantities of the crude

Table 6.2 Physical and chemical properties of some aflatoxins.

Aflatoxin	Molecular formula	Molecular weight	Melting point	UV absorption maxima (ε) in methanol	
				265 nm	360–362 nm
B$_1$	$C_{17}H_{12}O_6$	312	268–269	12 400	21 800
B$_2$	$C_{17}H_{14}O_6$	314	286–289	12 100	24 000
G$_1$	$C_{17}H_{12}O_7$	328	244–246	9600	17 700
G$_2$	$C_{17}H_{14}O_7$	330	237–240	8200	17 100
M$_1$	$C_{17}H_{12}O_7$	328	299	14 150	21 250 (357)
M$_2$	$C_{17}H_{14}O_7$	330	293	12 100 (264)	22 900 (357)
Aflatoxicol	$C_{17}H_{14}O_6$	314	230–234	10 800 (261)	14 100 (325)

product formed under natural conditions are presented. The reason is that secondary fungal metabolism often leads to a cocktail of different metabolites, which may act additively or synergistically. Accumulation of toxicological data, and information on occurrence and food consumption, has enabled risk assessment to be carried out and regulations introduced for the most important mycotoxins. The latest worldwide regulations for mycotoxins in food and feed were reviewed in 2003 (FAO, 2004). The presence of mycotoxins in raw commodities only becomes of concern for human health if they survive storage, processing and preparation of the food item as consumed or if they give rise to toxic metabolites of the parent compound. Similarly, their occurrence in animal feedingstuffs only presents a risk for human health if they, or a toxic metabolite, are transferred in significant amounts to meat or animal products such as eggs, milk or dairy products, although they may impair animal health and affect productivity. A comprehensive review 'Mycotoxins: Risks in Plant, Animal, and Human Systems' (CAST, 2003) provides an excellent overview of the mycotoxin problem while a reference work addressing the subject of mycotoxins in animal feeds is that by Smith and Henderson (1991).

The diversity of chemical structures exhibited by mycotoxins (Figs. 6.1–6.8) results in a wide range of acute and chronic toxicological effects. The scourge of the Middle Ages in Europe, St Anthony's Fire, was caused by ergot alkaloids in cereals and many other animal and human diseases have been attributed, at least in part, to the presence of mycotoxins in food or the environment, although conclusive evidence for such associations is often difficult to obtain. Some mycotoxins are proven or suspected carcinogens, mutagens or teratogens, while others have been shown to challenge the immune systems of man and animals. This raises the possibility of increased susceptibility to other diseases often without suspicion of mycotoxin involvement.

It has also been demonstrated that intake of mycotoxins can occur as a result of their presence in fungal spores in the atmosphere (Douwes et al., 2003) This form of exposure may make a significant additional contribution to total intake under some circumstances, such as damp domestic housing (Flannigan et al., 1991).

6.1.2 Mycotoxins and fungi

Mycotoxins may be formed in the field before harvest following infection by any of a variety of fungal species, members of the genera *Fusarium*, *Alternaria* and *Aspergillus* being the most important. Crops are also susceptible to fungal attack after harvest, during transport or when in store, the main mycotoxigenic species being those of *Penicillium* and *Aspergillus*. There is often a close relationship between fungal species and the secondary metabolites produced but this relies on accurate taxonomic information based on correct identification of species according to criteria agreed internationally. The most important mycotoxins found in food are listed in Table 6.3 together with the principal fungal species responsible for their production.

Information about the fungi involved has sometimes been contradictory due to misidentification of fungi at the species level or disagreement over the names applied to specific species. This has been especially true of the genus *Penicillium* (Frisvad, 1989). Extreme caution is required in attempting to relate fungal infection with the presence of mycotoxins as these substances can only form if toxigenic strains of fungi are present and conditions are suitable to trigger fungal metabolism. Conversely, apparently mould-free products can still be contaminated with mycotoxins because the populations of fungal flora change throughout food production and mycotoxins concentrations are subsequently affected by processing.

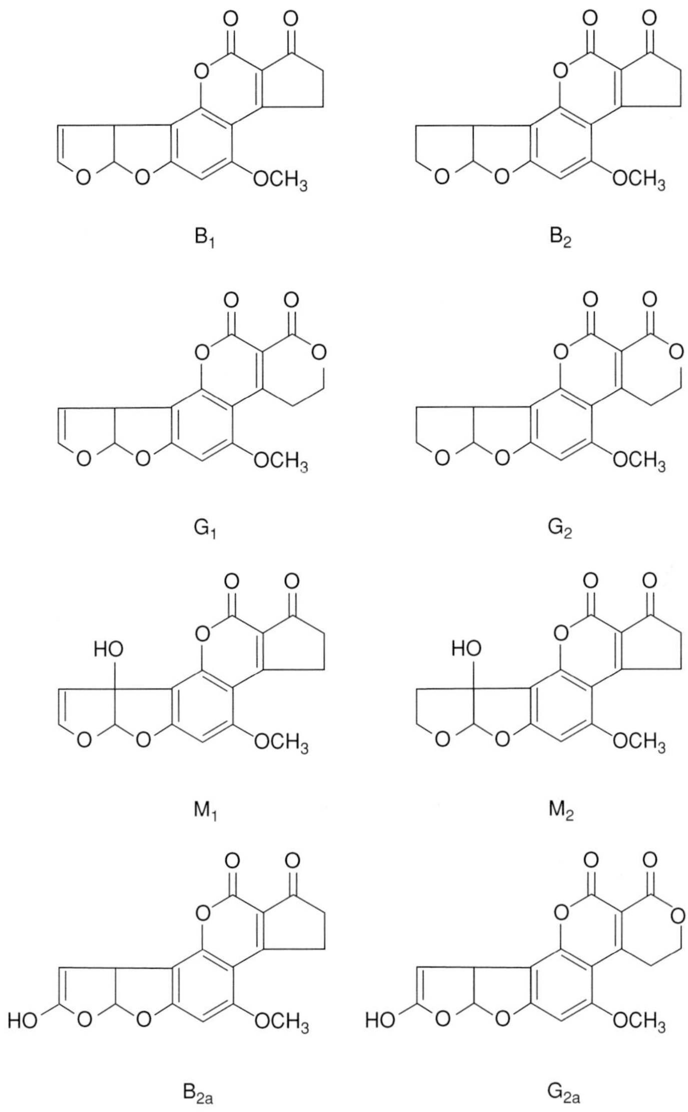

Fig. 6.1   Structures of aflatoxins.

The development of mycotoxins depends on many factors, including temperature, available moisture, oxygen levels, nutrients, trace elements, genetic variation between strains of specific species or competition with other organisms that may be present.

6.1.3   Sampling and detection

The way in which fungi colonise and grow in crops in the field or infect food commodities during transport or storage, results in a highly heterogeneous distribution of any mycotoxins that subsequently develop. This has important implications when sampling: statistically based protocols are required to ensure that a representative sample is drawn. The results of studies of

Table 6.3 Important mycotoxins that have been found in some food commodities and the fungi responsible.

Mycotoxin	Main fungal species	Foods infected
Aflatoxins B_1, B_2, G_1, G_2	A. flavus, A. parasiticus[a]	Nuts, figs, dried fruit, spices, rice bran, maize
Aflatoxins M_1, M_2	Metabolic products of aflatoxins B_1 and B_2	Milk and dairy products
Ochratoxin A	P. verrucosum, A. ochraceus	Cereals, coffee beans, field beans, beer, nuts
Deoxynivalenol, nivalenol	F. graminearum, F. culmorum, F. crookwellense	Cereals
T-2 toxin, HT-2 toxin	F. poae, F sporotrichioides	Cereals
Zearalenone	F. graminearum, F. culmorum, F. crookwellense	Cereals
Fumonisins B_1, B_2, B_3	F. moniliforme, F. proliferatum	Maize, maize products
Patulin	P. expansum	Apple juice, fruits, silage
Citrinin	P. verrucosum	Cereals
Cyclopiazonic acid	A. flavus, P. commune	Cereals, pulses, nuts, cheese
Sterigmatocystin	A. versicolor, A. nidulans	Cereals, cheese
Moniliformin	Fusarium species	Cereals
Alternariol, alternariol monomethyl ether, tenuazonic acid	Alternaria alternata, A. tenuis	Fruit, tomatoes, oil seeds, cereals

[a] Aflatoxins B_1 and B_2 only.

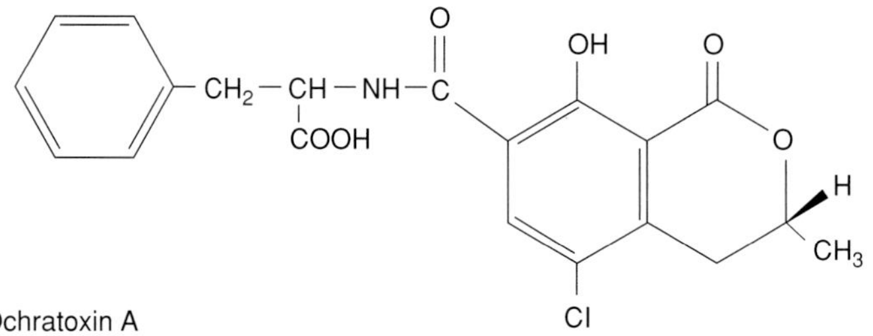

Ochratoxin A

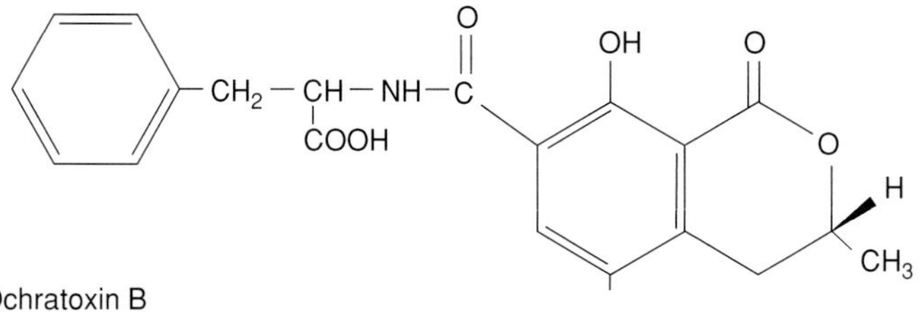

Ochratoxin B

Fig. 6.2 Structures of ochratoxin A and B, related compounds and citrinin.

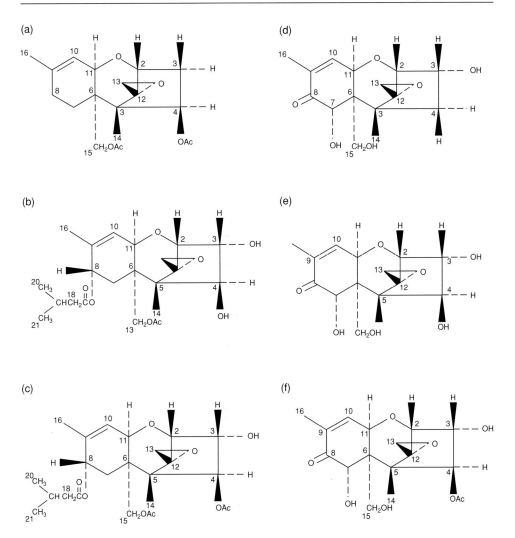

Fig. 6.3 Structures of some trichothecene mycotoxins: (a) Diacetylverrucarol; (b) HT-2 toxin; (c) T-2 toxin; (d) 4-Deoxynivalenol; (e) Nivalenol; (f) Fusarenon-X.

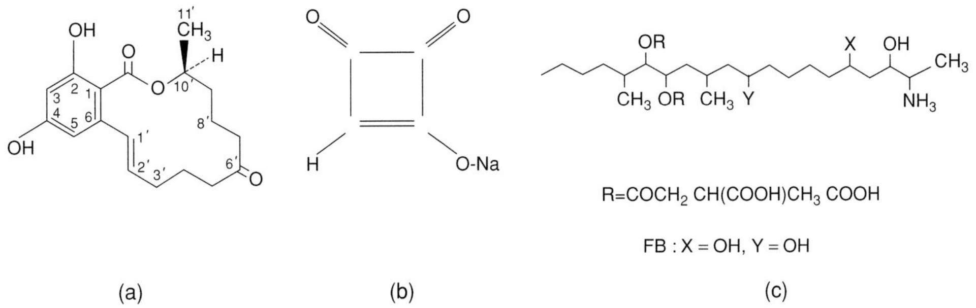

R=COCH$_2$ CH(COOH)CH$_3$ COOH

FB : X = OH, Y = OH

(a)   (b)   (c)

Fig. 6.4 Structures of zearalenone, fumonisins and moniliformin.

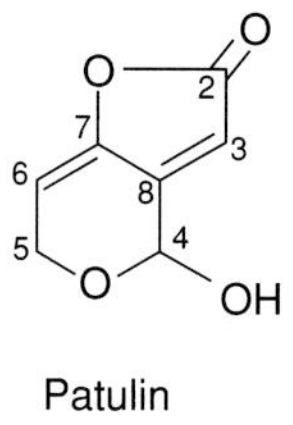

Patulin

Fig. 6.5 Structure of patulin.

the distribution of toxins within a consignment of material are used to determine the number and size of samples required to be taken. However, sometimes this may conflict with the practicalities involved in acquiring samples. Processed food and drink usually presents less of a sampling problem, as mycotoxins are likely to be distributed in a more homogeneous manner. However, sound sampling is critical if a true assessment of the amounts present is to be achieved.

Detection of low µg/kg amounts of fungal compounds within a raw material presents a major challenge to the analytical chemist. Hence, specific and precise analytical methods have been developed and are discussed later for each mycotoxin in turn with emphasis on those methods that have been subject to collaborative testing. Gilbert and Anklam (2002) have discussed the validation of analytical methods for the determination of mycotoxins in foodstuffs. Significant advances in methodology have been made recently such as the use of immunoaffinity column and molecular imprint technology for clean-up and various configurations of HPLC/MS for separation and detection (Zöllner and Mayer-Helm, 2006). Because methods are being continually improved, to enable access to the most up-to-date techniques it is now accepted that for regulatory control purposes a method should meet agreed performance parameters (criteria approach) rather than specifying a mandatory method. For quality assurance purposes, a limited number of certified reference materials or test materials are available for different matrices and these can be used to allow the analyst to check how methods perform. These materials can be obtained from organisations such as the IRMM (Institute for Reference Materials and Measurements), UK FAPAS® (Food Analysis

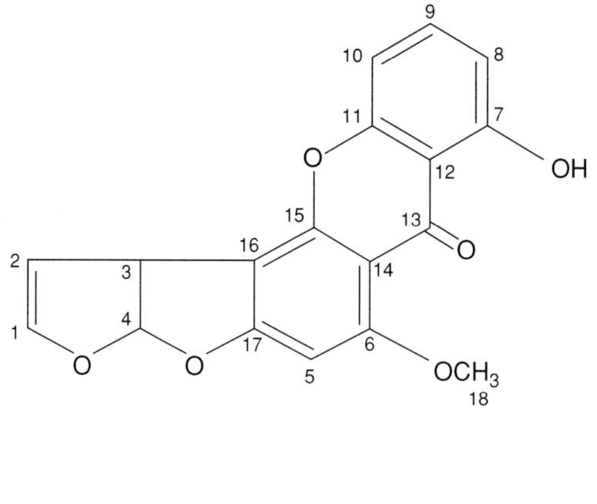

Fig. 6.6 Structures of sterigmatocystin and related compounds.

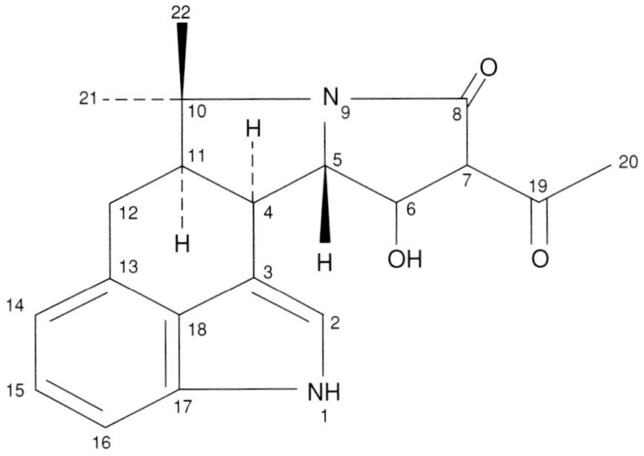

Fig. 6.7 Structure of cyclopiazonic acid.

Performance Assessment Scheme) and PROMEC Unit, Medical Research Council, South Africa.

Antibodies have been raised to most of the important mycotoxins and a range of immuno-logically based methods now complement, or has been incorporated into, chemically based methods such as thin-layer chromatography (TLC), high performance liquid chromatography (HPLC), gas chromatography and tandem mass spectrometric methods. Antibodies bound to activated Sepharose are commercially available as highly selective clean-up columns.

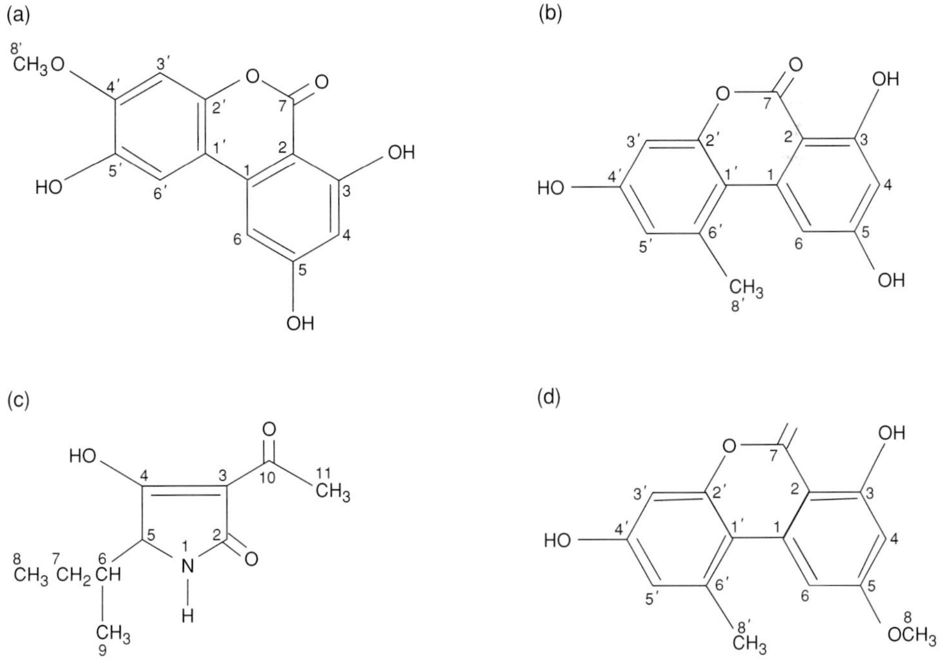

Fig. 6.8 Structures of *Alternaria* mycotoxins: (a) Altenuisiol; (b) Alternariol; (c) Tenuazonic acid; (d) Alternariol monomethyl ether.

6.1.4 Significance and control

The occurrence of mycotoxins in food is intermittent and the concentrations variable, at least in developed areas of the world. However, the potential for wide-scale problems exists should the appropriate circumstances arise. Low-level contamination of foods by highly biologically active chemicals presents a major difficulty in assessing their true significance for man. Development of methods for mycotoxins in body fluids provides additional evidence and information on intake. For example, Gilbert *et al.* (2001) assessed dietary exposure to ochratoxin A using a duplicate diet approach and analysis of urine and plasma samples that offers the opportunity of using ochratoxin A in urine as a simple and reliable biomarker to estimate exposure to this mycotoxin.

Fungi and mycotoxins, intimately associated with growing crops and stored foods, are difficult to eliminate, especially when they develop prior to harvest. Good farming and storage practice are important factors in reducing the occurrence and amounts of mycotoxins. It is unfortunate however that the less developed nations tend to have climatic conditions that encourage mould growth and mycotoxin formation and hence are faced with much greater problems, while at the same time having fewer resources to detect, control and train operators in the means to reduce the extent of contaminated food.

In summary, the aim during production of food crops and consumer food products should be to reduce levels of mycotoxins to the lowest that can be technologically achieved within the existing economic constraints and with regard to risk assessment, where this has been carried out. The existence of a statutory limit requires systematic inspection, sampling and analysis at agreed key points in trade and marketing.

Concerns of the food industry and other bodies include the means of observing legal limits, instigation of appropriate monitoring and control protocols, representative sampling, identification of products likely to be susceptible and the ability to respond to any public or media concern as, and if, the need arises. The mycotoxin field requires a multi-disciplinary approach involving, for example, chemists, mycologists, toxicologists, veterinarians and statisticians.

6.2 AFLATOXINS

6.2.1 Chemical properties

The aflatoxins consist of a group of approximately 20 related fungal metabolites, although only aflatoxins B_1, B_2, G_1, G_2 and M_1 are normally found in crops and animal feed. The chemical structures of the most important aflatoxins and their derivatives are shown in Fig. 6.1. Aflatoxins B_2 and G_2 are the dihydro derivatives of the parent compounds. They are produced by *Aspergillus species* such as *A. flavus*, *A. parasiticus* and *A. nominus* and can occur in a wide range of important raw food commodities such as cereals, nuts, spices, figs and dried fruit. Although the highest concentrations are usually formed in food crops grown and stored in the warmer areas of the world, the international trading of these important commodities ensures that aflatoxins are not only a problem for the producing nations but are also of concern for importing countries.

Aflatoxins M_1 and M_2 are the hydroxylated metabolites of aflatoxins B_1 and B_2 and are produced when cows and other ruminants ingest feed contaminated with these mycotoxins. They are then excreted in milk and may subsequently contaminate other dairy products such as cheese and yoghurt.

The chemistry of the aflatoxins was reviewed (Roberts, 1974). More recently the chemistry and biology of aflatoxin B_1 has been updated (Smela *et al.*, 2001) and the role of aflatoxins in food safety comprehensively appraised (Abbas, 2005). Some of their important physical and chemical properties are given in Table 6.2. Aflatoxins are crystalline substances, freely soluble in moderately polar solvents such as chloroform, methanol and dimethyl sulphoxide and they dissolve in water to the extent of 10–20 mg/L.

Crystalline aflatoxins are extremely stable in the absence of light and UV radiation, even at temperatures in excess of $100°C$. Solutions prepared in chloroform or benzene are stable for years if kept cold and in the dark. The purity and concentration of reference solutions can be calibrated using molar absorptivity data (Scott, 1990). The lactone ring makes aflatoxins susceptible to alkaline hydrolysis and processes involving ammonia or hypochlorite have been investigated as means for their removal from food commodities, although questions concerning the toxicity of the breakdown products have restricted the use of this means of eradicating aflatoxins from food and animal feeds. If alkaline treatment is mild, acidification will reverse the reaction to reform the original aflatoxin. In acid solution, aflatoxins B_1 and G_1 are converted to aflatoxins B_{2a} and G_{2a} by acid catalytic addition of water across the double bond of the furan ring. Oxidising reagents react and the molecules lose their fluorescence.

Aflatoxins are quite stable in many foods and are fairly resistant to degradation in variety of processes. Smith *et al.* (1994) and Park (2002) and Scudamore (2004) among others have summarised the stability of aflatoxins in some food processes. The stability can be affected by many factors such as the temperature, pH, length and severity of treatment and presence of other ingredients.

6.2.2   Analytical methods

Sampling remains a considerable source of error in the analytical determination of aflatoxins and other mycotoxins. Failure to take representative samples may invalidate subsequent analysis. Specific plans have been developed for commodities such as corn and peanuts and have been included in directives within the European Union (EU). The distribution of aflatoxin M_1 in liquid milk is fairly homogeneous making sampling of this material less onerous.

Analytical methods have been developed based principally on TLC, HPTLC, HPLC, HPLC/Mass Spectrometry and enzyme-linked immunosorbent assay. Aflatoxins fluoresce under UV irradiation, although aflatoxin B_1 and aflatoxin G_1 need derivatisation to enhance the fluorescence to a similar level to that of aflatoxins B_2 and G_2. This forms the basis for their detection by TLC, the original method of choice. On TLC plates the four substances are distinguished on the basis of their fluorescent colour and R_f values, B standing for blue and G for green with subscripts relating to the relative chromatographic mobility. Aflatoxin B_1 is usually, but not always, found in the highest concentration. TLC can be used to identify and quantify aflatoxins at levels as low as 1 µg/kg and it still remains the method of choice where resources are limited. This technique can be further refined by introducing two separation steps (two dimensions) for example developing using a mixture of ether, methanol and water for the first stage and chloroform and acetone for the second step or by using HPTLC.

Extraction of aflatoxins from foods and feeds usually involves the use of solvent systems such as aqueous acetonitrile, methanol or acetone, or mixtures. The use of solid-phase extraction (SPE) with C_{18} cartridges, or immunoaffinity columns (IAC) is now well established for subsequent clean-up in aflatoxin analysis. Mycosep™ columns, which remove matrix components efficiently and can produce a purified extract within a very short time, are also available. Conventional clean-up with silica columns has also been used. Pre- or post-column derivatisation is usually performed for low-level detection, of a wide range of foods and animal feedingstuffs. Derivatisation can take place using Br_2, I_2 (for B and G aflatoxins) or trifluoro-acetic acid (TFA for aflatoxin M_1). The emitted light is detected at 435 nm after

Table 6.4 Selected, collaboratively tested methods for aflatoxins.

Aflatoxins	Commodities	Reference/method
Aflatoxins B_1, B_2, G_1, G_2	Corn, almonds, Brazil nuts, peanuts, and pistachio nuts	Trucksess et al., 1994, IA/HPLC
Aflatoxins B_1, B_2, G_1, G_2	Peanut butter, pistachio paste, fig paste and paprika powder	Stroka et al., 2000, IA/HPLC/Br
B_1	Baby food	Stroka et al., 2001, IA/HPLC
M_1	Milk	Dragacci et al., 2001, IA/HPLC
M_1	Milk	Grosso et al., 2004, IA/TLC

IA, immunoaffinity.

excitation at 365 nm. Small amounts of water tend to give higher extraction efficiencies. The determination of aflatoxins by liquid chromatography with fluorescence detection in food analysis has been reviewed (Jaimez et al., 2000).

Dunne et al. (1993) have described a multi-mycotoxin method, which uses hydrochloric acid and dichloromethane for extraction although halogenated solvents are now undesirable on environmental grounds. Limits of detection below 1 µg/kg can be routinely achieved with an analytical precision of ±30%. Antibodies have been raised to the aflatoxins and a growing number of immunological test systems based on these have been developed in different formats for both quantitative and monitoring purposes (Patel, 2004). Commercially available tests are available to identify and measure aflatoxins in food within a very short time. The user should however be fully aware of any limitations involved.

Automated analysis using immunoaffinity column clean-up and high-performance liquid chromatographic determination was introduced for routine analysis of foods and animal feeds as early as 1991 (Sharman and Gilbert, 1991). A large number of analytical methods have been published for the aflatoxins and the reader should consult the literature further. A selection of those methods examined in inter-laboratory testing is given in Table 6.4. Identification with mass spectrometry (LC/MS) is becoming increasingly popular, particularly for difficult matrices and has been reviewed (Zöllner and Mayer-Helm, 2006). These methods are sensitive enough to operate in the concentration ranges where legal limits have been established for commodities like maize, peanuts, figs and spices. Nevertheless, the application of these methods is expensive and requires expert knowledge. HPLC-MS/MS methods have been reported for the determination of aflatoxins in peanuts, maize feed and whole milk (Pazzi et al., 2005), for aflatoxins in a range of foods (Takino et al., 2004) and for seeking a range of mycotoxins in animal offal food products (Driffield et al., 2003).

6.2.3 Occurrence in raw materials and processed foods

Governments of most developed nations carry out surveillance on a regular basis to monitor the intake of aflatoxins by the human population so that action can be taken if this should become necessary. Aflatoxins in peanuts moving in international trade have on occasions been found at levels of 1000 µg/kg or more, and products such as peanut butter may be contaminated by smaller amounts. Other nuts particularly prone to contamination are pistachios and brazils. Climate plays a crucial part in the conditions that encourage aflatoxin, so that the problem varies in severity from year-to-year. Drought leading to crop stress followed by rain is particularly unwelcome for cereals producers. Maize amongst the cereals is particularly prone to aflatoxins. As cereals are used in the brewing industry beer may also be contaminated (Mably et al., 2005).

Table 6.5 Regulations for aflatoxins in the EU, 2004.

µg/kg	B_1	$B_1 + B_2 + G_1 + G_2$	M_1
Groundnuts, nuts, dried fruit and processed products thereof for direct human consumption or as a food ingredient	2	4	
Groundnuts to be subjected to sorting, or other physical treatment, before human consumption or as a food ingredient	8	15	
Nuts and dried fruit to be subjected to sorting, or other physical treatment, before human consumption or as a food ingredient	5	10	
Cereals and processed products thereof for direct human consumption or as a food	2	4	
Chillies, chilli powder, cayenne pepper, paprika, white and black pepper, nutmeg, ginger and turmeric ingredient	5	10	
Milk (raw milk, milk for the manufacture of milk-based products and heat-treated milk)			0.05
Capsicum; dried fruits whole or ground including chillies, chilli powder, cayenne and paprika	5	10	
Piper; fruits including white and black pepper	5	10	
Nutmeg, ginger and turmeric	5	10	

In practice, any growing crop in which *A. flavus* or *A. parasiticus* can develop is potentially at risk from aflatoxins contamination. This is reflected by regulations in the European Commission (EC) so that in addition to nuts and cereals, dried figs (e.g. Şenyuva *et al.*, 2005), dates, spices, herbs, manioc, cottonseed, copra and even melon seeds (Bankole *et al.*, 2004) are all recognised sources of aflatoxins exposure now subject to regulation (Table 6.5). There are many references to the occurrence of aflatoxins to these and other commodities. Aflatoxins are also found in medicinal herbs and plant extracts (Reif and Metzger, 1995).

6.2.4 Toxicology

Aflatoxins are both acutely and chronically toxic. Aflatoxin B_1 is one of the most potent hepatocarcinogens known (Fishbein, 1979) and hence the long-term chronic exposure of extremely low levels of aflatoxins in the diet is an important consideration for human health. Aflatoxins have been implicated in sub-acute and chronic effects in humans. These effects include primary liver cancer, chronic hepatitis, jaundice, hepatomegaly, cirrhosis through repeated ingestion of low levels of aflatoxin and can also affect the immune system (Pier, 1991). Aflatoxin B_1 is a potent mutagen causing chromosomal aberrations in a variety of plant, animal and human cells.

In the temperate, developed areas of the world, acute poisoning in animals is rare and in man is now extremely unlikely. The outbreak of so-called 'turkey-X disease', which caused the deaths of 100 000 turkeys and other poultry in the UK in 1960, was caused by extremely high concentrations of aflatoxins in imported groundnut meal (Allcroft and Carnaghan, 1962). This alerted industry and governments to the potentially devastating effects of mycotoxins, particularly the aflatoxins.

Acute aflatoxin toxicity has been demonstrated in a wide range of mammals, fish and birds; rabbits, dogs, primates, ducks, turkeys and trout are all highly susceptible. For most species the LD_{50} is between 0.5 and 10 mg/kg body weight (bw). The liver is the principal target organ although the site of the hepatic effect varies with species. Effects on the lungs, myocardium and kidneys have also been observed and aflatoxin can accumulate in the brain. Teratogenic effects following administration of high doses of aflatoxin have been reported in some species (Elis and Di Paola, 1967).

Acute poisoning of man by aflatoxins still occurs occasionally in some areas of the world. Cases of human aflatoxicosis have been reported sporadically, mainly in Africa and Asia. The majority of cases involve consumption of contaminated cereals, most frequently maize, rice or cassava, or cereal products such as pasta or peanut meal (see, for example, Oyelami *et al.*, 1996). A classic case occurred during a Chinese festival in Malaysia in which approximately 40 persons were affected and 13 children died (Chao *et al.*, 1991) after eating noodles highly contaminated with aflatoxin and boric acid. One of the largest recent outbreaks occurred in rural Kenya in April 2004 (Lewis *et al.*, 2005), resulting in 317 recognised cases and 125 deaths. Home-grown maize was the source of the outbreak.

6.2.5 Regulation and control

The complete elimination of aflatoxins in human and animal food, while desirable, is extremely difficult as they have the potential to arise in a wide range of agricultural products. Risk assessments have been carried out for aflatoxin (see, for example, IARC, 1993). Because aflatoxin B_1 is a genotoxic carcinogen, most agencies, including the Joint Expert Committee on Food Additives (JECFA) and the US Food and Drug Administration, have not set a tolerable daily intake (TDI) figure. In common with other dietary carcinogens, it is generally accepted that amounts in food should be reduced to the lowest levels that are technologically possible. Regulations have been set for human food and animal feed in many countries (FAO, 2004). In the EC, aflatoxin is strictly controlled by maximum permissible limits (Table 6.5) and by recommendations for sampling. As an example of what can be achieved over time, regulation of aflatoxin B_1 in animal feedingstuffs in the UK since the 1980s (Anon, 1982) has been effective in steadily reducing amounts of aflatoxin M_1 in milk, as shown by regular surveillance (Ministry of Agriculture, Fisheries and Food, 1980, 1987, 1993).

6.3   OCHRATOXIN A

6.3.1 Chemical properties

The ochratoxins are a group of structurally related compounds of which ochratoxin A is the most important and most commonly occurring. They consist of a polyketide-derived dihydroiso-coumarin moiety linked through the 12-carboxy group to phenylalanine (Fig. 6.2). Ochratoxin A is a colourless crystalline compound, exhibiting blue fluorescence under UV light. It crystallises from benzene to give a product melting at $90°C$ containing one molecule of benzene. This can be removed under vacuum at $120°C$ to give a solid melting at $168°C$. It crystallises in a pure form from xylene. The sodium salt is soluble in water. In the acid form it is moderately soluble in polar organic solvents such as chloroform, methanol and acetonitrile, and dissolves in dilute aqueous sodium bicarbonate. It yields phenylalanine and an optically active lactone acid, ochratoxin α on acid hydrolysis. Reaction in methanol and hydrochloric acid yields the methyl ester, while methylation with diazomethane gives the *o*-methyl methyl ester. It can be stored in ethanol for at least a year under refrigeration and protected from light.

Ochratoxin A is a moderately stable molecule and will survive most food processing to some extent. This has been reviewed by Scott (1996) and the accompanying volume contains a series of papers covering the fate of ochratoxin A during malting and brewing (Baxter, 1996), during bread making (Subirade, 1996), as a result of processing in cereals (Alldrick, 1996), during processing of coffee (Viani, 1996), during processing of meat products (Gareis, 1996) and during processing in animal feed (Scudamore, 1996). During a study of extrusion no more than 40% of ochratoxin A was destroyed under the harshest treatments employed (Scudamore *et al.*, 2004). In a study of the fate of ochratoxin A in the processing of whole wheat grains during milling and bread production the mycotoxin concentrated in the bran and was reduced in the flour while most ochratoxin A survived subsequent baking into bread (Scudamore *et al.*, 2003).

In biological systems, ochratoxin A will bind to serum albumin and for the majority of its lifetime within the human body, OTA remains bound to the plasma protein human serum albumin (Perry *et al.*, 2003).

6.3.2   Analytical methods

Sensitive and reliable analytical methods have been developed with detection limits better than 1 μg/kg. Extraction solvents used for ochratoxin A have included mixtures of chloroform plus orthophosphoric acid, ethyl acetate plus phosphoric acid and acetonitrile plus water. Thin layer chromatography has been widely used (e.g. Krogh and Nesheim, 1982) but this has now been mainly replaced by HPLC- based methods. Clean-up of acid extracts can be achieved by back-partition into sodium bicarbonate although immunoaffinity columns are now often used as the main clean-up stage.

A number of methods have been examined by inter-laboratory testing often as part of the AOAC testing programme. Those developed and tested for cereals include determination of ochratoxin A in barley, wheat bran and rye (Larsson and Möller, 1996), in barley using immunoaffinity columns (Entwisle *et al.*, 2000), in wheat (Scudamore and McDonald, 1998) and in baby foods (Burdaspal *et al.*, 2001). Methods for roasted coffee have also used im-munoaffinity clean-up (Entwisle *et al.*, 2001). The discovery of the presence of ochratoxin A in wine and beer has led to the development and testing of methods specific for these beverages (Visconti *et al.*, 2001a; Leitner *et al.*, 2002). McDonald *et al.* (2003) tested a method for determination of ochratoxin A in currants, raisins, sultanas, mixed dried fruit and dried figs using acidified methanol for extraction. Post-column pH shift by the addition of 1.1 M ammonia solution to the column eluant enhances fluorescence.

Methods have also been developed to determine the occurrence of ochratoxin A in biological fluids such as plasma and urine (Gilbert *et al.*, 2001) and human milk (Miraglia *et al.*, 1995) and this has enabled estimated dietary intakes for babies to be calculated.

6.3.3   Occurrence in raw materials and processed foods

Ochratoxin A is produced by certain strains of *Aspergillus ochraceus* and related species, and by *Penicillium verrucosum*. There are literature reports of other *Penicillia* producing ochratoxin A but most of these are probably as the result of mis-identification. *A. ochraceus* occurs principally in tropical climates while *P. verrucosum* is a common storage fungus in temperate areas such as Canada, Eastern Europe, Denmark, parts of South America and the UK. The main ochratoxin A producing fungus occurring in grapes has shown to be *Aspergillus carbonarius* (Valero *et al.*, 2005).

Ochratoxin A often occurs in stored cereals and has been found in other foods including coffee (Tsubouchi *et al.*, 1988), beer (Payen *et al.*, 1983), dried fruit (Ozay and Alperden, 1991), wine (Zimmerli and Dick, 1996), cocoa (Ministry of Agriculture, Fisheries and Food, 1980), nuts (Cooper *et al.*, 1982), spices (Thirumala-Devi *et al.*, 2001; Fazekas *et al.*, 2005) and liquorice (Bresch *et al.*, 2000). A comprehensive review of the literature on the worldwide occurrence of ochratoxin has been carried out by Speijers and van Egmond (1993). In Europe an assessment of dietary intake of ochratoxin A has been undertaken under the 'SCOOP' programme (European Commission, 2002a). This involved collection of data on occurrence in food using methodology considered as soundly based. This together with consumer consumption data enabled the dietary intake of ochratoxin A to be calculated. Estimate of dietary intake on the basis of ochratoxin A level in serum/plasma has also been carried out.

6.3.4 Toxicology

Ochratoxin A is a potent carcinogenic mycotoxin that can affect kidneys, the immune system and the nervous system. The kidney is the most sensitive target organ. However, its dechloro derivative, ochratoxin B, is non-toxic. A nephrotoxic effect has been demonstrated in all mammalian species tested to date (Harwig *et al.*, 1983). In acute toxicity studies LD_{50} values vary greatly in different species, the dog and pig being especially susceptible.

The European Food Safety Authority (EFSA) published an opinion on ochratoxin A (EC, 2006a). They concluded that 'although some early epidemiological data had suggested that ochratoxin A might be involved in the pathogenesis of distinct renal diseases and otherwise rare tumours of the kidneys in certain endemic regions of the Balkan Peninsula these epidemiological data were incomplete and did not justify the classification of ochratoxin A as a human renal carcinogen'. Ochratoxin A has been found to be a potent renal toxin in all of the animal species tested, the dog being the most sensitive. The extent of renal injury is dose-dependent, but also associated with the duration of exposure, as ochratoxin A accumulates in renal tissue.

Recent scientific evidence indicates that the site-specific renal toxicity as well as the DNA damage and genotoxic effects of ochratoxin A, measured in various in vivo and in vitro studies, are most likely attributable to cellular oxidative damage. Furthermore, advanced chemical analytical procedures have failed to demonstrate the existence of specific ochratoxin A–DNA adducts.

Human exposure to ochratoxin A has been clearly demonstrated by its detection in blood (e.g. Bauer *et al.*, 1986; Breitholtz *et al.*, 1991; Palli *et al.*, 1999; Filali *et al.*, 2002; Sangare-Tigori *et al.*, 2006) and breast milk (Miraglia *et al.*, 1993).

6.3.5 Regulation and control

The presence of ochratoxin A in foodstuffs is clearly undesirable, although few countries have introduced statutory control to date (Food and Agriculture Organisation of the United Nations, 2004). Most attention to legislation for ochratoxin has been within the EU and a summary of current limits for ochratoxin A in cereals and other products is given (Table 6.6) although limits are currently being introduced and modified as more information on occurrence and toxicology is acquired. A full description of limits, recommended sampling protocols and analytical methods is provided (EC, 2005a,b).

The prevention and control of ochratoxin A in cereals depends on the crop being dried promptly at harvest and good store hygiene. In general ochratoxin A develops slowly in grain

Table 6.6 Maximum limits for ochratoxin A in raw cereal grain and finished products intended for human consumption, in the EU as of October 2006.

Product	Ochratoxin A (μg/kg)
Raw cereal grains (including raw rice and buckwheat)	5.0
All products derived from cereals (including processed cereal products and cereal grains intended for direct human consumption)	3.0
Baby foods and processed cereal-based foods for infants and young children	0.5
Dried vine fruits (currants, raisins, sultanas)	10.0
Roasted coffee beans and ground roasted coffee except soluble coffee	5.0
Wine and other wine and/or grape must based beverages	2.0
Green coffee, dried fruit other than vine fruit, beer, cocoa, cocoa products, liqueur wines, meat products, spices, liquorice	No current limits

below 18% moisture content but this depends on temperature, grain water activity, extent of infection with P. verrucosum and the length of time the cereal remains under such conditions. When grain is dried by ambient air the bottom layer dries first while the upper layers may remain damp for much longer. The availability of adequate drying capacity is important, hot air facilities being preferable. Even after drying, cereal bulks should be monitored regularly for temperature, moisture content and insect activity and action taken if it becomes necessary. For grapes, the critical control points for the control of ochratoxin A in the grape-wine chain have been established (Battilani *et al.*, 2003).

In addition, contamination of animal feeds with ochratoxin A may result in the presence of residues in edible offal and blood serum, whereas the ochratoxin A contamination in meat, milk and eggs is negligible.

6.4 DEOXYNIVALENOL AND THE TRICHOTHECENES

6.4.1 Chemical properties

The 12,13-epoxytrichothecenes (Fig. 6.3) are a group of related and biologically active mycotoxins produced by certain species of *Fusarium*, such as *F. poae, F. sporotrichioides, F. culmorum* and *F. graminearum*. They are often classified as Group A and Group B compounds depending on whether they have a side chain on the C-7 atom. The most commonly reported Group A trichothecenes include T-2 toxin, HT-2 toxin, neosolaniol, monoacetoxyscirpenol and diacetoxyscirpenol. These are highly soluble in ethyl acetate, acetone, chloroform, methylene chloride and diethyl ether.

Deoxynivalenol (commonly called DON or vomitoxin) is the most widely occurring Group B trichothecene, together with nivalenol, 3-acetyldeoxynivalenol, 15-acetyldeoxynivalenol, fusarenone-X, scirpentriol and T-2 tetraol. These compounds are highly hydroxylated and thus relatively polar, being soluble in methanol, acetonitrile and ethanol.

Another group of trichothecenes that is generally more acutely toxic than T-2 toxin is known as the macrocyclic trichothecenes. These are produced by mould species such as *Stachybotrys atra* and include the satratoxins, verrucarins and roridins, which may be produced in hay and straw stored under unsatisfactory conditions. However, there is little evidence that these compounds occur in human food, although the presence of macrocyclic trichothecenes in airborne fungal spores may contribute to some forms of sick building syndrome (Croft *et al.*, 1986).

All trichothecenes containing an ester group are hydrolysed to their respective parent alcohols when treated with alkali. A dilute solution of potassium carbonate, sodium hydroxide or ammonium hydroxide hydrolyses T-2 toxin and neosolaniol to T-2 tetraol and diacetoxy- and monoacetoxy-scirpenol to scirpentriol. Many of the alcohols are unaffected, even by hot dilute alkali. Trichothecenes are thus chemically stable and can persist for long periods once formed. Prolonged boiling in water or under highly acidic conditions causes a skeletal rearrangement due to opening of the epoxide ring. Owing to the hindered nature of the epoxide and stability of the ring system, reactions of the trichothecenes usually proceed in a manner predictable from sound chemical principles. For example, primary and secondary hydroxyl groups are easily oxidised to the aldehyde and ketone derivatives by reagents such as CrO_3–H_2SO_4 in acetone, CrO_3–pyridine and CrO_3–acetic acid.

6.4.2 Analytical methods

The determination of trichothecene mycotoxins is complicated by the number of closely related compounds that can occur together and the low UV absorbance properties of the molecules. Combinations of solvents, usually acetonitrile/water and methanol/water, are used for the extraction of grain, food, and feeds. A variety of solid-phase materials such as, silica gel, florisil, cyano and C_{18} SPE-cartridges are used for subsequent clean-up. DON and T-2 toxin immunoaffinity columns are also now available.

Methods based on TLC are still common especially where resources are limited and, separation efficiency and precision have been increased with the introduction of high-performance TLC (HPTLC) and scanning densitometers. Concentrated sulphuric acid and *p*-anisaldehyde and other derivatisation reagents are often used to give characteristic colours, which aid detection (Mirocha *et al.*, 1977).

Several HPLC methods have been published for the determination of trichothecenes in food and cereals such as that for baby foods and animal feed (Stroka *et al.*, 2006). However, different methods of analysis may be required for type A and B trichothecenes, e.g. HPLC with UV detection is not usually applicable to type A trichothecenes as they lack a keto group at the C-8 position. In these circumstances MS-based methods may be the preferred methods for the determination of type A trichothecenes. However, low wavelength UV detection is used in DON HPLC methods although sensitivity is poor, lack of specificity being the main problem. The earlier GC methods are based on derivatisation procedures such as trimethylsilylation or fluoroacylation to increase volatility and sensitivity with electron capture detection or the use of mass spectrometric or tandem mass spectrometric detection (MS/MS). The choice of derivatisation reagent depends on the type of trichothecene and the method of detection.

A number of studies have been carried out into improving and verifying trichothecene analysis. Pettersson and Langseth (2002) carried out inter-comparisons of GC/MS methods and studied the factors that influence method performance as a pre-requisite for producing certified calibrants. Mateo *et al.* (2001) published a critical study of, and improvements in, chromatographic methods for the analysis of type B trichothecenes. In a study in which laboratories used their own in-house methods for the determination of the *Fusarium* mycotoxins zearalenone and DON in common wheat and maize samples, the results were deemed acceptable for DON at the 100 µg/kg level (Josephs *et al.*, 2001).

LC/MS with either atmospheric pressure chemical ionisation (APCI) or electrospray ionisation or tandem mass spectrometry (MS/MS) have been successfully employed for the determination and identification of trichothecenes at trace levels (Biselli and Hummert, 2005;

Klotzel *et al.*, 2005; Klotzel *et al.*, 2006). Trace mycotoxin analysis in complex biological and food matrices using LC/MS has been comprehensively reviewed (Zöllner and Mayer-Helm, 2006).

A variety of test kits are now marketed for the rapid detection of individual trichothecenes, mostly for DON and T-2 toxin. These are particularly useful for screening purposes and may be set to provide a yes/no answer at a pre-set level or in ELISA formats. If necessary, samples can then be checked using fully quantitative methods. Rapid methods for DON and other trichothecenes have been reviewed (Schneider *et al.*, 2004).

6.4.3   Occurrence in raw materials and processed foods

Trichothecenes occur in cereals and other food commodities. There are many reports of DON and nivalenol in cereals although T-2 toxin and HT-2 toxin are found much less frequently except in oats. Surveillance commonly targets DON only although other trichothecenes are likely to be present.

A collection of occurrence data for *Fusarium toxins* was carried out in food in the EC (European Commission, 2003) and enabled a preliminary assessment of dietary intake by the population of member states to be carried out (Schothorst and van Egmond, 2004). By far most of the occurrence data were obtained for DON in wheat. Among cereals, corn (maize) showed the highest level of contamination with trichothecenes. There was a significant lack of consumption data in some countries. In particular, information on baby and children's food was generally not available. Wheat and wheat containing products (like bread and pasta) represented the major source of intake for the four trichothecenes, DON, nivalenol, HT-2 toxin and T-2 toxin. DON and related compounds were found in a survey of Soy products in Germany using an MS method (Schollenberger *et al.*, 2007).

Studies have suggested that DON may sometimes be present in cereals in a bound forms and an acid solvolysis procedure may be required to release this bound mycotoxin (Liu *et al.*, 2005b).

There are occasional reports of trichothecenes in commodities other than cereals and in view of the widespread distribution of *Fusarium* spp., this remains a possibility.

6.4.4   Toxicology

The acute toxicity of the trichothecenes varies considerably and LD_{50} values for mice (intraperitoneal route) for some trichothecenes are given in Table 6.7 LD_{50} (mg/kg bw). The LD_{50} value for DON is about ten times that for nivalenol, T-2 toxin and HT-2 toxin that are in turn about ten times greater than for the macrocyclic mycotoxins. Fortunately, these have rarely been reported in food. Acute trichothecene toxicity is characterised by gastrointestinal

Table 6.7 LD_{50} values for mice (ip) for some trichothecenes.

Trichothecene	LD50 (mg/kg bw)
Deoxynivalenol	70
Diacetoxyscirpenol	23
Neosolaniol	14.5
HT-2 toxin	9.0
T-2 toxin	5.2
Nivalenol	4.1
Verrucarin A	0.5

Table 6.8 EU Regulations for DON, EC as from 1st July 2007.

Cereal product	Limit (μg/kg)
Unprocessed cereals other than durum wheat, oats and maize	1250
Unprocessed durum wheat and oats	1750
Unprocessed maize with the exception of unprocessed maize intended to be processed by wet milling	1750
Pasta (dry). Cereals intended for direct human consumption, cereal flour, bran and germ as end product marketed for direct human consumption with the exception of foodstuffs listed below	750
Milling fractions of maize with particle size >500 micron falling within CN code 1103 13 or 1103 20 40	750
Milling fractions of maize with particle size ≤500 micron falling within CN code 1102 20	1250
Bread (including small bakery wares), pastries, biscuits, cereal snacks and breakfast cereals	500
Processed cereal-based food for infants and young children and baby food	200

disturbances, such as vomiting, diarrhoea and inflammation, dermal irritation, feed refusal, abortion, anaemia and leukopenia. This group of toxins is acutely cytotoxic and strongly immunosuppressive. The trichothecenes have not been shown to be mutagenic or carcinogenic but do inhibit DNA and protein synthesis.

DON is a common contaminant of cereals and causes vomiting in pigs at relatively low concentrations. However, pigs are very sensitive to its presence and will reject contaminated feed, effectively limiting any further toxic effects. DON is, however, immunosuppressive in low concentrations and this may be more important than its low acute toxicity. Because of the number of closely related metabolites likely to occur in combination in foods or animal feeds, the toxicology is complex with both synergistic and antagonistic effects. This has been discussed by Miller (1995).

Alimentary toxic aleukia (ATA) has probably been the most common human trichothecene mycotoxicosis. T-2 toxin is thought to have contributed to the epidemiology of ATA in Russia, which was responsible for widespread disease and many deaths. Continuous exposure to trichothecenes results in skin rashes, which may proceed to necrotic lesions.

6.4.5 Regulation and control

Legislation for DON is set in a number of countries (FAO, 2004) and the introduction of internationally agreed legislation is expected to continue. Legislation for DON in cereals that was introduced during 2006 and 2007 within the EU is summarised (Table 6.8). In addition, Statutory or Guideline limits for T-2 toxin and HT-2 toxin in cereals are expected to be introduced during 2008. Codes of practice for minimising the occurrence of DON and the trichothecenes are being developed by international bodies such as CODEX/WHO/FAO and within the EU. Some basic rules can be summarised; avoid growing maize as the previous crop, reduce crop debris on soil surface, avoid susceptible varieties, apply a T3 fungicide against *Fusarium* and avoid delays in drying at harvest.

6.5   ZEARALENONE

6.5.1   Chemical properties

Zearalenone (Fig. 6.4) is a phenolic resorcyclic acid lactone produced by a number of species of *Fusarium* including *F. culmorum*, *F. graminearum* and *F. crookwellense*. In fungal cultures a number of closely related metabolites are formed but there is only limited evidence that these occur in foodstuffs, although there is experimental evidence for some transmission of zearalenone and α- and β-zearalenols into the milk of sheep, cows and pigs fed high concentrations (Mirocha *et al.*, 1981).

Zearalenone is a white crystalline compound that exhibits blue-green fluorescence when excited by long wavelength UV light (360 nm) and a more intense green fluorescence when excited with short wavelength UV light (260 nm). In methanol, UV absorption maxima occur at 236 nm ($\varepsilon = 29\,700$), 274 nm ($\varepsilon = 13\,909$) and 316 nm ($\varepsilon = 6020$). Maximum fluorescence in ethanol occurs with irradiation at 314 nm and with emission at 450 nm. Solubility in water is about 0.002 g/100 mL. Zearalenone is slightly soluble in hexane and progressively more so in benzene, acetonitrile, methylene chloride, methanol, ethanol and acetone. It is also soluble in aqueous alkali. It appears to be very stable in many processes.

6.5.2   Analytical methods

Ground cereal samples are usually extracted using a suitable solvent such as acetonitrile, chloroform, ethyl acetate or methanol and water or acidic solutions. TLC initially provided an acceptable method in laboratories with limited instrumentation and a recent improvement has been introduced by using HPTLC (Ostry and Skarkova, 2003). Liquid–liquid partitioning or solid-phase extraction is necessary to provide clean extracts for separation and detection. Recently commercially available clean-up columns have become a widely used clean-up medium (Visconti and Pascale, 1998; Kruger *et al.*, 1999). An HPLC immunoaffinity column method for barley, maize and wheat flour, polenta, and maize-based baby food has shown to meet required analytical criteria in an interlaboratory study (MacDonald *et al.*, 2005).

In a comparative study, mixtures of methanol and 1% aqueous NaCl were found to be the best extraction solvent while immunoaffinity columns were very effective for clean-up although of lower capacity than solid phase extraction columns (Llorens *et al.*, 2002).

There have been several recent developments for the determination of low concentrations of zearalenone in foods and body liquids or tissues including an automated flow-through immunosensor method for the determination of zearalenone in pig feed (Urraca *et al.*, 2005). Among a number of LC/MS methods, Pallaroni and von Holst (2003) determined zearalenone in wheat and corn using pressurised liquid extraction and liquid chromatography–electrospray mass spectrometry, a method based on LC-APCI-MS was developed for the determination of zearalenone and its metabolites in urine, plasma and faeces of horses (Songsermsakul *et al.*, 2006) and for bovine milk by LC/MS/MS (Sorensen and Elbaek, 2005).

Molecular imprint technology produced by developing a polymer template that binds the zearalenone molecule has been used to provide a clean-up method for zearalenone and alpha-zearalenol in cereal and swine feed sample extracts (Urraca *et al.*, 2006). Krska *et al.* (2005) has reviewed recent advances in zearalenone methodology.

6.5.3 Occurrence in raw materials and processed foods

Zearalenone occurs in cereal grains especially maize and can survive processing into cereal products (Williams, 1985) such as maize beer (Lovelace and Nyathi, 1977), wheat flour (Tanaka *et al.*, 1985). It has also been reported in walnuts (Jemmali and Mazerand, 1980). The presence of zearalenone in whole plants and parts of maize used for silage making has been investigated in Germany (Oldenberg, 1993). In that work, zearalenone was detected at concentrations up to 300 µg/kg and mainly formed at the end of the ripening process, with subsequent contamination of the silage. In a survey in Germany, zearalenone has been found in small amounts in vegetables, fruits, oilseeds and nuts (Schollenberger *et al.*, 2005). Information on the occurrence of zearalenone within the EU has been collated and summarised (Gareis *et al.*, 2003).

Zearalenone is only partly decomposed by heat. Approximately 60% of zearalenone remained unchanged in bread while about 50% survived the production of noodles (Matsuura and Yoshizawa, 1981). In dry milling of maize, concentrations in the main food-producing fractions, including flour and grits, were reduced by 80–90% although increased concentrations were found in bran and germ (Bennett *et al.*, 1976). Its distribution in wet-milled corn products is reviewed by Bennett and Anderson (1978).

6.5.4 Toxicology

The most important effect of zearalenone is on the reproductive system, its acute toxicity being low. The ability of zearalenone to cause hyperestrogenism, particularly in swine, has been known for many years. Two comprehensive reviews of this have been published by Mirocha *et al.* (1971) and Mirocha and Christensen (1974). In New Zealand, zearalenone in pasture grass is a recognised cause of infertility in sheep (Towers and Sprosen, 1992). Trial feeding of female pigs demonstrated that a concentration of 0.25 mg/kg, or even less, produced changes in the reproductive organs (Bauer *et al.*, 1987).

In a review by the International Agency for Research on Cancer (IARC, 1993), it was concluded that there was limited evidence in experimental animals for the carcinogenicity of zearalenone. Evidence for genotoxicity has been contradictory but Pfohl-Leskowicz *et al.* (1995) showed that zearalenone is genotoxic in mice.

A risk assessment of the mycotoxin has been carried out (Kuiper-Goodman *et al.*, 1987) and that paper provides a comprehensive review of the literature up to that date. A recent comprehensive review of the toxicity, occurrence, metabolism, detoxification, regulations and intake of zearalenone has been published (Zinedine *et al.*, 2006). An opinion on zearalenone was published in 2000 (European Commission, 2000) and for animals in 2004 (EC, 2006b).

6.5.5 Regulation and control

Initial legislation for zearalenone in cereals was introduced from July 2006 within the EU. Prior to this legislation existed in about 12 countries worldwide (FAO, 2004). These limits were modified and introduced in July 2007 (EC, 2007). In summary, these limits range from 100 µg/kg in unprocessed cereals other than maize, to 350 µg/kg in maize, 75 µg/kg in cereals intended for direct human consumption, cereal flour, bran and germ marketed for direct human consumption, down to 20 µg/kg in processed cereal foods for infants and young children. The reader should consult the EC document above for full details of these limits. Because zearalenone often occurs along with other *Fusarium* mycotoxins codes of practice to prevent its formation are similar, e.g. to DON.

6.6   FUMONISINS

6.6.1   Chemical properties

The fumonisins (Fig. 6.4) are a group of at least 15 closely related mycotoxins that often occur in maize and less commonly in other products, the most important compound being fumonisin B_1. They are polar metabolites formed by several species of *Fusarium*, but mainly *F. verticilliodes* (Sacc.) Nirenberg (= *F. moniliforme* Sheldon). Their structures are based on a long hydroxylated hydrocarbon chain. Two hydroxyl groups are esterified to two propane-1,2,3-tricarboxylic acids. Fumonisin B_1 differs from fumonisin B_2 in that it has an extra hydroxyl group at the 10-position.

Fumonisins contain four free carboxyl groups and an amino group, which accounts for their solubility in water and some polar organic solvents. Fumonisin B_1 is stable in acetonitrile–water (1:1) over a 6-month period at 25°C (Visconti *et al.*, 1993) and in methanol if stored at −18°C but steadily degrades at 25°C and above. Fumonisin B_1 is hydrolysed under some conditions and may bind to food constituents. The pure substance is a white hygroscopic powder. The insolubility in many organic solvents partly explains the difficulty in their original identification.

A review of the occurrence and toxicity of the fumonisins is that of Dutton (1996) while the history and discovery of this group of mycotoxins has been described (Marasas, 2001).

6.6.2   Analytical methods

TLC, HPLC, MS/MS and immunochemical methods have been reported for the fumonisins. Lack of a suitable chromophore in the molecule means that fumonisins must be derivatised with reagents such as *p*-anisaldehyde, fluorescamine or *o*-phthaldialdehyde to allow detection by TLC or HPLC. Currently, HPLC methods, mainly employing fluorescence detection together with sample pre-treatment, are the methods of choice for routine analysis, and an overview has been presented for the determination of the fumonisin mycotoxins in various matrices (Arranz *et al.*, 2004). Reliable immunoaffinity clean-up columns are available and becoming used widely.

The first collaborative study of a method for liquid chromatographic determination of fumonisins B_1, B_2 and B_3 in corn was a joint AOAC–IUPAC collaborative trial (Sydenham *et al.*, 1996). Visconti *et al.* (2001b) presented an in-house and inter-laboratory validation study for the determination of fumonisin B_1 and B_2 in corn-based foodstuffs including cornflakes by HPLC using immunoaffinity clean-up.

Stability and problems in the recovery of fumonisins added to corn-based foods has been known to be a problem for many years (Scott and Lawrence, 1994; Avantaggiato *et al.*, 2003). The analysis of heat-processed corn foods such as cornflakes, corn-based breakfast cereals, tortilla chips and corn chips can in addition present further special problems in analysis as hydrolysed fumonisin B_1 as well as protein- and total-bound fumonisin B_1 can be present as well as the parent fumonisins. Bound (hidden) fumonisins cannot be detected by conventional analysis and improved methods for the determination of bound fumonisin B_1 were developed (Park *et al.*, 2004). A number of studies have been carried out recently developing LC/MS methods (e.g. Cavaliere *et al.*, 2005; Faberi *et al.*, 2005; Paepens *et al.*, 2005).

Determination of total fumonisins in corn by competitive direct ELISA (Bird *et al.*, 2002) and commercial rapid test kits are suitable for use in monitoring situations.

6.6.3   Occurrence in raw materials and processed foods

In the years since fumonisins were first identified there have been many surveys reporting fumonisins in harvested maize sometimes in very high concentrations. Worldwide occurrence of fumonisins in maize has been reviewed (Shepherd et al., 1996) while a more recent review of their occurrence in foods has been carried out (Soriano and Dragacci, 2004). While maize is by far the most important commodity in which fumonisins occur, there have been reports of their occurrence in other commodities, usually in much lower concentrations. These include wheat, barley and soybean in Spain (Castella et al., 1999) and other warm growing areas. However, a recent re-examination of wheat from South Africa claimed to contain fumonisins as high as 1.7 mg/kg failed to find any evidence of their presence (Shepherd et al., 2005).

Fumonisins have been reported to be quite stable and not broken down even by moderate heat (Bordson et al., 1995). However, no fumonisins were detected in tortilla flour made by treatment with calcium hydroxide (nixtamilisation) and it has been suggested that this process degrades fumonisins (Sydenham et al., 1991). Scott and Lawrence (1994) succeeded in obtaining about 80% reduction in fumonisin B_1 by heating at higher temperatures. However, even when fumonisin appears to be destroyed a breakdown product (hydrolysed fumonisin) can be formed and is reported to be just as toxic as the parent compound (Hopkins and Murphy, 1993).

Fumonisins have also been reported in tea and medicinal plants in Portugal (Martins et al., 2001) and Turkey (Omurtag and Yazioglu, 2004), and in asparagus in China (Liu et al., 2005a)

6.6.4   Toxicology

Although fumonisins were not formally identified until 1988 (Bezuidenhout et al., 1988; Gelderblom et al., 1988), the effects of these compounds were observed in a sporadic fatal disease in horses and related species called equine leucoencephalomalacia (ELEM). Affected animals commonly lose appetite, become lethargic and develop neurotoxic effects after a period of ingesting contaminated feed. Autopsy shows oedema in the brain and liquefaction of areas within the cerebral hemispheres.

On a weight-for-weight basis, fumonisins are however far less acutely toxic than the aflatoxins. In contrast, fumonisins commonly occur in concentrations of mg/kg (parts per million) in maize (Shephard et al., 1996), and up to 300 mg/kg has been reported (Fazekas and Tothe, 1995), whereas aflatoxins are usually measured at concentrations of μg/kg (parts per billion in foods).

Fumonisins are considered toxic due to their effects on sphingolipid synthesis (Riley et al., 1993). Alteration in sphingolipid base ratios occurs almost immediately after exposure because fumonisin inhibits ceramide synthetase (Wang et al., 1992). This property is indicative of fumonisin exposure in a number of species including horses and pigs. Animal studies with ^{14}C-labelled fumonisin B_1 generally show the uptake to be poor and elimination rapid. The effect of fumonisins on mammals appears to be species related. In pigs, fumonisins induce pulmonary oedema and hydrothorax, with thoracic cavities filled with a yellow liquid. There may also be respiratory problems and foetal mortality.

In rats, fed material from *F. moniliforme* cultures, primary hepatocellular carcinomas were produced (Gelderblom et al., 1988; Gelderblom and Snyman, 1991). These results were reproduced using purified fumonisins B_1, B_2 and B_3 (Gelderblom et al., 1993, 1994). However, experimental carcinogenicity studies have been hampered by lack of pure standards.

Lebepe-Mazur *et al.* (1995a,b) showed that fumonisin B_1 affected the foetus in pregnant rats, causing low litter weights and foetal bone development as compared with controls. Šegvić and Pepeljnjak (2001) have reviewed their effects on the health of livestock.

In humans, there appears to be a link in some maize-consuming areas of the world between fumonisin toxicity and the occurrence of oesophageal cancer. Further epidemiological studies are required to define more precisely the role of *F. moniliforme* and its metabolites in oesophageal cancer. The incidence of this form of cancer is high in the Transkei, China, and in northern Italy. Many studies of the toxicology of fumonisins are in progress. The full significance of fumonisins in maize for human and animal health still remains to be determined.

6.6.5   Regulation and control

In 2003, only four countries had limits for fumonisins of either 1000 or 3000 μg/kg (FAO, 2004). After consideration of all the available information, the EC introduced a comprehensive range of limits for the total of fumonisin B_1 and fumonisin B_2 (combined) in maize and maize products during 2007. In summary, limits set range from 4000 μg/kg in unprocessed maize (but not maize destined for wet milling), to 1000 μg/kg in maize intended for direct human consumption, 800 μg/kg in breakfast cereals, down to 200 μg/kg in processed cereal foods for infants and young children. The reader should consult the appropriate document for full details of these limits and regulations (EC, 2007).

6.7   PATULIN

6.7.1   Chemical properties

Patulin is a polyketide lactone (Fig. 6.5) produced by certain species of *Penicillium, Aspergillus* and *Byssochlamys*. It can be isolated as colourless to white crystals from ethereal extracts that have no optical activity, melts at about 110°C and sublimes in high vacuum at 70–100°C. It is soluble in water, methanol, ethanol, acetone and ethyl or amyl acetate, and less soluble in diethyl ether and benzene. It undergoes all the reactions expected of a secondary alcohol, reduces warm Fehling's solution and decolourises potassium permanganate. It is stable in acid solutions but can be decomposed by boiling in 2 N sulphuric acid for 6 hours. It is susceptible to alkaline hydrolysis, reduced by sulphur dioxide and by fermentation.

6.7.2   Analytical methods

Analysis of patulin is usually by HPLC using extraction with ethyl acetate and UV detection and there are many methods described in the literature. An earlier ISO standard for determination of patulin in apple juice, apple juice concentrates and drinks containing apple juice is based on TLC (ISO, 1993). Chloroform is preferred for storage of patulin solutions, as it tends to decompose in distilled water.

An HPLC method for the determination of patulin in clear and cloudy apple juices and apple puree has been subject to collaborative study (MacDonald *et al.*, 2000). HORRAT values for patulin ranged from 0.5 to 1.3 that indicates an acceptable precision and these were comparable to or better than values reported in the then AOAC–IUPAC Official First

Action Method (AOAC International, 1995). This paper also provides a brief review of other intercomparison trials carried out for determination of patulin. An interlaboratory study of an HPLC method for quantitation of patulin at 10 ng/mL in apple-based products intended for infants showed acceptable within-laboratory and between-laboratory precision for each matrix, as required by current European legislation (Arranz *et al.*, 2005). Another recent HPLC method is that of Iha and Sabino (2006). 5-Hydroxymethylfurfural (HMF) often occurs in products susceptible to patulin contamination and care is needed to ensure this compound is separated from patulin during HPLC to avoid possible misidentification.

Mass spectrometric methods are increasingly being developed using APCI (Sewram *et al.*, 2000; Ito *et al.*, 2004).

Attempts to develop rapid tests suitable for routine monitoring of juices based on antibody technology have proved difficult because the small molecular size of patulin hinders the raising of the highly specific antibodies and this remains a challenge yet to be met.

6.7.3 Occurrence in raw materials and processed foods

An assessment of the dietary intake of patulin by the population of the EU Member States has been carried out (European Commission, 2002b). Patulin is found in fruits, including apples, pears, grapes and melons. *P. expansum* appears to be the mould usually responsible for patulin in apple juice. In whole fruits, visual inspection will usually identify poor quality items. The principal risk arises when unsound fruit is used for the production of juices and other processed products. Patulin has also been reported in vegetables, cereal grains and silage (Beretta *et al.*, 2000). Apples are by far the most important fruit contaminated by patulin and this can transfer to apple-based foods. However, when sulphur dioxide is used as a food preservative in fruit juice or other foods, patulin will be broken down. It is not usually found in alcoholic beverages or vinegar because it is destroyed during fermentation but has been found in 'sweet' cider in which unfermented apple juice is added to the cider. In acid conditions, it is relatively stable to heat processes up to about 100°C.

6.7.4 Toxicology

Patulin possesses wide-spectrum antibiotic properties and has been tested in humans to evaluate its ability to treat the common cold. However, its effectiveness has never been proven and its use to treat medical conditions has not been pursued because it irritates the stomach, causing nausea and vomiting. In acute and short-term studies patulin causes gastrointestinal problems, haemorrhaging and ulceration. For patulin, the LD_{50} for the rat has been reported as 15 mg/kg bw (Broom *et al.*, 1944) and 25 mg/kg after subcutaneous injection (Katzman *et al.*, 1944). Death was usually caused by pulmonary oedema. Patulin injected in large amounts over a 2-month period was carcinogenic, resulting in induction of sarcomas at the injection site (Dickens and Jones, 1961). In long-term studies at lower dose levels these effects were not observed. Patulin has also been shown to be immunotoxic and neurotoxic. The International Agency for Research on Cancer (IARC, 1986) concluded that no evaluation could be made of the carcinogenicity of patulin to humans and that there was inadequate evidence in experimental animals. Based on reproduction and long-term carcinogenicity studies in rats and mice JECFA allocated a Provisional Tolerable Weekly Intake of 7 µg/kg bw.

6.7.5 Regulation and control

Patulin exposure is often controlled by the setting of a 'guideline' or a 'recommended' maximum concentration, e.g. typically 50 μg/L or μg/kg. Statutory regulation for patulin is still restricted to a few countries but is expected soon within the EU. A code of practice for the prevention and reduction of patulin contamination in apple juice and apple juice ingredients in other beverages has been published by CODEX (CA/RCP 50-2003). One of its many recommendations not surprisingly is to ensure the use of sound good quality apples.

6.8   OTHER MYCOTOXINS

There is an extensive range of fungal metabolites that have been detected in laboratory cultures, but few of these have been found in food commodities. Smith *et al.* (1994) reviewed mycotoxins in human health and the reader is recommended to consult this and later works for more information on the following lengthy, but incomplete, list: beauvericin, citreoviridin, enniatins, ergot alkaloids, fusaric acid, fusarin C, fusaproliferin,gliotoxin, lolitrem B, mycophenolic acid, 3-nitropropionic acid, penicillic acid, PR-toxin, roquefortine C, rubratoxin A, satratoxins G and H, sporidesmin, viomellein, vioxanthin and xanthomegnin and toxins from *Aspergillus fumigatus* and *A. clavatus*.

To present a risk for human health such compounds must survive processing and be present in food ingested by the consumer, in toxicologically significant amounts. Animals and livestock are at greater potential risk as feedstuffs are often fed directly or with less processing than is normal for humans. The possibility of yet undetected mycotoxins of significance should not be dismissed. The fumonisins, for example escaped identification until the mid-1980s although their effects had been known for a long time.

6.8.1   Citrinin

Citrinin was first isolated as a pure compound from a culture of *Penicillium citrinum* in 1931. Its structure is shown in Fig. 6.2. Many species of *Penicillium* have been reported to produce citrinin (Frisvad, 1989) including *P. verrucosum*. Hence, citrinin often occurs together with ochratoxin A, but is more readily lost in analytical procedures and is sought much less frequently. *Aspergillus terreus, A. carneus* and *A. niveus* may also produce this mycotoxin. Citrinin has mainly been found in rice and other cereals.

Citrinin crystallises as lemon-coloured needles melting at 172°C. It is sparingly soluble in water but soluble in dilute sodium hydroxide, sodium carbonate or sodium acetate, in methanol, acetonitrile, ethanol and most other polar organic solvents. Some photodecomposition occurs in fluorescent light both in solution and in the solid state. It can be degraded in acid, or alkaline solution. Colour reactions include brown with ferric chloride, green with titanium chloride and deep wine-red with hydrogen peroxide followed by alkali. Mono-acetate, diethyl, methyl ester and dihydro derivatives can be prepared.

There is little information on the fate of citrinin during processing but it is likely to be degraded by heat and alkali. However, even though citrinin may be destroyed, toxic breakdown products have been demonstrated (Trivedi *et al.*, 1993).

Citrinin causes kidney damage and mild liver damage in the form of fatty infiltration. A review of citrinin toxicity is given by Scott (1977). Citrinin often co-occurs with ochratoxin and has been implicated in mycotoxic nephropathy of pigs (Krogh *et al.*, 1973). It seems unlikely that citrinin presents much risk to humans.

6.8.2 Sterigmatocystin

Sterigmatocystin (Fig. 6.6) is a toxic metabolite that is closely related in structure and toxicology to the aflatoxins (Ohtsubo *et al.*, 1978; Ueno and Ueno, 1978) and is mainly produced by the fungi *Aspergillus nidulans* and *A. versicolor*. The physical properties of sterigmatocystin are summarised (Table 6.1) together with those of several other mycotoxins reviewed in subsequent parts of this chapter. It crystallises as pale yellow needles and is readily soluble in methanol, ethanol, acetonitrile, benzene and chloroform. It reacts with hot ethanolic KOH and is methylated by methyl sulphate and methyl iodide. Methanol or ethanol in acid produces dihydro-ethoxysterigmatocystin.

The acute toxicity, carcinogenicity and metabolism of sterigmatocystin have been compared with those for aflatoxin and several other hepatotoxic mycotoxins (Wannemacher *et al.*, 1991). Cattle exhibiting bloody diarrhoea, loss of milk production and in some cases death were found to have ingested feed containing *A. versicolor* and high levels of sterigmatocystin of about 8 mg/kg (Vesonder and Horn, 1985).

Sterigmatocystin has been found in mouldy grain, green coffee beans and cheese, although information on its occurrence in foods is limited. It occurs much less frequently than the aflatoxins, although analytical methods for its determination are less sensitive. A survey of 1580 samples including wheat, corn and rice taken from 12 provinces located in east-north, west-north, west-south, east and central China showed the average contaminated content of sterigmatocystin in wheat, corn and rice to be 68.9 μg/kg, 32.2 μg/kg and 13.9 μg/kg with a contamination rate of 98%, 89% and 72%, respectively (Tian and Liu, 2004).

An analytical method for the determination of sterigmatocystin in cheese, bread and corn products, using HPLC with atmospheric pressure ionisation mass spectrometric detection has been developed (Scudamore *et al.*, 1996). Sterigmatocystin is now included in several MS/MS methods so that sensitive detection should be possible in the future.

6.8.3 Cyclopiazonic acid

Cyclopiazonic acid is a toxic indole tetramic acid (Fig. 6.7) that was first isolated from *Penicillium cyclopium* (Holzapfel, 1968) and subsequently from other *Penicillium* species, *Aspergillus flavus* (Luk *et al.*, 1977) and *A. versicolor* (Ohmono *et al.*, 1973). Cyclopiazonic acid is an optically active colourless crystalline compound that is soluble in chloroform, dichloromethane, methanol, acetonitrile and sodium bicarbonate. It reacts with 0.1 N sulphuric acid in methanol, gives a blue-violet Ehrlich colour reaction and an orange-red reaction with ferric chloride.

It has been detected in naturally contaminated mixed feeds, maize, peanuts, milk, cheese and other foods and feeds. Methods for its detection are less sensitive than for mycotoxins such as aflatoxin B_1 and ochratoxin A. Cyclopiazonic acid together with tenuazonic acid has been found in tomatoes in Brazil (da Motta *et al.*, 2001).

Cyclopiazonic acid only appears to be toxic when present in high concentrations. Cyclopiazonic acid was found to be a neurotoxin when injected intraperitoneally into rats and lesions in the liver, kidney, spleen and other organs were observed (Purchase, 1971). It can co-occur with aflatoxins (Takashi *et al.*, 1992) and may enhance the overall toxic effect when this happens (Cole, 1986). There is a dearth of human exposure data. However, 'Kodua' poisoning in India resulting from ingestion of contaminated millet seeds has been linked to this toxin.

The main method development for cyclopiazonic at this time has been into methods for its determination in milk (Losito *et al.*, 2002) and cheese (Zambonin *et al.*, 2001) although it may potentially occur with aflatoxins, for example in nuts.

6.8.4 Moniliformin

Moniliformin is formed by a number of *Fusarium* species and occurs as the sodium or potassium salt of 1-hydroxycyclobut-1-ene-3,4-dione (Fig. 6.4). It is soluble in water and polar solvents. On heating to 360°C moniliformin decomposes without melting. UV maxima are 229 nm and 260 nm in methanol.

Data on the occurrence of moniliformin in food are scarce. Thiel *et al.* (1982) showed that levels up to 12 mg/kg occurred in maize intended for human consumption in the Transkei. Maize-milled products destined for incorporation into animal feedingstuffs in the UK showed that samples were contaminated with concentrations up to 4.6 mg/kg (Scudamore *et al.*, 1998). Moniliformin has also been shown to occur in other cereals such as wheat and rice.

There is a limited data on the effects of moniliformin on mammalian species. The oral LD_{50} in rodents is approximately 50 mg/kg and for day-old cockerels 4 mg/kg. The ip LD_{50} values were 21 and 29 mg/kg, respectively, for female and male mice. In acute studies, the main lesion appears to be intestinal haemorrhage but in sub-acute and chronic studies, in a variety of avian species and laboratory rodents, the principal target was the heart. Moniliformin is a potent inhibitor of mitochondrial pyruvate and ketoglutarate oxidation. Data on the effects of moniliformin on reproduction and the foetus, and its mutagenic and carcinogenic potential are negative but extremely limited. In humans, moniliformin has casually been linked with Keshan disease, which is endemic in some areas of China.

Contaminated maize may contain a cocktail of toxic *Fusarium*-derived residues such as fumonisins, zearalenone and trichothecenes (Thiel *et al.*, 1982; Scudamore *et al.*, 1998). Thus some of the published data in which maize-containing moniliformin has been fed to animals may need re-evaluation.

6.8.5 Alternaria toxins

Alternaria is a common genus that usually invades crops at the pre-harvest stage and under suitable conditions may lead to production of certain mycotoxins (Magan *et al.*, 1984). The mycotoxins that occur most frequently are tenuazonic acid, alternariol monomethyl ether and alternariol (Fig. 6.8). Altenuene, iso-altenuene and altertoxins I and II may occur occasionally. Tenuazonic acid is a colourless, viscous oil and is a monobasic acid with pK_a 3.5. It is soluble in methanol and chloroform. On standing, heating or treatment with a base, optical activity is lost and crystallisation may occur as a result of formation of isotenuazonic acid. It forms complexes with calcium, magnesium, copper, iron and nickel ions. Tenuazonic acid is usually stored as its copper salt. Alternariol and alternariol monomethyl ether crystallise from ethanol as colourless needles, with melting points with decomposition of 350°C and 267°C, respectively. They sublime in a high vacuum without decomposing at 250°C and 180–200°C. They are soluble in most organic solvents and give a purple colour reaction with ethanolic ferric chloride. Altenuene crystallises as colourless prisms melting at 190–191°C.

Altertoxin I is an amorphous solid melting at 180°C and fluoresces bright yellow under UV light.

Fruits, vegetables and oilseeds are the food commodities that are most affected (Stinson *et al.*, 1981). Occurrence of *Alternaria* species and their mycotoxins in oilseeds has been reported by a number of workers, e.g. in sunflower seed (Chulze *et al.*, 1995) and in oilseed rape and sunflower seed meal (Nawaz *et al.*, 1997). Mycotoxins produced by *Alternaria* have also been reported in apples, olives, tomato products and cereals such as sorghum, wheat and rye (Visconti *et al.*, 1986; Grabarkiewicz-Szczesna *et al.*, 1989; Ansari and Shrivastava, 1990; Bottalico and Logrieco, 1992). Tenuazonic acid has been found in tomatoes in Brazil (da Motta *et al.*, 2001).

Toxicology: *Alternaria* toxins exhibit both acute and chronic effects. The LD_{50} values for alternariol monomethyl ether, alternariol, altenuene and altertoxin I in mice are 400, 400, 50 and 0.2 mg/kg bw, respectively. Those for tenuazonic acid are 162 and 115 mg/kg bw (by the intravenous route) for male and female mice, respectively.

Alternaria toxins have been implicated in animal and in human health disorders (Woody and Chu, 1992). Deaths in rabbits and poultry have been reported as a result of toxic action of *Alternaria* species found in the fodder and feed (Forgacs *et al.*, 1958; Wawrzkiewicz *et al.*, 1989). *Alternaria* was also detected in cereal samples in which *Fusarium* was implicated as the likely cause for an outbreak of ATA in Russia (Joffe, 1960).

Tenuazonic acid has been most studied of the *Alternaria* toxins. Its principal mode of action appears to be the inhibition of protein synthesis by suppressing the release of newly formed proteins from ribosomes into supernatant fluid (Shigeura and Gordon, 1963). It exhibits anti-tumour, antiviral and antibacterial activity. Alternariol and alternariol monomethyl ether show foetotoxic and teratogenic effects in mice, including a synergistic effect when a combination of the toxins was administered (Pero *et al.*, 1973). Most *Alternaria* mycotoxins exhibit considerable cytotoxic activity, including mammalian toxicity. The altertoxins are of particular concern because of their mutagenic activity (Scott and Stoltz, 1980; Chu, 1981; Stack and Prival, 1986). The latter showed that altertoxin III exhibits mutagenic activity that is approximately one-tenth that of aflatoxin B_1, while altertoxins I and II showed less mutagenicity.

References

Abbas, H.K. (2005) *Aflatoxin and Food Safety*. CRC Press, Boca Raton, FL, pp. 1–427.

Allcroft, R. and Carnaghan, R.B.A. (1962) Groundnut toxicity. *Aspergillus flavus* toxin (aflatoxin) in animal products: Preliminary communication. *Veterinary Record*, **74**, 863–864.

Alldrick, A.J. (1996) The effects of processing on the occurrence of ochratoxin A in cereals. *Food Additives and Contaminants*, **13**(suppl), 27–28.

Anon. (1982) *The Feeding Stuffs (Sampling and Analysis) Regulations 1982*. Statutory Instrument 1982, No. 1144. HMSO, London.

Ansari, A.A. and Shrivastava, A.K. (1990) Natural occurrence of *Alternaria* mycotoxins in sorghum and ragi from North Bihar, India. *Food Additives and Contaminants*, **7**, 815–820.

AOAC International (1995) *Official Methods of Analysis*, 16th edn. AOAC International, Gaithersburg, MD, Method 995.10.

Arranz, I., Baeyens, W.R., Van der Weken, G., De Saeger, S. and Van Peteghem, C. (2004) Review: HPLC determination of fumonisin mycotoxins. *Critical Reviews of Food Science and Nutrition*, **44**, 195–203.

Arranz, I., Derbyshire, M., Kroeger, K., Miscke, C., Stroka, J. and Anklam, E. (2005) Liquid chromatographic method for the determination of patulin at 10 ng/L in Apple-based products intended for infants: Interlaboratory study. *Journal of the Association of Official Analytical Chemists International*, **88**, 518–525.

Avantaggiato, G., De La Campa, R., Miller, J.D. and Visconti, A. (2003) Effects of muffin processing on fumonisins from 14C-labeled toxins produced in cultured corn kernels. *Journal of Food Protection*, **66**, 1873–1878.

Bankole, S.A., Ogunsanwo, B.M. and Mabekoje, O.O. (2004) Natural occurrence of moulds and aflatoxin B1 in melon seeds from markets in Nigeria. *Food and Chemical Toxicology*, **42**, 1309–1314.

Battilani, P., Pietri, A., Silva, A. and Giorni, P. (2003) Critical control points for the control of ochratoxin A in the grape-wine chain. *Journal of Plant Pathlogy*, **85**(suppl), 285.

Bauer, J., Gareis, M. and Gedek, B. (1986) Incidence of ochratoxin A in blood serum and kidneys of man and animals, in *Proceedings of the 2nd World Congress on Foodborne Infections and Intoxications*, Berlin, 26–30 May 1986, p. 907.

Bauer, J., Heinritzi, K., Gareis, M. and Gedek, B. (1987) Veranderungen am Genitaltrakt des weiblichen Schweines nach Verfätterung praxisrelevanter Zearalenonmengen. *Tierärztliche Praxis*, **15**, 33–36.

Baxter, E.D. (1996) The fate of ochratoxin A during malting and brewing. *Food Additives and Contaminants*, **13**(suppl), 23–24.

Bennett, G.A., Peplinski, A.J., Brekke, O.L. and Jackson, L.K. (1976) Zearalenone: Distribution in dry-milled fractions of contaminated corn. *Cereal Chemistry*, **53**, 299–307.

Bennett, G.A. and Anderson, R.A. (1978) Distribution of aflatoxin and/or zearalenone in wet-milled corn products: A review. *Journal of Agricultural Food Chemistry*, **26**, 1055–1060.

Beretta, B., Gaiaschi, A., Galli, C.L. and Restani, P. (2000) Patulin in apple-based foods: Occurrence and safety evaluation. *Food Additives and Contaminants*, **17**, 399–406.

Bezuidenhout, S.C., Gelderblom, W.C.A., Gorst-Allman, C.P., Horak, R.M., Marasas, W.F.O., Spiteller, G. and Vleggaar, R. (1988) Structure elucidation of the fumonisins, mycotoxins from *Fusarium moniliforme*. *Journal of the Chemical Society, Chemical Communications*, **82**, 743–745.

Bird, C.B., Malone, B., Rice, L.G., Ross, P.F., Eppley, R. and Abouzied, M.M. (2002) Determination of total fumonisins in corn by competitive direct enzyme-linked immunosorbent assay. *Journal of the Association of Official Analytical Chemists International*, **85**, 404–410.

Biselli, S. and Hummert, C. (2005) Development of a multicomponent method for *Fusarium* toxins using LC-MS/MS and its application during a survey for the content of T-2 toxin and deoxynivalenol in various feed and food samples. *Food Additives and Contaminants*, **22**, 752–760.

Bordson, G.O., Meerdink, G.L., Bauer, J.K. and Tumbleson, M.E. (1995) Effects of drying temperature on fumonisin recovery from feeds. *Journal of the Association of Official Analytical Chemists International*, **78**, 1183–1188.

Bottalico, A. and Logrieco, L. (1992) *Alternaria* plant disease in Mediterranean countries and associated mycotoxins, in *Alternaria – Biology, Plant Disease and Metabolites* (eds. J. Chelkowski and A. Visconti). Elsevier, Amsterdam, pp. 209–232.

Breitholtz, A., Olsen, M., Dahlbäck, Å. and Hult, K. (1991) Plasma ochratoxin A levels in three Swedish populations surveyed using an ion-pair HPLC technique. *Food Additives and Contaminants*, **8**, 183–192.

Bresch, H., Urbanek, M. and Nusser, M. (2000) Ochratoxin A in food containing liquorice. *Nahrung*, **44**, 276–278.

Broom, W.A., Bulbring, E., Chapman, C.J., Hampton, J.W.F., Thomson, A.M., Ungar, J., Wien, R. and Woolfe, G. (1944) The pharmacology of patulin. *British Journal of Experimental Pathology*, **25**, 195–207.

Burdaspal, P., Legarda, T.M. and Gilbert, J. (2001) Determination of ochratoxin A in baby food by immunoaffinity column clean up with liquid chromatography: Inter-laboratory study. *Journal of the Association of Official Analytical Chemists International*, **84**, 1445–1452.

CAST (2003) *Mycotoxins: Risks in Plant, Animal, and Human Systems, Council for Agricultural Science and Technology*. Ames, Iowa, USA. Task Force Report No. 139, January 2003, ISBN 1-887383-22-0.

Castellá, G., Bragulat, M.R. and Cabañes, F.J. (1999) Surveillance of fumonisins in maize-based feeds and cereals from Spain. *Journal of Agricultural Food Chemistry*, **47**, 4707–4710.

Cavaliere, C., Foglia, P., Pastorini, E., Samperi, R. and Lagana, A. (2005) Development of a multiresidue method for analysis of major Fusarium mycotoxins in corn meal using liquid chromatography/tandem mass spectrometry. *Rapid Communications in Mass Spectrometry*, **19**, 2085–2093.

Chao, T.C., Maxwell, S.M. and Wong, S.Y. (1991) An outbreak of aflatoxicosis and boric acid poisoning in Malaysia. *Journal of Pathology*, **164**, 225–233.

Chu, F.S. (1981) Isolation of altenuisol and altertoxin I and II. Minor mycotoxins elaborakt by *Alternaria*. *Journal of the American Oil Chemists Society*, **58**, 1006A–1008A.

Chulze, S.N., Torres, A.M., Dalcero, A.M., Etcheverry, M.G., Ramirez, M.L. and Farnochi, M.C. (1995) *Alternaria* mycotoxins in sunflower seeds: Incidence and distribution of the toxins in oil and meal. *Journal of Food Protection*, **58**, 1133–1135.

CODEX ALIMENTARIUS (2003) Code of practice for the prevention and reduction of patulin contamination in apple juice and apple juice ingredients in other beverages, RCP 50.

Cole, R.J. (1986) Etiology of turkey-X disease in retrospect: A case for the involvement of cyclopiazonic acid. *Mycotoxin Research*, **2**, 3.

Cooper, S.J., Wood, G.M., Chapman, W.B. and Williams, A.P. (1982) Mycotoxins occurring in mould-damaged foods, in *Proceedings of the Fifth International IUPAC Symposium on Mycotoxins and Phycotoxins*, 1–3 September 1982, Vienna, Austria, IUPAC, pp. 64–67.

Croft, W.A., Jarvis, B.B. and Yatawara, C.S. (1986) Airborne outbreak of trichothecene toxicosis. *Atmospheric Environment*, **20**, 549–552.

da Motta, S., Lucia, M. and Soares, V. (2001) Survey of Brazilian tomato products for alternariol, alternariol monomethyl ether, tenuazonic acid and cyclopiazonic acid. *Food Additives and Contaminants*, **18**, 630–634.

Dickens, F. and Jones, H.E.H. (1961) Carcinogenic activity of a series of reactive lactones and related substances. *British Journal of Cancer*, **15**, 85–100.

Driffield, M., Hird, S.J. and MacDonald, S.J. (2003) The occurrence of a range of mycotoxins in animal offal food products by HPLC-MS/MS. *Aspects of Applied Biology*, **68**, 205–210.

Douwes, J., Thorne, P., Pearce, N. and Heederik, D. (2003) Review bioaerosol health effects and exposure assessment: Progress and prospects. *Annals of Occupational Hygiene*, **47**, 187–200.

Dragacci, S., Grosso, F. and Gilbert, J. (2001) Immunoaffinity column cleanup with liquid chromatography for determination of aflatoxin M_1 in liquid milk: Collaborative study. *Journal of the Association of Official Analytical Chemists International*, **84**, 437–443.

Dunne, C., Meaney, M. and Smyth, M. (1993) Multi-mycotoxin detection and clean-up for aflatoxins, ochratoxin and zearalenone in animal feed ingredients using high performance liquid chromatography and gel permeation chromatography. *Journal of Chromatography*, **629**, 229–235.

Dutton, M. (1996) Fumonisins, mycotoxins of increasing importance: Their nature and their effects. *Pharmacology and Therapeutics*, **70**, 137–161.

Elis, J. and Di Paola, J.A. (1967) Aflatoxin B_1. *Archives of Pathology*, **83**, 53–57.

Entwisle, A.C., Williams, A.C., Mann, P.J., Slack, P.T. and Gilbert, J. (2000) Liquid chromatographic method with immunoaffinity column clean up for determination of ochratoxin A in barley: Collaborative study. *Journal of the Association of Official Analytical Chemists*, **83**, 1377–1383.

Entwisle, A.C., Williams, A.C., Mann, P.J., Russell, J., Slack, P.T. and Gilbert, J. (2001) Combined phenyl silane and immunoaffinity column cleanup with liquid chromatography for determination of ochratoxin A in roasted coffee, Collaborative study. *Journal of the Association of Official Analytical Chemists*, **84**, 444–450.

European Commission (2000) *Opinion of the Scientific Committee on Food, Fusarium Toxins, Part 2: Zearalenone*, 22 June 2000, European commission Health & Consumer Protection Directorate-General, 12pp.

European Commission (2002a) Assessment of dietary intake of ochratoxin A. EUR Report 17523 – *Reports on tasks for Scientific Cooperation*. Office of Official Publications of the EC, L-2985, Luxembourg.

European Commission (2002b) Assessment of dietary intake of patulin by the population of EU Member States. *Report on tasks for Scientific Cooperation*. Report of experts participating in task 3.2.8, March 2002.

European Commission (2003) Collection of occurrence data of *Fusarium* toxins in food and assessment of dietary intake by the population of EU Member States. *Report on tasks for Scientific Cooperation*. Report of experts participating in task 3.2.10, April 2002, 606 pp.

European Commission (2005a) *Summary Minutes of the Meeting of the Standing Committee on the Food Chain and Animal Health*. Animal Nutrition Section, Brussels, 28–29 June 2005, 7 pp.

European Commission (2005b) Commission Regulation (EC) No 123/2005, 28.1.2005 EN. *Official Journal of the European Union*, **L 25**, 3–5.

European Commission (2006a) Opinion of the scientific panel on contaminants in the food chain on a request from the commission related to ochratoxin A in food. *The EFSA Journal*, **365**, 1–56. Available from: www.efsa.eu.int.

European Commission (2006b) Commission recommendation of 17 August 2006 on the presence of deoxynivalenol, zearalenone, ochratoxin A, T-2 and HT-2 and fumonisins in products intended for animal feeding *Official Journal of the European Union*, **L 229**, 7–9.

European Commission (2007) Commission Regulation (EC) No 1126/2007, 28.7.2007 EN. Amending Regulation (EC) No 1881/2006 Setting maximum limits for certain contaminants as regards *Fusarium* toxins in maize and maize products. *Official Journal of the European Union*, **L 255/ 14-17**.

Faberi, A., Foglia, P., Pastorini, E. and Samperi, R. (2005) Determination of type B fumonisin mycotoxins in maize and maize-based products by using a QqQlinear ion trap mass spectrometer. *Rapid Communications in Mass Spectrometry*, **19**, 275–282.

Fazekas, B. and Tothe, H.E. (1995) Incidence of fumonisin B_1 in maize cultivated in Hungary. *Magy Állatorv Lapja*, **50**, 515–518.

Fazekas, B., Tar, A. and Kovács, M. (2005) Aflatoxin and ochratoxin A content of spices in Hungary. *Food Additives and Contaminants*, **22**, 856–863.

Filali, A., Betbeder, A.M., Baudrimont, I., Benayada, A., Soulaymani, R. and Creppy, E.E. (2002) Ochratoxin A in human plasma in Morocco: A preliminary survey. *Human and Experimental Toxicology*, **21**, 241–245.

Fishbein, L. (1979) Range of potency of carcinogens in animals, in *Potential Industrial Carcinogens and Mutagens*. Elsevier Scientific, Amsterdam, Oxford and New York, p. 1.

Flannigan, B., Mccabe, E.M. and Mcgarry, F. (1991) Allergenic and toxigenic microorganisms in houses, in *Pathogens in the Environment* (ed. B. Austin). *Journal of Applied Bacteriology Symposium Supplement*, **70**, 61S–73S.

Food and Agriculture Organisation of the United Nations (2004) Worldwide regulations for mycotoxins in food and feed was reported in 2003 (FAO 2004); FAO Food and Nutrition Paper No. 81; 2004.

Forgacs, J., Koch, H., Carll, W.T. and White-Stevens, R.H. (1958) Additional studies on the relationship of mycotoxicoses to poultry haemorrhagic syndrome. *American Journal of Veterinary Research*, **19**, 744–753.

Frisvad, J.C. (1989) The connection between the *Penicillia* and *Aspergilli* and mycotoxins with emphasis on misidentified isolates. *Archives of Environmental Contamination and Toxicology*, **18**, 452–467.

Gareis, M. (1996) Fate of ochratoxin A on processing of meat products. *Food Additives and Contaminants*, **13**(suppl), 35–38.

Gareis, M., Schothorst, R.C., Vidnes, A., Bergsten, C., Paulsen, B., Brera, C. and Miraglia, M. (2003) *Collection of Occurrence Data of Fusarium Toxins in Food and Assessment of Dietary Intake by the Population of EU Member States*. Report of Experts Participating in SCOOP Task 3.2.10. Available from: http://europa.eu.int/comm/food/fs/scoop/task3210.pdf.

Gelderblom, W.C.A., Jaskiewicz, K., Marasas, W.F.O., Thiel, P.G., Horak, R.M., Vleggaar, R. and Kriek, N.P.J. (1988) Fumonisins—novel mycotoxins with cancer-promoting activity produced by *Fusarium moniliforme*. *Applied Environmental Microbiology*, **54**, 1806–1811.

Gelderblom, W.C.A. and Snyman, S.D. (1991) Mutagenicity of potentially carcinogenic mycotoxins produced by *Fusarium moniliforme*. *Mycological Research*, **7**, 46–50.

Gelderblom, W.C.A., Cawood, M.E., Snyman, S.D., Vleggaar, R. and Marasas, W.F. (1993) Structure–activity relationship in short-term carcinogenesis and cytotoxicity assays. *Food Chemistry and Toxicology*, **31**, 407–414.

Gelderblom, W.C.A., Cawood, M.E., Snyman, S.D. and Marasas, W.F. (1994) Fumonisin B_1 dosimetry in relation to cancer initiation in rats. *Carcinogenesis*, **15**, 209–214.

Gilbert, J., Brereton, P. and MacDonald, S. (2001) Assessment of dietary exposure to ochratoxin A in the UK using a duplicate diet approach and analysis of urine and plasma samples. *Food Additives and Contaminants*, **18**, 1088–1093.

Gilbert, J. and Anklam, E. (2002) Validation of analytical methods for determining mycotoxins in foodstuffs. *Trends in Analytical Chemistry*, **21**, 468–486.

Grabarkiewicz-Szczesna, J., Chelkowski, J. and Zajkowski, P. (1989) Natural occurrence of *Alternaria* mycotoxins in the grain and chaff of cereals. *Mycotoxin Research*, **5**, 77–80.

Grosso, F., Fremy, J.M., Bevis, S. and Dragacci, S. (2004) Joint IDF-IUPAC-IAEO (FAO). *Food Additives and Contaminants*, **21**, 348–357.

Harwig, J., Kuiper-Goodman, T. and Scott, P.M. (1983) Microbial food toxicants: Ochratoxins, in *Handbook of Foodborne Diseases of Biological Origin* (ed. M. Recheigl, Jr). CRC Press, Boca Raton, FL, pp. 193–238.

Holzapfel, C.W. (1968) The isolation and structure of cyclopiazonic acid, a toxic metabolite of *Penicillium cyclopium*. *Tetrahedron*, **24**, 2101–2119.

Hopkins, E.C. and Murphy, P.A. (1993) Detection of fumonisins B_1, B_2, B_3 and hydrolyzed fumonisin B_1 in corn-containing foods. *Journal of Agricultural and Food Chemistry*, **41**, 1655–1658.

IARC (1986) Patulin. *IARC Monographs*, **40**, 83–98.

IARC (1993) *Some Naturally Occurring Substances: Food Items and Constituents, Heterocyclic Aromatic Amines and Mycotoxins*. IARC Monographs on the Evaluation of Carcinogenic Risks to Humans, Vol. 56, IARC, Lyon, France.

Iha, M.H. and Sabino, M. (2006) Determination of patulin in apple juice by liquid chromatography. *Journal of the Association of Official Analytical Chemists International*, **89**, 139–143.

International Standard ISO/DIS 8128-1(E) (1993) Apple juice, apple juice concentrates and drinks containing apple juice. Determination of patulin content.

Ito, R., Yamazaki, H., Inoue, K., Yoshimura, Y., Kawaguchi, M. and Nakazawa, H. (2004) Development of liquid chromatography–electrospray mass spectrometry for the determination of patulin in apple juice: Investigation of its contamination levels in Japan. *Journal of the Association of Official Analytical Chemists International*, **52**, 7464–7468.

Jaimez, J., Fente, C.A., Vazquez, B.I., Franco, C.M., Cepeda, A., Mahuzier, G. and Prognon, P. (2000) Review: Application of the assay of aflatoxins by liquid chromatography with fluorescence detection in food analysis. *Journal of Chromatography A*, **882**, 1–10.

Jemmali, M. and Mazerand, C. (1980) Présence de zéaralénone ou F_2 dans les noix de commerce. *Annules de Microbiologie (Paris)*, **B131**, 319–321.

Joffe, A.Z. (1960) Mycoflora of overwintered cereals and its toxicity. *Bulletin of the Research Council of Israel, Section D*, **9**, 101–126.

Josephs, R.D., Schuhmacher, R. and Krska, R. (2001). International inter-laboratory study for the determination of the Fusarium mycotoxins zearalenone and deoxynivalenol in agricultural commodities. *Food Additives and Contaminants*, **18**, 417–430.

Katzman, P.A., Hays, E.E., Cain, C.K., Van Wyk, J.J., Reithel, F.J., Thayer, S.A., Doisy, E.A., Gaby, W.L., Carroll, C.J., Muir, R.D., Jones, L.R. and Wade, N.J. (1944) Clavacin, an antibiotic substance from *Aspergillus clavatus*. *Journal of Biological Chemistry*, **154**, 475–486.

Klotzel, M., Gutsche, B., Lauber, U. and Humpf, H.U. (2005) Determination of 12 type A and B trichothecenes in cereals by liquid chromatography–electrospray ionization tandem mass spectrometry. *Journal of Agricultural Food Chemistry*, **53**, 8904–8910.

Klotzel, M., Lauber, U. and Humpf, H.U. (2006) A new solid phase extraction clean-up method for the determination of 12 type A and B trichothecenes in cereals and cereal-based food by LC-MS/MS. *Molecular Nutrition and Food Research*, **50**, 261–269.

Krogh, P., Hald, B. and Pedersen, E.J. (1973) Occurrence of ochratoxin A and citrinin on cereals associated with porcine nephropathy. *Acta Pathologica et Microbiologica Scandanavica, Section B*, **81**, 689–695.

Krogh, P. and Nesheim, S. (1982) *Environmental Carcinogens. Selected Methods of Analysis. Vol. 5: Some Mycotoxins*, No. 44 (eds. L. Stoloff, M. Castegnaro, P.M. Scott, I.K. O'Neill and H. Bartsch). IARC Scientific Publications, Lyon, France.

Krska, R., Welzig, E., Berthiller, F., Molinelli, A. and Mizaikoff, B. (2005) Advances in the analysis of mycotoxins and its quality assurance [Review]. *Food Additives and Contaminants*, **22**, 345–353.

Kruger, S.C., Kohn, B., Ramsey, C.S. and Prioli, R. (1999) Rapid immunoaffinity-based method for determination of zearalenone in corn by fluorometry and liquid chromatography. *Journal of the Association of Official Analytical Chemists International*, **82**, 1364–1368.

Kuiper-Goodman, T., Scott, P.M. and Watanabe, H. (1987) Risk assessment of the mycotoxin zearalenone. *Regulatory Toxicology and Pharmacology*, **7**, 253–306.

Larsson, K. and Möller, T. (1996) Liquid chromatographic determination of ochratoxin A in barley, wheat bran and rye by the AOAC/IUPAC/NMKL method: NMKL collaborative study. *Journal of the Association of Official Analytical Chemists*, **79**, 1102–1105.

Lebepe-Mazur, S., Bal, H., Hopmans, E., Murphy, P. and Hendrich, S. (1995a) Fumonisin B_1 is fetotoxic in rats. *Veterinary and Human Toxicology*, **37**, 126–130.

Lebepe-Mazur, S., Wilson, T. and Hendrich, S. (1995b) *Fusarium proliferatum* fermented corn stimulates development of placental glutathione S-transferase-positive altered hepatic foci in female rats. *Veterinary and Human Toxicology*, **37**, 39–45.

Leitner, A., Zollner, P., Paolillo, A., Stroka, J., Papadopoulou-Bouraoui, A., Jaborek, S., Anklem, E. and Lindner, W. (2002) Comparison of methods for the determination of ochratoxin A in wine. *Analytica Chimica Acta*, **453**, 33–41.

Lewis, L., Onsongo, M., Njapau, H., Schurz-Rogers, H., Luber, G., Kieszak, S., Nyamongo, J., Backer, L., Dahiye, A.M., Misore, A., De Cock, K. and Rubin, C. (2005) Aflatoxin contamination of commercial maize products during an outbreak of acute aflatoxicosis in Eastern and Central Kenya. *Environmental Health Perspectives*, **113**, 1763–1767.

Liu, C., Liu, F., Xu, W., Kofoet, A., Hnas-Ulrch Humpf, H.-U. and Jiang, S. (2005a) Occurrence of fumonisins B1 and B2 in asparagus from Shandong province, P.R. China. *Food Additives and Contaminants*, **22**, 673–676.

Liu, Y., Walker, F., Hoeglinger, B. and Buchenauer, H. (2005b) Solvolysis procedures for the determination of bound residues of the mycotoxin deoxynivalenol in fusarium species infected grain of two winter wheat cultivars preinfected with barley yellow dwarf virus. *Journal of Agricultural Food Chemistry*, **53**, 6864–6869.

Llorens, A., Mateo, R., Mateo, J.J. and Jimenez, M. (2002) Comparison of extraction and clean-up procedures for analysis of zearalenone in corn, rice and wheat grains by high performance liquid chromatography with photodiode array and fluorescence detection. *Food Additives and Contaminants*, **19**, 272–281.

Losito, I., Monaci, L., Aresta, A. and Zambonin, C.G. (2002) LC-ion trap electrospray MS-MS for the determination of cyclopiazonic acid in milk samples. *Analyst*, **127**, 499–502.

Lovelace, C.E.A. and Nyathi, C.B. (1977) Estimation of the fungal toxins zearalenone and aflatoxin, contaminating opaque maize beer in Zambia. *Journal Science Food and Agriculture*, **28**, 288–292.

Luk, K.C., Kobbe, B. and Townsend, J.M. (1977) Production of cyclopiazonic acid by *Aspergillus flavus* Link. *Applied Environmental Microbiology*, **33**, 211–212.

Mably, M., Mankotia, M., Cavlovic, P., Tam, J., Wong, L., Pantazopoulos, P., Calway, P. and Scott, P.M. (2005) Survey of aflatoxins in beer sold in Canada. *Food Additives and Contaminants*, **22**, 1252–1257.

MacDonald, S.J., Long, M. and Gilbert, J. (2000) Liquid chromatographic method for the determination of patulin in clear and cloudy apple juices and apple puree: Collaborative study. *Journal of the Association of Official Analytical Chemists*, **86**, 1387–1394.

MacDonald, S.J., Anderson, S., Brereton, P. and Wood, R. (2003) Determination of ochratoxin A in currants, raisins, sultanas, mixed dried fruit, and dried figs by immunoaffinity column cleanup with liquid chromatography: Inter-laboratory study. *Journal of the Association of Official Analytical Chemists International*, **86**, 1164–1171.

MacDonald, S.J., Anderson, S., Brereton, P., Wood, R. and Damant, A. (2005) Determination of zearalenone in barley, maize and wheat flour, polenta, and maize-based baby food by immunoaffinity column cleanup with liquid chromatography: Interlaboratory study. *Journal of the Association of Official Analytical Chemists International*, **88**, 1733–1740.

Magan, N., Cayley, G.R. and Lacey, J. (1984) Effect of water activity and temperature on mycotoxin production by *Alternaria alternata* in cultures and on wheat grain. *Applied Environmental Microbiology*, **47**, 1113–1117.

Marasas, W.F.O. (2001) Discovery and occurrence of the fumonisins: A historical perspective. *Environmental Health Perspectives*, **109**(suppl 2), 239–243.

Martins, M.L., Martins, H.M. and Bernardo, F. (2001) Fumonisins B1 and B2 in black tea and medicinal plants. *Journal of Food Protection*, **64**, 1268–1270.

Mateo, J.J., Llorens, A., Mateo, R. and Jimenez, M. (2001) Critical study of and improvements in chromatographic methods for the analysis of type B trichothecenes. *Journal of Chromatography A*, **918**, 99–112.

Matsuura, Y. and Yoshizawa, T. (1981) Effect of food additives and heating on the decomposition of zearalenone in wheat flour. *Journal of the Food Hygiene Society of Japan*, **22**, 293–298.

Miller, J.D. (1995) Fungi and mycotoxins in grain: Implications for stored product research. *Journal of Stored Product Research*, **31**, 1–16.

Ministry of Agriculture, Fisheries and Food (1980) *Surveillance of Mycotoxins in the United Kingdom*. The fourth report of the Steering Group on Food Surveillance, The Working Party on Mycotoxins, Food Surveillance Paper No. 4. HMSO, London.

Ministry of Agriculture, Fisheries and Food (1987) *Mycotoxins*. The eighteenth report of the Steering Group on Food Surveillance, The Working Party on Naturally Occurring Toxicants in Food: Sub-Group on Mycotoxins, Food Surveillance Paper No. 18. HMSO, London.

Ministry of Agriculture, Fisheries and Food (1993) *Mycotoxins: Third Report*. The thirty-sixth report of the Steering Group on Chemical Aspects of Food Surveillance, Sub-Group on Mycotoxins, Food Surveillance Paper No. 36. HMSO, London.

Miraglia, M., Brera, C., Corneli, S. and De Dominics, R. (1993) Ochratoxin A in Italy: Status of knowledge and perspectives, in *Human Ochratoxicosis and its Pathologies* (eds. E.E. Creppy, M. Castegnaro and G. Dirheimer). John Bibbey Eurotext, Montrouge, France, pp. 129–140.

Miraglia, M., de Dominicis, A., Brera, C., Corneli, S., Cava, E., Menghetti, E. and Miraglia, E. (1995) Ochratoxin A levels in human milk and related food samples: An exposure assessment. *Natural Toxins*, 3, 436–444.

Mirocha, C.J., Christensen, C.M. and Nelson, G.H. (1971) F-2 (zearalenone), estrogenic mycotoxin from Fusarium, in *Microbial Toxins*. Academic Press, New York.

Mirocha, C.J. and Christensen, C.M. (1974) Oestrogenic mycotoxins synthesised by *Fusarium*, in *Mycotoxins* (ed. I.F.H. Purchase). Elsevier, Amsterdam, pp. 129–148.

Mirocha, C.J., Pathre, S.V. and Christensen, C.M. (1977) Chemistry of *Fusarium* and *Stachybotrys* mycotoxins, in *Mycotoxic, Fungi, Mycotoxins and Mycotoxicoses*, Vol. 1 (eds. T.D. Wyllie and L.G. Morehouse). Marcel Dekker, New York.

Mirocha, C.J., Pathre, S.V. and Robison, T.S. (1981) Comparative metabolism of zearalenone and transmission into bovine milk. *Food and Cosmetics Toxicology*, **19**, 25–30.

Nawaz, S., Scudamore, K.A. and Rainbird, S.C. (1997) Mycotoxins in ingredients of animal feeding stuffs: I. Determination of *Alternaria* mycotoxins in oilseed rape meal and sunflower seed meal. *Food Additives and Contaminants*, **14**, 249–262.

Ohmono, S., Sugita, M. and Abe, M. (1973) Isolation of cyclopiazonic acid, cyclopiazonic acid imine and bissecodehydrocyclopiazonic acid from the cultures of *Aspergillus versicolor*. *Journal of the Agricultural Chemical Society of Japan*, **47**, 57–63.

Ohtsubo, K., Saito, M., Kimura, H. and Tsuruta, O. (1978) High incidence of hepatic tumours in rats fed mouldy rice contaminated with *Aspergillus versicolor* and sterigmatocystin. *Food Cosmetics, Toxicology*, **16**, 143–150.

Oldenberg, E. (1993) Occurrence of zearalenone in maize. *Mycotoxin Research*, **9**, 72–78.

Omurtag, G.Z. and Yazicioglu, D. (2004) Determination of fumonisins B_1 and B_2 in herbal tea and medicinal plants in Turkey by high-performance liquid chromatography. *Journal of Food Protection*, **67**, 1782–1786.

Ostry, V. and Skarkova, J. (2003) A HPTLC method for the determination of the mycotoxin zearalenone in cereal products. *Mycotoxin Research*, **19**, 64–68.

Oyelami, O.A., Maxwell, S.M. and Adeoba, E. (1996) Aflatoxins and ochratoxin A in the weaning food of Nigerian children. *Annals of Tropical Paediatrics*, **16**, 137–140.

Ozay, G. and Alperden, I. (1991) Aflatoxin and ochratoxin A contamination of dried figs (*Ficus carina* L.) from the 1988 crop. *Mycotoxin Research*, **7**, 85–91.

Paepens, C., De Saeger, S., Van Poucke, C., Dumoulin, F., Van Calenbergh, S. and Van Peteghem, C. (2005) Development of a liquid chromatography/tandem mass spectrometry method for the quantification of fumonisin B1, B2 and B3 in cornflakes. *Rapid Communications in Mass Spectrometry*, **19**, 2021–2029.

Pallaroni, L. and von Holst, C. (2003) Determination of zearalenone from wheat and corn by pressurized liquid extraction and liquid chromatography–electrospray mass spectrometry. *Journal of Chromatography A*, **993**, 39–45.

Palli, D., Miraglia, M., Saieva, C., Masala, G., Cava, E., Colatosti, M., Corsi, A.M., Russo, A. and Brera, C. (1999) Serum levels of ochratoxin A in healthy adults in tuscany: Correlation with individual characteristics and between repeat measurements. *Cancer Epidemiology Biomarkers and Prevention*, **8**, 265–269.

Park, D.L. (2002) Effect of processing on aflatoxin. *Journal of Experimental Medicine and Biology*, **504**, 173–179.

Park, J.W., Scott, P.M., Lau, B.P. and Lewis, D.A. (2004) Analysis of heat-processed corn foods for fumonisins and bound fumonisins. *Food Additives and Contaminants*, **21**, 168–178.

Patel, P. (2004) Mycotoxin analysis: Current and emerging technologies, in *Mycotoxins in Food detection and control* (eds. N. Magan and M. Olsen). Woodhead Publishing Ltd, Cambridge, UK, pp. 88–110.

Payen, J., Girard, T., Gaillardin, M. and Lafont, P. (1983) Sur la présence de mycotoxines dans des bières. *Microbiologie Aliments Nutrition*, **1**, 143–146.

Pazzi, M., Medana, C., Brussino, M. and Baiocchi, C. (2005) Determination of aflatoxins in peanuts, maize feed and whole milk by HPLC-MS2 and MS3 tandem mass spectrometry. *Annali di Chemica*, **95**, 803–811.

Pero, R.W., Porner, H., Blois, M., Harvan, D. and Spalding, J.W. (1973) Toxicity of metabolites produced by the '*Alteria*'. *Environmental Health Perspectives*, **June**, 87–94.

Perry, J.L., Yuri, V., Il'ichev, V., Kempf, R., McClendon, J., Park, G., Richard, A., Manderville, R.A., Rüker, F., Dockal, M. and Simon, J.D. (2003) Binding of ochratoxin A derivatives to human serum albumin. *Journal of Physical Chemistry B*, **107**, 6644–6647.

Pettersson, H. and Langseth, W. (2002) *Inter-comparison of Trichothecene Analysis and Feasibility to Produce Certified Calibrants. Method Studies*. BCR information, EU Report EUR 20285/1 EN, 82pp.

Pfohl-Leszkowicz, A., Chekir-Ghedir, L. and Bacha, H. (1995) Genotoxicity of zearalenone, an oestrogenic mycotoxin: DNA adduct formation in female mouse tissues. *Carcinogenesis*, **16**, 2315–2320.

Pier, A.C. (1991) The influence of mycotoxins on the immune system, in *Mycotoxins and Animal Foods* (eds. J.E. Smith and R.S. Henderson). CRC Press, Boca Raton, FL, pp. 489–497.

Purchase, I.F.H. (1971) The acute toxicity of the mycotoxin cyclopiazonic acid to rats. *Toxicology and Applied Pharmacology*, **18**, 114–123.

Reif, K. and Metzger, W. (1995) Determination of aflatoxins in medicinal herbs and plant extracts. *Journal of Chromatography A*, **692**, 131–136.

Riley, R.T., An, S., Showker, A.L., Yoo, H.S., Norred, W.P., Chamberlain, W.J., Wang, E., Merrill, A.H., Motelin, G., Beasley, V.R. and Haschek, W.M. (1993) Alteration of tissue and serum sphinganine to sphingosine ratio: An early biomarker of exposure to fumonisin containing feeds in pigs. *Toxicology and Applied Pharmacology*, **118**, 105–112.

Roberts, J.C. (1974) Aflatoxins and sterigmatocystin. *Fortschritte der Chemie Organischer Naturstoffe*, **31**, 119–151.

Sangare-Tigori, B., Moukha, S., Kouadio, J.H., Dano, D.S., Betbeder, A.M., Achour, A. and Creppy, E.E. (2006) Ochratoxin A in human blood in Abidjan, Cote d'Ivoire. *Toxicon*, **47**, 894–900.

Schneider, E., Curtui, V., Seidler, C., Dietrich, R., Usleber, E. and Martlbauer, E. (2004) Rapid methods for deoxynivalenol and other trichothecenes. *Toxicology Letters*, **153**, 113–121.

Schollenberger, M., Muller, H.M., Rufle, M., Suchy, S., Planck, S. and Drochner, W. (2005) Survey of Fusarium toxins in foodstuffs of plant origin marketed in Germany. *International Journal of Food Microbiology*, **97**, 317–326.

Schollenberger, M., Muller, H.M., Rufle, M., Terry-Jara, H., Suchy, S., Plank, S. and Drochner, W. (2007) Natural occurrence of *Fusarium* toxins in soy food marketed in Germany. *International Journal of Food Microbiology*, **113**, 142–146.

Schothorst, R.C. and van Egmond, H.P. (2004) Report from SCOOP task 3.2.10 "collection of occurrence data of Fusarium toxins in food and assessment of dietary intake by the population of EU member states". Subtask: Trichothecenes. *Toxicology Letters*, **53**, 133–143.

Scott, P.M. (1977) *Penicillium* mycotoxins, in *Mycotoxic, Fungi, Mycotoxins and Mycotoxicoses*, Vol. 1 (eds. T.D. Wyllie and L.G. Morehouse). Marcel Dekker, New York, pp. 283–356.

Scott, P.M. (1990) Natural poisons, 971.22 standards for aflatoxins. *AOAC Official Methods of Analysis*. AOAC, Gaithersburg, MD, pp. 1186–1187.

Scott, P.M. (1996) Effects of processing and detoxification treatments on ochratoxin A: Introduction. *Food Additives and Contaminants*, **13**(suppl), 19–22.

Scott, P.M. and Stoltz, D.R. (1980) Mutagens produced by *Alternaria alternata*. *Mutation Research*, **78**, 33–40.

Scott, P.M. and Lawrence, G.A. (1994) Stability and problems in recovery of fumonisins added to corn-based foods. *Journal of the Association of Official Analytical Chemists*, **77**, 541–545.

Scudamore, K.A. (1996) The effects of processing on the occurrence of ochratoxin A in cereals. *Food Additives and Contaminants*, **13**(suppl), 39–42.

Scudamore, K.A. (2004) Control of mycotoxins: Secondary processing, in *Mycotoxins in Food Detection and Control* (eds. N. Magan and M. Olsen). Woodhead Publishing Ltd, Cambridge, UK, pp. 228–243.

Scudamore, K.A., Hetmanski, M.T., Clarke, P.T., Barnes, K.A. and Startin, J.R. (1996) An analytical method for the determination of sterigmatocystin in cheese, bread and corn products using HPLC with atmospheric pressure ionisation mass spectrometric detection. *Food Additives and Contaminants*, **13**, 343–358.

Scudamore, K.A. and Macdonald, S.J. (1998) A collaborative study of an HPLC method for the determination of ochratoxin A in wheat using immunoaffinity column clean-up. *Food Additives and Contaminants*, **15**, 401–410.

Scudamore, K.A., Nawaz, S. and Hetmanski, M.T. (1998) Mycotoxins in ingredients of animal feeding stuffs: II. Determination of mycotoxins in maize and maize products. *Food Additives and Contaminants*, **15**, 30–55.

Scudamore, K.A., Banks, J. and Macdonald, S.J. (2003) The fate of ochratoxin A in the processing of whole wheat grains during milling and bread production. *Food Additives and Contaminants*, **20**, 1153–1163.

Scudamore, K.A., Banks, J. and Guy, R. (2004) The fate of ochratoxin A in the processing of whole wheat grains during extrusion. *Food Additives and Contaminants*, **21**, 488–497.

Šegvić, M.S. and Pepeljnjak, S. (2001) Fumonisins and their effects on animal health: A brief review. *Veterinarski Arhiv*, **71**, 299–323.

Şenyuva, H.Z., Gilbert, J., Ozcan, S. and Ulken, U. (2005) Survey for co-occurrence of ochratoxin A and aflatoxin B1 in dried figs in Turkey by using a single laboratory-validated alkaline extraction method for ochratoxin A. *Journal of Food Protection*, **68**, 1512–1515.

Sewram, V., Nair, J.J., Nieuwoudt, T.W., Leggott, N.L. and Shephard, G.S. (2000) Determination of patulin in apple juice by high-performance liquid chromatography-atmospheric pressure chemical ionization mass spectrometry. *Journal of Chromatography A*, **897**, 365–374.

Sharman, M. and Gilbert, J. (1991) Automated aflatoxin analysis of foods and animal feeds using immunoaffinity column clean-up and high-performance liquid chromatographic determination. *Journal of Chromatography*, **543**, 220–225.

Shephard, G.S., Thiel, P.G., Stockenström, S. and Sydenham, E.W. (1996) Worldwide survey of fumonisin contamination of corn and corn-based products. *Journal of the Association of Official Analytical Chemists International*, **79**, 671–687.

Shephard, G.S., van der Westhuizen, L., Gatyeni, P.M., Katerere, D.R. and Marasas, W.F.O. (2005) Do fumonisin mycotoxins occur in wheat? *Journal of Agricultural and Food Chemistry*, **53**, 9293–9296.

Shigeura, H.T. and Gordon, C.N. (1963) The biological activity of tenuazonic acid. *Biochemistry*, **2**, 1132–1137.

Smela, M.E., Curier, S.S., Bailey, E.A. and Essingmann, J.M. (2001) The chemistry and biology of aflatoxin B: From mutational spectrometry to carcinogenesis. *Carcinogenesis*, **22**, 535–545.

Smith, J.E. and Henderson, R.S. (1991) *Mycotoxins and Animal Foods*. CRC Press, Boca Raton, FL.

Smith, J.E., Lewis, C.W., Anderson, J.G. and Solomons, G.L. (1994) A literature review carried out on behalf of the agro-industrial division, E2, of the European Commission Directorate-General XII for scientific research and development, in *Mycotoxins in Human Nutrition and Health*. European Commission.

Songsermsakul, P., Sontag, G., Cichna-Markl, M., Zentek, J. and Razzazi-Fazeli, E. (2006) Determination of zearalenone and its metabolites in urine, plasma and faeces of horses by HPLC-APCI-MS. *Journal of Chromatography B Analytical Technology Biomedical Life Science*, **843**, 252–261.

Sorensen, L.K. and Elbaek, T.H. (2005) Determination of mycotoxins in bovine milk by liquid chromatography tandem mass spectrometry. *Journal of Chromatography B Analytical Technology Biomedical Life Science*, **820**, 183–196.

Soriano, J.M. and Dragacci, S. (2004) Intake, decontamination and legislation of fumonisins in foods. *Food Research International*, **37**, 985–1000, 1010.

Speijers, G.J.A. and van Egmond, H.P. (1993) World-wide ochratoxin A levels in food and feeds, in *Human Ochratoxicosis and its Pathologies* (eds. E.E. Creppy, M. Castegnaro and G. Dirheimer). John Bibbey Eurotext, Montrouge, France, pp. 85–100.

Stack, M.E. and Prival, M.J. (1986) Mutagenicity of the *Alternaria* metabolites altertoxins I, II and III. *Applied Environmental Microbiology*, **52**, 718–722.

Stinson, E.E., Osman, S.F., Heisler, E.G., Siciliano, J. and Bills, D.D. (1981) Mycotoxin production in whole tomatoes, apples, oranges and lemons. *Journal of Agricultural Food Chemistry*, **29**, 790–792.

Stroka, J., Anklam, E., Jorissen, U. and Gilbert, J. (2000) Immunoaffinity column cleanup with liquid chromatography using post-column bromination for determination of aflatoxins in peanut butter, pistachio paste, fig paste, and paprika powder: Collaborative study. *Journal of the Association of Official Analytical Chemists International*, **83**, 320–340.

Stroka, J., Anklam, E., Jorissen, U. and Gilbert, J. (2001) Determination of aflatoxin B_1 in baby food (infant formula) by immunoaffinity column cleanup liquid chromatography with postcolumn bromination: Collaborative study. *Journal of the Association of Official Analytical Chemists International*, **84**, 1116–1123.

Stroka, J., Derbyshire, M., Mischke, C., Ambrosio, M., Kroeger, K., Arranz, I., Sizoo, E. and van Egmond, H. (2006) Liquid chromatographic determination of deoxynivalenol in baby food and animal feed: Interlaboratory study. *Journal of the Association of Official Analytical Chemists International*, **89**, 1012–1020.

Subirade, I. (1996) Fate of ochratoxin A during bread making. *Food Additives and Contaminants*, **13**(suppl), 285.

Sydenham, E.W., Shephard, G.S., Thiel, P.G., Marasas, W.F.O. and Stockenström, S. (1991) Fumonisin contamination of commercial corn-based human foodstuffs. *Journal of Agricultural Food Chemistry*, **39**, 2014–2018.

Sydenham, E.W., Shephard, G.S., Thiel, P.G., Stockenstrom, S, Snijman, W. and Van Schalkwik, D.J. (1996) Liquid chromatographic determination of fumonisins B_1, B_2, and B_3 in corn: AOAC–IUPAC collaborative study. *Journal of the Association of Official Analytical Chemists International*, **79**, 688–696.

Takashi, U., Trucksess, M.W., Beaver, W.R., Wilson, D.M., Dorner, J.W. and Dowell, F.E. (1992) Co-occurrence of cyclopiazonic acid and aflatoxin in corn and peanuts. *Journal of the Association of Official Analytical Chemists*, **75**, 838.

Takino, M., Tanaka, T., Yamaguchi, K. and Nakahara, T. (2004) Atmospheric pressure photo-ionization liquid chromatography/mass spectrometric determination of aflatoxins in food. *Food Additives and Contaminants*, **21**, 76–84.

Tanaka, T., Hasegawa, A., Matsuki, Y., Lee, U.-S. and Ueno, Y. (1985) Rapid and sensitive determination of zearalenone in cereals by high-performance liquid chromatography with fluorescence detection. *Journal of Chromatography*, **328**, 271–278.

Thiel, P.G., Meyer, C.J. and Mararas, W.F.O. (1982) Natural occurrence of moniliformin together with deoxynivalenol and zearalenone in Transkeian corn. *Journal of Agricultural Food Chemistry*, **30**, 308–312.

Thirumala-Devi, K., Mayo, M.A., Reddy, G., Emmanuel, K.E., Larondelle, Y. and Reddy, D.V.R. (2001) Occurrence of ochratoxin A in black pepper, coriander, ginger and turmeric in India. *Food Additives and Contaminants*, **18**, 830–835.

Tian, H. and Liu, X. (2004) Survey and analysis on sterigmatocystin contaminated in grains in China [Article in Chinese]. *Wei Sheng Yan Jiu*, **33**, 606–608.

Towers, N.R. and Sprosen, J.M. (1992) *Fusarium* mycotoxins in pastoral farming: Zearalenone induced infertility in ewes, in *Recent Advances in Toxicology Research*, Vol. 3 (eds. P. Gopalakrishnakone and C.K. Tan). National University of Singapore, Singapore, pp. 272–284.

Trivedi, A.D., Hiroto, M., Dol, E. and Kitabatake, N. (1993) Formation of a new toxic compound, citrinin Hl, from citrinin on mild heating in water. *Journal of the Chemical Society, Perkin Transactions*, **1**, 2167–2171.

Trucksess, M.W., Stack, M.E., Nesheim, S., Page, S.W., Albert, R.H. and Romer, T.R. (1994) Multifunctional column coupled with liquid chromatography for determination of aflatoxins B1, B2, G1, G2 in corn, almonds, Brazil nuts, peanuts and pistachio nuts: Collaborative study. *Journal of the Association of Official Analytical Chemists*, **77**, 1512–1521.

Tsubouchi, H., Terada, H., Yamamoto, K., Hisada, K. and Sakabe, Y. (1988) Ochratoxin A found in commercial roast coffee. *Journal of Agricultural Food Chemistry*, **36**, 540–542.

Ueno, Y. and Ueno, I. (1978) Toxicology and biochemistry of mycotoxins, in *Toxicology. Biochemistry and Pathology of Mycotoxins* (eds. K. Uraguchi and M. Yamazaki). John Wiley, New York, pp. 107–155.

Urraca, J.L., Benito-Pena, E., Perez-Conde, C., Moreno-Bondi, M.C. and Pestka, J.J. (2005) Analysis of zearalenone in cereal and Swine feed samples using an automated flow-through immunosensor. *Journal of Agricultural Food Chemistry*, **53**, 3338–3344.

Urraca, J.L., Marazuela, M.D. and Moreno-Bondi, M.C. (2006) Molecularly imprinted polymers applied to the clean-up of zearalenone and alpha-zearalenol from cereal and swine feed sample extracts. *Analytical Bioanalytical Chemistry*, **385**, 1155–1161.

Valero, A., Marín, S., Ramos, A.J. and Sanchis, V. (2005) Ochratoxin A-producing species in grapes and sun-dried grapes and their relation to ecophysiological factors. *Letters in Applied Microbiology*, **41**, 196.

Vesonder, R.F. and Horn, B.W. (1985) Sterigmatocystin in dairy cattle feed contaminated with *Aspergillus versicolor*. *Applied Environmental Microbiology*, **49**, 234–235.

Viani, R. (1996) Fate of ochratoxin A (OTA) during processing of coffee. *Food Additives and Contaminants*, **13**(suppl), 29–34.

Visconti, A., Logrieco, A. and Bottalico, A. (1986) Natural occurrence of *Alternaria* mycotoxins in olives—their production and possible transfer into the oil. *Food Additives and Contaminants*, **3**, 323–330.

Visconti, A., Doko, M.B., Bottalico, C., Schurer, B. and Boenke, A. (1993) The stability of fumonisins (fumonisin B_1 and fumonisin B_2) in solution, in *Proceedings of the UK Workshop on Occurrence and Significance of Mycotoxins, Slough* (ed. K.A. Scudamore). Central Science Laboratory, MAFF, London, 21–23 April 1993, pp. 196–199.

Visconti, A. and Pascale, M. (1998) Determination of zearalenone in corn by means of immunoaffinity clean up and high performance liquid chromatography with fluorescence detection. *Journal of Chromatography A*, **815**, 133–140.

Visconti, A., Pascale, M. and Centonze, G. (2001a) Determination of ochratoxin A in wine and beer by immunoaffinity column clean up and liquid chromatographic analysis with fluorometric detection: Collaborative study. *Journal of the Association of Official Analytical Chemists International*, **84**, 1818–1827.

Visconti, A., Solfrizzo, M. and De Girolamo, A. (2001b) Determination of fumonisins B_1 and B_2 in corn and corn flakes by liquid chromatography with immunoaffinity column clean-up: Collaborative study. *Journal of the Association of Official Analytical Chemists International*, **84**, 1828–1837.

Wang, E., Ross, F., Wilson, T.M., Riley, R.T. and Merrill, A.H. (1992) Increases in serum sphingosine and sphinganine and decreases in complex sphingolipids in ponies given feed containing fumonisins, mycotoxins produced by *Fusarium moniliform*. *Journal of Nutrition*, **122**, 1706–1716.

Wannemacher, R.W., Bunner, D.L. and Neufeld, H.A. (1991) Toxicity of trichothecenes and other related mycotoxins in laboratory animals, in *Mycotoxins and Animal Foods* (eds. J.E. Smith and R.S. Henderson). CRC Press, Boca Raton, FL, pp. 499–552.

Watson, D.H. (1985) Toxic fungal metabolites in food. *CRC Critical Reviews in Food Science and Nutrition*, **22**, 177–198.

Wawrzkiewicz, K., Gluch, A., Rubaj, B. and Wrobel, M. (1989) *Alternaria* spp., an opportunist pathogen. *Medycyna Weterynary*, **45**, 27–30.

Williams, B.C. (1985) Mycotoxins in foods and feedstuffs, in *Mycotoxins: A Canadian Perspective* (eds. P.M. Scott, H.L. Trenholm and M.D. Sutton). National Research Council Canada, Ottawa, pp. 49–53.

Woody, M.A. and Chu, F.S. (1992) Toxicology of *Alternaria* mycotoxins, in *Alternaria – Biology. Plant Disease and Metabolites* (eds. J. Chelkowski and A. Visconti). Elsevier, Amsterdam, pp. 409–433.

Zambonin, C.G., Monaci, L. and Aresta, A. (2001) Determination of cyclopiazonic acid in cheese samples using solid-phase microextraction and high performance liquid chromatography. *Food Chemistry*, **75**, 249–254.

Zimmerli, B. and Dick, R. (1996) Ochratoxin A in table wine and grape-juice: Occurrence and risk assessment. *Food Additives and Contaminants*, **13**, 655.

Zinedine, A., Soriano, J.M., Molto, J.C. and Manes, J. (2006) Review on the toxicity, occurrence, metabolism, detoxification, regulations and intake of zearalenone: An oestrogenic mycotoxin. *Food and Chemical Toxicology*, **114**, 25–29.

Zöllner, P. and Mayer-Helm, B. (2006) Trace mycotoxin analysis in complex biological and food matrices by liquid chromatography–atmospheric pressure ionisation mass spectrometry. *Journal of Chromatography A*, **1136**, 123–169.

7 Phytoestrogens

Don Clarke and Helen Wiseman

Summary

Phytoestrogens are plant-derived compounds, which commonly occur in a wide range of foodstuffs. Phytoestrogens can exert biological activity by interacting with the mammalian hormone system, mimicking the effect of mammalian steroidal estrogens. This oestrogenicity may have a variety of potential health benefits. This chapter reviews the expanding list of known phytoestrogens, the increasing range of foods known to contain oestrogenic compounds, advances in measurement techniques and the outcome of clinical investigations of health effects.

7.1   INTRODUCTION

Phytoestrogens are plant-derived compounds, which commonly occur in a wide range of foodstuffs. The interest in phytoestrogens is that they can exert biological activity by interacting with the mammalian hormone system, mimicking the effect of mammalian steroidal estrogens such as 17β-oestradiol. This activity may have potential benefits to health and therefore phytoestrogens have been a rapidly expanding and fast moving field of study over the last decade. This chapter reviews the reasons for this growing interest such as the expanding list of known phytoestrogens, the increasing range of foods known to contain oestrogenic compounds, advances in measurement techniques and the outcome of clinical investigations.

Oestrogenicity in the environment is a field of study in its own right. There are an ever-increasing number of xenoestrogens being reported, these are man-made chemicals which interact with oestrogen receptors. These are classified as endocrine disruptors and are treated separately from both the natural mammalian steroids and natural plant-based oestrogens. The diet can therefore contain many different classes of chemical, which are all oestrogenic, ranging from natural constituents to environmental contaminants. This chapter is however restricted to the plant-based naturally occurring phytoestrogens.

7.2   THE STRUCTURE OF PHYTOESTROGENS

It is immediately apparent from the chemical structures that the isoflavone phytoestrogens and the main human steroidal oestrogen have a very similar three-dimensional shape and size. While simplistic, this shape and distance between the two hydroxyl groups, which provide the binding points, is fundamental to the structure activity relationship. This feature can be seen

(a) Steroidal estrogens

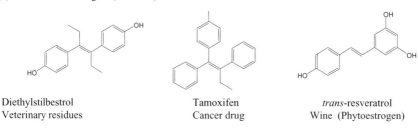

      B-Estradiol   17A-ethynylestradiol

(b) Non-steroidal estrogens (Stilbenes)

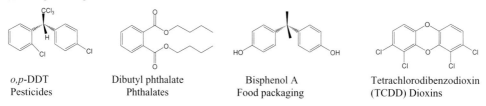

Diethylstilbestrol   Tamoxifen   *trans*-resveratrol
Veterinary residues   Cancer drug   Wine (Phytoestrogen)

(c) Estrogen distruptors

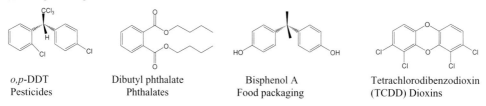

o,p-DDT   Dibutyl phthalate   Bisphenol A   Tetrachlorodibenzodioxin
Pesticides   Phthalates   Food packaging   (TCDD) Dioxins

Fig. 7.1   Structures of (a) steroidal oestrogens, (b) non-steroidal oestrogens and (c) endocrine disruptors.

in the synthetic hormone diethyl stilbestrol, a veterinary growth promoter, but becomes more difficult to visualise in the other synthetic xenoestrogens such as DDT pesticides, dioxins, phthalates, polybrominated flame retardants and bisphenol A (Fig. 7.1). It is not possible to accurately predict oestrogenicity from chemical structures alone, although pointers to activity have been presented (Katzenellenbogen *et al.*, 1996). This is further complicated by metabolism, which can make low activity or inactive compounds oestrogenic. The qualitative structure activity relationship approach (QSAR) has been successfully applied to predict receptor binding of similar structures (Aizawa and Hu, 2003), and this topic has been recently reviewed (Vaya and Tamir, 2004).

7.3   OCCURRENCE AND LEVELS IN PLANT MATERIALS

7.3.1   Defining phytoestrogens by class and structure

The phytoestrogens are generally described as broad classes of compounds based on the underlying chemical ring structures as shown in Fig. 7.2. The definition of which are oestrogenic and which have more potent biological activity is often disputed, as the many different *in vivo*

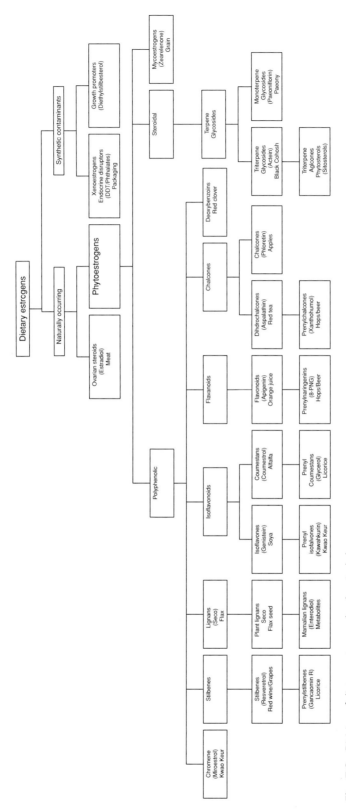

Fig. 7.2 Division of dietary estrogens by chemical class.

and *in vitro* assays can give contradictory outcomes. As a consequence of this the literature describing natural products often gives the impression that all such similar compounds with closely related polyphenolic structures are phytoestrogens, although this is not often the case.

The range of chemicals that are proven or suspected of being phytoestrogens has increased rapidly especially, as anecdotal cures and traditional Chinese herbal remedies are re-examined with modern techniques for measuring oestrogenicity and identifying the bioactive principles. Phytoestrogens can also be described on the basis of their major source, soya isoflavones (genistein, daidzein, glycitein), red clover isoflavones (biochanin A, formononetin), kudzu isoflavones (daidzein, puerarin), flax seed lignans (matairesinol, secoisolariciresiniol), hop prenylnaringenins (8-prenylnaringenin, 6-prenylnaringenin, 6,8-diprenylnaringenin). There is a huge overlap as soya also generally contains detectable quantities of biochanin A and formononetin, while red clover contains moderate quantities of daidzein and genistein as shown in Fig. 7.3.

7.3.2 Compositional databases

The occurrence of isoflavones, lignans and coumestrol are now well studied in most common foods and this is described adequately elsewhere. Databases have been prepared from these original source documents, to allow the diet consumed as recorded by food frequency questionnaires to be converted into dietary intake figures. As an example, the objective of the VENUS project (Kiely *et al.*, 2003) was to evaluate existing data on dietary exposure to compounds with oestrogenic and anti-oestrogenic effects present in plant foods as constituents or contaminants, and thereby to permit the assessment of exposure to e.g. isoflavones in European populations. Data on the isoflavone (genistein and daidzein) content of 791 foods, including almost 300 foods commonly consumed in Europe, were collected. Levels of coumestrol, formononetin, biochanin A and lignans in a limited number of foods were also included. Databases are often compilations of previously published analyses and this database also contains information on the references sourced for the compositional data, on the analytical methods used by each author and on the number of foods analysed in each reference (Kiely *et al.*, 2003). In order to demonstrate the performance of a database it must be able to successfully predict phytoestrogen intake from dietary records without conducting any actual analyses. This is achieved through filling out a food frequency questionnaire each day, or through a single dietary recall interview and to correlate these theoretical intake values with actual experimentally derived food, plasma and urine phytoestrogen concentrations (Valsta *et al.*, 2003; Milder *et al.*, 2005; Park *et al.*, 2005; Horn-Ross *et al.*, 2006; Ritchie *et al.*, 2006; Blitz *et al.*, 2007).

7.3.3 Factors affecting isoflavone content in soya

Isoflavones are phenolic secondary metabolites found mostly in legumes. These compounds play key roles in many plant–microbe interactions. Because of their biological activities, metabolic engineering of isoflavonoid biosynthesis in legume and non-legume crops has significant agronomic and nutritional impact. This has the aim of producing crops with enhanced plant disease resistance and providing dietary isoflavones for the improvement of human health (Yu and McGonigle, 2005).

The levels of phytoestrogens in plants vary enormously. One of the earliest studies found isoflavone concentrations in different soybean varieties to range from 1160 to 3090 µg/g, almost a threefold range (Eldridge and Kwolek, 1983) This was confirmed a decade later

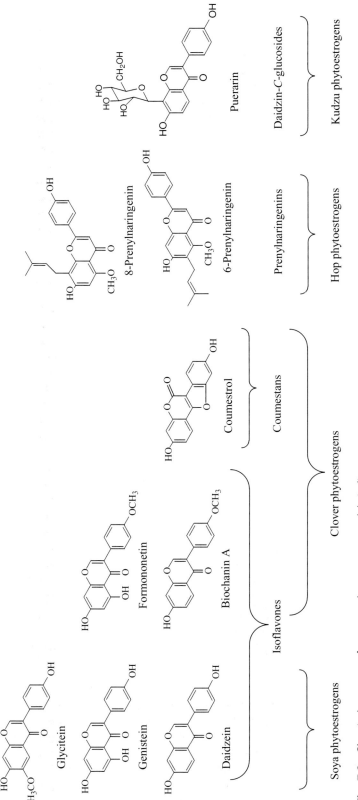

Fig. 7.3 Chemical structures of common phytoestrogens and their dietary sources.

with similar isoflavone concentrations being reported from within a single cultivar of soybean (Wang and Murphy, 1994). Concentrations of isoflavones found in 1996 were almost twice those found in 1995, for six soybean cultivars. The effect of eight different growing locations with different soil types also produced a twofold range in concentrations within a given years crop. This was interpreted as a significant genotype × environmental interaction (Hoeck *et al.*, 2000). Similar interactions were observed in a 21 cultivar, 5 location trial (MacDonald *et al.*, 2005). The application of fertilisers at various rates was found to have no effect on isoflavone content (Sequin *et al.*, 2006). Wounding and pathogen attack (stem rot fungus) play a key role in elevating isoflavone content, with a similar range of effect to that observed previously and attributed to environmental causes. Cultivars differ in ability to accumulate isoflavones following wounding and/or infection (Wegulo *et al.*, 2005). The date of planting had a great influence on levels of isoflavones, with early planted soybeans having lower concentrations of isoflavones. High temperatures during seed development significantly reduce concentrations of isoflavones (Tsukamoto *et al.*, 1995; Carrao-Panizzi *et al.*, 1999). Care must be taken to select commercially viable strains. From a practical viewpoint, the isoflavone content has been shown to increase with increased seed yield, while oil and protein content appear less responsive to seed yield. This suggests that both higher and lower isoflavone content need not affect any other of the commercial critical seed quality traits (Yin and Vyn, 2005; Primomo *et al.*, 2005b).

Population phenotyping and genotyping is underway, mapping the quantitative trait loci (QTL mapping) that associate with the isoflavone content of soya seeds. This marker-assisted selection is being used to develop soya strains with a genetic predisposition to produce higher isoflavone levels independent of the environmental conditions (Primomo *et al.*, 2005a,b).

7.3.4 Other possible phytoestrogens

There are a number of other classes of phytoestrogen that are less well understood. The majority of phytoestrogens are polyphenols, there is little linking with the chemical structures of polyphenols, steroidal triterpenes and monoterpene glucosides. The phytoestrogenic potency varies enormously and a number of the lesser-known phytoestrogens are considered more potent oestrogens than the better-known soya isoflavone phytoestrogens.

There are four factors involved in exposure to dietary phytoestrogens. Firstly, consumers are now better informed about the potential benefits of phytochemicals and can therefore choose to consciously increase exposure by adding soya products into their diets. Secondly, at the same time soya usage has been increasing in significance, in processed foods largely outside the control of the consumer, with this secondary exposure being the driving force behind the recent rise in the UK daily exposure to isoflavones. Thirdly, there are still a number of uncharacterised phytoestrogen sources in the diet. With the large cultural changes in recent decades, few Westerners now eat solely their own traditional diet. Many include other, especially Asian foods. Lastly, the ever-increasing scientific basis for health benefits has led to an influx of dietary supplements of soya, flaxseed and red clover. More recently this has expanded to cover further plant-based phytoestrogens, new to the UK diet, which are supplements rather than foods. These are generally deliberately taken as herbal remedies. These are increasing in popularity and represent a significant exposure to phytoestrogens. Often the health claims of these products are based solely around oestrogenic potency, i.e., that a known phytoestrogen is present and that it is more oestrogenically potent, than for instance genistein, thereby inferring this is more beneficial, without regard to quantity and dose. Many of the claims associated with these products should be treated with caution; however, they do currently represent the front line in commercial phytoestrogen research.

Many of these drugs and supplements contain potent phytoestrogens, but the dose and efficacy and relevance to a particular biological endpoint are generally not proven to rigorous scientific standards. The better-characterised products, based on soya isoflavones, red clover isoflavones and black cohosh triterpenes are gaining credibility as front line treatments of vasomotor symptoms and other menopausal complaints. Despite the far larger financial investment in researching traditional dietary exposure and the many potential chronic disease health benefits, it is somewhat surprising that the various supplements for the treatment of menopausal symptoms are now considered scientifically well validated and these simplistic substitutes to hormone replacement therapy (HRT) currently remain the only proven therapeutic uses for phytoestrogens (Howes *et al.*, 2006; Nachtigall *et al.*, 2006; Thompson Coon *et al.*, 2007).

The structures of a number of selected minor phytoestrogens are given in Fig. 7.4 with the chemical classes and a common dietary source.

7.3.4.1 Black cohosh

While still relatively unknown in the UK, black cohosh has become one of the most important herbal products in the US dietary supplements market. It is manufactured from roots and rhizomes of *Cimicifuga racemosa* (Ranunculaceae) and is widely used in the treatment of menopausal symptoms and menstrual dysfunction. This formulation is perhaps the best example of a well-studied and clinically proven use of phytoestrogens delivering a clear health benefit (Geller and Studee, 2005). Black cohosh is believed not to exert an oestrogenic effect on the breast (Ruhlen *et al.*, 2007). However, it prevents proliferation of breast cancer cells (Einbond *et al.*, 2006). Products sometimes contain the isoflavone formononetin, but as extracts of these products that contain no isoflavones remain active, and the oestrogenic effect is correlated to the triterpene glycoside fraction, these products and components are considered as non-isoflavone supplements (Onorato and Henion, 2001; Jiang *et al.*, 2006). Formulations are generally standardised against two triterpene glycosides 27-deoxyacetin and acetin, but contain many related compounds (Lai *et al.*, 2005) (Fig. 7.4).

Phytosterols. Plant sterols (phytosterols) are structurally similar to cholesterol and are found in bread, vegetable fats, fruit and vegetables and vegetable-based fats (Normén *et al.*, 2001). β-sitosterol is the most prevalent, and can be consumed at *ca.* 200 mg/day (Jimenez-Escriq *et al.*, 2006). As β-sitosterol has been suspected of being a phytoestrogen for some time, an intake of 200 mg/day would be a considerable exposure (Fig. 7.4). These compounds are generally more widely know as additives in margarine where they are used to achieve a reduction in serum cholesterol when they are consumed as part of a low fat, low cholesterol diet.

Toxicological studies have found that phytosterols did not bind to rat uterine oestrogen receptors (ER), or stimulate transcriptional activity of human ER in a recombinant yeast assay. Furthermore, there was no evidence of oestrogenicity in the immature rat uterotrophic assay. With negative results in the benchmarking uterotrophic assay, by definition phytosterols could not therefore be considered as phytoestrogens in rodents (Baker *et al.*, 1999). However, the egg yolk protein formed via the vitellogin gene is expressed in fish, providing unequivocal evidence that in aquatic systems phytosterols and β-sitosterol are oestrogenic (Mellanen *et al.*, 1996, 1999; Nakari, 2005). Differences in the primary sequences of rodent, human and fish oestrogen receptors and metabolism may explain the discrepancy between fish and

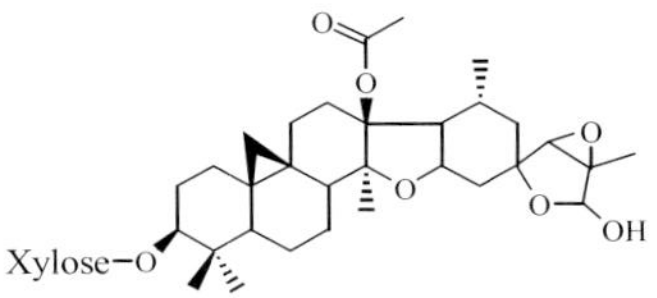

i) Acetin
Triterpene glycoside

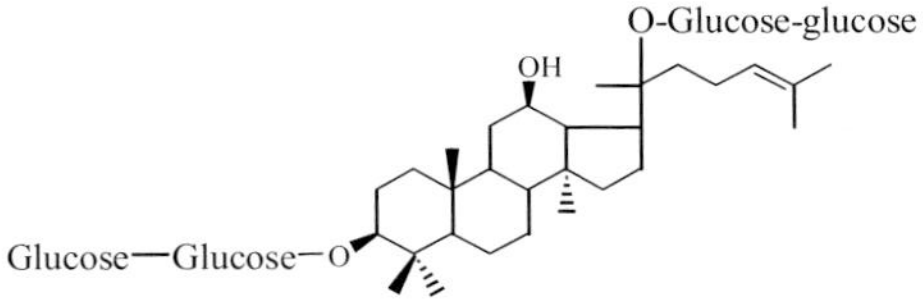

ii) Ginsenoside-Rb1
Triterpene glycoside

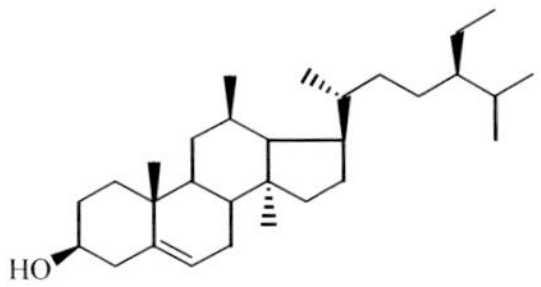

iii) β-sitosterol
Triterpene aglycone

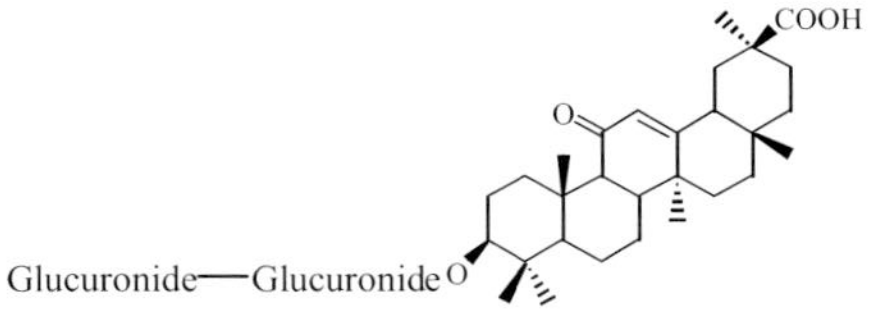

iv) Glycyrrhizin
Tritepene glycoside

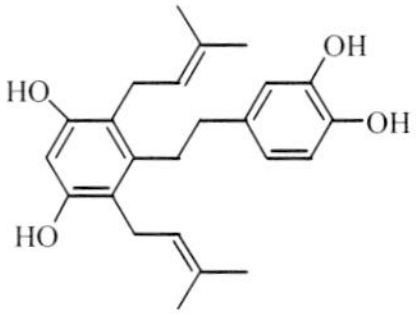

v) Gancaonin R
Prenyl dihydrostilbene

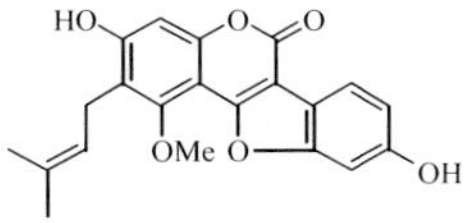

vi) Glycyrol
Prenyl coumestan

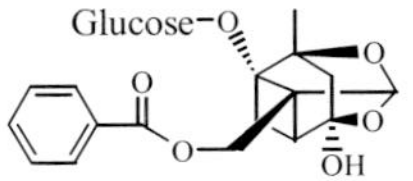

vii) Paeoniflorin
monoterpene glucoside

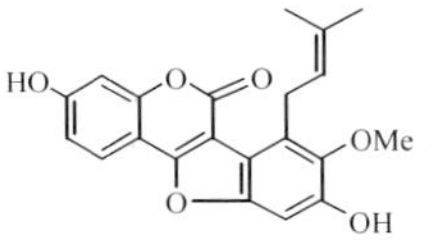

viii) Mirificoumestan
Prenyl coumestan

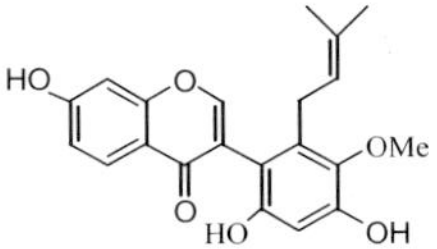

ix) Kawakhurin
Prenyl isoflavone

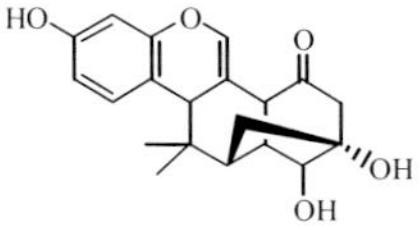

x) Deoxymiroestrol
Chromene

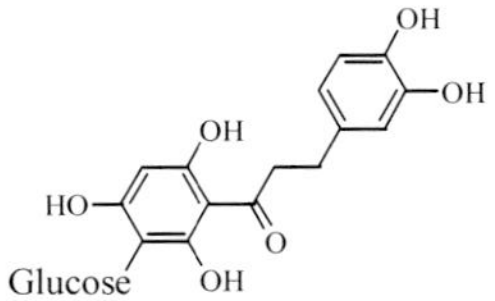

xi) Aspalathin
C-glucoside dihydrochalcone

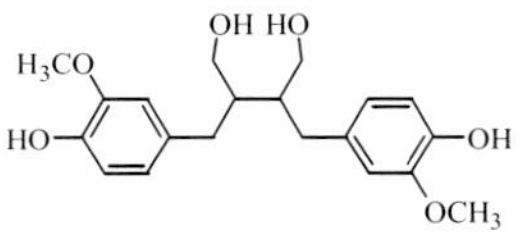

xii) Secoisolariciresinol
lignan

Fig. 7.4 Chemical structures of probable minor dietary phytoestrogens and their primary dietary sources: (i) acetin from black cohosh, (ii) ginsenoside-Rb1 from ginseng, (iii) β-sitosterol from vegetable oils, (iv) glycyrrhizin from licorice, (v) gancaonin R from licorice, (vi) glycyrol from licorice, (vii) paeoniflorin from paeony, (viii) mirificoumestan from Kwao keur, (ix) kawakhurin from Kwao Keur, (x) deoxymiroestrol from Kwao Keur, (xii) aspalathin from red tea, (xiii) secoisolariciresiniol from flax seed.

mammalian assays. In rats, β-sitosterol was found to reduce sperm count and testicular weight (Moghadasian, 2000), and to produce elevated plasma oestradiol and testosterone levels in field voles (Nieminen *et al.*, 2003). Recent evidence suggests that while β-sitosterol can exert a range of biological activity, modulating the growth of human breast cancer cells is a recognised oestrogenic role in humans (Ju *et al.*, 2004). This suggests that there can be some oestrogenic effects in mammals and phytosterols should continue to be viewed with caution. More importantly, the presence of β-sitosterol and related compounds in the waste stream from the Kraft pulp and paper process caused serious environmental pollution resulting in the feminisation of fish (Mellanen *et al.*, 1996, 1999; Nakari, 2005).

7.3.4.2 Miroestrol

Miroestrol was isolated in 1940, and the structure assigned in 1960 from *Pueraria mirifica* (Cain, 1960). This still relatively unknown phytoestrogen and more recently deoxymiroestrol have been isolated from the Thai drug based on this plant root. More recently it has been suggested that deoxymiroestrol is the actual oestrogenic principle found in *Pueraria mirifica* and that miroestrol is formed as an oxidation artefact (Chansakaow *et al.*, 2000). The relative binding of deoxymiroestrol and miroestrol to the ER of a human breast cancer cell line has been found to be 50 and 260 times respectively, the molar excess needed for 50% inhibition of ^{3}H oestradiol (Matsumura *et al.*, 2005). Genistein required a 1000-fold excess showing the potency of miroestrol and deoxymiroestrol as 4 and 20 times more potent than genistein, respectively. Deoxymiroestrol and 8-prenylnaringenin have comparable potency, in this assay, although deoxymiroestrol is generally considered the more potent overall when other assays are taken into account.

7.3.4.3 Prenylflanonoids

Prenylated flavonoids are perhaps the second most potent class of phytoestrogens (Milligan *et al.*, 1999), although the oestrogenic activity is still less than 1% of that of 17 β-oestradiol (Milligan *et al.*, 2000). While potent and found in a number of plants, prenylnaringenins are perhaps of limited effect as hops (*Humulus lupulus*) are to date the only identified dietary source of prenyl naringenins (Fig. 7.3). Beer is the most important source. These compounds are introduced through the use of hops as the bitter flavouring agent. Further prenylnaringenins are formed during the brewing process. Recent studies suggest that isoxanthohumol is also converted into 8-prenylnaringenin in the human distal colon (Possemiers *et al.*, 2006). After consumption, human intestinal microbiota convert further isoxanthohumol from the beer into 8-prenylnaringenin. Depending on inter-individual differences this could increase the intake of 8-prenylnaringenin 10-fold from beer consumption (Possemiers *et al.*, 2005). Despite the extremely high potency and the subtle increase in exposure through post-dietary-intake biotransformations, the actual daily intake level of prenylnaringenins is considered so low as to represent no benefit or detriment to health (Stevens and Page, 2004).

Phase I enzymes, for example, Cyp1 A2, are thought to be inhibited by xanthohumol and 8-prenylnaringenin (Miranda *et al.*, 2000b). Phase II enzymes by contrast have been shown to be induced in vitro. For example, NAD(P)H:quinone reductase has been shown to be induced by xanthohumol (Miranda *et al.*, 2000a). Xanthohumol could thus have beneficial effects on the detoxification of carcinogenic compounds by the inhibition of phase I and the induction of phase II enzymes.

Selectivity, as well as oestrogenic potency, differs for 8-prenylnaringenin binding to oe-strogen receptors in comparison with isoflavonoids. 8-Prenylnaringenin was found to be a 100 times more potent ERα agonist than genistein, but a much weaker agonist of ERβ in further oestradiol-competition assays for receptor binding (Schaefer *et al.*, 2003). The prenyl group of 8-prenylnaringenin is critical to its observed oestrogenicity as can be seen by the comparative lack of oestrogenicity of naringenin. Recent studies have investigated substituting the prenyl group (at C8) with alkyl chains of varying lengths and branching patterns (Roelens *et al.*, 2006). The new alkyl naringenins were found to have an activity spectrum ranging from partial to full agonism to antagonism. Interestingly, 8-(2,2-dimethylpropyl)naringenin showed full agonist characteristics using ERα activity assays, but pronounced antagonist characteristics using ERβ activity assays. This clearly showed the potential for the chemical optimisation of phytoestrogens, in particular the flavonoids, as potential Selective Oestrogen Modulators (SERMs) of the future. The search for new natural prenylflavonoids also continues, in particular evaluating prenylflavonoids from tonics and herbal medicines for their oestrogenic properties (Wang *et al.*, 2006).

What is lacking in comparing prenylated flavonoids with isoflavonoids and flavonoids is epidemiological and feeding trial data. Until this is forthcoming it is unknown whether or not they, as supplements, will be regarded as such potential health maintaining agents as the isoflavonoids. In view of the relatively high oestrogenicity of prenylated flavonoids, concern has been expressed over their potential adverse effect on health. The presence of 'feminising' agents in beer is considered a sensitive topic and is not something most brewers wish to promote.

7.3.4.4 Stilbenes

Resveratrol is widely distributed in the plant kingdom, but the principal dietary sources are grapes (*Vitis* spp.), peanuts (*Arachis* spp.), berries (blue berries and cranberries, *Vaccinium* spp.) and rhubarb (*Rheum* spp.) (Sigreorelli and Ghidoni, 2005). Recently resveratrol has been produced in transgenic apple fruit so in the future it may be present in even greater quantities in the diet (Ruhmann *et al.*, 2006). By far the most widely researched dietary stilbene is resveratrol or 3,5,4-t-hydroxystilbene which is found in plants mainly in the *trans* form. Like most polyphenols, resveratrol is generally found conjugated, principally as 3-O-β-D-glucosides called piceids. Other minor conjugated forms contain 1 or 2 methyl groups (e.g. pterostilbene), a sulfate group, or a fatty acid. As with isoflavonoids, absorption of resveratrol in the human intestine is principally via the aglycone form and therefore there is a requirement for glycosidases, including bacterial types, in the intestine. As regards metabolism and excretion, it is interesting that flavonoids such as quercetin inhibit the glucuronidation of resveratrol and may therefore increase its bioavailability (Sigreorelli and Ghidoni, 2005).

The principal health benefits of resveratrol ingestion are seen as the prevention of cardiovascular disease and cancer. Proposed mechanisms for the promotion of cardiovascular health include, in common with other phytoestrogens, antioxidant action and the induction of nitrous oxide to maintain vasodilation (Orallo *et al.*, 2002). Interestingly, resveratrol was found to inhibit nitric oxide production and iNOS (inducible nitric oxide synthase) expression in cancer cells, in contrast to its vasodilatory function (Roman *et al.*, 2002).

There has been particular interest, however, in resveratrol for its potential prevention of cancer. As for isoflavonoids and 8-prenylnaringenin, proposed mechanisms include the down regulation of phase I enzymes and up-regulation of phase II enzymes (Szaefer *et al.*, 2004). Sirtuins are a nicotinamide adenosine dinucleotide (NAD)-dependent class of deacetylases responsible for regulating the response to DNA damage and gene silencing process of ageing and survival. Resveratrol was found to activate human sirtuin 1 (SIRT 1) and sensitised cells to apoptosis (Yeung *et al.*, 2004).

In addition to interfering with cell cycle control, resveratrol, like other phytoestrogens, is potentially able to overcome drug resistance of tumours, for example breast cancer, that express multi-drug resistance associated proteins (ATP-dependent pumps that remove chemotherapeutics out of cells) (Cooray *et al.*, 2004). There appears therefore great potential for the use of resveratrol as a chemopreventative agent, although as for many other phytoestrogens further specific feeding trials need to be carried out to determine more precisely the potential health benefits. Resveratrol is antagonistic on both ERα and ERβ at high concentrations (Mueller *et al.*, 2004). Resveratrol is not a potent phytoestrogen (Mueller *et al.*, 2004; Harris *et al.*, 2005; Matsumura *et al.*, 2005). It would appear therefore that oestrogenicity is not the most significant property of resveratrol in terms of human health and that properties such as the activation of SIRT 1 will prove more significant for this compound in the future.

7.3.4.5   Licorice

Glycyrrhizin, a triterpene in licorice root, is 50 times sweeter than sugar leading to its use as a sweetener. Glycyrrhizin is believed to be oestrogenic as is the aglycone glycyrrhetic acid (Sharaf *et al.*, 1975). About 90 phenolic compounds have been isolated from the different species of licorice plants (*Glycyrrhiza*). Six of these are oestrogenic. The dihydrostilbene gancaonin R has a higher binding affinity than genistein (Fig. 7.4). While five others, liquiritigenin (hydroxychalcone), isobavachin (prenylfavanone), sigmoidin B (prenylflavanone), glycyrol (prenylcoumestan), glabrene (pyranoisoflavene) have binding affinities similar to genistein and daidzein (Nomura *et al.*, 2002). Licorice is especially important in Japan, where good health and phytoestrogen intake are believed to be closely linked. Licorice is frequently used in Japan as an over-the-counter medicine and most elderly Japanese choose licorice over synthetic medicine. Glycyrrhizin has preventative effects of oestrogen-related endometrial carcinogenesis in mice (Niwa *et al.*, 2007).

7.3.4.6   Ginseng

In the USA, ginseng (*Panax ginseng*) is used to alleviate menopausal symptoms. Ginseng contains a range of ca. 30 related ginsenosides, steroidal saponins, which comprise some 3–6% of ginseng. Two of the more active constituents are ginsenosides Rb1 and Rh1 (Fig. 7.4) (Punnonen and Lukola, 1980; Chan *et al.*, 2002; Lee *et al.*, 2003; Cho *et al.*, 2004). The most recent work suggested that ginsenoside Rg1 is an extremely potent phytoestrogen in the human breast cell proliferation assay, and it is more potent than e.g. coumestrol. As there is no specific binding to ER, ginsenosides may activate ER via a ligand-independent pathway (Chen *et al.*, 2006).

7.3.4.7   Peony

Peony is an ancient, traditional Chinese herbal medicine. The active component paeoniflorin (Fig. 7.4) is commonly used to treat dysmenorrhea (painful menses), polycystic ovary syndrome and pre-menstrual syndrome (PMS). Peony shows some weak oestrogen-like effects, acting like a very weak anti-oestrogen, particularly as part of the formula shakuyaku-kanzo-to.

In a preliminary study, this formula was shown to improve fertility in women affected by polycystic ovary syndrome (Takahashi and Kitao, 1994).

Paeoniflorin, glycyrrhetic acid and glycyrrhizin may affect the conversion between delta 4-androstenedione and testosterone to inhibit testosterone synthesis and stimulate aromatase activity to promote oestradiol synthesis by the direct action on the proestrous ovary (Takeuchi *et al.*, 1991).

7.3.4.8   Red tea phytoestrogens

The polyphenols from *Aspalathus linearis* have recently been introduced into the Western diet as red tea, or rooibos, a South African beverage containing no caffeine and low levels of tannins. This drink is promoted for its high antioxidant content, these same antioxidant polyphenols are also phytoestrogenic. Nothofagin is of similar oestrogenic potency to genistein, while vitexin, isovitexin, luteolin, luteolin-7-glucoside, hemiphlorin, aspalathin (Fig. 7.4) have more moderate oestrogenicity (Shimamura *et al.*, 2006).

7.3.4.9   Deoxybenzoins

As these diphenolics are considered to be the final intermediates in the synthesis of isoflavones they are also often considered to be extraction artefacts. These compounds are now being investigated for their oestrogenic activity (Fokialakis *et al.*, 2004). It has been known for some time that the oestrogenicity of red clover (*Trifolium subterraneum*) extracts increases markedly after treatment with ethanolic alkali and that this is probably due to as yet uncharacterised deoxybenzoins (Beck *et al.*, 1966). They have been found in several plant species, including licorice (*Glycyrrhiza*) and spiny restharrow (*Oononis spinosa*) another trifolium plant, which is used in homeopathic medicine, although the oestrogenic activity of such plants is often mistakenly attributed to isoflavones. The affinity of deoxybenzoins for ERα and ERβ show some bias towards ERβ (Fokialakis *et al.*, 2004) as is typified by the isoflavone genistein and as such deoxybenzoins represent a new class of ERβ selective phytoestrogen.

7.4   MEASUREMENT OF INDIVIDUAL PHYTOESTROGENS

The measurement of phytoestrogens is widely recognised as being highly challenging. Success requires an extremely high degree of expertise in both wet laboratory and instrument operation to achieve consistently accurate and precise results. With the continued absence of recognised reference materials it continues to be difficult to generate sufficient quality control to allow an objective and independent comparison of data generated by different laboratories conducting similar investigations.

7.4.1   Traditional approaches

The measurement of phytoestrogens has changed considerably in the last decade. A relatively recent review provides in depth information on methods (Wu *et al.*, 2004). Mass spectrometry (MS) methods are under continuous development, while many of the other methods are of more academic interest. Measurement of very high levels of phytoestrogens, such as in soya products has always been readily achieved by straightforward high performance liquid chromatography (HPLC) with ultraviolet (UV) detection. This approach has changed little in

Table 7.1   Main methods that may be used to quantify concentrations of phytoestrogens.

Method	Detector	Reference
GC	MS	Grace et al. (2003)
HPLC methods	UV	Kim et al. (2007)
	FLD	Richelle et al. (2002)
	Electrochemical	Penalvo and Nurmi (2006)
	NMR	Fritche et al. (2002)
Capillary electrophoresis	Capillary zone electrophoresis	Peng and Ye (2006)
	Micellar electrokinetic chromatography	Micke (2006)
	Capillary electrochromatography	Starkey et al. (2002)
Immunoassays	Radioimmunoassay	Lapcik et al. (2003)
	Enzyme-linked immunosorbent assays	Vitkova et al. (2004)
	Time-resolved fluoroimmunoassay	Brouwers et al. (2003)
	ER-CALUX	Legler et al. (2002)

the last decade and is a reliable and robust technique, which continues to be routinely used in niche applications, particularly measuring high concentrations of isoflavones in soya foods. It is generally combined with an acid hydrolysis step to convert all the glucoside forms into a lesser number of aglycones to further simplify the quantitation aspects of the analysis. This approach has the advantage of being low-cost and readily available, and has been extensively applied in support of food production and labelling claims. A small number of research groups have left out the acid hydrolysis step and directly measured the intact glucosides to establish the ratios of aglycone, primary glucosides, malonyl and acetyl glucosides. As the form in which the core phytoestrogen aglycone is ingested and presented internally affects bioavailability, these food conjugate analyses are conducted in support of clinical trials (Setchell et al., 2001; Wiseman et al., 2002). There are now a wide range of complementary liquid chromatography techniques with a range of detectors that can all be used to measure hydrolysed aglycones in food. Most of these techniques have relatively little merit for the analysis of complex samples with multiple analytes. Fluorescence is limited to a few analytes such as coumestrol and equol where fluorimetric detection is significantly more sensitive than UV (Richelle et al., 2002). Other approaches such as electrochemical detection with the coulometric electrode array detector lack the specificity to be used in any but the simplest applications (Penalvo and Nurmi, 2006). These are summarised in Table 7.1.

The bulk of analyses and the addition of new compositional data for individual foods is carried out by LC-MS/MS, where it is useful to have the higher degree of specificity for uniquely identifying various structurally similar phytoestrogens. For analysis of well-characterised foods where the analyte distribution is unambiguous HPLC-UV is still considered to be of adequate specificity.

7.4.2   Mass spectrometry

The low-level analysis of biological samples for phyoestrogens has always been more technically challenging. Clinically relevant concentrations of phytoestrogens in food are measured in mg/kg, while the human circulatory concentrations derived from these foods are two orders of magnitude lower in urine (μg/mL) and three orders lower in plasma (ng/mL).

The initial analytical methods were derived directly from traditional steroid analyses. For low-level pharmacokinetic studies involving the analysis of 1–1000 ng/mL of analytes in urine and plasma samples, the standard approach involved multiple chromatographic clean-up steps and derivatisation to trimethyl silyl ethers, to increase volatility before analysis by GC-MS. This approach was technically difficult, time-consuming and prone to numerous quality control failures. The advent of triple quadrupole LC-MS/MS instruments has resulted in these methods being quickly abandoned and replaced by simpler and more reproducible methods where the phytoestrogens can be analysed directly without derivatisation. While still technically challenging, when conducted by LC-MS operators with the same degree of technical expertise, these data sets are of vastly improved quality and more importantly are not prone to batch failures. Urine can be hydrolysed in situ with glucuronidase and analysed directly by LC-MS/MS, with much greater precision. Current LC-MS/MS instrumentation can now deliver better sensitivity than traditional GC-MS-based analyses working from the same sample size. Plasma analyses continued to be GC-MS-based until very recently when it became feasible to conduct a simple protein precipitation with acetonitrile and to dry this down and reconstitute for LC-MS/MS analysis. This approach now performs well at 1 ng/mL using 1 mL of human plasma. Many researchers now consider that food analyses with all the intrinsic problems with solvent selection and optimising extraction efficiencies for each food and analyte combination to be far more challenging than the biological analyses or MS issues. To summarise, the current best practise is either, HPLC-UV (Kim *et al.*, 2007), or LC-MS/MS for food analyses, LC-MS/MS for urine (Clarke *et al.*, 2002) and either LC-MS/MS, or GC-MS (Grace *et al.*, 2003) for plasma analyses.

The most significant recent advance in the underlying analytical chemistry has been the synthesis of a range of stable isotope standards of phytoestrogens. Isoflavones, plant and mammalian lignans, coumestrol as well as some glucosides and glucuronides have been prepared containing three carbon-13 atoms (Clarke *et al.*, 2002; Al-Maharik and Botting, 2004; Oldfield *et al.*, 2004; Fryatt and Botting, 2005; Haajanen and Botting, 2006). All modern mass spectrometers, using both GC and LC front-end sample introduction and separation techniques benefit from a procedure known as isotope dilution mass spectrometry (IDMS). In HPLC-UV methods an internal standard (IS) is added in a known quantity and all analyte peak sizes compared to this (normalisation). The IS must elute at a significantly different retention time to all the quantified analytes and thus by definition is chemically different from all of the analytes being measured. In MS, by using the chemically identical analytes that have undergone synthesis to contain three atoms of carbon-13 in place of carbon-12, the internal standard can be considered identical at the physical level, behaving exactly like the unlabelled analogue, co-eluting with the unlabelled version. This results in an exact match with identical experimental losses throughout any extraction procedure, and also lessens matrix effects and ionisation differences within the final mass spectral quantitation step. Having a different mass allows a labelled standard to be treated as a separate entity in the detector, although both are measured simultaneously. It is then the ratio of unlabelled to labelled versions that is used to calculate concentrations. Initially labelled standards were produced with deuterium (hydrogen-2), as the heavier isotope, these are less stable than carbon-13 versions as the deuterium can often be exchanged back out and replaced with hydrogen. This effect was minimal but observable during the acid hydrolysis of glycones and glucuronides to aglycones.

LC-MS/MS has also been used to probe questions of metabolism and pharmacokinetics. It is now possible to measure phytoestrogens in their biological, circulatory and excretory forms, glucuronides and sulfates (Clarke *et al.*, 2002), as well as cellular metabolites such as nitro derivatives. Anti-inflammatory agents are used in chemopreventive strategies. The inflammatory response involves the production of cytokines and proinflammatory oxidants such as hypochlorous acid (HOCl) and peroxynitrite (ONO_2-) produced by neutrophils and macrophages, respectively. The aromatic nature of polyphenols makes them potential targets of oxidation. Both chlorinated and nitrated genistein are formed by human neutrophils. These data imply a potential role for modified forms of genistein that would be produced in the inflammatory environment in and around a tumour (D'Alessandro *et al.*, 2003).

7.4.3 Immunoassays

As immunoassays are based on specific antibodies, these tests measure concentrations of discrete analytes. For a coumestrol immunoassay 3-O-carboxymethylcoumestrol was prepared as the hapten, which was conjugated to bovine serum albumin and used to immunise rabbits. The resultant rabbit polyclonal antiserum and I^{125} labelled hapten-tyrosine methyl ester conjugate was used as the radioligand to produce a quantitative assay. This was very specific to coumestrol with negligible cross-reactivity to isoflavones (Lapcik *et al.*, 2003). A different assay is needed for each phytoestrogen, which limits application. Immunoassays are often produced as competitive enzyme-linked immunosorbent assays (ELISA). A set of ELISA assays have been prepared for daidzein, genistein and biochanin A (Vitkova *et al.*, 2004). ELISA assays developed for steroidal oestrogens are often used to measure phytoestrogenicity as these assays are non-selective and phytoestrogens cross-react (Shimamura *et al.*, 2006). The available immunoassays, including enzyme immunoassay (CIA), fluoroimmunoassay (FIA), chemiluminescence immunoassay (CLIA) have been recently reviewed (Zhao *et al.*, 2007).

7.4.4 Online bioactivity assays

A new screening technology that combines biological analysis with the resolution power of HPLC is being developed, and this is referred to here as high-resolution screening (HRS) (Schobel *et al.*, 2001). The interactions of the analyte with the oestrogen receptor proceed at high speed in a closed, continuous-flow reaction detection system, which is coupled directly to the outlet of an HPLC column. The reaction products of this homogeneous fluorescence enhancement-type assay were detected online using a flow-through fluorescence detector, which was operated in combination with MS. This HRS system with biochemical detection can be described as LC-BCD-FLD/MS. This system dramatically enhances the speed of biologically active compound characterisation in natural product extracts compared to traditional bioassay directed fractionation approaches.

Online biochemical detection coupled to MS (LC-BCD-MS) was applied to profile the oestrogenic activity in a pomegranate peel extract. After an HPLC separation, the biologically active compounds were detected by an online beta-oestrogen receptor (β-ER) bioassay. Using this approach three oestrogenic compounds, i.e. luteolin, quercetin and kaempferol, were detected in pomegranate (van Elswick *et al.*, 2004). There are major limitations in that these techniques can only be based on the simplest in vitro assays and that at the technical level, in coupling enzyme assays to MS the choice and compatibility of solvent and buffer is severely limiting. Non-volatile buffers contaminate the spectrometer ion source and cause ionisation suppression. These online techniques (HRS by LC-BCD-MS) represent the potential future of phytoestrogen analytical chemistry.

7.5 MEASUREMENT OF OESTROGENICITY

The strength of traditional chromatographic analytical chemistry is the accurate quantitation of known compounds. The shortcoming is that only well-characterised analytes can be quantified and this requires considerable prior knowledge of the sample, to measure appropriate analytes. Phytoestrogen research by definition is derived from these compounds having biological effects. The oestrogenic potential of phytoestrogens is the defining property of these compounds. It is not however at all straightforward to measure oestrogenicity. The gold standard is the *in vivo* uterotrophic assay where immature rodents are fed the phytoestrogen under study and when mature, the weight of the uterus is then measured, the biological endpoint being growth in response to phytoestrogens, in comparison to control animals not exposed to oestrogens. The potency of a phytoestrogen is expressed as that required to produce an equivalent change in weight. A phytoestrogen would generally be expected to have increased the weight of the uterus. While this is ultimately the final test of whether a compound is a phytoestrogen, it is a lengthy and expensive process. A transgenic test has recently been developed – the blue mouse test in which tissues stain blue where oestrogen activity is present. This has the benefit of looking at the whole body rather than a single tissue such as the uterus. There are many endpoints studied, age of puberty, cell differentiation and proliferation, gene expression, oestrus cycle length, hormone concentrations etc. The majority of these tests can only be applied in animal models. A summary of the main techniques is given in Table 7.2. These bioassays quantify a simple biological effect (endpoint) and do not determine a concentration of the causative chemical (analyte).

Many other simpler tests are conducted in order to screen a wider range of samples and to support other studies. One of the major problems with in vitro oestrogenicity tests is that the chemical under test may be inactive as it requires metabolic activation before acquiring oestrogenicity. Classic examples are the increase in potency between the soya isoflavone daidzein and one of its metabolites equol, or the conversion of inactive plant lignans matairesinol and secoisolariciresiniol into the phytoestrogenic mammalian lignans, enterodiol and enterolactone by the action of gut microflora. The reverse is also seen where 8-prenylnaringenin is extremely potent in vitro and relatively inactive *in vivo*. Simple tests which do not take metabolism and transport mechanisms into account (in vitro) can therefore quite legitimately give widely differing results to more complex *in vivo* tests for the same chemicals. As there are many relevant mechanisms that are under study, it is difficult to make an accurate overall comparison of which phytoestrogens are more active. The current consensus taking into account both *in vitro* and *in vivo* studies ranks oestrogenic potency of the most widely studied analytes as: oestradiol ≥ deoxymiroestrol ≥ coumestrol ≥ genistein and equol > glycitein > 8-prenylnaringenin > daidzein > formononetin, biochanin A, 6-prenylnaringenin, xanthohumol, isoxanthohumol.

The different types of oestrogenic assay can be divided into three groups corresponding to their position in the oestrogen-dependent metabolic pathway (Figure 7.5). The first, cell proliferation measures a physiological response resulting from production of functional proteins, the second receptor-dependent gene expression measures the ability to activate the oestrogen responsive element to produce a protein product, the third receptor-binding assays measure the strength of the interaction with oestrogen and the receptor. Ligand-independent

Table 7.2 Main methods that may be used to quantify the bioactivity of phytoestrogens.

Method	Comments
In vitro	
Receptor-binding assay	Assay is easy to perform
Measures the affinity between oestrogens and oestrogen receptors	Offers a choice of receptor source, e.g. possibility to select subtype and isoform
	Assay measures affinity to oestrogen receptors
	Not possible to distinguish agonists from antagonists
Cell proliferation assay	Assay is easy to perform
Indirect measure of oestrogenic activity via ability of a compound to stimulate proliferation in an oestrogen-responsive cell line (e.g. MCF-7 cells)	Potency estimates can be derived
	Can distinguish agonists from antagonists
	Cell proliferation may occur independent of oestrogen receptors
	Cell lines can differ in their response to an oestrogen
Reporter gene assay	Provides estimation of potency
Indirect measurement of oestrogenic activity via expression of a reporter gene engineered into a cell line	Can distinguish agonists from antagonists
	Artificial system and dependent on cell line, response element and reporter gene used
Analysis of changes in gene expression	Provides estimation of potency
Estimation of oestrogen induced gene expression	Can be used to investigate tissue-specific effects
	Does not inform about functional response
In vivo	
Uterotrophic assay	Provides estimation of potency
Measurement of uterine growth in response to phytoestrogens in a rodent model low in endogenous oestrogens	Absorption, metabolism and excretion are incorporated in the assessment
	Only measures oestrogenic effects in one tissue
Transgenic mouse assay	Provides estimation of potency
Measures expression of an oestrogen sensitive reporter gene engineered into transgenic mice	Absorption, distribution, metabolism and excretion are incorporated in the assessment
	Identifies and measures effects in oestrogen-responsive tissues

processes can produce protein expression without binding to the standard oestrogen receptor.

One of the more interesting approaches is the CALUX assay. The *in vitro* oestrogen receptor-mediated chemically activated luciferase gene expression (CALUX) assay uses cells transfected with an ER-mediated luciferase gene construct. This expresses by emitting light when exposed to oestrogens. This luciferase construct has also been introduced to Zebra fish and the difference in reporter gene activation studied between the two systems. The fish will glow when exposed to endocrine disruptors (Legler *et al.*, 2002). Little work has been done using CALUX to measure phytoestrogens as there is currently no cost benefit in CALUX versus LC-MS.

To conduct a hazard or risk assessment of xenoestrogens and phytoestrogens requires the building of a complex equivalency factor model. This starts from knowledge of endocrine-disrupting chemicals. The relative potency (RP) of individual compounds relative to a standard (e.g. 17β-oestradiol) needs to be determined for several receptor-mediated responses.

The oestrogenic equivalent (EQ) of a mixture is defined as EQ = Sigma$[E-i]x$ RPi, where $E-i$ are concentrations of individual ER agonists in any mixture and RPi are the

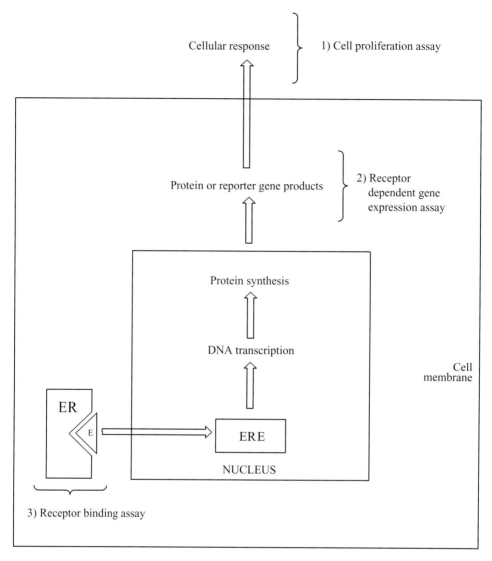

Fig. 7.5 Simplified schematic representing the mechanism of (1) the oestrogen-dependent cell proliferation assay and the location of (2) reporter gene and (3) ligand-binding assays. E, oestrogen; ER, oestrogen
receptor; ERE, oestrogen responsive element. (Reproduced from Thomson, B.M., Cressey, P.J. and Shaw,
I.C. Dietary exposure to xenoestrogens in New Zealand. *Journal of Environmental Monitoring,* **5**, 229–235,
Figure 2. Copyright 2003 by permission of the Royal Society of Chemistry.)

relative potencies (Safe, 1998). There are a number of obvious problems associated with
implementing this assessment. Not least of which will be deciding which actual responses
are to be measured. Oestrogenic responses are a complex integration of cell signalling and
agreeing the potency factors will take some time. The model assumes additivity, but a number
of non-additive interactions (synergies), both greater than and less than additive are being
reported for mixtures (Borgert *et al.*, 2003). There are few examples of the model in use
comparing low-dose exposures to xeno and natural (dietary) endocrine disrupters. While

problematic, this type of assessment is fundamental to future progress in the field and has already been used to confirm that phytoestrogens have a greater overall contribution to dietary oestrogenicity than synthetic chemicals (endocrine disruptors) (Thomson *et al.*, 2003).

7.6 OESTROGEN RECEPTORS

7.6.1 Introduction

As phytoestrogens are polyphenols, they exhibit numerous other biological effects. The key activities generally proceed by mimicking oestrogens, by interaction with the oestrogen receptors. Oestrogens play a vital role in the growth, development and homeostasis of oestrogen responsive tissues. Oestrogens bind to the ligand-binding domain of oestrogen receptors resulting in either the activation or repression of target genes (Katzenellenbogen *et al.*, 1996; Brosens and Parke, 2003; Koehler *et al.*, 2005; Dahlman-Wright *et al.*, 2006). Oestrogen receptors mediate the biological activity of oestrogens and are ligand-inducible nuclear transcription factors that belong to the 'nuclear hormone receptor superfamily' (NR). In humans this is composed of 48 genes with diverse roles in metabolic homeostasis, development and detoxification. Uniquely, oestrogen receptors are the only steroid receptors in the 'nuclear hormone receptor superfamily' which lack specificity and in addition when binding oestrogens, are able to interact with a large number of non-steroidal compounds. These endocrine disruptors (ED) frequently show a structural similarity to the steroid nucleus of the oestrogen, including phytoestrogens, and drug and environmental xenoestrogens (Brosens and Parker, 2003). Structural features of these EDs enable them to bind to oestrogen receptors to elicit responses ranging from agonism to antagonism of the endogenous hormone-ligand (Miksicek, 1995).

7.6.2 Receptors – ERα and ERβ

Originally it was believed that only one oestrogen receptor existed. This is in contrast to other members of the nuclear receptor superfamily where multiple forms of each receptor have been found. The first receptor that could bind 17β-oestradiol was identified in the late 1950s and was cloned in 1986 (Greene *et al.*, 1986) and a second oestrogen receptor was reported in 1996 (Kuiper *et al.*, 1996). Today, these two oestrogen receptors are now known as ERα and ERβ (Koehler *et al.*, 2005; Dahlman-Wright *et al.*, 2006). The amino acid sequence identity between ERα and ERβ is very high ($\sim$97%) in the DNA-binding domain, lower ($\sim$56%) in the ligand-binding domain and poor ($\sim$24%) in the N terminus (Koehler *et al.*, 2005; Dahlman-Wright *et al.*, 2006).

ERβ was identified from rat prostate and ovary tissues (Kuiper *et al.*, 1996) and it is of considerable interest that ERβ shows a different tissue distribution to ERα. ERβ was first reported to be strongly expressed in ovary, uterus, brain, bladder, testis, prostate, lung (Kuiper *et al.*, 1997) and normal human breast tissue (Crandall *et al.*, 1998). Normal and cancerous breast tissues display distinct ERα and ERβ profiles, some data shows regulation of ERβ by promoter methylation (frequently observed in cancer) and indicates that the ERβ transcript is down-regulated in breast tumours consequently it has been suggested that ERβ is a possible tumour suppressor gene. However, ERβ is expressed in the majority of breast tumours and a similar percentage of tumours express ERβ as ERα (Dahlman-Wright *et al.*, 2006). ERβ has been shown to be expressed in the cardiovascular system (Makela *et al.*, 1999) where results from oestrogen receptor knockout mice indicate that ERα is important in the pathophysiology of the vessel wall, whereas ERβ knockout mice display abnormalities in ion channel function and an age-related hypertension (Dahlman-Wright *et al.*, 2006). Expression of ERβ appears to occur at different sites in the brain from ERα (Kuiper *et al.*, 1997), in rodents ERα is mostly found in regions involved in control of reproductive function such as the hypothalamus, whereas ERβ is more widely distributed including in the hippocampus and cortex. ERβ has also been found to be expressed in bone (Arts *et al.*, 1997; Onoe *et al.*, 1997; Stossi *et al.*, 2004) where studies of female ERβ knockout mice show that ERβ is responsible for repression of the ERα-mediated growth stimulating effect on oestrogen on bone (Dahlman-Wright *et al.*, 2006).

Genistein is a much better ligand for ERβ than for the ERα (~30-fold higher binding affinity) (Kuiper *et al.*, 1997; Koehler *et al.*, 2005), it can also act as an oestrogen agonist via both ERα and ERβ in some test systems (Kuiper *et al.*, 1998; Mueller *et al.*, 2004). The selective oestrogen receptor antagonist raloxifene and the related anticancer drug tamoxifen (Wiseman, 1994) can inhibit the mitogenic effects of oestrogen in reproductive tissues, while maintaining the beneficial effects of oestrogen in other tissues. Genistein also behaves as a partial oestrogen agonist in human kidney cells transiently expressing ERβ, suggesting that it may be a partial oestrogen antagonist in some cells expressing ERβ (Barkhem *et al.*, 1998). Binding studies to ERβ variants have demonstrated that coumestrol and genistein bind to ERβ2 with a weaker affinity than to ERβ1 (Petersen *et al.*, 1998).

7.6.3 Oestrogen receptor-binding assays

Oestrogen receptor-binding assays provide a quantitative indication of the oestrogenic activity of a phytoestrogen relative to that of oestradiol at the receptor level. This is the most studied effect of phytoestrogens and this in vitro test rather than *in vivo* testing is routinely taken as the first proof of phytoestrogenicity. Experiments examine the binding affinity of phytoestrogens to oestrogen receptors, which is termed the relative-binding affinity (RBA). RBAs are usually expressed as the concentration of a compound required to inhibit oestradiol binding by 50% (IC_{50}; Blair *et al.*, 2000) (see Fig. 7.6). RBAs provide a simple measurement of the affinity of a compound to the receptor but do not indicate whether a compound will behave as an agonist, antagonist or partial agonist.

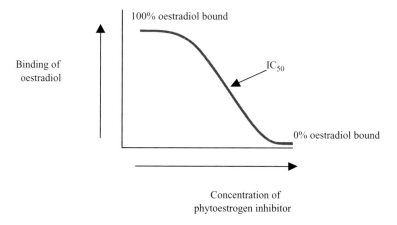

Fig. 7.6 Determination of IC_{50} from a competitive ligand-binding assay.

Conclusions

Phytoestrogens are becoming increasingly prevalent in the Western diet, through intake of soya products and through increasing exposure to ethnic foods and the acceptance of dietary supplements and traditional plants. Many known compounds are now being re-examined and found to have oestrogenic properties and new classes of increasingly potent phytoestrogens are being discovered.

Significant improvements have been made in the analytical techniques such as LC-MS for measuring phytoestrogens and in the many bioassays for detecting their biological properties. There has been an enormous body of published work generated on the actions of phytoestrogens. The studies showing no adverse effects such as on human reproduction are of major importance. It is still early days for the long job of documenting and fully investigating the numerous promising findings for potential health benefits. The investigations proving any long-term health benefits for phytoestrogens while of most interest are proving to be more challenging than expected.

References

Aizawa, T. and Hu, J.-Y. (2003) Quantitative structure–activity relationships for estrogen receptor binding affinity of phenolic chemicals. *Water Research*, **37**, 1213–1222.

Al-Maharik, N. and Botting, N.P. (2004) A new short synthesis of coumestrol and its application for the synthesis of [6,6a,11a-C-13(3)]coumestrol. *Tetrahedron*, **60**(7), 1637–1642.

Arts, J., Kuiper, G.G.J.M. and Janssen, J.M.M.F. (1997) Differential expression of estrogen receptors α and β mRNA during differentiation of human osteoblast SV-HFO cells. *Endocrinology*, **138**, 5067–5070.

Baker, V.A., Hepburn, P.A., Kennedy, S.J., Jones, P.A., Lea, L.J., Sumpter, J.P. and Ashby, J. (1999) Safety evaluation of phytosterol esters. Part 1. Assessment of oestrogenicity using a combination of *in vivo* and *in vitro* assays. *Food and Chemical Toxicology*, **37**(1), 13–22.

Barkhem, T., Carlsson, B. and Nilsson, Y. (1998) Differential response of estrogen receptor α and estrogen receptor β to partial estrogen agonists/antagonists. *Molecular Pharmacology*, **54**, 105–112.

Beck, A.B., Kowala, C. and Lamberton, J.A. (1966) 2,4-dihydroxy-4'-methoxybenzil, a probable artefact from formononetin in subterranean clover extracts. *Australian Journal of Chemistry*, **19**, 1755–1757.

Blair, R.M., Fang, H., Branham, W.S., Hass, B.S., Dial, S.L., Moland, C.L., Tong, W., Shi, L., Perkins, R. and Sheehan, D.M. (2000) The estrogen receptor relative binding affinities of 188 natural and xenochemicals: Structural diversity of ligands. *Toxicological Sciences*, **54**, 138–153.

Blitz, C.L., Murphy, S.P. and Au, D.L.M. (2007) Adding lignan values to a food composition database. *Journal of Food Composition and Analysis*, **20**(2), 99–105.

Borgert, C.J., LaKind, J.S. and Witorsch, R.J. (2003) A critical review of methods for comparing estrogenic activity of endogenous and exogenous chemicals in human milk and infant formula. *Environmental Health Perspectives*, **111**(8), 1020–1036.

Bronsens, J.J. and Parker, M.G. (2003) Oestrogen receptor hijacked. *Nature*, **423**, 487–488.

Brouwers, E., L'homme, R., Al-Maharik, N., Lapcik, O., Hampl, R., Wahala, K., Mikola, H. and Adlercreutz, H. (2003) Time-resolved fluoroimmunoassay for equol in plasma and urine. *Journal of Steroid Biochemistry and Molecular Biology*, **84**(5), 577–588.

Cain, J.C. (1960) Miroestrol: An oestrogen from the plant *Pueraria mirifica*. *Nature*, **188**, 774–777.

Carrao-panizzi, M.C., Beleia, A.D. and Oliveira, M.C.N. (1999) Effects of genetics and environment on isoflavone content of soybean from different regions of Brazil. *Pesquisa Agropecuaria Brasileira*, **34**(10), 1787–1795.

Chan, R.Y., Chen, W.F., Dong, A., Guo, D. and Wong, M.S. (2002) Estrogen-like activity of ginsenoside Rg1 derived from *Panax notoginseng*. *Journal Clinical Endocrinology Metabolism*, **87**, 3691–3695.

Chansakaow, S., Ishikawa, T., Seki, H., Sekine, K., Okada, M. and Chaichantipyuth, C. (2000) Identification of deoxymiroestrol as the actual rejuvenating principle of 'Kwao Keur', *Pueraria mirifica*. The known miroestrol may be an artefact. *Journal of Natural Products*, **63**, 173–175.

Chen, W.-F., Lau, W.-S., Cheung, P.-K., Guo, D.-A. and Wong, M.-S. (2006) Activation of insulin-like growth factor I receptor-mediated pathway by ginsenoside Rg1. *British Journal of Pharmacology*, **147**, 542–551.

Cho, J., Park, W., Lee, S., An, W. and Lee, Y. (2004) Ginsenoside-Rb1 from *Panax ginseng* C. A. Meyer activates estrogen receptor-α and $-\beta$, independent of ligand binding. *Journal of Clinical Endocrinology and Metabolism*, **89**(7), 3510–3515.

Clarke, D.B., Lloyd, A.S., Botting, N.P., Oldfield, M.F., Needs, P.W. and Wiseman, H. (2002) Measurement of intact sulfate and glucuronide phytoestrogen conjugates in human urine using isotope dilution liquid chromatography–tandem mass spectrometry with [$^{13}C_3$]-isoflavone internal standards D. *Analytical Biochemistry*, **309**, 158–172.

Cooray, H.C., Janvilisri, T., van Ween, H.W., Hladky, S.B. and Barrand, M.A. (2004) Interaction of breast cancer resistance protein with plant polyphenols. *Biochemical and Biophysical Research Communications*, **317**, 269–275.

Crandall, D.L., Busler, D.E., Novak, T.J., Weber, R.V. and Kral, J.G. (1998) Identification of estrogen receptor RNA in human breast and abdominal subcutaneous adipose tissue. *Biochemical and Biophysical Research Communications*, **248**, 523–526.

Dahlman-Wright, K., Cavailles, V., Fugua, S.A., Jordan, V.C., Katzenellenbogen, J.A., Korach, K.S., Maggi, A., Muramatsu, M., Parker, M.G. and Gustafsson, J.-A. (2006) International Union of Phamacology. LXIV. Estrogen receptors. *Pharmacological Reviews*, **58**, 773–781.

D'Alessandro, T., Prasain, J., Benton, M.R., Botting, N., Moore, R., Darley-Usmar, V., Patel, R. and Barnes, S. (2003) Polyphenols, inflammatory response, and cancer prevention: Chlorination of isoflavones by human neutrophils. *Journal of Nutrition*, **133**(11, suppl 1), 3773S–3777S.

Einbond, L.S., Shimizu, M., Nuntanakorn, P., Seter, C., Cheng, R., Jiang, B., Kronenberg, F., Kennelly, E.J. and Weinstein, I.B. (2006) Actein and a fraction of black cohosh potentiate antiproliferative effects of chemotherapy agents on human breast cancer cells. *Planta Medica*, **72**(13), 1200–1206.

Eldridge, A.C. and Kwolek, W.F. (1983) Soybean isoflavones: Effect of environment and variety on composition. *Journal of Agriculture and Food Chemistry*, **31**, 394–396.

Fokialakis, N., Lambrinidis, G., Mitsiou, D.J., Aligiannis, N., Mitakou, S., Skaltsounis, A.L., Pratsinis, H., Mikros, E. and Alexis, M.N. (2004) A new class of phytoestrogens; evaluation of the oestrogenic activity of deoxybenzoins. *Chemistry and Biology*, **11**, 397–406.

Fritche, J., Angoelal, M. and Dachtler, M. (2002) On-line liquid-chromatography–nuclear magnetic resonance spectroscopy–mass spectrometry coupling for the separation and characterization of secoisolariciresinol diglucoside isomers in flaxseed. *Journal of Chromatography A*, **972**(2), 195–203.

Fryatt, T. and Botting, N.P. (2005) The synthesis of multiply C-13-labelled plant and mammalian lignans as internal standards for LC-MS and GC-MS analysis. *Journal of Labelled Compounds and Radiopharmaceuticals*, **48**(13), 951–969.

Geller, S.E. and Studee, L. (2005) Botanical and dietary supplements for menopausal symptoms: What works, what does not. *Journal of Womens Health*, **14**(7), 634–649.

Grace, P.B., Taylor, J.I., Botting, N.P., Fryatt, T., Oldfield, M.F. and Bingham, S.A. (2003) Quantitation of isoflavones and lignans in urine using gas chromatography/mass spectrometry. *Analytical Biochemistry*, **315**(1), 114–121.

Greene, G.L., Gilna, P., Waterfield, M., Baker, A., Hort, A. and Shine, J. (1986) Sequence and expression of human estrogen receptor complementary DNA. *Science*, **231**, 1150–1154.

Haajanen, K. and Botting, N.P. (2006) Synthesis of multiply C-13-labeled furofuran ligands using C-13-labelled cinnamyl alcohols as building blocks. *Steroids*, **71**(3), 231–239.

Harris, D.M., Besselink, E., Henning, S.M., Go, V.L.W. and Heber, D. (2005) Phytoestrogens induce differential estrogen receptor alpha- or beta-mediated responses in transfected breast cancer cells. *Experimental Biology and Medicine*, **230**, 558–568.

Hoeck, J.A., Fehr, W.R. and Welke, G.A. (2000) Influence of genotype and environment on isoflavone contents of soybean. *Crop Science*, **40**, 48–51.

Horn-Ross, P.L., Barnes, S., Lee, V.S., Collins, C.N., Reynolds, P., Lee, M.M., Stewart, S.L., Canchola, A.J., Wilson, L. and Jones, K. (2006) Reliability and validity of an assessment of usual phytoestrogen consumption (United States). *Cancer Causes and Control*, **17**(1), 85–93.

Howes, L.G., Howes, J.B. and Knight, D.C. (2006) Isoflavone therapy for menopausal flushes: A systematic review and meta-analysis. *Matauritas*, **55**, 203–211.

Jiang, B., Kronenberg, F., Balick, M.J. and Kennelly, E.J. (2006) Analysis of formononetin from black cohosh (*Actaea racemosa*). *Phytomedicine*, **13**(7), 477–486.

Jimenez-Escrig, A., Santos-Hidalgo, A.B. and Saura-Calixto, F. (2006) Common sources and estimated intake of plant sterols in the Spanish diet. *Journal of Agricultural and Food Chemistry*, **54**(9), 3462–3471.

Ju, Y.H., Clausen, L.M., Allred, K.F., Almada, A.L. and Helferich, W.G. (2004) Beta-sitosterol, beta-sitosterol glucoside, and a mixture of beta-sitosterol and beta-sitosterol glucoside modulate the growth of estrogen-responsive breast cancer cells in vitro and in ovariectomized athymic mice. *Journal of Nutrition*, **134**(5), 1145–1151.

Katzenellenbogen, J.A., O'Malley, B.W. and Katzenellenbogen, B.S. (1996) Tripartite steroid hormone receptor pharmacology: Interaction with multiple effector sites as a basis for the cell- and promoter-specific action of these hormones. *Molecular Endocrinology*, **10**, 119–131.

Kiely, M., Faughnan, M., Wahala, K., Brants, H. and Mulligan, A. (2003) Phyto-oestrogen levels in foods: The design and construction of the VENUS database. *British Journal of Nutrition*, **89**(suppl 1), S19–S23.

Kim, J.-A., Hong, S.-B., Jung, W.-S., Yu, C.-Y., Ma, K.-H., Gwag, J.-G. and Chung, I.-M. (2007) Comparison of isoflavones composition in seed, embryo, cotyledon and seed coat of cooked-with-rice and vegetable soybean (*Glycine max* L) varieties. *Food Chemistry*, **102**, 738–744.

Koehler, K.F., Helguero, L.A., Haldosen, L.-A, Warner, M. and Gustafsson, J.-A. (2005) Reflections on the discovery and significance of estrogen receptor β. *Endocrine Reviews*, **26**, 465–478.

Kuiper, G.G.J.M., Enmark, E., Pelto-Huikko, M., Nilsson, S. and Gustafsson, J.-A. (1996) Cloning of a novel estrogen receptor expressed in rat prostate and ovary. *Proceedings of the National Academy of Sciences USA*, **93**, 5925–5930.

Kuiper, G.G.J.M., Carlsson, B. and Grandien, K. (1997) Comparison of the ligand binding specificity and transcript tissue distribution of estrogen receptors α and β. *Endocrinology*, **138**, 863–870.

Kuiper, G.G.J.M., Lemmen, J.G. and Carlsson, B.O. (1998) Interaction of estrogenic chemicals and phytoestrogens with estrogen receptor β. *Endocrinology*, **139**, 4252–4263.

Lai, G.F., Wang, Y.F., Fan, L.M., Cao, J.X. and Luo, S.D. (2005) Triterpenoid glycoside from Cimicifuga racemosa. *Journal of Asian Natural Products Research*, **7**(5), 695–699.

Lapcik, O., Stursa, J., Kleinova, T., Vitkova, M., Dvorakova, H., Klejdus, B. and Moravcoca, J. (2003) Synthesis of hapten and conjugates of coumestrol and development of immunoassay. *Steroids*, **68**(14), 1147–1155.

Lee, Y., Jin, Y., Lim, W., Ji, S., Choi, S., Jang, S. and Lee, S. (2003) A ginsenoside-Rh1, a component of ginseng saponin, activates oestrogen receptor in human breast carcinoma MCF-7 cells. *The Journal of Steroid Biochemistry and Molecular Biology*, **84**(4), 463–468.

Legler, J., Zeinstra, L.M., Schuitemaker, F., Lanser, P.H., Bogerd, J., Brouwer, A., Vethaak, A.D., De Voogt, P., Murk, A.J. and van der Burg, B. (2002) Comparison on *in vivo* and in vitro reporter gene assays for short term screening of estrogenic activity. *Environmental Science and Technology*, **36**(20), 4410–4415.

MacDonald, R.S., Guo, J.Y., Copeland, J., Browning, J.D., Sleper, D., Rottinghaus, G.E. and Berhow, M.A. (2005) Environmental influences on isoflavones and saponins in soybeans and their role in colon cancer. *The Journal of Nutrition*, **135**, 1239–1242.

Makela, S., Savolainen, H. and Aavik, E. (1999) Differentiation between vasculoprotective and uterotrophic effects of ligands with different binding affinities to estrogen receptors α and β. *Proceedings of the National Academy of Sciences USA*, **96**, 7077–7082.

Matsumura, A., Ghosh, A., Pope, G.S. and Darbre, P.D. (2005) Comparative study of oestrogenic properties of eight phytoestrogens in MCF7 human breast cancer cells. *The Journal of Steroid Biochemistry and Molecular Biology*, **94**, 431–443.

Mellanen, P., Petänen, T., Lehtimäki, J., Mäkelä, S., Bylynd, G., Holmbom, B., Mannila, E., Oikkari, A. and Santti, R. (1996) Wood-derived estrogens: Studies *in vitro* with breast cancer cell lines and *in vivo* in trout. *Toxicology and Applied Pharmacology*, **136**, 381–388.

Mellanen, P., Soimasua, M., Holmbom, B., Oikari, A. and Santti, R. (1999) Expression of the vitellogenin gene in the liver of juvenile whitefish (coregonus lavaraetus L s.l) exposed to effluents from pulp and paper mills. *Ecotoxicology and Environmental Safety*, **43**(2), 133–137.

Micke, G.A., Fujiya, N.M., Tonin, F.G., Costa, A.C.D. and Tavares, M.F.M. (2006) Method development and validation for isoflavones in soy germ pharmaceutical capsules using micellar electrokinetic chromatography. *Journal of Pharmaceutical and Biomedical Analysis*, **41**(5), 1625–1632.

Miksicek, R.J. (1995) Estrogenic flavonoids. Structural requirements for biological activity. *Proceedings of the Society for Experimental Biology and Medicine*, **208**, 44–50.

Milder, I.E.J., Arts, I.C.W., van de Putte, B., Venema, D.P. and Hollman, P.C.H. (2005) Lignan contents of Dutch plant foods: A database including lariciresinol, pinoresinol, secoisolariciresinol and matairesinol. *British Journal of Nutrition*, **93**(3), 393–402.

Milligan, S.R., Kalita, J.C., Heyerick, A., Rong, H., De Cooman, L. and De Keukeleire, D. (1999) Identification of a potent phytoestrogen in hops (*Humulus lupulus* L.) and beer. *Journal of Clinical Endocrinology and Metabolism*, **84**, 2249–2252.

Milligan, S.R., Kalita, J.C., Pocock, V., Van De Kauter, V., Stevens, J.F., Deinzer, M.L., Rong, H. and De Keukeleire, D. (2000) The endocrine activities of 8-prenylnaringenin and related hop (*Humulus lupulus* L.) flavonoids. *Journal of Clinical Endocrinology and Metabolism*, **85**, 4912–4915.

Miranda, C.L., Aponso, G.L., Stevens, J.F., Deinzer, M.L. and Buhler, D.R. (2000a) Prenylated chalcones and flavanones as inducers of quinone reductase in mouse Hepa 1c1c7 cells. *Cancer Letters*, **149**, 21–29.

Miranda, C.L., Yang, Y.H., Henderson, M.C., Stevens, J.F., Santana-Rios, G., Deinzer, M.L. and Buhler, D.R. (2000b) Prenylated flavonoids from hops inhibit the metabolic activation of the carcinogenic heterocyclic amine 2-amino-3-methylimidazo[4,5-*f*]quinoline, mediated by cDNA-expressed human CYP1A2. *Drug Metabolism and Disposition*, **28**, 1297–1302.

Moghadasian, M.H. (2000) Pharmacological properties of plant sterols. *In vivo* and *in vitro* observations. *Life Sciences*, **67**, 605–615.

Muelle, S.O., Simon, S., Chae, K., Metzier, M. and Korach, K.S. (2004) Phytoestrogens and their human metabolites show distinct agonistic and antagonistic properties on estrogen receptor alpha (ERalpha) and ERbeta in human cells. *Toxicological Sciences*, **80**, 14–25.

Nachtigall, L.E., Baber, R.J., Barentsen, R., Durand, N., Panay, N., Pitkin, J., van de Weijer, P.H.M. and Wysocki, S. (2006) Complementary and hormonal therapy for vasomotor symptom relief: A conservative clinical approach. *Journal of Obstetrics and Gynaecology Canada*, **28**(4), 279–289.

Nakari, T. (2005) Estrogenicity of phytosterols evaluated *in vitro* and *in vivo*. *Environmental Science*, **12**(2), 87–97.

Nieminen, P., Mustonen, A.-M., Lindstrom-Seppa, P., Karkkainen, V., Musalo-Rauhamaa, H. and Kukkonen, J.V.K (2003) Phytosterols affect endocrinology and metabolism of the field vole (*Microtus agrestis*). *Experimental Biology and Medicine*, **228**, 188–193.

Niwa, K., Lian, Z., Onogi, K., Yun, W., Tang, L., Mori, H. and Tamaya, T. (2007) Preventive effects of glycyrrhizin on estrogen-related endometrial carcinogenesis in mice. *Oncology Reports*, **17**(3), 617–622.

Nomura, T., Fukai, T. and Akiyama, T. (2002) Chemistry of phenolic compounds of licorice (*Glycyrrhiza* species) and their estrogenic and cytotoxic activities. *Pure and Applied Chemistry*, **74**(7), 1199–1206.

Normén, A.L., Brants, H.A.M., Voorrips, L.E., Andersson, H.A., van den Brandt, P.A. and Goldbohm, R.A. (2001) Plant sterol intakes and colorectal cancer risk in the Netherlands Cohort Study on Diet and Cancer. *American Journal of Clinical Nutrition*, **74**(1), 141–148.

Oldfield, M.F., Chen, L.R. and Botting, N.P. (2004) Synthesis of [3,4,8-C-13(3)]daidzein. *Tetrahedron*, **60**(8), 1887–1893.

Onoe, Y., Miyaura, C., Ohta, H., Nozawa, S. and Suda, T. (1997) Expression of estrogen receptor β in rat bone. *Endocrinology*, **138**, 4509–4512.

Onorato, J. and Henion, J.D. (2001) Evaluation of triterpene glycoside estrogenic activity using LC/MS and immunoaffinity extraction. *Analytical Chemistry*, **73**(19), 4704–4710.

Orallo, F., Alvarez, E., Camina, M., Leiro, J.M., Gomez, E. and Fernandez, P. (2002) The possible implication of trans-resveratrol in the cardioprotective effects of long-term moderate wine consumption. *Molecular Pharmacology*, **61**, 294–302.

Park, M., Paik, H.Y. and Joung, H. (2005) Establishment of isoflavone database of Korean foods and estimation of isoflavone intake of Korean adolescent girls. *Journal of the Federation of American Societies for Experimental Biology*, **19**(4, part 1, suppl S), A88–A88.

Penalvo, J.L. and Nurmi, T. (2006) Application of coulometric electrode array detection to the analysis of isoflavonoids and lignans. *Journal of Pharmaceutical and Biomedical Analysis*, **41**(5), 1497–1507.

Peng, Y.Y. and Ye, J.N. (2006) Determination of isoflavones in red clover by capillary electrophoresis with electrochemical detection. *Fitoterapia*, **77**(3), 171–178.

Petersen, D.N., Tkalcevic, G.T., Koza-Taylor, P.H., Turi, T.G. and Brown, T.A. (1998) Identification of estrogen receptor beta2, a functional variant of estrogen receptor beta expressed in normal rat tissues. *Endocrinology*, **139**, 1082–1092.

Possemiers, S., Heyerick, A., Robbens, V., De Keukeleire, D. and Verstraete, W. (2005) Activation of proestrogens from hops (*Humulus lupulus* L.) by intestinal microbiota; conversion of isoxanthohumol into 8-prenylnaringenin. *Journal of Agriculture and Food Chemistry*, **53**, 6281–6288.

Possemiers, S., Bolca, S., Grootaert, C., Heyerick, A., Decroos, K., Dhooge, W., De Keukeleire, D., Rabot, S., Verstraete, W. and Van de Wiele, T. (2006) The prenylflavonoid isoxanthohumol from hops (*Humulus lupulus* L) is activated to the potent phytoestrogen 8-prenylnaringenin in vitro and in the human intestine. *Journal of Nutrition*, **136**, 1862–1867.

Primomo, V.S., Poysa, V., Ablett, G.R., Jackson, C.J., Gijzen, M. and Rajcan, I. (2005a) Mapping QTL for individual and total isoflavone content in soybean seeds. *Crop Science*, **45**(6), 2454–2464.

Primomo, V.S., Poysa, V., Ablett, G.R., Jackson, C. and Rajcan, I. (2005b) Agronomic performance of recombinant inbred line populations segregating for isoflavone content in soybean seeds. *Crop Science*, **45**(6), 2203–2211.

Punnonen, R. and Lukola, A. (1980) Oestrogen-like effect of ginseng. *British Medical Journal*, **281**, 1110.

Richelle, M., Pridmore-Merten, S., Bodenstab, S., Enslen, E. and Offord, E.A. (2002) Hydrolysis of isoflavone glycosides to aglycones by β-glycosidase does not alter plasma and urine pharmacokinetics in post-menopausal women. *The Journal of Nutrition*, **132**, 2587–2592.

Ritchie, M.R., Cummings, J.H., Morton, M.S., Steel, M.C., Bolton-Smith, C. and Riches, A.C. (2006) A newly constructed and validated isoflavone database for the assessment of total genistein and daidzein intake. *British Journal of Nutrition*, **95**(1), 204–213.

Roelens, F., Heldring, N., Dhooge, W., Bengtsson, M., Comhaire, F., Gustafsson, J.A., Treuter, E. and De Keukeleire, D. (2006) Subtle side-chain modifications of the hop phytoestrogen 8-prenylnaringenin result in distinct agonist/antagonist activity profiles for oestrogen receptors alpha and beta. *Journal of Medicinal Chemistry*, **49**, 7357–7365.

Roman, V., Billard, C., Kern, C., Ferry-Dumazet, H., Izard, J.C., Mohammad, R., Mossalayi, D.M. and Kolb, J.P. (2002) Analysis of resveratrol-induced apoptosis in human B-cell chronic leukaemia. *British Journal of Haematology*, **117**, 842–851.

Ruhlen, R.L., Tracy, J.K., Haubner, J., Zhu, W.Z., Qin, W.Y., Lamberson, W.R., William, R., Rottinghaus, G.E. and Sauter, E.R. (2007) Black cohosh lacks an estrogenic effect in breast. *Journal of Nutrition*, **137**(1, suppl S),288S–288S.

Ruhmann, S., Treutter, D., Fritsche, S., Briviba, K. and Szankowski, I. (2006) Piceid (resveratrol glucoside) synthesis in stilbene synthase transgenic apple fruit. *Journal of Agriculture and Food Chem*istry, **54**, 4633–4640.

Safe, S.H. (1998) Hazard and risk assessment of chemical mixtures using the toxic equivalency factor approach. *Environmental Health Perspectives*, **106**(suppl 4), 1051–1058.

Schaefer, O., Humpel, M., Fritzemeier, K.H., Bohlmann, R. and Schleuning, W.D. (2003) 8-Prenyl naringenin is a potent ERalpha selective phytoestrogen present in hops and beer. *Journal of Steroid Biochemistry and Molecular Biology*, **84**, 359–360.

Schobel, U., Frenay, M., van Elswick, D.A., McAndrews, J.M., Long, K.R., Olson, L.M., Bobzin, S.C. and Irth, K. (2001) High resolution screening of plant natural product extracts for estrogen receptor alpha and beta binding activity using an online HPLC-MS biochemical detection system. *Journal of Biomolecular Screening*, **6**(5), 291–303.

Sequin, P. and Zheng, W. (2006) Potassium, phosphorus, sulfur, and boron fertilization effects on soybean isoflavone content and other seed characteristics. *Journal of Plant Nutrition*, **29**(4), 681–698.

Setchell, K.D.R., Brown, N.M., Desai, P., Zimmer-Nechemias, L., Wolfe, B.E., Brashear, W.T., Kirschner, A.S. and Cassidy, A. (2001) Bioavailability of pure isoflavones in healthy humans and analysis of commercial soy isoflavone supplements. *The Journal of Nutrition*, **131**, 1362S–1375S.

Sharaf, A., Gomaa, N. and El Gamal, M.H.A. (1975) Glycyrrhetic acid as an active oestrogenic substance separated from *Glycyrrhiza glabra* (liquorice). *Egyptian Journal of Pharmaceutical Sciences*, **16**(2), 245–251.

Shimamura, N., Miyase, T., Umehara, K., Warashina, T. and Fujii, S. (2006) Phytoestrogens form *Aspalathus linearis*. *Biological and Pharmaceutical Bulletin*, **29**(6), 1271–1274.

Signorelli, P. and Ghidoni, R. (2005) Resveratrol as an anticancer nutrient: Molecular basis, open questions and promises. *The Journal of Nutritional Biochemistry*, **16**, 449–466.

Starkey, J.A., Mechref, Y., Byun, C.K., Steinmetz, R., Fuqua, J.S., Pescovitz, O.H. and Novotny, M.V. (2002) Determination of trace isoflavone phytoestrogens in biological materials by capillary electrochromatography. *Analytical Chemistry*, **74923**, 5998–6005.

Stevens, J.F. and Page, J.E. (2004) Xanthohumol and related prenylflavonoids from hops to beer: To your good health! *Phytochemistry*, **65**, 1317–1330.

Stossi, F., Barnett, D.H. and Frasor, J. (2004) Transcriptional profiling of oestrogen-regulated gene expression via estrogen receptor (ER) α or ERβ in human osteosarcoma cells: Distinct and common target genes for these receptors. *Endocrinology*, **145**, 3473–3486.

Szaefer, H., Cichocki, M., Brauze, D. and Baer-Bubowska, W. (2004) Alteration in phase I and phase II enzyme activities and polycyclic aromatic hydrocarbons – DNA adduct formation by plant phenolics in mouse epidermis. *Nutrition and Cancer – an International Journal*, **48**, 70–77.

Takahashi, K. and Kitao, M. (1994) Effect of TJ-68 (shakuyaku-kanzo-to) on polycystic ovarian disease. *International Journal of Fertility and Menopausal Studies*, **39**, 69–76.

Takeuchi, T., Nishii, O., Okamura, T. and Yaginuma, T. (1991) Effect of paeoniflorin, glycyrrhizin and glycyrrhetic acid on ovarian androgen production. *American Journal of Chinese Medicine*, **19**(1), 73–78.

Thomson, B.M., Cressey, P.J. and Shaw, I.C. (2003) Dietary exposure to xenoestrogens in New Zealand. *Journal of Environmental Monitoring*, **5**, 229–235.

Thompson Coon, J., Pittler, M.H. and Ernst, E. (2007) *Trifolium pratense* isoflavones in the treatment of menopausal hot flushes: A systematic review and meta-analysis. *Phytomedicine*, **14**, 153–159.

Tsukamoto, C., Shimada, S., Igita, K., Kudou, S., Kokubun, M., Okubo, K. and Kitamura, K. (1995) Factors affecting isoflavone content in soybean seeds: Changes in isoflavones, saponins, and composition of fatty acids at different temperatures during seed development. *Journal of Agriculture and Food Chemistry*, **43**, 1184–1192.

Valsta, L.M., Kilkkinen, A., Mazur, W., Nurmi, T., Lampi, A.M., Ovaskainen, M.L., Korhonen, T., Adlercreutz, H. and Pietinen, P. (2003) Phyto-oestrogen database of foods and average intake in Finland. *British Journal of Nutrition*, **89**(suppl 1), S31–S38.

van Elwijk, D.A., Schobel, U.P., Lansky, E.P., Irth, H. and vab der Greef, J. (2004) Rapid dereplication of estrogenic compounds in pomegranate (*Punica granatum*) using on-line biochemical detection coupled to mass spectrometry. *Phytochemistry*, **65**(2), 233–241.

Vaya, J. and Tamir, S. (2004) The relation between the chemical structure of flavonoids and their estrogen-like activities. *Current Medicinal Chemistry*, **11**(10), 1333–1343.

Vitkova, M., Mackova, Z., Fukal, L. and Lapcik, O. (2004) Enzyme immunoassay for the determination of isoflavones. *Chemické Listy*, **98**, 1135–1139.

Wang, H.J. and Murphy, P.A. (1994) Isoflavone composition of American and Japanese soybeans in Iowa: Effects of variety, crop year, and location. *Journal of Agriculture and Food Chemistry*, **42**, 1674–1677.

Wang, Z.Q., Weber, N., Lou, Y.J. and Proksch, P. (2006) Prenylflavonoids as nonsteroidal phytoestrogens and related structure–activity relationships. *ChemMedChem*, **1**, 482–488.

Wegulo, S.N., Yang, X.B., Martinson, C.A. and Murphy, P.A. (2005) Effects of wounding and inoculation with *Sclerotinia sclerotiorum* on isoflavone concentration in soybeans. *Canadian Journal of Plant Sciences*, **85**(4), 749–760.

Wiseman, H. (1994) *Tamoxifen: Molecular Basis of use in Cancer Treatment and Prevention*. John Wiley, Chichester.

Wiseman, H., Casey, K., Clarke, D., Barnes, K. and Bowey, E. (2002) The isoflavone aglycone and gluco-conjugate content of commercial and "home-prepared" high soy and low soy foods and food dishes in the UK selected for use in chronic controlled nutritional studies. *Journal of Agricultural and Food Chemistry*, **50**, 1404–1419.

Wu, Q., Wang, M. and Simon, J.E. (2004) Analytical methods to determine phytoestrogenic compounds. *Journal of Chromatography B*, **812**, 325–355.

Yeung, F., Hoberg, J.E., Ramsey, C.S., Keller, M.D., Jones, D.R., Frye, R.A. and Mayo, M.W. (2004) Modulation of NF-kappaB-dependent transcription and cell survival by the SIRT 1 deacetylase. *The European Molecular Biology Organization Journal*, **23**, 2369–2380.

Yin, X.H. and Vyn, T.J. (2005) Relationships of isoflavone, oil, and protein in seed with yield of soybean. *Agronomy Journal*,**97**(5), 1314–1321.

Yu, O. and McGonigle, B. (2005) Metabolic engineering of isoflavone biosynthesis. *Advances in Agronomy*, **86**, 147–190.

Zhao, M.P., Zhou, S., Yan, J. and Li, L. (2007) Immunochemical analysis of endogenous and exogenous estrogens. *Current Pharmaceutical Analysis*, **3**(1), 25–28.

8   β-Carboline Alkaloids

Tomás Herraiz

Summary

Tetrahydro-β-carbolines (THβCs) and β-carbolines (βCs) are naturally and food-occurring indole alkaloids produced through a Pictet–Spengler condensation from indoleamines and carbonyl compounds. This chapter covers their analysis, formation, and occurrence in foods as well as their biological and toxicological activity. Sample preparation procedures based on solid phase extraction (SPE) followed by selective and sensitive GC-MS and RP-HPLC coupled with fluorescence and mass detection (including tandem MS) are currently used for analysis. A number of THβC and βC molecules occur in foods under a varying range of concentration and distribution patterns from ng/kg to mg/kg. Among THβCs are tetrahydro-β-carbolines, tetrahydro-β-carboline-carboxylic acids, and polyol- or phenolic-tetrahydro-β-carbolines; whereas among aromatic βCs are norharman, harman, and carbohydrate-derived βCs. Formation of these compounds occurs during food production, processing, and storage depending on several biological, chemical, and technological factors. Transformation among compounds also exists and the fully aromatic βCs (norharman and harman) arise in foods from oxidation of tetrahydro-β-carboline-3-carboxylic acid. THβCs and βCs are bioactive substances exhibiting a broad range of biological and pharmacological actions including enzyme inhibition and binding to several body and brain receptors. They are also antioxidants and free radical scavengers. Various toxicological aspects such as their involvement as slow acting neurotoxins or their possible co-mutagenic or cytotoxic properties are of current interest. We are exposed to THβCs and βCs through the diet and their exogenous intake from foods is likely contributing to the occurrence of these compounds in the human body where they act as bioactive substances.

8.1   INTRODUCTION

THβCs and βCs form a class of indole alkaloids which appear in several families of natural products (Allen and Holmstedt, 1980) and exhibit a broad range of pharmacological and biological activities. Molecules of this kind show neuroactive, antimicrobial, antiviral, antioxidant, cytotoxic, and anticarcinogenic actions. Good examples are reserpine, an antihypertensive agent from *Rauwolfia serpentina*, yohimbine from Yohimbe bark (*Pausinystalia yohimbe*), the anticancer-related vinca alkaloids like ajmalicine from *Catharanthus roseus*, the psychoactive substances harmine, harmaline, and tetrahydroharmine from *Peganum harmala*, and from *Banisteriopsis caapi* which is used to prepare Ayahuasca, an hallucinogenic beverage used in rituals by the Amazonian tribes, and the recently discovered antitumor brominated β-carbolines called eudistomins and eudistomidins isolated from the tunicate *Eudistoma* sp. and manzamines isolated from marine sponges.

A number of simple THβCs and βCs have been detected so far in biological tissues and fluids of mammals (Airaksinen and Kari, 1981a; Melchior and Collins, 1982; Rommelspacher *et al.*, 1991; Brossi, 1993; Fekkes and Bode, 1993; Matsubara *et al.*, 1998; Parker *et al.*, 2004). These findings have triggered many studies and speculations on a possible function or biochemical role of these *mammalian* alkaloids *in vivo*. They might function as neuro-modulators via effects on monoamine oxidase (MAO), monoamine uptake and binding to several brain receptors such as serotonin, benzodiazepine, or imidazoline (Buckholtz, 1980; Braestrup *et al.*, 1980; Airaksinen and Kari, 1981b; Rommelspacher *et al.*, 1994; Robinson *et al.*, 2003; Herraiz and Chaparro, 2005). β-carbolines have also attracted notable scientific interest from a toxicological approach because they act as co-mutagens or precursors of mutagens (Higashimoto *et al.*, 1996; Totsuka *et al.*, 1998; Boeira *et al.*, 2002), cause neuronal cell death (Brenneman *et al.*, 1993; Bringmann *et al.*, 2006), and are bioactivated to produce endogenous neurotoxins (Matsubara *et al.*, 1992, 1998; Hamann *et al.*, 2006). Those are only a few examples but enough to conclude that the biological activity and toxicity of these compounds appear to be complex and its complete delineation is still needed and awaiting future elucidation.

In the last few years, it has become clear that β-carboline alkaloids occur in many commercial foods of different origin, and that they are produced during food production processing and storage (Herraiz *et al.*, 1993; Herraiz, 1996, 1998, 1999a,b, 2000a,b,c,d, 2002, 2004a,b; Gutsche and Herderich, 1997a,b; Pfau and Skog, 2004). The diet may somehow contribute to the ultimate presence of those alkaloids in the human biological tissues and fluids. Therefore, the availability and exposure to xenobiotic β-carbolines are currently a matter of interest. These compounds could exert biological and toxicological actions when ingested from foods, which will be potentiated further if they are accumulated within body tissues. This chapter will deal with the chemistry, formation, analysis, as well as the occurrence and factors influencing the content of THβCs and βCs in foods. In addition, it will address the potential biological activity and toxicity expected for these substances.

8.2 CHEMISTRY, SYNTHESIS, AND FORMATION OF β-CARBOLINE ALKALOIDS

The carbolines constitute a group of isomeric heterocyclic compounds classified as pyrido(b)indoles (Fig. 8.1) and which differ on the position of N-atom within the pyrido ring. THβCs and βCs contain a tricyclic pyrido(3,4-b)indole structure with a reduced and a fully aromatic pyridine ring, respectively. The nitrogen of βCs is at N-2 compared to mutagenic and carcinogenic α- and γ-carbolines detected in well-cooked meats, which have the nitrogen atom in other positions of the pyridine. Molecular diversity of β-carbolines appears with distinct substituents both at the pyridine and benzenic ring, and also through different degrees of saturation in the pyridine ring producing THβCs, dihydro-β-carbolines (DHβCs), and the fully aromatic βCs.

The most effective and commonly used route for synthetic preparation of THβCs is the so-called Pictet–Spengler reaction between indolethylamines (i.e., tryptamine and tryptophan derivatives) and aldehydes or α-ketoacids (Brossi, 1993; Cox and Cook, 1995). The mechanism of this reaction proceeds via formation of an iminium cation intermediate (Schiff base) and cyclization probably through spiroindolenium species to give THβCs (Fig. 8.2).

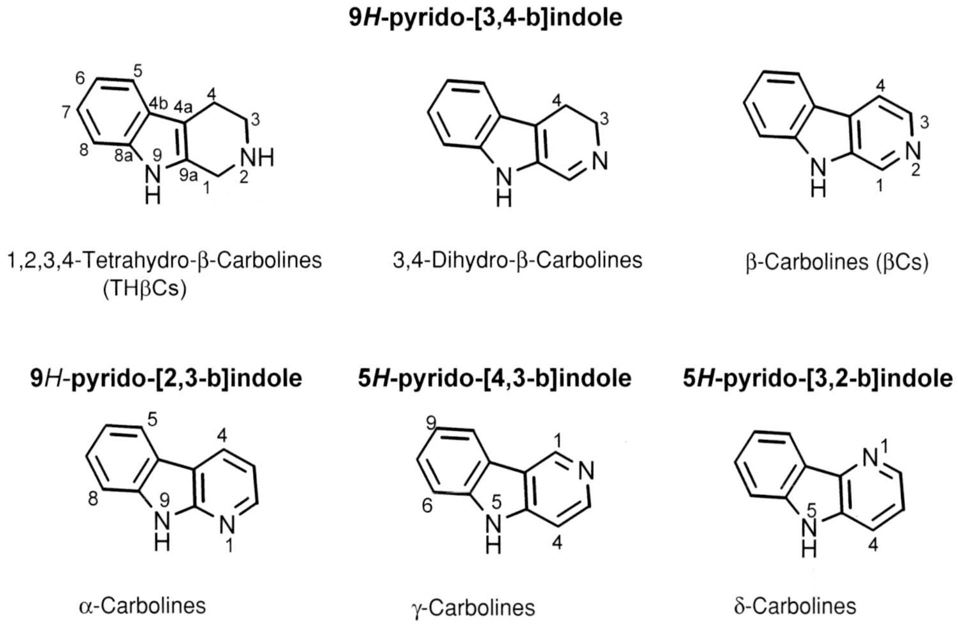

9*H*-pyrido-[3,4-b]indole

1,2,3,4-Tetrahydro-β-Carbolines
(THβCs)

3,4-Dihydro-β-Carbolines

β-Carbolines (βCs)

9*H*-pyrido-[2,3-b]indole

5*H*-pyrido-[4,3-b]indole

5*H*-pyrido-[3,2-b]indole

α-Carbolines

γ-Carbolines

δ-Carbolines

Fig. 8.1 Structures and nomenclature of carbolines (pyrido-b-indoles).

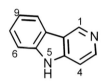
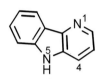

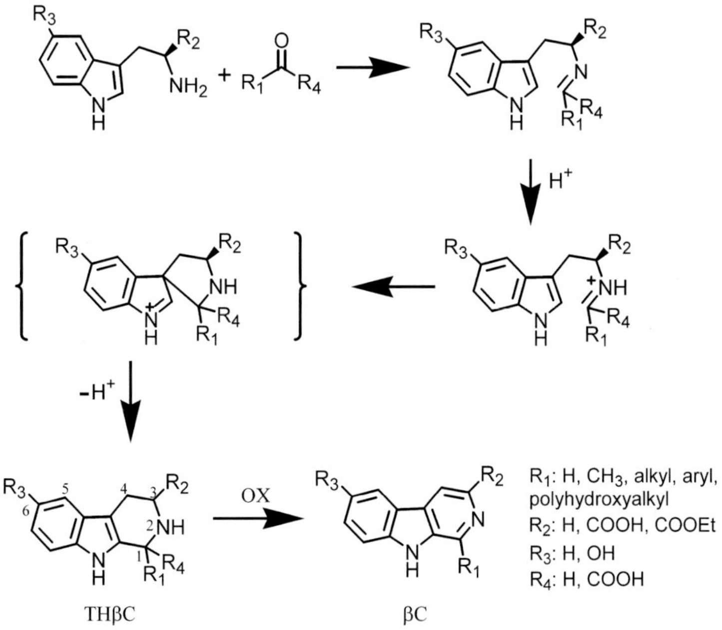

R₁: H, CH₃, alkyl, aryl, polyhydroxyalkyl
R₂: H, COOH, COOEt
R₃: H, OH
R₄: H, COOH

THβC

βC

Fig. 8.2 Pictet–Spengler reaction between indolethylamines or amino acids and aldehydes or α-ketoacids to give THβCs which can be oxidized to aromatic βCs.

This reaction is catalyzed by Brønsted acids and in a lesser extent by Lewis acids. Each specific THβC molecule produced in this reaction depends on the indolealkylamine precursor and carbonyl compound involved. Thus, from tryptophan, it produces tetrahydro-β-carboline-3-carboxylic acid (THβC-3-COOH), whereas THβCs arise from tryptamine. Serotonin (5-hydroxy-tryptamine) gives the corresponding 6-hydroxy-THβCs. α-Ketoacids and aldehydes other than formaldehyde provide THβCs with a chiral carbon at C-1 (two enantiomers) whereas two diastereoisomers (*cis* and *trans*) occur from L-tryptophan-derived 1,3-disubstituted THβC-3-COOH ($1S,3S$ and $1R,3S$). The stereochemistry of this reaction has been studied and it may be controlled under specific conditions (Cox and Cook, 1995). Oxidation of the THβC pyridine ring produces dihydro (DHβCs) or the fully aromatic βCs, which are two important subclasses of these alkaloids with singular biological properties. This oxidation can be accomplished with several chemical reagents such as palladium on carbon, elemental sulfur, DDQ, MnO_2, SeO_2. Another commonly employed method for synthetic preparation of βCs is the Bischler–Napieralski reaction. It also uses tryptophan or tryptamine derivatives as starting materials and involves the acylation of the amine followed by the treatment with strong dehydrating agents (phosphorus pentoxide and phosphorus oxychloride) resulting in the cyclization to give DHβCs. The latter can be oxidized to give the fully aromatic βCs or reduced to give THβCs. This reaction usually requires more vigorous reaction conditions than those necessary for the Pictet–Spengler condensation.

Under the relatively mild conditions existing in foods compared to chemical synthesis, formation of THβCs occurs as illustrated in Fig. 8.2 through a cyclocondensation reaction between indolealkylamines (tryptamine, tryptophan, or serotonin and their derivatives) and carbonyl compounds (aldehydes or α-ketoacids) (Herraiz *et al.*, 1993; Herraiz, 1996, 1997, 1998, 1999a,b, 2000a,b,c,d, 2002, 2004b). Both reactants occur in foods and a number of chemical, technological, and biological factors during preparation, processing, and storage may determine the presence of βCs. A subsequent stepwise oxidation of THβCs gives DHβCs and those aromatic βCs occurring in foods. Although THβCs, DHβCs, and βCs may exhibit different chemical and biological properties, and consequently each one is worth of being considered on its own, they will be considered together here for practical reasons and to provide a whole picture of these compounds in foods.

8.3   ANALYSIS OF β-CARBOLINE ALKALOIDS

THβCs and βCs are usually present at relatively low concentrations in foods, thus a clean-up step including purification and trace enrichment is necessary for their analysis. This analysis requires a careful isolation, separation, and chemical identification (for review, see Herraiz, 2000a). High-resolution chromatographic techniques providing efficient and selective separations (i.e., GC and HPLC) should be used to avoid co-eluting and interfering peaks. For detection, UV-absorption, fluorescence, and mass spectrometry have been used with good results.

8.3.1   Sample preparation

As for many other analytes in foods, SPE has become a very useful tool for sample preparation and isolation of THβCs and βCs (Herraiz, 2000a). The use of SPE with different cartridges and mechanisms has important advantages over liquid–liquid extraction. SPE allows rapid clean-up providing extracts free of co-extractives, thus reducing possible artifacts, which have been a traditional concern for high sensitivity assays of THβCs. Indeed, generation of artifacts during sample preparation may occur by a Pictet–Spengler reaction arising from traces of naturally occurring indoleamines and aldehydes or α-ketoacids (see Fig. 8.2). This is a critical point and needs to be carefully considered during sample preparation. Two approaches have been used: trapping aldehydes with reagents such as semicarbazide (Bosin *et al.*, 1983), and removing traces of indoleamines with fluorescamine (Tsuchiya *et al.*, 1994) or methyl chloroformate (Bosin and Jarvis, 1985; Herraiz, 1996, 2000a). Additional precautions are the use of appropriate controls and blanks which may include labeled precursors, direct chromatographic analysis, and mass detection without sample preparation, and preparation of the sample at different working conditions (pH, solvents, temperature). Early quantitative methods using solvent extractions that did not take into account the possible formation of artifacts should be regarded with caution.

SPE of THβCs alkaloids can be accomplished by using different cartridges and mechanisms based on nonpolar and ionic-exchange interactions (Herraiz, 2000a). C_{18}-SPE has been used for the isolation of THβCs from biological samples and foods (Schouten and Bruinvels, 1985; Musshoff *et al.*, 1993; Herraiz and Sanchez, 1997; Gutsche and Herderich, 1997a). A fast and reliable ion-exchange–SPE procedure used for THβCs is based on benzenesulfonic cation exchange columns (SCX). In this procedure after loading acidified samples, cartridges are washed with HCl and water, the pH adjusted with phosphate buffer pH 9, and the THβCs extracted from the column with a mixture of the same buffer and methanol (Adachi *et al.*, 1991a; Herraiz *et al.*, 1993). This procedure was used for extraction of 3-carboxylated THβCs and THβCs from many commercial foods with good results (Herraiz and Ough, 1993; Herraiz, 1996, 1998, 1999a,b, 2000b,c,d, 2004b; Herraiz and Galisteo, 2003; Herraiz and Papavergou, 2004).

Aromatic βCs can be isolated by liquid–liquid extraction (Bosin and Faull, 1988a,b), and by using SPE procedures working on reversed phase or ion-exchange mechanisms. As compared with THβCs, the formation of artifacts of aromatic βCs during sample preparation is not considered a major problem. However, artifact formation should not be ruled out because the most abundant THβC-3-COOHs and 1,3-dicarboxylic-THβCs are oxidized under relative mild conditions to afford the less abundant aromatic βCs (Herraiz, 2000a,b,c, 2004a,b). A reliable SPE procedure is based on PRS (propylsulfonic acid-derivatized silica) sorbents (Adachi *et al.*, 1991b; Herraiz, 2000a, 2002). In this method, acidified samples are loaded onto PRS sorbents and after washing the columns with diluted acid and water, βCs are eluted with a mixture (1:1) of methanol–phosphate buffer pH 9. With this method, βCs (norharman and harman) were recently reported in coffee brews (Herraiz, 2002), and many other foodstuffs and tobacco smoke (Herraiz, 2004a).

8.3.2 Chromatographic analysis and detection

Both, GC and HPLC have been used for separation of β-carbolines (Herraiz, 2000a). Analysis of THβCs by GC (detection by FID or ECD) and GC-MS requires derivatization of THβCs to obtain volatile derivatives (i.e., *N*-trifluoroacetyl, *N*-pentafluoropropionyl or *N*-heptafluorobutyryl or other acyl reagents) (Beck *et al.*, 1983; Matsubara *et al.*, 1986; Bringmann *et al.*, 1996; Herraiz and Ough, 1994; Herraiz, 1997, 2000a). Halogen derivatives are good electron acceptors providing a high sensitivity with ECD detectors. GC-MS under negative ion chemical ionization mass spectrometry (NICI-MS) was a sensitive method to analyze THβCs (Johnson *et al.*, 1984; Matsubara *et al.*, 1986; Hayashi *et al.*, 1990; Tsuchiya *et al.*, 1996). GC-MS under electron ionization (EI) gives abundant molecular ions ($M^{+•}$) along with characteristic mass fragments of THβCs that can be used for quantification; for this purpose, the samples often incorporate deuterated standards. GC-MS (EI) allowed the identification of 3-carboxylated THβCs (THβC-3-COOHs), which are the most abundant THβCs in foods. These THβCs were found in many foods including blue cheese, yogurt, cider, soy and Tabasco sauces, orange juice, vinegar and in some alcoholic beverages such as wine, cider and beer, and smoked foods (Herraiz and Sanchez, 1997; Papavergou and Herraiz, 2003). For that, N-methoxycarbonyl methyl ester derivatives of THβC-3-COOHs were obtained and separated (including $1S,3S$ and $1R,3S$ diastereoisomers) into nonpolar capillary columns (Fig. 8.3). Derivatization of THβC-3-COOH requires two derivatization steps, though the use of methyl chloroformate reagent allowed derivatization of the amine and carboxylic group in one step (in presence of pyridine and methanol) (Hušek, 1991; Herraiz

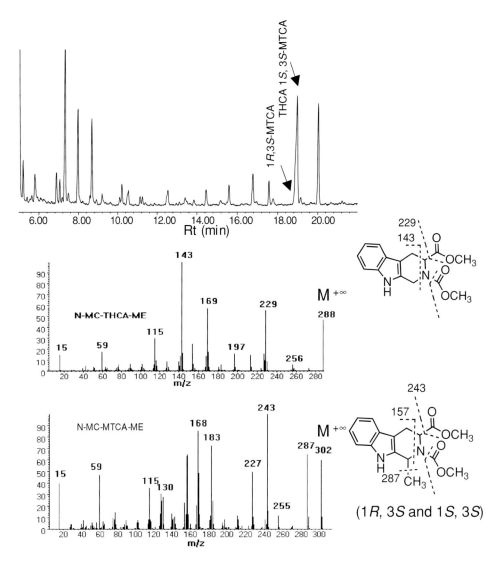

Fig. 8.3   GC-MS (electron ionization) and mass fragmentation of two main THβC-3-COOH in foods as N-methoxycarbonylmethyl ester derivatives. Chromatogram corresponding to sauce (Tabasco).

and Sanchez, 1997). By using those derivatives, THβCs gave good and relative clean total ion chromatograms and their EI-spectra and mass fragmentation was provided and assigned (Herraiz, 1997). Fragmentation is dominated by retro Diels-Alder rearrangement providing the ions $C_{10}H_9N^+$ (m/z 143) for THβCs and $C_{11}H_{11}N^+$ (m/z 157) for 1-methyl-THβCs, respectively. THβCs also exhibit abundant molecular ions ($M^{+\bullet}$) and N-methoxycarbonyl derivatives suffer loss of $COOCH_3$. GC-MS (EI) was used to study 3-carboxylic THβCs in wines as their N-trifluoroacetyl methyl ester derivatives (Herraiz and Ough, 1994), and to analyze the presence of N-nitroso THβCs derivatives in nitrite-treated foods by monitoring the $(M-NO)^+$ and molecular ions ($M^{+\bullet}$) (Sen *et al.*, 1995).

Simple molecules of aromatic βCs such as norharman ($9H$-pyrido-(3,4-b)-indole) and harman (1-methyl-$9H$-pyrido-(3,4-b)indole) can be also analyzed by GC. Harman was derivatized with pentafluorobenzyl bromide (Bosin and Faull, 1988a,b) and the resulting 9-pentafluorobenzylharman analyzed under negative ion chemical ionization (NICI) GC-MS that provided a single ion at m/z 181 corresponding to the indole anion. Under specific chromatographic conditions, harman and norharman can be analyzed by GC-MS without chemical derivatization by using nonpolar stationary phases. This approach allowed their determination in cooked meats and meat extracts, cigarette smoke, and oxidation of THβCs (Skog *et al.*, 1998; Herraiz and Galisteo, 2002a; Herraiz and Chaparro, 2005). However, chromatographic adsorption might be a drawback of this analysis.

HPLC combined with fluorescence or mass detection is generally the best choice for the analysis of both THβC and βC alkaloids (Herraiz, 2000a,c). Disadvantages of GC, such as the need for chemical derivatization including possible artifact generation are currently solved by HPLC-fluorescence and HPLC-MS of underivatized THβCs or βCs. The analysis of THβCs can be accomplished by using C_{18}-RP-HPLC with fluorescence detection under appropriate wavelengths (e.g., 270 nm excitation and 343 nm emission). This method has been applied to the analysis of THβCs in foods following SCX sample preparation (Herraiz, 1996, 1998, 1999a,b,c,d, 2004b). Detection by fluorescence is both sensitive and selective offering information about THβCs through the spectrum of chromatographic peaks (Herraiz, 1996, 2000a,b). HPLC-fluorescence also allows the analysis of THβC derivatives containing different substituents at the tetrahydropyrido ring such as alkyl, polyhydroxyalkyl, and benzenic or phenolic substituents (Herraiz and Galisteo, 2002b; Herraiz *et al.*, 2003). Nevertheless, some compounds with specific substituents may lose the fluorescence (quenching) of the tetrahydropyrido indole ring.

Chemical identification of THβCs in foods has improved greatly by using RP-HPLC coupled to mass spectrometry. HPLC-MS (electrospray ionization – ESI) provides good protonated molecular ions $(M + H)^+$ while using low energy collision induced dissociation (CID) gives the neutral loss of iminoacetic acid moiety $C_2H_3NO_2$ ($-73 \ u$) for THβC-3-COOH and imine moiety $CH_2 = NH$ ($-29 \ u$) for THβCs, which arise from retro Diels-Alder fragmentation, as mentioned above for electron ionization (IE) (Herraiz, 1997; Gutsche and Herderich, 1997a,b, 1998; Herraiz and Galisteo, 2002b; Herraiz *et al.*, 2003). Based on this fragmentation in combination with product ion experiments, HPLC/MS/MS allowed a successful substructure-specific identification (profiling) of isomeric and co-eluting THβCs (Gutsche and Herderich, 1997a,b, 1998). HPLC-MS (ESI) has been employed to identify THβCs in foods, including cooked and smoked foods, wines, sauces, and fruit products (Adachi *et al.*, 1991a; Sen *et al.*, 1995; Gutsche and Herderich, 1997a,b, 1998; Herraiz, 2000a,b,c,d, 2004b; Herraiz and Galisteo, 2002b, 2003; Herraiz and Papavergou, 2004), and also to identify N-nitroso-THβC in foods (Wakabayashi *et al.*, 1983; Sen *et al.*, 1995). Figure 8.4 illustrates the identification of THβCs in a tomato juice. In addition, novel

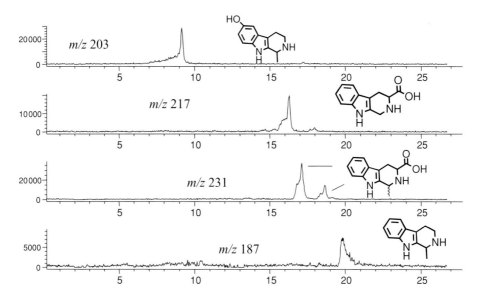

Fig. 8.4 Identification by HPLC-MS (ESI, positive ionization) of THβCs isolated from commercial tomato juice. (Reprinted with permission from Herraiz and Galisteo in *Journal of Agricultural and Food Chemistry*, **51**, 7156–7161. Copyright 2003 with permission from the American Chemical Society.)

molecules such as phenolic THβCs (derived from reaction of tryptophan and phenolic aldehydes) and carbohydrate-derived THβCs have been identified in foods and food-related samples by HPLC-MS (Herraiz and Galisteo, 2002b; Herraiz *et al.*, 2003).

Aromatic βCs are usually analyzed by RP-HPLC-fluorescence and MS. βCs exhibit high native fluorescence and fluorescent detection is a suitable and sensitive method for quantitative analysis with spectra providing qualitative information (Herraiz, 2000a,b). Under acidic eluents, excitation is set at 245 or 300 nm and emission around 433–445 nm. The amount of βCs in foods is relatively low (usually at ng/g) and often they appear under interfering chromatographic peaks being usually necessary its identification by MS. Mild ESI gives simple mass spectra in which the main signal is due to protonated molecular ions $(M + H)^+$ (e.g., *m/z* 169 for norharman, *m/z* 183 for harman) (Herraiz, 2000a, 2002, 2004a,b). Although βCs are more stable toward the ionization process than THβCs, characteristic fragments (i.e., HCN) can be obtained by increasing fragmentation by CID or tandem MS/MS (Toribio *et al.*, 2002). Figure 8.5 illustrates the identification of aromatic βCs norharman and harman in coffee brews by HPLC-MS (electrospray, positive ionization) (Herraiz, 2002). They were also identified in many other processed foods (Herraiz, 2000a, 2004a). Similarly, HPLC with CID and tandem MS is a useful tool for identification of novel carbohydrate-derived aromatic βCs in foods arising from a reaction of tryptophan and glucose (Diem and Herderich, 2001a,b; Herraiz, unpublished results).

8.4 OCCURRENCE OF β-CARBOLINE ALKALOIDS IN FOODS AND EXPOSURE

8.4.1 THβCs in foods

A number of THβCs resulting from a Pictet–Spengler condensation between indolethylamines and aldehydes or α-ketoacids have been detected in foods (Fig. 8.6). From tryptophan

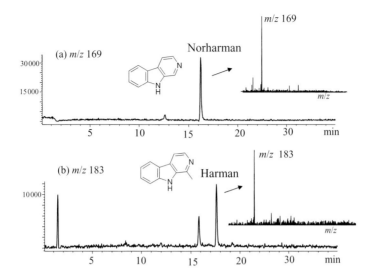

Fig. 8.5 Identification by HPLC-MS (ESI, positive ionization) of β-carbolines norharman and harman in coffee brews. [Reproduced from Herraiz, T. Identification and occurrence of the bioactive β-carbolines norharman and harman in coffee brews. *Food Additives and Contaminants*, **19**, 748–754. Copyright 2002 with permission from Taylor & Francis Ltd, http://www.tandf.co.uk/journals (http://www.informaworld.com).]

as precursor THβC-3-carboxylic acids (**1a–1h**) are produced whereas tryptamine or serotonin afford THβCs (**2a–d**) or 6-hydroxy-THβCs (**3**), respectively. Reactions of tryptophan with glucose or pentoses provide carbohydrate-derived THβCs (**5ab**) (Rönner *et al.*, 2000; Diem and Herderich, 2001b; Herraiz and Galisteo, 2002b), and reactions with phenolic aldehydes give phenolic-derived THβCs (**6**) (Herraiz *et al.*, 2003). Nitrosated foods may contain *N*-nitroso-THβCs **4ab** (Sen *et al.*, 1991). Table 8.1 lists the concentration ranges of several THβCs detected in foods. THβC-3-COOHs **1a–c** are generally the major THβCs, which may appear up to in several mg/kg (mg/L), and reaching exceptionally the hundred of mg/kg (e.g., soy sauces). THβC-3-COOHs **1a–c** are widespread in commercial foods and beverages. They have been found in fruits and fruit processed products (fruit juices, jams, and purees) (Herraiz, 1996, 1998, 1999a, 2004b; Herraiz and Galisteo, 2003), in food of animal origin such as cooked fish, meats, and sausages (Herraiz, 2000b; Herraiz and Papavergou, 2004), in smoked foods (cheese, fish, and meats) (Papavergou and Clifford, 1992; Sen *et al.*, 1995; Papavergou and Herraiz, 2003), fermented products such as chocolate and cocoa (Herraiz, 2000d), and alcoholic fermented drinks, sauces and vinegars (Adachi *et al.*, 1991a; Herraiz *et al.*, 1993). The THβC ester **1d** (diastereoisomers 1*S*,3*S* and 1*R*,3*S*) was found in alcoholic fermented beverages in concentrations up to 0.5 mg/L (Herraiz, 1999b). 1,3-Dicarboxylic-THβCs **1ef** derived from α-ketoacids such as pyruvic acid, and glyoxylic acid have been detected in alcoholic beverages and sauces (Gutsche and Herderich, 1997a,b, 1998). Polyol THβCs have been found in foods, and specifically 1-pentahydroxypentyl-1,2,3,4-tetrahydro-β-carboline-3-carboxylic acid **5ab**, a glucose tetrahydro-β-carboline derivative, was recently identified and quantified in commercial fruit juices and jams reaching up to 6.5 µg/g (Herraiz and Galisteo, 2002b). As seen in Table 8.1, small amounts of *N*-nitroso THβC-3-COOHs **4ab** were also detected in nitrite-treated foods (Wakabayashi *et al.*, 1983; Sen *et al.*, 1995) suggesting that, if preserved with nitrite, foods rich in THβC-3-COOHs could produce the corresponding *N*-nitroso compounds. Experiments with model reactions simulating foods

Structure	R_1	R_2	R_3	Comp.
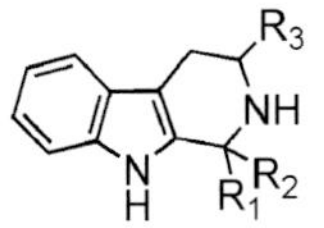	H	H	COOH	1a
	CH_3	H	COOH	1b (1S,3S), 1c (1R,3S)
	CH_3	H	COOEt	1d (1S,3S; 1R,3S)
	COOH	H	COOH	1e
	COOH	CH_3	COOH	1f
	CH_2OH	H	COOH	1g
	$(CH_2)_2COOH$	H	COOH	1h
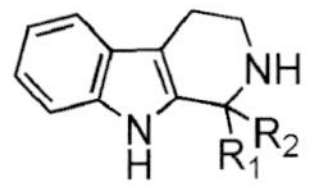	H	H		2a
	CH_3	H		2b
	COOH	H		2c
	COOH	CH_3		2d
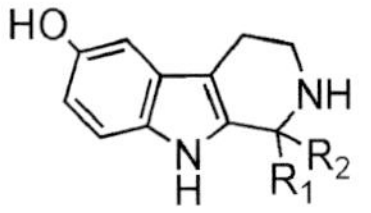	H	H		3a
	CH_3	H		3b
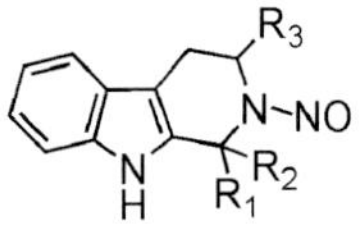	H	H	COOH	4a
	CH_3	H	COOH	4b
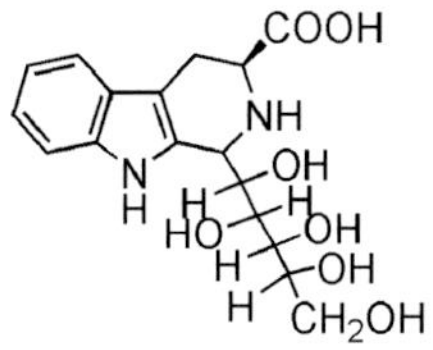				5a (cis) 5b (trans)
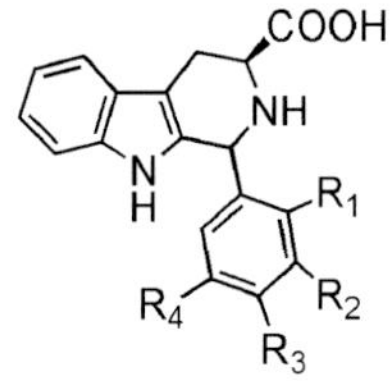	$(R_1, R_2, R_3, R_4 : H$ $OH, OCH_3)$			6

Fig. 8.6 THβCs found in foods and food-related samples. Compounds listed in Table 8.1 are **1a**: 1,2,3,4-tetrahydro-β-carboline-3-carboxylic acid; **1bc**: 1-methyl-1,2,3,4-tetrahydro-β-carboline-3-carboxylic acid (1S,3S; **b** and 1R,3S; **c**); **1d**: 1-methyl-1,2,3,4-tetrahydro-β-carboline-3-carboxylic acid ethyl ester (1S,3S; 1R,3S); **1e**: 1,2,3,4-tetrahydro-β-carboline-1,3-dicarboxylic acid; **1f**: 1-methyl-1,2,3, 4-tetrahydro-β-carboline-1,3-dicarboxylic acid; **1g**: 1-hydroxymethyl-1,2,3,4-tetrahydro-β-carboline-3-carboxylic acid; **2a**: 1,2,3,4-tetrahydro-β-carboline (tryptoline); **2b**: 1-methyl-1,2,3,4-tetrahydro-β-carboline; **3b**: 6-hydroxy-1-methyl-1,2,3,4-tetrahydro-β-carboline; **4a**: 2-N-nitroso-1,2,3,4-tetrahydro-β-carboline-3-carboxylic acid; **4b**: 2-N-nitroso-1-methyl-1,2,3,4-tetrahydro-β-carboline-3-carboxylic acid; **5**: 1-pentahydroxypentyl-1,2,3,4-tetrahydro-β-carboline-3-carboxylic acid (cis, **a** and trans, **b**).

also indicate that additional aldehydes appearing in foods, both aliphatic and aromatic, may afford new THβCs as well. Of interest, are phenolic aldehydes (syringaldehyde, salicylaldehyde, anisaldehyde, vanillin, and benzaldehyde) giving phenolic THβCs (Herraiz et al., 2003). Those THβCs have been proposed as a new type of antioxidants containing phenolic and indolic rings (Herraiz and Galisteo, 2002c). Novel THβC-3-COOHs containing a pyrrolidinethione at C-1 were recently detected in fermented radish up to 10 mg kg^{-1} (Ozawa et al., 1999).

Table 8.1   Concentration ranges (mg/L, µg/g) of tetrahydro-β-carbolines (THβCs) in foods and beverages.

Food	1a	1b	1c	1d	1e	1f	1g	2a	2b	3b	4a	4b	5ab
Wine	<0.65ᵃ	0.3–13.7	<4.0	0.003–0.53	>0.01	>0.01	<0.01	0.0004–0.001	0.008–0.10	<0.0002			
Beer	<0.8	0.3–13.1	0.08–3.9	<0.010	>0.01	>0.01	<0.07	<0.012	0.003–0.10	0.04–0.08			
Distillate	<0.04	<0.8	<0.2	<0.028				0.009–0.012	0.0003				
Liquour	<0.23	<5.7	<1.6	<0.054						<0.002			
Soft drink	<0.09	<0.35	<0.07										
Cider	<0.02	0.05–0.14	0.01–0.04										
Fruit	<0.05	<6.6	<1.8						<1.9	<3.9			<3.85
Fruit juices	0.01–1.45	0.04–9.1	0.03–2.5						<0.4	<3.7			<0.37
Jam	0.06–0.42	0.1–2	0.02–0.7										<5.2
Fruit purees	0.03–0.11	0.03–1.18	0.01–0.36										
Fish (raw)	<2.5												
Fish (cooked)	0.03–6.4	0–0.16											
Smoked fish	0.03–12.2	<0.01					<0.04						
Meat (raw)	<0.01												
Meat (cooked)	<0.05	<0.05										<0.05	
Smoked meat	<0.7						<0.03						
Sausages	<1.8	<2.9	<0.6				<0.5	<0.8	<9.9				
Smoked sausage	0.09–15.8	<16.6	<4				<0.5	<22.6	<29		<0.34		
Soy sauce	2.1–69.6	72–360.5	20–87.6		>10	>10	0.2–5.1		>0.01		0.03		
Seasonings	0.03–3.18	0.6–24	0.15–5.6		>10	>10			>0.01			<0.25	
Vinegar	0.01–0.12	3.9–7.5	0.84–2.06		>0.01	>0.01		>0.01					<2.3
Bread	0.12–2.51	0.03–0.32	0.01–0.21										
Flour	0.13–2.5	0.018	0.005										
Yogurt	0.01–0.02	0.04–0.11	0.01–0.02										
Chocolate	0.1–0.7	0.5–2	0.2–0.9						<0.21	0.4–4.0			
Cheese	<0.87	<2.06	<0.5							<0.028			
Smoked cheese	0.07–6.1						<0.035						

1a–c (Herraiz et al., 1993; Herraiz, 1996, 1998, 1999, 2000b,c,d, 2004b; Herraiz and Galisteo, 2003; Papavergou and Herraiz, 2003; Herraiz and Galisteo, 2003; Papavergou and Herraiz, 2004); **1d** (Herraiz, 1999b); **1e, 1f, 2a** (Gutsche and Herderich, 1997a,b, 1998; Herraiz, unpublished); **1g** (Papavergou and Clifford, 1992; Sen et al., 1995; Herraiz and Papavergou, 2004); **2ab** (Beck and Holmstedt, 1981; Matsubara et al., 1986; Tsuchiya et al., 1996; Gutsche and Herderich, 1998; Herraiz, 2000d; Herraiz and Galisteo, 2003; Herraiz and Papavergou, 2004); **3b** (Beck et al., 1983; Herraiz, 2000d; Herraiz and Galisteo, 2003); **4ab** (Sen et al., 1991, 1995; Herraiz, unpublished); **5ab** (Herraiz and Galisteo, 2002b).
ᵃ This value reflects a range from less than limit of detection (nd) to the given figure.
Compounds are as in Fig. 8.6.

Compound	Structure	Name
7	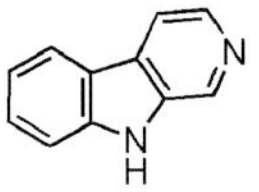	Norharman (9H-pyrido[3,4-b]indole)
8	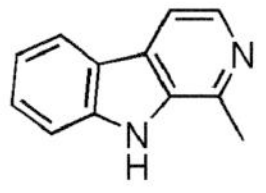	Harman (1-methyl-9H-pyrido[3,4-b]indole)
9		1-(1,3,4,5-tetrahydroxypent-1-yl)--9H-pyrido[3,4-b]indole R₁, R₂; H or OH **9a, 9b**
10		1-(1,4,5-trihydroxypent-1-yl)--9H-pyrido[3,4-b]indole R₁, R₂; H or OH **10a, 10b**
11	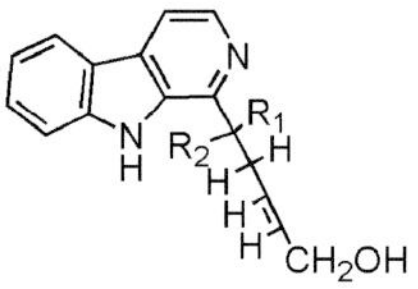	1-(1,5-dihydroxypent-3-en-1-yl)--9H-pyrido[3,4-b]indole R₁, R₂; H or OH **11a, 11b**

Fig. 8.7 Structures of fully aromatic βCs found in foods.

Apart from those THβCs arising from tryptophan, THβCs are also generated in foods from tryptamine and serotonin as indoleamine precursors (Fig. 8.6 and Table 8.1). We have reported the presence of acetaldehyde-condensation products of tryptamine and serotonin **2b** and **3b** in several fruits and juices (Herraiz and Galisteo, 2003) and chocolate (Herraiz, 2000d). Interestingly, the occurrence of one or another specific THβC appears to be dependent on the corresponding indoleamine precursor of each type of fruit or juice. The occurrence of **2a** and **2b** has also been reported in fermented, smoked, and matured sausages (Papavergou and Herraiz, 2003; Herraiz and Papavergou, 2004) and detected in some wines and sauces (Tsuchiya et al., 1996; Gutsche and Herderich, 1998; Herraiz, unpublished).

The ranges of concentration determined for each THβC, given in Table 8.1, show a large variability among foods. Nevertheless, considering the high and widespread occurrence of several of these compounds, the exposure to THβCs coming from the diet may easily reach the amount of several mg/person/day. This exposure will increase or otherwise decrease greatly depending on the type and individual foods ingested during the diet.

8.4.2 βCs in foods

In general, less attention has been given to aromatic βCs in foods (Fig. 8.7). Table 8.2 lists the occurrence of these compounds in foods. With a few exceptions, the level of the βCs

Table 8.2 β-Carbolines norharman (**7**), harman (**8**), and gluco-β-carbolines (**9,10**) in foodstuffs (ng/g or µg/L) and cigarette smoke (ng/cigarette).

Food	7 X	7 Range	8 X	8 Range	9a Range	9b Range	10 Range
Dairy products	<0.03ᵃ		<0.03				
Soft drinks	<0.03		<0.03				
Fruit juice	0.8	<5.5ᵇ	0.59	<1.8	15–395	12–347	2.8–89
Jam	0.27	<1.07	0.08	<0.41			
Wine	0.16	<0.56	1.35	<8.04			
Fortified wines (sherry and porto)	0.5	<1.3	8.6	0.76–39.2			
Distillate	0.73	<3.1	0.95	<5.3			
Alcoholic beverages (total)	0.36	<3.1	3.4	<39.2			
Vinegar	0.88	<3.5	5.8	<14.3	<249	<204	<35.6
Soy sauce	44.0	32–52	187.6	162.3–200.6	980–1819	769–1751.8	187.4–643.5
Seasoning	12.6	3.4–24.7	12.7	0.24–52.1	151–1921.6	121.4–1577.0	46.1–222.0
Flour	<0.03		0.2	<0.8			
Bread	22.1	<65.4	0.94	<2.07			
Toasted bread	91.0	41.7–164.2	<0.03				
Cookies/biscuit	16.5	7.7–34	2.14	0.4–4.6			
Breakfast cereals	46.5	<187.4	20.1	<91.7			
Fish/meat soups	28.2	2.6–83.9	3.73	0.5–11.0			
Meat (uncooked)ᶜ	<0.03		<0.03				
Cooked meat (medium)ᶜ	0.19	<0.4	<0.03				
Cooked meat ("well-done")ᶜ	82.3	36.4–128.1	26.4	20.7–32.1			
Fresh fish (uncooked)ᶜ	<0.03		<0.03				
Cooked fishᶜ	24.7	6.4–48.11	2.02	0.7–4.3			
Sausage	2.31	<7.4	1.24	<3.7			
Smoked fish	1.3	0–4.5	0.04	<0.2			
Chocolate	6.8	4.14–10.9	9.18	5.7–16.5			
Nut/dry fruit	5.2	<13.7	1.7	<6.9			
Instant coffeeᵈ	2100	1400–3500	420	270–680			
Brewed coffeeᵈ	1430	1000–2380	335	230–550			
Espresso/brewed coffee	91.4	23.6–165.9	22.38	5.06–40.8			
Cigarette smoke (ng/cigarette)ᵉ	1242.8	152–1966	500.8	55–814			

ᵃ Less than limit of detection: 0.03 ng/g (undetected).
ᵇ This value reflects the range from less than limit of detection to the given figure.
ᶜ Cooked meats (beef and pork fillet) or fish (hake, salmon, swordfish) were pan-fried and were cooked medium or well-done (prolonged cooking).
ᵈ Expressed as nanogram of βs in brewed coffee per gram of ground coffee used to make filtered or instant coffee.
ᵉ As nanogram of βs in mainstream smoke per cigarette. X : mean (ng/g or **g**/L, except when mentioned otherwise).
Data summarized from Herraiz (2004a) (**7** and **8**) and from Diem and Herderich (2001a) (**9,10**).

norharman **7** and harman **8** is low (i.e., in the low μg/L or ng/g order) (Herraiz, 2004a), although large variations occur among foods ranging from undetectable to relatively high levels. They were mostly undetectable in dairy products and soft drinks, and generally a low amount was found in fruit products. Variable levels of βCs, particularly harman **8**, were found in fermented alcoholic beverages such as red, white, and fortified wines. With only a few exceptions, high alcohol distillates showed very low levels or undetectable βCs. A relatively high level of both norharman **7** and harman **8** appeared in soy sauce and other seasonings and **8** was the major βC in vinegar. Regarding those processed and cooked foods, βCs particularly **7**, appeared in breads, toasted breads, breakfast cereals, and cookies. Both, **7** and **8** occurred in "well-done" cooked fish and meat (levels up to 160 ng/g), but not in the corresponding uncooked samples. It is noticeable that there is a high occurrence of both **7** and **8** in brewed coffee (ready to take coffee), irrespective of the variety and type (instant, ground, decaffeinated) (Herraiz, 2002). Indeed, coffee is comparatively the foodstuff affording the highest relative amount of these alkaloids. βCs **7** and **8** are present in a high concentration in cigarette smoke (from hundred to several thousands of ng/cigarette) (Herraiz, 2004a), and smoking is an important source of exposure to these compounds in addition to the diet.

Aromatic βC derivatives with polyhydroxyalkyl or a furan moiety have been also reported in foods (Nakatsuka *et al.*, 1986; Diem and Herderich, 2001a,b). Polyol βCs (**9–11**) arising from a reaction of tryptophan with carbohydrates have been identified and quantified in several sauces, vinegars, and fruit juices in a concentration range even higher than that determined for norharman and harman (Diem and Herderich, 2001a,b). They are also present in other processed foods (Herraiz, unpublished results).

Although the level of aromatic βCs in food and beverages is rather low compared to THβCs, successive ingestion of foodstuffs containing these βCs might raise substantially human exposure to these compounds. As shown in Table 8.2, brewed coffee, cooked and highly processed foods such as those "well-done" or broiled, seasonings (such as soy sauce), vinegar, and fermented alcoholic beverages along with cigarette smoke, are the main sources of exposure to **7** and **8**. Concerning carbohydrate-derived carbolines, exposure to these carbolines will come mainly from processed foods such as fruit juices, sauces, and others. Thus, ingestion of compounds **7–11** may easily reach the level of hundreds or even thousands μg/person/day. Exposure to these xenobiotic βCs may explain a correlation found among plasma βCs with alcoholism and smoking. Indeed, the presence of βCs in foods and smoke should be considered in biomedical studies seeking relationships between smoking or alcohol drinking and βCs in the body (Spijkerman *et al.*, 2002; Fekkes *et al.*, 2004).

8.5   FACTORS INFLUENCING THE PRESENCE OF β-CARBOLINES IN FOODS

Foods show a large variability in the content of β-carbolines, suggesting that many factors are involved. Few attempts have been made to rationalize those effects. Factors influencing the presence of βCs are biological, chemical, and technological, including the amount of precursors, pH, temperature, storage time, oxidants, antioxidants, preservatives, microorganisms, and processing conditions (e.g., fermentation, smoking, heating, cooking, and ripening) (Herraiz and Ough, 1993; Herraiz, 1996, 1998, 1999a,b, 2000a,b,c,d, 2002, 2004a,b). Both, carbonyl compounds and indoleamines and amino acids occur in foods, and a cyclocondensation producing THβCs (see Fig. 8.2) may take place during food production, processing, and storage. In the presence of reactants, the pH of the media is a determinant factor of the reaction rate with formation favored at low pH. However, at the pH found in many beverages and foods (pH values 3–5), generation of tryptophan-acetaldehyde and/or formaldehyde-derived carbolines **1a–c** can readily progress. Reaction with formaldehyde is faster and less pH-dependent than with acetaldehyde. Formation of THβCs increases with temperature whereas it decreases with carbonyl trapping. During alcoholic fermentation, yeasts were a significant biological factor because they produced acetaldehyde able to react with tryptophan giving **1bc** (Herraiz and Ough, 1993; Herraiz et al., 1993). Reactions involving other aliphatic and aromatic aldehydes, α-ketoacids, or carbohydrates may behave in a similar way, and glyco-THβCs **5ab** and phenolic THβCs **6** increased exponentially at low pH and with high processing temperatures (Herraiz and Galisteo, 2002b; Herraiz et al., 2003).

The relative presence and content of THβCs in foods appear to be determined by those specific indolethylamines and aldehydes involved. Thus, tryptophan-derived 1-methyl-THβC-3-carboxylic acid **1bc** was generally Found as a major carboline in fermented beverages and fruits, owing to the highest amount of free acetaldehyde compared to formaldehyde in those products. This compound increased with storage and aging of fruits (Herraiz, 1999a; Ichikawa et al., 2004). In the same manner, other THβCs appear resulting from the involvement of a different indolethylamine (tryptamine, serotonin) (Herraiz, 1999a, 2004b; Herraiz and Galisteo, 2003; Herraiz and Papavergou, 2004). Thus, fruits containing serotonin may afford 6-hydroxy-1-methyl-THβC **3b** and those with tryptamine give 1-methyl-THβC **2b**. THβC-3-carboxylic acid **1a** was a major THβC in smoked foods (Papavergou and Herraiz, 2003) arising from a reaction of tryptophan with formaldehyde that is present in wood smoke (Potthast and Eigner, 1985). This finding was evidenced by analyzing the outer and interior parts of smoked products. Thus, the outer part of smoked sausage, fish, and cheese, which is in direct contact with smoke, contained much higher amount of **1a** (up to eightfold) than the inner counterpart (Papavergou and Herraiz, 2003). Again, the occurrence of several THβCs **1a–c**, **2** in sausages of different types (cooked, fermented, ripened, smoked, and unsmoked), depended on the relative presence of indoleamines and the technological process involved (smoking or not) (Herraiz and Papavergou, 2004). In this case, meat microorganisms may also release tryptamine through amino acid decarboxylases, which will be able to react with aldehydes generated during ripening or otherwise added from wood smoke. The concentration of THβCs may depend on a variety of factors during elaboration process such as smoking, fermentation and microorganisms involved, ripening and storage, artisanal production, raw meats, ingredients used for flavoring, spices, etc.

The origin of the fully aromatic βCs, norharman **7** and harman **8**, was attributed to pyrolysis. It is becoming clear that βCs in foods arise from THβC-3-COOH through a decarboxylative oxidation (Herraiz, 2000b,c, 2004a,b) (Fig. 8.8). Factors such as storage time, heating (e.g., cooking), presence of oxidants and free radicals such as hydrogen peroxide (H_2O_2), hydrogen peroxide (H_2O_2)–transition metals (Fe^{2+}) (Fenton reaction), and sodium nitrite may accelerate the formation of βCs from the corresponding THβC-3-COOHs (Herraiz, 2000b,c; Herraiz and Galisteo, 2002a; Herraiz, 2004b). Under the same conditions THβCs lacking a 3-COOH did not provide appreciable amounts of aromatic βCs, whereas tetrahydro-β-carboline-1,3-dicarboxylic acids **1e**, **1f** also afforded aromatic βCs at the highest temperature of heating (80°C) or in presence of oxidants (Herraiz, 2004b). As seen in Tables 8.1 and 8.2, the occurrence of βCs **7** and **8** in foods seems to correlate with the levels of THβC-3-COOHs **1a–c**. Thus, sauces, vinegars, and alcoholic beverages contain more harman **8** and also more **1bc**; whereas processed foods, smoked, and cooked meats and fish contain more norharman **7** and also more **1a**. The formation of **7** and **8** in well-cooked or broiled foods such as fish and meats may result from heating and/or pyrolysis of tryptophan and its

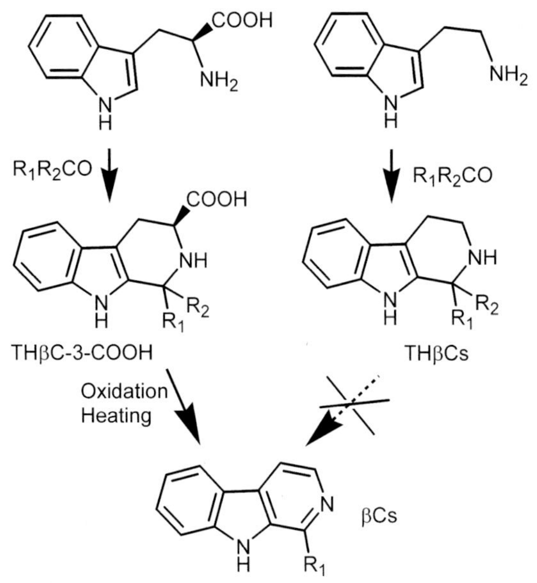

Fig. 8.8 Formation of βCs from THβC-3-COOHs in foodstuffs.

subsequent condensation with aldehydes generated in the process to give THβC-3-COOH, which are further oxidized (Herraiz, 2000b). The absence of **7** and **8** in fresh or slightly processed foods compared with cooked (well done) samples might result from the absence of THβC-3-COOH and/or processing conditions needed to afford βCs (i.e., the involvement of oxidants and heating) (Herraiz, 2000b, 2004a). In a similar way, βCs in brewed coffee result from the roasting process of coffee beans (Herraiz, 2002). In the same manner, the formation of carbohydrate-derived βCs (**9–11**) during food processing may arise from oxidation of *glyco*-THβC-3-COOH **5ab** and both species usually appear in the same products (Tables 8.1 and 8.2) (Diem and Herderich, 2001a,b; Herraiz and Galisteo, 2002b).

8.6 BIOLOGICAL ACTIVITY AND TOXICOLOGY OF β-CARBOLINES

Molecules with structure of β-carbolines exhibit a broad spectrum of biological and pharmacological activities including neuroactive, antithrombotic, antimicrobial, antiparasiticidal, antimalarial, antiviral, and antitumor actions (Cao *et al.*, 2005; Di Giorgio *et al.*, 2004; Kusurkar and Goswami, 2004; Zhao *et al.*, 2006). The pharmacology of these substances is complex and they act at different sites within mammalian tissues. A primary target is the central nervous system and several βCs (i.e., norharman **7**, harman **8**, harmine, and harmaline) are considered psychoactive indoles. In fact, the hallucinogenic beverage Ayahuasca or yage, used by Indian tribes, contains these substances besides N, N-dimethyltryptamine. βCs (i.e., **7**, **8**, and some THβCs) are known to cross the blood–brain barrier, distributing along the brain (Fekkes and Bode, 1993; Anderson *et al.*, 2006). They may exert as neuromodulators on the serotonin neurotransmitter system, producing neuroendocrine and behavioral effects consistent with a serotoninergic involvement (Buckholtz, 1980; Robinson *et al.*, 2003). They are inhibitors of MAO and serotonin uptake, and bind to benzodiazepine-GABA receptor show-ing several physiological effects (Airaksinen and Kari, 1981b; Rommelspacher et al., 1994; Adell et al., 1996). The finding that β-carboline-3-carboxylate esters were potent ligands of benzodiazepine receptor (Braestrup et al., 1980) pushed research to achieve new drugs with increased binding affinity to this receptor. Most are inverse agonist showing anxiogenic and convulsant properties (Cox and Cook, 1995).

Most effects of βCs are attributed to inhibition of MAO enzymes (Rommelspacher et al., 1994; Herraiz and Chaparro, 2005, 2006a,b). Indeed, βCs can modulate monoamine levels in vivo (Adell et al., 1996; Baum et al., 1996), and increase extracellular dopamine and serotonin levels, probably by inhibition of MAO-A. Interestingly, βCs present in foods and the environment are able to inhibit MAO-A and -B. Thus, norharman **7** and harman **8** isolated from tobacco smoke and coffee brews proved to be potent and reversible inhibitors of MAO-A (harman and norharman) and MAO-B (norharman) with Ki in the low μM or nM range (Herraiz and Chaparro, 2005, 2006a,b). Other food-occurring THβCs such as tryptoline **2a** and 1-methyltryptoline **2b** were also MAO-A inhibitors but weaker than the fully aromatic ones. Inhibition of MAO-A and -B has implications in behavior conditions and pathologies such as addiction, depression, and Parkinson's disease. βCs could exert MAO inhibitory actions if accumulated in the body following their ingestion via foods (coffee and others) or environmental sources (smoking).

Hudson et al. (1999) have proposed that aromatic βCs could be endogenous ligands of imidazoline binding sites (IBS) assigning a new role to these alkaloids. βCs bind with high affinity to I_1-BS and I_2-BS imidazoline receptors and this may explain some aspects of their pharmacology (Robinson et al., 2003). βCs (harman **7**) induce hypotensive effects following central administration and change cardiovascular parameters (Shi et al., 2000), induce hy-pothermic response (Adell et al., 1996), modulate food intake (Robinson et al., 2003), induce insulin secretion (Morgan et al., 2003) as well as anxiolysis and antidepressant-like effects in rats (Aricioglu and Altunbas, 2003). Some of these physiological responses are likely related to MAO inhibition, but binding to imidazoline sites may play a role (Husbands et al., 2001; Robinson et al., 2003). βCs (harman **8**, norharman **7**, harmalan, and tetrahydroharman **2b**) displaying high affinity for IBS might work as endogenous ligands of imidazoline receptors (Parker et al., 2004; Miralles et al., 2005). Nevertheless, some of these compounds are also xenobiotics (Tables 8.1 and 8.2) and an exclusive endogenous formation is unlikely.

THβCs occurring in foods and biological systems act as antioxidants and radical scav-engers (Herraiz and Galisteo, 2002a, 2003; Ichikawa et al., 2004). Their antioxidant capacity against $ABTS^{+\bullet}$ (total antioxidant capacity assay) was stronger than ascorbic acid and Trolox (a water-soluble form of vitamin E). THβCs contain an indole ring that would afford an indolyl cation or neutral radical through single electron transfer while acting as radical scavengers (Fig. 8.9). Those indolyl radicals might be further oxidized to aromatic βCs such as **7** and **8**, as occurs with THβC-3-COOH (Herraiz and Galisteo, 2002a; Herraiz, 2004b), or to still un-known compounds. Interestingly, the actions of THβCs as antioxidants may follow a similar trend and mechanism to that of neurohormone melatonin (N-acetyl-5-methoxytryptamine), which has been extensively characterized as a potent in vitro and in vivo antioxidant (Reiter et al., 2002). Dietary or endogenously formed THβC alkaloids, after being absorbed and/or accumulated in tissues and fluids, might play a role as potential antioxidants by protect-ing against harmful radicals during oxidative stress. Nevertheless, the contribution of these compounds to total antioxidant activity (TAC) in foods is small owing to their low concen-tration compared with vitamins, carotenoids, and phenols. However, many other indoles may act as radical scavengers as well, suggesting that indolic compounds may constitute a new

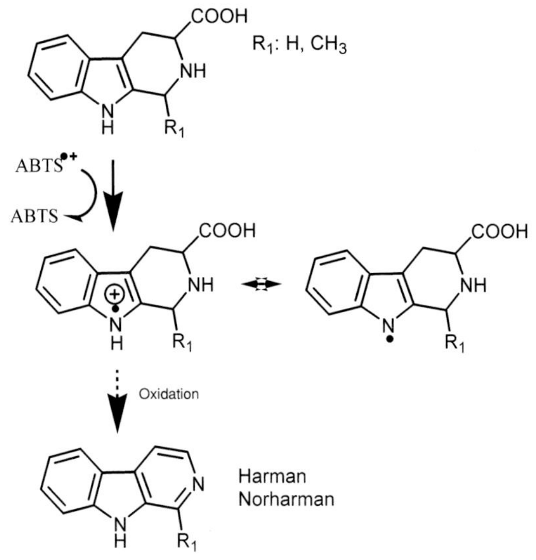

Fig. 8.9 THβCs act as antioxidants and radical scavengers (antioxidative capacity against ABTS[•+]). THβC-3-COOH are oxidized to the fully aromatic βCs.

class of antioxidants (Herraiz and Galisteo, 2004). Besides its antioxidant capacity, THβCs and βCs also show anti-aggregation activity of platelets (Tsuchiya *et al.*, 1999; Zhao *et al.*, 2006).

Numerous reports suggest the involvement of βCs in addiction, drug withdrawal and/or pathological states. THβCs (**2a**) and βCs (harman **8**) were implicated in alcoholism because administration of these substances to rats significantly altered alcohol consumption (Myers, 1989). βCs (norharman **7**) may attenuate drug withdrawal (Fekkes *et al.*, 2004). Specific βCs may exhibit or participate in toxicological actions. In the 90s, it was suggested that THβCs impurities might be involved in the etiology of the eosinophilia-myalgia syndrome (EMS) associated with the ingestion of impure L-tryptophan that occurred in the USA. One impurity identified as 1,1'-ethylidene-bis-tryptophan (EBT) was suggested to cause the disease (Mayeno *et al.*, 1990). EBT is converted to 1-methyl-1,2,3,4-tetrahydro-β-carboline-3-carboxylic acid **1bc** in acidic solution such as gastric fluid. Brenneman *et al.* (1993) reported that **1b** (1*S*,3*S* diastereoisomer) affected neuronal survival in vitro suggesting a role for this THβC in the etiology of some of the neuropathic features of L-Trp-EMS.

A growing interest is currently given to THβCs and βCs as neurotoxins (Matsubara *et al.*, 1992, 1998; Östergren *et al.*, 2004; Bringmann *et al.*, 2006). Endogenous or xenobiotics β-carbolines after bioactivation in the brain may afford *N*-methylcarbolinium cations which are toxic to cellular respiratory chain (complex I) (Fig. 8.10). These substances might act as protoxins in idiopathic Parkinson's disease (Hamman *et al.*, 2006). Indeed, THβCs and β-carbolinium cations are structural analogues of MPTP (1-methyl-4-phenyl-1,2,3,6-tetrahydropyridine), a well-known neurotoxin, and MPP[+] (1-methyl-4-phenylpyridinium), its corresponding active metabolite. Bioactivation may occur with the participation of *N*-methyltransferases in mammalian brain, which catalyze 2*N*-, and 9*N*-methylation of βCs (Gearhart *et al.*, 2000). Recently, we have shown that cytochrome P450 2D6 and heme

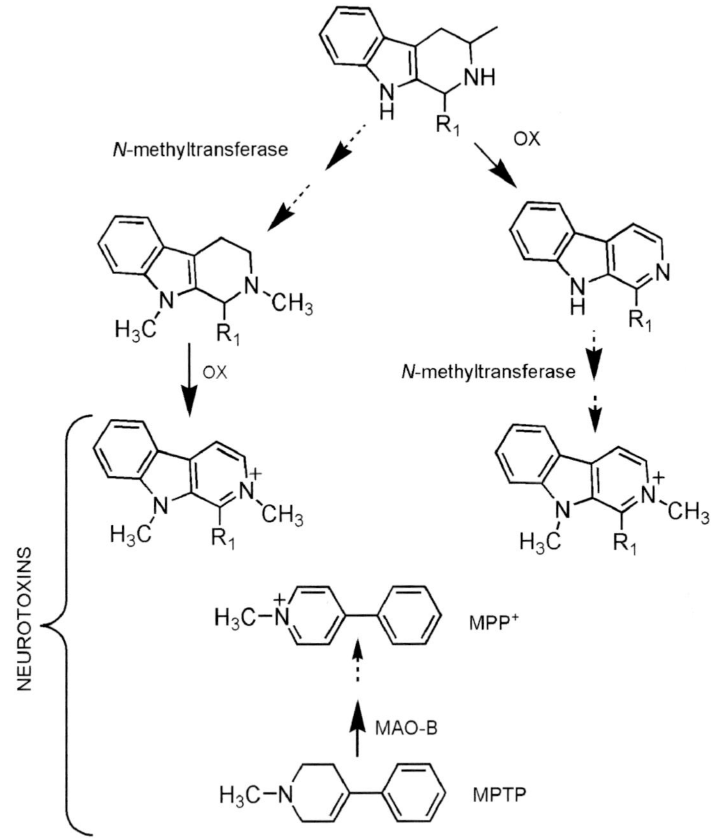

Fig. 8.10 Bioactivation route of THβCs and βCs to N-methyl-β-carbolinium cations (MβC+) which are neurotoxins equivalent to N-methyl-4-phenylpyridinium (MPP+) (a parkinsonian neurotoxin) arising from MPTP (N-methyl-4-phenyl-1,2,3,6-tetrahydropyridine).

peroxidases participate in the metabolism of MPTP and N-methyltetrahydro-β-carbolines (Herraiz et al., 2006, 2007). The possible implications of these or related substances in neurotoxicity as well as the involvement of the polymorphic cytochrome P-450 2D6 in their metabolism are subjects of current interest (Yu et al., 2003; Herraiz et al., 2006).

THβCs are precursors of N-nitroso compounds (see Table 8.1), and may react with nitrite in foods, mouth, or stomach producing mutagenic compounds (Wakabayashi et al., 1983; Higashimoto et al., 1996; Diem et al., 2001; Masuda et al., 2005; Herraiz and Galisteo, in preparation). Aromatic βCs have been also studied in relation to their mutagenicity and toxicity (Boeira et al., 2002) and norharman **7** and harman **8** are described as co-mutagenic compounds in presence of aromatic amines such as aniline or toluidine (Totsuka et al., 1998; Nii, 2003). The mutagenicity and cytotoxicity of βCs are also linked to their interaction and intercalation with DNA, action as photosensitizers and DNA cleavers (Cao et al., 2005; Guan et al., 2006), interaction with cytochrome P450 (Nii, 2003) and topoisomerase I inhibition (Sobhani et al., 2002).

Conclusions and future directions

THβC and βC alkaloids occur in foods (and subsequently in biological systems) under a wide range of concentrations and distribution patterns. Advances in their identification and analysis by using specific sample preparation protocols (SPE) and separation techniques (GC-MS, HPLC-MS, and tandem MS) are rapidly contributing to improve our knowledge on these bioactive compounds occurring in foods. From pharmacological studies, we currently know that they exhibit a broad range of biological activities including antimicrobial, antiviral, antiparasitic, antioxidant, neuroactive, cytotoxic, and neurotoxic actions. Consequently, their presence in foods and the diet is an interesting matter. These alkaloids form under mild conditions in foods and resulting from a nonenzymatic Pictet–Spengler cyclization during food production, processing, and storage. Their formation depends on an array of biological, chemical, and technological factors, which only now start to be explained. These alkaloids can be defined as xenobiotics, and therefore the origin of βCs found in biological systems might arise from both endogenous and exogenous sources. In conclusion, the diet contributes to the ultimate presence of these alkaloids in the human biological tissues and fluids and the ingestion of foods containing high level of THβCs and βCs may increase the level of these compounds in the body.

For the future, in addition to increase our knowledge about these compounds, it should be clarified whether these alkaloids, at the level ingested from foods (exposure), may contribute and/or exhibit, whatever in the short- or long-term, any of the diverse biological and pharmacological actions assigned to them. This subject will continue to be the object of scientific interest. Research dealing with the biological significance of THβCs or βCs has focused on their effects on the central nervous system such as inhibition of serotonin uptake, MAO inhibition, and binding to benzodiazepine-GABA or imidazoline receptors. Other pharmacological effects are also being considered such as antimicrobial, antiviral, cytotoxic, or antitumor actions. New reports have recently appeared on the promising actions of THβCs occurring in foods as antioxidants and free radical scavengers against oxidative stress or as antithrombotic agents. Concerning food safety, the possible toxicological actions of some of these compounds need to be considered in detail, and paying special attention to them as new lines of research suggest remarkable actions of THβC and βC alkaloids as potential protoxins, neurotoxicants and/or co-mutagens.

Acknowledgments

The author is grateful to Spanish government (MEC), projects AGL2006-02414 and AGL2003-01233 for financial support.

References

Adachi, J., Mizoi, Y., Naito, T., Ogawa, Y., Uetani, Y. and Ninomiya, I. (1991a) Identification of tetrahydro-β-carboline-3-carboxylic acid in foodstuffs, human urine and human milk. *Journal of Nutrition*, **121**, 646–652.

Adachi, J., Mizoi, Y., Naito, T., Yamamoto, K., Fujiwara, S. and Ninomiya, I. (1991b) Determination of β-carbolines in foodstuffs by high-performance liquid-chromatography and high-performance liquid-chromatography mass-spectrometry. *Journal of Chromatography*, **538**, 331–339.

Adell, A., Biggs, T.A. and Myers, R.D. (1996) Action of harman (1-methyl-β-carboline) on the brain: Body temperature and in vivo efflux of 5-HT from hippocampus of the rat. *Neuropharmacology*, **35**, 1101–1107.

Airaksinen, M.M. and Kari, I. (1981a) β-Carbolines, psychoactive compounds in the mammalian body. 1. Occurrence, origin and metabolism. *Medical Biology*, **59**, 21–34.

Airaksinen, M.M. and Kari, I. (1981b) β-Carbolines, psychoactive compounds in the mammalian body. 2. Effects. *Medical Biology*, **59**, 190–211.

Allen, J.R.F. and Holmstedt, B.R. (1980) The simple β-carboline alkaloids. *Phytochemistry*, **19**, 1573–1582.

Anderson, N.J., Tyacke, R.J., Husbands, S.M., Nutt, D.J., Hudson, A.L. and Robinson, E.S.J. (2006) In vitro and ex vivo distribution of ^{3}H harmane, an endogenous β-carboline, in rat brain. *Neuropharmacology*, **50**, 269–276.

Aricioglu, F. and Altunbas, H. (2003) Harmane induces anxiolysis and antidepressant-like effects in rats. *Annals of the New York Academy of Sciences*, **1009**, 196–200.

Baum, S.S., Hill, R. and Rommelspacher, H. (1996) Harman-induced changes of extracellular concentrations of neurotransmitters in the nucleus accumbens of rats. *European Journal of Pharmacology*, **314**, 75–82.

Beck, O. and Holmstedt, B. (1981) Analysis of 1-methyl-1,2,3,4-tetrahydro-β-carboline in alcoholic beverages. *Food and Cosmetics Toxicology*, **19**, 173–177.

Beck, O., Bosin, T.R. and Lundman, A. (1983) Analysis of 6-hydroxy-1-methyl-1,2,3,4-tetrahydro-β-carboline in alcoholic beverages and food. *Journal of Agricultural and Food Chemistry*, **31**, 288–292.

Boeira, J.M., Viana, A.F., Picada, J.N. and Henriques, J.A.P. (2002) Genotoxic and recombinogenic activities of the two β-carboline alkaloids harman and harmine in *Saccharomyces cerevisiae*. *Mutation Research*, **500**, 39–48.

Bosin, T.R., Holmstedt, B., Lundman, A. and Beck, O. (1983) Presence of formaldehyde in biological media and organic solvents. Artifactual formation of tetrahydro-β-carbolines. *Analytical Biochemistry*, **128**, 287–293.

Bosin, T.R. and Jarvis, C.A. (1985) Derivatization in aqueous solution, isolation and separation of tetrahydro-β-carbolines and their precursors by liquid chromatography. *Journal of Chromatography*, **341**, 287–293.

Bosin, T.R. and Faull, K.F. (1988a) Harman in alcoholic beverages. Pharmacological and toxicological implications. *Alcoholism – Clinical and Experimental Research*, **12**, 679–682.

Bosin, T.R. and Faull, K.F. (1988b) Measurement of β-carbolines by high performance liquid chromatography with fluorescence detection. *Journal of Chromatography*, **428**, 229–236.

Braestrup, C., Nielsen, M. and Olsen, C.E. (1980) Urinary and brain β-carboline-3-carboxylates as potent inhibitors of brain benzodiazepine receptors. *Proceedings of the National Academy of Sciences of the United States of America*, **77**, 2288–2292.

Brenneman, D.E., Page, S.W., Schultzberg, M., Thomas, F.S., Zelazowski, P., Burnet, P., Avidor, R. and Sternberg, E.M. (1993) A decomposition product of a contaminant implicated in L-tryptophan eosinophilia-myalgia-syndrome affects spinal-cord neuronal cell-death and survival through stereospecific, maturation and partly interleukin-1-dependent mechanisms. *Journal of Pharmacology and Experimental Therapeutics*, **266**, 1029–1035.

Bringmann, G., Friedrich, H., Birner, G., Koob, M., Sontag, K.H., Heim, C., Kolasiewicz, W., Fahr, S., Stablein, M., God, R. and Feineis, D. (1996) Endogenous alkaloids in man. 26. Determination of the dopaminergic neurotoxin 1-trichloromethyl-1,2,3,4-tetrahydro-β-carboline (TaClo) in biological samples using gas chromatography with selected ion monitoring. *Journal of Chromatography B*, **687**, 337–348.

Bringmann, G., Feineis, D., Bruckner, R., God, R., Grote, C. and Wesemann, W. (2006) Synthesis of radiola-belled 1-trichloromethyl-1,2,3,4-tetrahydro-β-carboline (TaClo), a neurotoxic chloral-derived mammalian alkaloid, and its biodistribution in rats. *European Journal of Pharmaceutical Sciences*, **28**, 412–422.

Brossi, A. (1993) Mammalian alkaloids II, in *The Alkaloids. Chemistry and Pharmacology*, Vol. 43 (ed. G.A. Cordell). Academic Press, San Diego, pp. 119–183.

Buckholtz, N.S. (1980) Neurobiology of tetrahydro-β-Carbolines. *Life Sciences*, **27**, 893–903.

Cao, R., Chen, H., Peng, W., Ma, Y., Hou, X., Guan, H., Liu, X. and Xu, A. (2005) Design, synthesis and in vitro and in vivo antitumor activities of novel β-carboline derivatives. *European Journal of Medicinal Chemistry*, **40**, 991–1001.

Cox, E.D. and Cook, J.M. (1995) The Pictet–Spengler condensation – a new direction for an old reaction. *Chemical Reviews*, **95**, 1797–1842.

Di Giorgio, C., Delmas, F., Ollivier, E., Elias, R., Balansard, G. and Timon-David, P. (2004) In vitro activity of the β-carboline alkaloids harmane, harmine, and harmaline toward parasites of the species *Leishmania infantum*. *Experimental Parasitology*, **106**, 67–74.

Diem, S., Gutsche, B. and Herderich, M. (2001) Degradation of tetrahydro-β-carbolines in the presence of nitrite: HPLC-MS analysis of the reaction products. *Journal of Agricultural and Food Chemistry*, **49**, 5993–5998.

Diem, S. and Herderich, M. (2001a) Reaction of tryptophan with carbohydrates: Identification and quantitative determination of novel β-carboline alkaloids in food. *Journal of Agricultural and Food Chemistry*, **49**, 2486–2492.

Diem, S. and Herderich, M. (2001b) Reaction of tryptophan with carbohydrates: Mechanistic studies on the formation of carbohydrate-derived β-carbolines. *Journal of Agricultural and Food Chemistry*, **49**, 5473–5478.

Fekkes, D. and Bode, W.T. (1993) Occurrence and partition of the β-carboline norharman in rat organs. *Life Sciences*, **52**, 2045–2054.

Fekkes, D., Bernard, B.F. and Cappendijk, S.L.T. (2004) Norharman and alcohol-dependency in male Wistar rats. *European Neuropsychopharmacology*, **14**, 361–366.

Gearhart, D.A., Collins, M.A., Lee, J.M. and Neafsey, E.J. (2000) Increased β-carboline 9N-methyltransferase activity in the frontal cortex in Parkinson's disease. *Neurobiology of Disease*, **7**, 201–211.

Guan, H., Liu, X., Peng, W., Cao, R., Ma, Y., Chen, H. and Xu, A. (2006) β-Carboline derivatives: Novel photosensitizers that intercalate into DNA to cause direct DNA damage in photodynamic therapy. *Biochemical and Biophysical Research Communications*, **342**, 894–901.

Gutsche, B. and Herderich, M. (1997a) HPLC-MS/MS profiling of tryptophan-derived alkaloids in food: Identification of tetrahydro-β-carbolinedicarboxylic acids. *Journal of Agricultural and Food Chemistry*, **45**, 2458–2462.

Gutsche, B. and Herderich, M. (1997b) High-performance liquid chromatography electrospray ionisation – tandem mass spectrometry for the analysis of 1,2,3,4-tetrahydro-β-carboline derivatives. *Journal of Chromatography A*, **767**, 101–106.

Gutsche, B. and Herderich, M. (1998) HPLC-MS/MS identification of tryptophan-derived tetrahydro-β-carboline derivatives in food. *Fresenius Journal of Analytical Chemistry*, **360**, 836–839.

Hamann, J., Rommelspacher, H., Storch, A., Reichmann, H. and Gille, G. (2006) Neurotoxic mechanisms of 2,9-dimethyl-β-carbolinium ion in primary dopaminergic culture. *Journal of Neurochemistry*, **98**, 1185–1199.

Hayashi, T., Todoriki, H. and Iida, Y. (1990) Highly sensitive method for the determination of 1-methyl-1,2,3,4-tetrahydro-β-carboline using combined capillary gas-chromatography and negative-ion chemical ionization mass-spectrometry. *Journal of Chromatography*, **528**, 1–8.

Herraiz, T. (1996) Occurrence of tetrahydro-β-carboline-3-carboxylic acids in commercial foodstuffs. *Journal of Agricultural and Food Chemistry*, **44**, 3057–3065.

Herraiz, T. (1997) Analysis of tetrahydro-β-carbolines and their precursors by electron ionization mass spectrometry. Identification in foodstuffs by gas chromatography mass spectrometry. *Rapid Communications in Mass Spectrometry*, **11**, 762–768.

Herraiz, T. (1998) Occurrence of 1,2,3,4-tetrahydro-β-carboline-3-carboxylic acid and 1-methyl-1,2,3,4-tetrahydro-β-carboline-3-carboxylic acid in fruit juices, purees, and jams. *Journal of Agricultural and Food Chemistry*, **46**, 3484–3490.

Herraiz, T. (1999a) 1-methyl-1,2,3,4-tetrahydro-β-carboline-3-carboxylic acid and 1,2,3,4-tetrahydro-β-carboline-3-carboxylic acid in fruits. *Journal of Agricultural and Food Chemistry*, **47**, 4883–4887.

Herraiz, T. (1999b) Ethyl 1-methyl-1,2,3,4-tetrahydro-β-carboline-3-carboxylate: A novel β-carboline found in alcoholic beverages. *Food Chemistry*, **66**, 313–321.

Herraiz, T. (2000a) Analysis of the bioactive alkaloids tetrahydro-β-carboline and β-carboline in food. *Journal of Chromatography A*, **881**, 483–499.

Herraiz, T. (2000b) Tetrahydro-β-carboline-3-carboxylic acid compounds in fish and meat: Possible precursors of co-mutagenic β-carbolines norharman and harman in cooked foods. *Food Additives and Contaminants*, **17**, 859–866.

Herraiz, T. (2000c) Analysis of tetrahydro-β-carboline-3-carboxylic acids in foods by solid-phase extraction and reversed-phase high-performance liquid chromatography combined with fluorescence detection. *Journal of Chromatography A*, **871**, 23–30.

Herraiz, T. (2000d) Tetrahydro-β-carbolines, potential neuroactive alkaloids, in chocolate and cocoa. *Journal of Agricultural and Food Chemistry*, **48**, 4900–4904.

Herraiz, T. (2002) Identification and occurrence of the bioactive β-carbolines norharman and harman in coffee brews. *Food Additives and Contaminants*, **19**, 748–754.

Herraiz, T. (2004a) Relative exposure to β-carbolines norharman and harman from foods and tobacco smoke. *Food Additives and Contaminants*, **21**, 1041–1050.

Herraiz, T. (2004b) Tetrahydro-β-carboline bioactive alkaloids in beverages and foods. *ACS Symposium Series*, **871**, 405–426.

Herraiz, T., Huang, Z.X. and Ough, C.S. (1993) 1,2,3,4-Tetrahydro-β-carboline-3-carboxylic acid and 1-methyl-1,2,3,4-tetrahydro-β-carboline-3-carboxylic acid in wines. *Journal of Agricultural and Food Chemistry*, **41**, 455–459.

Herraiz, T. and Ough, C.S. (1993) Chemical and technological factors determining tetrahydro-β-carboline-3-carboxylic acid content in fermented alcoholic beverages. *Journal of Agricultural and Food Chemistry*, **41**, 959–964.

Herraiz, T. and Ough, C.S. (1994) Separation and characterization of 1,2,3,4-tetrahydro-β-carboline-3-carboxylic acids by HPLC and GC-MS. Identification in wine samples. *American Journal of Enology and Viticulture*, **45**, 92–101.

Herraiz, T. and Sanchez, F. (1997) Presence of tetrahydro-β-carboline-3-carboxylic acids in foods by gas chromatography mass spectrometry as their *N*-methoxycarbonyl methyl ester derivatives. *Journal of Chromatography A*, **765**, 265–277.

Herraiz, T. and Galisteo, J. (2002a) Tetrahydro-β-carboline alkaloids that occur in foods and biological systems act as radical scavengers and antioxidants in the ABTS assay. *Free Radical Research*, **36**, 923–928.

Herraiz, T. and Galisteo, J. (2002b) Identification and occurrence of the novel alkaloid pentahydroxypentyl-tetrahydro-β-carboline-3-carboxylic acid as a tryptophan glycoconjugate in fruit juices and jams. *Journal of Agricultural and Food Chemistry*, **50**, 4690–4695.

Herraiz, T. and Galisteo, J. (2002c) Phenolic tetrahydro-β-carbolines as antioxidants. Patent No ES 2211295.

Herraiz, T. and Galisteo, J. (2003) Tetrahydro-β-carboline alkaloids occur in fruits and fruit juices. Activity as antioxidants and radical scavengers. *Journal of Agricultural and Food Chemistry*, **51**, 7156–7161.

Herraiz, T., Galisteo, J. and Chamorro, C. (2003) L-tryptophan reacts with naturally occurring and food-occurring phenolic aldehydes to give phenolic tetrahydro-β-carboline alkaloids: Activity as antioxidants and free radical scavengers. *Journal of Agricultural and Food Chemistry*, **51**, 2168–2173.

Herraiz, T. and Galisteo, J. (2004) Endogenous and dietary indoles: A class of antioxidants and radical scavengers in the ABTS assay. *Free Radical Research*, **38**, 323–331.

Herraiz, T. and Papavergou, E. (2004) Identification and occurrence of tryptamine- and tryptophan-derived tetrahydro-β-carbolines in commercial sausages. *Journal of Agricultural and Food Chemistry*, **52**, 2652–2658.

Herraiz, T. and Chaparro, C. (2005) Human monoamine oxidase is inhibited by tobacco smoke: β-carboline alkaloids act as potent and reversible inhibitors. *Biochemical and Biophysical Research Communications* **326**, 378–386.

Herraiz, T. and Chaparro, C. (2006a) Analysis of monoamine oxidase enzymatic activity by reversed-phase high performance liquid chromatography and inhibition by β-carboline alkaloids occurring in foods and plants. *Journal of Chromatography A*, **1120**, 237–243.

Herraiz, T. and Chaparro, C. (2006b) Human monoamine oxidase enzyme inhibition by coffee and β-carbolines norharman and harman isolated from coffee. *Life Sciences* **78**, 795–802.

Herraiz, T., Guillén, H., Arán, V.J., Idle, J.R. and Gonzalez, F.J. (2006) Comparative aomatic hydroxylation and *N*-demethylation of MPTP neurotoxin and its analogs, *N*-methylated β-carboline and isoquinoline alkaloids, by human cytochrome P450 2D6. *Toxicology and Applied Pharmacology*, **216**, 387–398.

Herraiz, T., Guillen, H. and Galisteo, J. (2007) *N*-methyltetrahydro-β-carboline analogs of 1-methyl-4-phenyl-1,2,3,6-tetrahydropyridine (MPTP) neurotoxin are oxidized to neurotoxic β-carbolinium cations by heme peroxidases. *Biochemical and Biophysical Research Communications*, **356**, 118–123.

Higashimoto, M., Yamamoto, T., Kinouchi, T., Matsumoto, H. and Ohnishi, Y. (1996) Mutagenicity of 1-methyl-1,2,3,4-tetrahydro-β-carboline-3-carboxylic acid treated with nitrite in the presence of alcohols. *Mutation Research*, **367**, 43–49.

Hudson, A.L., Price, R., Tyacke, R.J., Lalies, M.D., Parker, C.A. and Nutt, D.J. (1999) Harmane, norharmane and tetrahydro-β-carboline have high affinity for rat imidazoline binding sites. *British Journal of Pharmacology*, **126**(suppl S), 2P, U6.

Husbands, S.M., Glennon, R.A., Gorgerat, S., Gough, R., Tyacke, R., Crosby, J., Nutt, D.J., Lewis, J.W. and Hudson, A.L. (2001) β-Carboline binding to imidazoline receptors. *Drug and Alcohol Dependence*, **64**, 203–208.

Hušek, P. (1991) Rapid derivatization and gas-chromatographic determination of amino acids. *Journal of Chromatography*, **552**, 289–299.

Ichikawa, M., Ryu, K., Yoshida, J., Ide, N., Yoshida, S., Sasaoka, T. and Sumi, S.I. (2004) Antioxidant effects of tetrahydro-β-carboline derivatives identified in aged garlic extract. *ACS Symposium Series*, **871**, 380–404.

Johnson, J.V., Yost, R.A. and Faull, K.F. (1984) Tandem mass-spectrometry for the trace determination of tryptolines in crude brain extracts. *Analytical Chemistry*, **56**, 1655–1661.

Kusurkar, R.S. and Goswami, S.K. (2004) Efficient one-pot synthesis of anti-HIV and anti-tumour β-carbolines. *Tetrahedron*, **60**, 5315–5318.

Masuda, S., Kanamori, H. and Kinae, N. (2005) Isolation of mutagenic β-carboline derivatives after nitrite treatment of Maillard reaction mixtures and analysis of these compounds from foodstuffs and human urine. *Bioscience Biotechnology and Biochemistry*, **69**, 2232–2235.

Matsubara, K., Fukushima, S., Akane, A., Hama, K. and Fukui, Y. (1986) Tetrahydro-β-carbolines in human urine and rat brain. No evidence of formation by alcohol drinking. *Alcohol and Alcoholism*, **21**, 339–345.

Matsubara, K., Neafsey, E.J. and Collins, M.A. (1992) Novel S-Adenosylmethionine-dependent indole-*N*-Methylation of β-Carbolines in brain particulate fractions. *Journal of Neurochemistry*, **59**, 511–518.

Matsubara, K., Gonda, T., Sawada, H., Uezono, T., Kobayashi, Y., Kawamura, T., Ohtaki, K., Kimura, K. and Akaike, A. (1998) Endogenously occurring β-carboline induces parkinsonism in nonprimate animals: A possible causative protoxin in idiopathic Parkinson's disease. *Journal of Neurochemistry*, **70**, 727–735.

Mayeno, A.N., Lin, F., Foote, C.S., Loegering, D.A., Ames, M.M., Hedberg, C.W. and Gleich, G.J. (1990) Characterization of peak-E, a novel amino acid associated with eosinophilia-myalgia-syndrome. *Science*, **250**, 1707–1708.

Melchior, C. and Collins, M.A. (1982) The route and significance of endogenous synthesis of alkaloids in animals. *CRC Critical Reviews in Toxicology*, **9**, 313–356.

Miralles, A., Esteban, S., Sastre-Coll, A., Moranta, D., Asensio, V.J. and Garcia-Sevilla, J.A. (2005) High-affinity binding of β-carbolines to imidazoline I-2B receptors and MAO-A in rat tissues: Norharman blocks the effect of morphine withdrawal on DOPA/noradrenaline synthesis in the brain. *European Journal of Pharmacology*, **518**, 234–242.

Morgan, N.G., Cooper, E.J., Squires, P.E., Hills, C.E., Parker, C.A. and Hudson, A.L. (2003) Comparative effects of efaroxan and β-carbolines on the secretory activity of rodent and human beta cells. *Annals of the New York Academy of Sciences*, **1009**, 167–174.

Musshoff, F., Daldrup, T. and Bonte, W. (1993) Gas-chromatographic mass-spectrometric screening procedure for the identification of formaldehyde-derived tetrahydro-β-carbolines in human urine. *Journal of Chromatography*, **614**, 1–6.

Myers, R.D. (1989) Isoquinolines, β-carbolines and alcohol drinking. Involvement of opioid and dopaminergic mechanisms. *Experientia*, **45**, 436–443.

Nakatsuka, S., Feng, B., Goto, T. and Kihara, K. (1986) Structures of Flazin and Ys, highly fluorescent compounds isolated from Japanese soy sauce. *Tetrahedron Letters*, **27**, 3399–3402.

Nii, H. (2003) Possibility of the involvement of 9H-pyrido(3,4-b)indole (norharman) in carcinogenesis via inhibition of cytochrome P450-related activities and intercalation to DNA. *Mutation Research*, **541**, 123–136.

Östergren, A., Annas, A., Skog, K., Lindquist, N.G. and Brittebo, E.B. (2004) Long-term retention of neurotoxic β-carbolines in brain neuromelanin. *Journal of Neural Transmission*, **111**, 141–157.

Ozawa, Y., Uda, Y., Matsuoka, H., Abe, M., Kawakishi, S. and Osawa, T. (1999) Occurrence of stereoisomers of 1-(2′-pyrrolidinethione-3′-yl)-1,2,3,4-tetrahydro-β-carboline-3-carboxylic acid in fermented radish roots and their different mutagenic properties. *Bioscience Biotechnology and Biochemistry*, **63**, 216–219.

Papavergou, E. and Clifford, M.N. (1992) Tetrahydro-β-carboline carboxylic acids in smoked foods. *Food Additives and Contaminants*, **9**, 83–95.

Papavergou, E. and Herraiz, T. (2003) Identification and occurrence of 1,2,3,4-tetrahydro-β-carboline-3-carboxylic acid: The main β-carboline alkaloid in smoked foods. *Food Research International*, **36**, 843–848.

Parker, C.A., Anderson, N.J., Robinson, E.S.J., Price, R., Tyacke, R.J., Husbands, S.M., Dillon, M.P., Eglen, R.M., Hudson, A.L., Nutt, D.J., Crump, M.P. and Crosby, J. (2004) Harmane and harmalan are bioactive components of classical clonidine-displacing substance. *Biochemistry*, **43**, 16385–16392.

Pfau, W. and Skog, K. (2004) Exposure to β-carbolines norharman and harman. *Journal of Chromatography B*, **802**, 115–126.

Potthast, K. and Eigner, G. (1985) Formaldehyde in smokehouse smoke and smoked meat products. *Fleischwirtschaft*, **65**, 1178–1186.

Reiter, R.J., Tan, D.X. and Allegra, M. (2002) Melatonin: Reducing molecular pathology and dysfunction due to free radicals and associated reactants. *Neuroendocrinology Letters*, **23**, 3–8.

Robinson, E.S.J., Anderson, N.J., Crosby, J., Nutt, D.J. and Hudson, A.L. (2003) Endogenous β-carbolines as clonidine-displacing substances. *Annals of the New York Academy of Sciences*, **1009**, 157–166.

Rommelspacher, H., May, T. and Susilo, R. (1991) β-Carbolines and tetrahydroisoquinolines. Detection and function in mammals. *Planta Medica*, **57**, S85–S92.

Rommelspacher, H., May, T. and Salewski, B. (1994) Harman (1-Methyl-β-carboline) is a natural inhibitor of monoamine oxidase type A in rats. *European Journal of Pharmacology*, **252**, 51–59.

Rönner, B., Lerche, H., Bergmüller, W., Freilinger, C., Severin, T. and Pischetsrieder, M. (2000) Formation of tetrahydro-β-carbolines and β-carbolines during the reaction of L-tryptophan and D-glucose. *Journal of Agricultural and Food Chemistry*, **48**, 2111–2126.

Schouten, M.J. and Bruinvels, J. (1985) High-performance liquid-chromatography of tetrahydro-β-carbolines extracted from plasma and platelets. *Analytical Biochemistry*, **147**, 401–409.

Sen, N.P., Seaman, S.W., Baddoo, P.A., Weber, D. and Malis, G. (1991) Analytical methods for the determination and mass spectrometric confirmation of 1-methyl-2-nitroso-1,2,3,4-tetrahydro-β-carboline-3-carboxylic acid and 2-nitroso-1,2,3,4-tetrahydro-β-carboline-3-carboxylic acid in foods. *Food Additives and Contaminants*, **8**, 275–290.

Sen, N.P., Seaman, S.W., Lan, B.P.Y., Weber, D. and Lewis, D. (1995) Determination and occurrence of various tetrahydro-β-carboline-3-carboxylic acids and the corresponding *N*-nitroso compounds in foods and alcoholic beverages. *Food Chemistry*, **54**, 327–337.

Shi, C.C., Chen, S.Y., Wang, G.J., Liao, J.F. and Chen, C.F. (2000) Vasorelaxant effect of harman. *European Journal of Pharmacology*, **390**, 319–325.

Skog, K., Solyakov, A., Arvidsson, P. and Jägerstad, M. (1998) Analysis of nonpolar heterocyclic amines in cooked foods and meat extracts using gas chromatography mass spectrometry. *Journal of Chromatography A*, **803**, 227–233.

Sobhani, A.M., Ebrahimi, S.A. and Mahmoudian, M. (2002) An in vitro evaluation of human DNA topoisomerase I inhibition by *Peganum harmala L.* seeds extract and its β-carboline alkaloids. *Journal of Pharmacy and Pharmaceutical Sciences*, **5**, 19–23.

Spijkerman, R., van den Eijnden, R., van de Mheen, D., Bongers, I. and Fekkes, D. (2002) The impact of smoking and drinking on plasma levels of norharman. *European Neuropsychopharmacology*, **12**, 61–71.

Toribio, F., Moyano, E., Puignou, L. and Galceran, M.T. (2002) Ion-trap tandem mass spectrometry for the determination of heterocyclic amines in food. *Journal of Chromatography A*, **948**, 267–281.

Totsuka, Y., Hada, N., Matsumoto, K., Kawahara, N., Murakami, Y., Yokoyama, Y., Sugimura, T. and Wakabayashi, K. (1998) Structural determination of a mutagenic aminophenylnorharman produced by the co-mutagen norharman with aniline. *Carcinogenesis*, **19**, 1995–2000.

Tsuchiya, H., Ohtani, S., Yamada, K., Takagi, N., Todoriki, H. and Hayashi, T. (1994) Quantitation of urinary 1,2,3,4-tetrahydro-β-carboline and 1-methyl-1,2,3,4-tetrahydro-β-carboline by high-performance liquid-chromatography. *Journal of Pharmaceutical Sciences*, **83**, 415–418.

Tsuchiya, H., Yamada, K., Tajima, K. and Hayashi, T. (1996) Urinary excretion of tetrahydro-β-carbolines relating to ingestion of alcoholic beverages. *Alcohol and Alcoholism*, **31**, 197–203.

Tsuchiya, H., Sato, M. and Watanabe, I. (1999) Antiplatelet activity of soy sauce as functional seasoning. *Journal of Agricultural and Food Chemistry*, **47**, 4167–4174.

Wakabayashi, K., Ochiai, M., Saito, H., Tsuda, M., Suwa, Y., Nagao, M. and Sugimura, T. (1983) Presence of 1-methyl-1,2,3,4-tetrahydro-β-carboline-3-carboxylic acid, a precursor of a mutagenic nitroso compound, in soy sauce. *Proceedings of the National Academy of Sciences of the United States of America*, **80**, 2912–2916.

Yu, A.M., Idle, J.R., Herraiz, T., Küpfer, A. and Gonzalez, F.J. (2003) Screening for endogenous substrates reveals that CYP2D6 is a 5-methoxyindolethylamine *O*-demethylase. *Pharmacogenetics*, **13**, 307–319.

Zhao, M., Bi, L.R., Wang, W., Wang, C., Baudy-Floc'h, M., Ju, J.F. and Peng, S.Q. (2006) Synthesis and cytotoxic activities of β-carboline amino acid ester conjugates. *Bioorganic and Medicinal Chemistry*, **14**, 6998–7010.

Plate 1 Wild plants of *Echium vulgare* in New Zealand. A rich monofloral source of nectar and pyrrolizidine alkaloids. Photo courtesy of Barrie Wills. (Reprinted with permission from Betteridge *et al.* in *Journal of Agricultural and Food Chemistry*, **53**, 1894–1902, Figure 4. Copyright 2005 with permission from the American Chemical Society.)

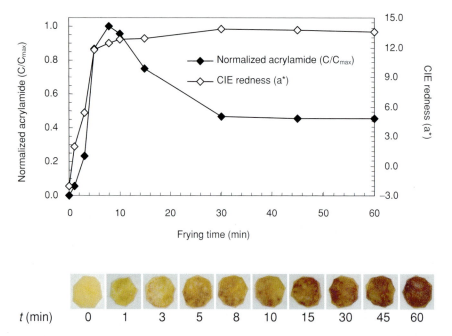

| t (min) | 0 | 1 | 3 | 5 | 8 | 10 | 15 | 30 | 45 | 60 |

Plate 2 Change of acrylamide concentration and CIE redness parameter a* in potato chips during frying at 170°C. (Reprinted from Gökmen, V., Şenyuva, H.Z., Dülek, B. and Çetin, A.E. Computer vision-based image analysis for the estimation of acrylamide concentrations of potato chips and french fries. *Food Chemistry*, **101**, 791–798. Copyright 2006b with permission from Elsevier.)

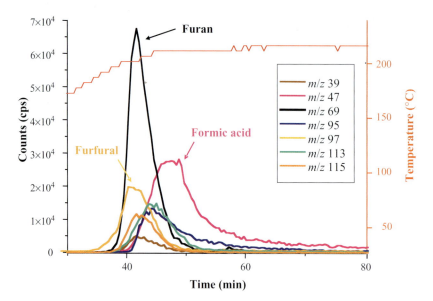

Plate 3 PTR-MS traces of selected ions obtained by heating ascorbic acid at 200°C. Some of these traces were identified by coupling with GC-MS after trapping the headspace on Tenax® tubes. (Reprinted with permission from Märk, J., Pollien, Ph., Lindinger, Ch., Blank, I. and Märk, T., Quantitation of furan and methylfuran formed in different precursor systems by proton transfer reaction mass spectrometry, *Journal of Agricultrual and Food Chemistry*, **54**, 2786–2793, copyright 2006, American Chemical Society.)

Part Two
Man-made components

9 Naturally Occurring Nitrates and Nitrites in Foods

Mari Reinik, Terje Tamme and Mati Roasto

Summary

Nitrates and nitrites can be found in a variety of plant-derived foods as naturally occurring compounds. Dietary exposure to both nitrates and nitrites are of interest from a human health perspective in terms of direct toxic effects (e.g. cyanosis) and also possible indirect effects as precursors of carcinogenic N-nitrosamines. Levels of nitrate in vegetables can range from below 10 mg/kg to as much as 10 000 mg/kg depending on many factors such as cultivar type, light intensity, soil composition, air temperature, growth density, moisture, maturity of plant, duration of growth period, harvesting time, size of the vegetable, storage time, edible plant portion and nitrogen sources. The levels of nitrates in fruit are low compared with the vegetables. It has been estimated that vegetables constitute a major source of human exposure to nitrates contributing up to 92% of the average daily intake. Natural levels of nitrites in food are low, usually remaining under the limit of detection.

9.1 INTRODUCTION

Nitrates and nitrites can be found in food as naturally occurring compounds. An interest in the dietary intakes of nitrates and nitrites has arisen mainly from the concern about their possible adverse effect on health. The natural occurrence of nitrates in plants is a consequence of the nitrogen cycle whereby mineral nitrogen is assimilated by the plant as nitrates to use them in the synthesis of plant proteins. Nitrates and nitrites are also used as food additives in cured meats due to their ability to protect products from *Clostridium botulinum* and other *Clostridium* species, and for their red colour-fixing properties. Nitrates and nitrites are found in drinking water due to both natural occurrence and contamination of water supplies, mostly from agricultural sources and municipal wastewater.

The concern over nitrates and nitrites in the diet has two aspects: they may create an excess of methaemoglobin possibly leading to toxic effects such as cyanosis and they may cause the endogenous formation of carcinogenic N-nitroso compounds.

Nitrate represents the stable oxidation state (V) of nitrogen and can be reduced to nitrite in the environment by micro-organisms and within human tissues. Nitrite represents a less stable oxidation state (III) of nitrogen and therefore can be further reduced to various compounds or oxidized to nitrate. Nitrite may endogenously react with secondary amines to form N-nitrosamines at low pH values, as is the case in the gastric environment of mammals. Nitrosoamines may also be pre-formed in foodstuffs during certain biological, chemical and physical processes in crops, industrial transformation or even at the time of consumption.

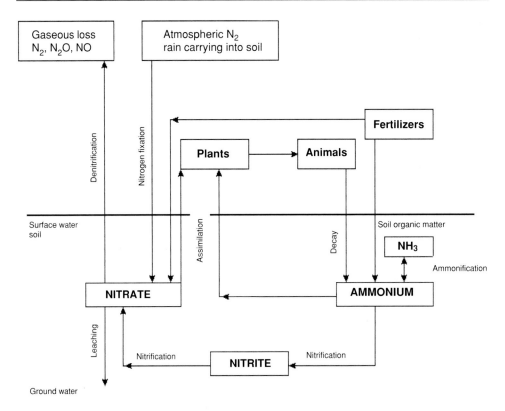

Fig. 9.1   The nitrogen cycle.

The nitrogen cycle, as shown in Fig. 9.1, is the pathway by which nitrogen is converted from the gaseous atmosphere to various inorganic and organic compounds and back again to the gaseous form (Nelson and Cox, 2000; Luf, 2002). This cycle is one of the most essential endlessly repeating processes in nature. All living organisms need nitrogen since it is an essential constituent of proteins, chlorophyll, nucleic acids and the building blocks of the genetic material as DNA and RNA.

In the biosphere, nitrogen alters continuously between the oxidation states +V (fully oxidized nitrogen – nitrate) and –III (fully reduced nitrogen – ammonia). The nitrogen cycle includes a number of redox reactions used either for assimilatory purposes or in respiratory processes for energy conservation. Prokaryotes play essential role in these reactions since only they have the enzymes carrying out these processes (Cabello *et al.*, 2004; González *et al.*, 2006).

The primary source of nitrogen is atmospheric air, from which molecular nitrogen makes up 78%. Incorporation of atmospheric nitrogen into terrestrial nitrogenous compounds takes place via a number of different pathways, including micro-organisms, plants, animals and humans through agricultural and industrial activities. Nitrogen in the air becomes a part of biological matter mostly through the actions of bacteria and algae in a process known as nitrogen fixation. The nitrogen-fixing bacteria take nitrogen from air and convert it into ammonia. The ammonia is further converted into nitrite, and consequently into nitrate by nitrifying bacteria, such as *Nitrosomonas* and *Nitrobacter*. Finally, the nitrate is taken up by plants and incorporated into tissues. Legume plants form nodules on the roots where symbiotic nitrogen-fixing bacteria take nitrogen from the air and convert it into ammonia. These plants can assimilate some nitrogen in the form of ammonium ions from the noodles (Lucinski *et al.*, 2002).

In plants, much of nitrogen is used in chlorophyll molecules, which are essential for the photosynthesis and further growth. Plant-eating animals use nitrates obtained from food to produce proteins. Carnivores obtain nitrogen compounds through eating herbivores. Humans receive nitrogen both from plants and animals.

Bacteria and plants can take up and readily reduce nitrate and nitrite to ammonia through the action of nitrate and nitrite reductases. The process requires energy provided by photosynthesis (Nelson and Cox, 2000). Nitrates accumulated in plants form a nitrogen reserve, which is needed for amino acid and protein synthesis (Elliott and Elliott, 2002).

Nitrate is returned to the environment through microbial degradation of plants and animal remains, as well as in animal faeces. Decomposition of dead organic matter by bacteria is the source of ammonia. Waste material from plants and animals returns nitrogen to the soils in which part of it is recycled and part returned after bacterial denitrification to the atmosphere to complete the nitrogen cycle. The denitrification of nitrate to nitrogen and nitrogen oxides takes place in the soil by bacteria but also in various natural water sources. Ideally, nitrogen flow into the system and out of it, would be balanced, while any excess import of nitrogen may lead to its accumulation, and/or losses in the form of gaseous nitrogen (N_2), nitrous oxide (N_2O), nitrogen monoxide (NO) and ammonia (NH_3) to the atmosphere, and as nitrate NO_3^- to the hydrosphere (Codispoti *et al.*, 2001; Oenema *et al.*, 2003). The balance of nitrogen cycle is often destroyed due to increasing use of nitrogen-based fertilizers. As a consequence more nitrogen is transformed to gaseous form (Hardisson and Gonzáez Padrón, 1996). Nitrate and nitrite are readily soluble in water and quite mobile in the environment. They have a high potential for entering surface water when it rains and groundwater through leaching. Because of the increased use of synthetic nitrogen fertilizers and livestock manure in intensive agriculture, vegetables and drinking water in agricultural areas may contain higher concentrations of nitrate than in the past.

9.2 METHODS OF ANALYSIS

A variety of analytical methods, including spectrophotometry, high-performance liquid chromatography (HPLC), ion chromatography (IC), gas chromatography (GC), polarography and capillary electrophoresis (CE), for the determination of nitrate and nitrite in food have been developed. Using of HPLC methods has gained more popularity in last decades since they are more rapid than classic methods based on reduction process followed by colorimetry. In a review paper by Moorcroft *et al.* (2001) the strategies employed to facilitate the determination of nitrate and nitrite, relevant analytical methodologies and various techniques have been presented. Similarly to the analytical determination of other unstable and unevenly distributed compounds in food, systematic bias may be introduced at all stages of the analytical procedure from sampling, storage, extraction to the end determination step. Tables 9.1 and 9.2 present an overview of the analytical methods used for the determination of nitrates and nitrites in recent years, their performance characteristics, advantages and limitations.

Table 9.1 Method performance characteristics in nitrate and nitrite analysis.

Method	LOQ (mg/kg)	Reproducibility RSD	Recovery	Reference
HPIC/HPLC, UV detection	Nitrate 2–50 mg/kg; nitrite 0.4–40 mg/kg	Nitrate 3.3–15.2%; nitrite 9.4%	Nitrate 96–109%; nitrite 96–108%	Vaessen and Schothorst, 1999; Merino et al., 2000; Chou et al., 2003; CEN, 2005b
HPIC/HPLC, conductivity detection	20–80 mg/kg; olive oil – 8.4–31 µg/kg (LOD)	2.4–10.2%	Nitrate 87–104%; nitrite 91–104%	CEN, 1997; De Martin and Restani, 2003; McMullen et al., 2005; Farrington et al., 2006; Dugo et al., 2007
IC/MS	0.1 mg/kg			Saccani et al., 2006
Cd reduction, spectrophotometric detection of summed nitrite and nitrate	Nitrate – 5–900 mg/kg (LOD); nitrite 0.3 mg/kg (LOD)	8.3–21%	80–117%	CEN, 1998; Petersen and Stoltze, 1999; AOAC, 2000
Flow-injection spectrophotometric method, through NO generation	Nitrate 20 mg/kg; nitrite 13 mg/kg	2.3–5.5%	Nitrate 93–110%; nitrite 88–97%	Andrade et al., 2003
Ion chromatographic separation, V(III)post-column reduction, spectrometric detection	25 mg/kg		82–107%	Casanova et al., 2006
Enzymatic reduction of nitrate to nitrite	Nitrite 10 mg/kg	7.7%		Cruz and Martins Loução, 2002; Zanardi et al., 2002; CEN, 2005a
Potentiometric methods	Nitrate 30 mg/kg	15%		Tamme et al., 2006
Electrophoresis	Nitrite 4 mg/kg (LOD)	Nitrate 2.54%; nitrite 4.51%	Nitrate 92–106%; nitrite 94–103%	Öztekin et al., 2002
Gas chromatography	100–200 µg/kg	3–13%	100%	Ross and Hotchkiss, 1985
Polarography	Nitrate 39 mg/kg	<7%	85–107%	Ximenes et al., 2000

Table 9.2 Advantages and limitations of various analytical methods for the determination of nitrates and nitrites in food.

Method	Advantages (+), limitations (−)
HPIC, HPL/UV	+ fast, sensitive, precise − possible positive interference due to co-eluting matrix components, equipment not available in all laboratories
HPIC, HPLC, conductivity detection	+ precise, rapid, multiple anions can be determined simultaneously − equipment not available in all laboratories
IC/MS	+ robust, sensitive − expensive equipment needed
Cd reduction, spectrophotometric detection of summed nitrite and nitrate	+ traditional method; no need for expensive equipment −carcinogenic Cd used, nitrite has to be determined separately, less precise for complicate matrixes and low concentrations
Ion chromatographic separation, V(III)post-column reduction, spectrometric detection	+ no Cd used, simultaneous determination of nitrite and nitrate
Enzymatic reduction of nitrate to nitrite	+ no Cd used − nitrite has to be determined separately; overestimation of nitrate values in meat samples
Potentiometric methods	+ inexpensive, rapid − suitable only for the determination of nitrates in vegetables, possible positive interference, lack of sensitivity
Electrophoresis	+ short analysis time, good alternative to IC methods
Gas chromatography	+ high sensitivity − derivatization needed, expensive equipment

9.2.1 Sampling and storage

Nitrate and nitrite are both potentially unstable, therefore all food and water samples should be analysed as soon as possible after collection. It is well established that nitrate is readily reduced by bacteria and may also be produced or utilized as a result of post-sampling metabolic processes that may occur in vegetables. Nitrite is chemically highly reactive and its concentration may drop rapidly during storage, particularly under mildly acidic conditions. In such circumstances it is essential to perform the analysis as soon as possible after sampling. If it is unavoidable to store the samples prior to starting analysis, the effects of the storage conditions on the stability of the analyte must be determined in advance.

9.2.2 Extraction

Many analytical methods for nitrite and nitrate determination in foodstuffs employ the same extraction procedure for both anions. In general, the sample is extracted into hot water or sodium tetraborate (Borax) and treated with protein precipitation reagents (Carrez reagents) prior to filtration and measurement. Several nitrate extraction methods from plant material have been evaluated by Farrington *et al.* (2006), the hot water extraction method described in European Standard EN 12014-2:1997 (CEN, 1997) was found to give most reliable results.

Extraction has to be performed under alkaline conditions, as nitrite may react with other matrix components if the extraction is carried out in even mildly acidic environment (Massey, 1996).

9.2.3 Spectrophotometric methods

Classic methods for the determination of nitrite are based on variations of the Griess di-azotization procedure, in which azo dye is produced in a reaction of diazonium salt with an aromatic amine or phenol. The formed diazonium ions are coupled to another aromatic compound in order to produce an azo dye (Kirk and Sawyer, 1991). This is the basis of many spectrophotometric methods and has also found in wide application for the analysis of nitrate after its prior reduction by spongy cadmium to nitrite (Ruiter and Bergwerff, 2005).

The most widely used colorimetric method is based on the reaction of nitrite, sulfanylamide and N-1-naphtylethylenediamine under acidic conditions to form a red azo dye. The red colour produced in the reaction is measured by a spectrophotometer at 538 nm. Nitrite present in the original sample is determined before nitrate reduction in cadmium column and nitrate content is obtained by the difference of the analysis of unreduced and reduced extract (Slack, 1987). Experimental conditions must be carefully controlled since factors such as pH, temperature, nature and concentration of the reagents affect the final colour intensity (Zanardi *et al.*, 2002).

Other spectrophotometric methods for nitrite are based on the catalytic effect on oxidation of gallocyanine (Ensafi and Bagherian-Dehaghei, 1999) or pyrogallol red (Kazemzadeh and Ensafi, 2001) by bromate. A novel method for the determination of nitrite and nitrate in food by spectrophotometric detection presented by Andrade *et al.* (2003) is based on the reduction of nitrite and nitrate to nitric oxide, with subsequent reaction with iron(II) and thiocyanate in an acid medium, forming $FeSCNNO^+$. The absorbance of the complex measured at 460 nm is proportional to the nitrite and nitrate concentrations. The use of cadmium columns to reduce nitrate to nitrite is widely applied in the analysis of cured meats (Oliveira *et al.*, 2004), cheese and baby food (AOAC, 2000), fruit and vegetables (CEN, 1998; Petersen and Stoltze, 1999).

In the older version of the colorimetric method various steps of analytical procedure are run manually whereas more recent methods apply automated equipment (Zanardi *et al.*, 2002; Oliveira *et al.*, 2004). Flow-injection spectrophotometric detection for simultaneous analysis of nitrite and nitrate has been used both for food and water samples by Petersen and Stoltze (1999), Monser *et al.* (2002), Andrade *et al.* (2003) and Ensafi *et al.* (2004). The injected sample is split into two streams. One of the streams is transported through a reductor microcolumn containing copperized cadmium, where nitrate is reduced to nitrite. The total nitrite concentration initially plus that produced in cadmium column is measured spectrophotometrically. The method is quick – according to the data by different authors, 10–25 samples can be analysed per hour. The parameters affecting the results of flow-injection analysis have been investigated by Gal *et al.* (2004). A continuous flow method for the analysis of vegetables and vegetable products using spectrophotometric detection has also been adopted as a European standard EN 12014-7:1998 (CEN, 1998).

Many attempts have been made to avoid using the toxic and carcinogenic metal cadmium. Other nitrate-reduction techniques include chemical treatment with vanadium(III) chloride for baby food samples (Casanova *et al.*, 2006), microbial reduction with *E. coli* for analysis of plants (Cruz and Martins Loução, 2002) and enzymatic reduction for the analysis of meat samples (CEN, 2005a).

Spectrophotometric assays are potentially vulnerable to interferences from a number of sources, most important of them being incomplete reduction of nitrate and over-reduction of nitrate via nitrite to lower oxides (Norwitz and Keliher, 1985). The close comparability of the results obtained for the nitrite content of cured meats by colorimetry and by HPLC suggests that this is not a widespread problem (Dennis *et al.*, 1990). The colorimetric assay may also be adversely affected by the turbidity of the measurement solution, which is likely to occur if the post-extraction precipitation and filtration steps are ineffective. To overcome the latter problem, Fox *et al.* (1982) have recommended charcoal or alkaline extraction.

Detection limits for nitrate and nitrite by reduction methods are generally around 1 mg/kg (Dennis *et al.*, 1990). For samples that contain appreciable quantities of ascorbate or other interferants the limit of detection may be an order of higher magnitude (Usher and Telling, 1975). However, at analyte concentrations below 20 mg/kg the agreement with other methods is not very good due to problems of chemical interference or turbidity. As with the exception of cured meats, nitrite is generally present at very low levels, reliable determination of it by colorimetric assay is often a problem. Zanardi *et al.* (2002) reported that three analytical methods based on the same principles (Cd manual and automated method, enzymatic method) for the determination of nitrite showed no significant differences. The results concerning nitrate showed a different pattern – the values obtained by the Cd method were considerably lower than those obtained by the chromatographic method (Alonso *et al.*, 1992). The difference has been attributed to a possible decline in efficiency of the Cd column due to possible interferences caused by other anions. Enzymatic test produces an overestimation of the nitrate values.

The AOAC (2000) method for meat samples involves oxidation of nitrite to nitrate with permanganate followed by acidification and treatment with *m*-xylenol. After nitration the nitroxylenol is removed from the samples by distillation and measured colorimetrically.

9.2.4   Chromatographic methods

The use of HPLC and IC for the analysis of nitrate and nitrite has increased substantially over recent years due to the number of favourable factors compared with colorimetric assays including the speed of the analysis, the fact that both anions can be determined at the same time, toxic reagents such as cadmium are not required and also that in some instances higher sensitivity and accuracy is obtainable.

HPLC/UV methods have been reported for the nitrate and nitrite in vegetables (Cheng and Tsang, 1998; Chou *et al.*, 2003; Tamme *et al.*, 2006), dairy products (Reece and Hird, 2000; Gapper *et al.*, 2004), cured meats (Dennis *et al.*, 1990; Reinik *et al.*, 2005) and beer (Massey *et al.*, 1990).

Reverse-phase (ion interaction/ion pair) HPLC has been used both in the analysis of water and food (Frolich, 1987; Mullins, 1987; Cheng and Tsang, 1998), although ion-exchange HPLC is preferred in foodstuff analysis (Pentchuk *et al.*, 1986; Dennis *et al.*, 1990; Siu and Henshall, 1998; Reece and Hird, 2000; Stalikas *et al.*, 2003; Gapper *et al.*, 2004). Normal phase ion-pair chromatography was used for water and vegetable samples by Butt *et al.* (2001). UV-detection is not suitable for multi-ion measurement due to the poor UV absorbance of chloride and phosphate. In such cases the analytes can be measured by indirect UV in which a UV-absorbing compound is included in the composition of mobile phase and anions owing a lower absorbance give a signal in the form of a negative peak on elution from the column (Mullins, 1987). Alternatively, the conductivity detector may be applied (Pentchuk *et al.*, 1986; De Martin and Restani, 2003; Kissner and Koppenaol, 2005; Masson *et al.*, 2005; McMullen *et al.*, 2005; Dugo *et al.*, 2007).

Achieving the stability of chromatographic performance is a problem in the analysis of complex food extracts, as components of the matrix may adsorb onto the analytical column and adversely affect subsequent analyses. Various types of solid-phase extraction cartridges have been tested to clean the sample extract prior to HPLC measurement: C18 SPE columns have been used by De Martin and Restani (2003) for analyses of leafy vegetables and by Vaessen and Schothorst (1999) for total diet samples; Dennis *et al.* (1990) have used cyclohexyl Bond Elut cartridges in the analysis of cured meats. Hunt and Seymour (1985) treated vegetable extracts with activated charcoal.

A number of authors have compared HPLC results with those obtained by other methods. In the case of water, good agreement with colorimetric results has been reported (Schroeder, 1987). Pentchuk *et al.* (1986) compared the determination of nitrate by HPLC and colorimetry of a number of types of vegetables and showed good agreement of the results. HPLC was found to give higher results than colorimetry in the analysis of nitrates in carrots, and this was thought likely to be due to incomplete colour development (Schuster and Lee, 1987). Dennis *et al.* (1990) have reported good agreement between HPLC and colorimetry for the analysis of nitrite and nitrate in cured meats. Better agreement between these methods has been achieved using reverse-phase HPLC than ion exchange (Massey, 1996). Three ion chromatographic methods, including two present European Standard methods EN 12014-2 (CEN, 1997) and EN 12014-4 (CEN, 2005b), were tested in an NMKL collaborative study (Merino *et al.*, 2000) and compared to spectrophotometric method. In the analysis of nitrite, no statistically significant difference was found between the spectrophotometric method and EN 12014-4. It was concluded that the use of a strong anion exchange column is necessary to ensure the reliability of results. Nitrite determination with the weak anion exchanger column used in standard method EN 12014-2 for the analysis of vegetable samples is not suitable for the determination of residual nitrate and nitrite in meat products. The method is also not reliable for the analysis of certain vegetables with low concentrations of nitrate.

The detection limits for HPLC-based methods are typically between 0.1 and 10 mg/kg for foodstuffs (Eggers and Cattle, 1986; Dennis *et al.*, 1990; Vaessen and Schothorst, 1999; Merino *et al.*, 2000; Chou *et al.*, 2003). In the validation of an ion chromatographic method for the determination of nitrates in some high-nitrate containing leafy vegetables, de Martin and Restani reported a limit of detection of 80 mg/kg.

Gas chromatographic methods for the measurement of nitrate and nitrite in water and foodstuffs involve the formation of a volatile derivative, extraction into organic solvent and measurement by GC using a selective detector. Wu *et al.* (1984) have developed a gas chromatographic electron-capture detector (GC/ECD) method for the pentafluorobenzyl derivative of nitrite. The procedure has a detection limit of 5 µg/L for aqueous samples and showed good agreement with a colorimetric assay for the analysis of cured meat, saliva and water. Similar performance characteristics were achieved by Funazo *et al.* (1980) in their method for nitrite involving GC/ECD detection of bromochlorobenzene formed by reaction of nitrite, bromoaniline and cupric chloride. Ross and Hotchkiss (1985) used the nitrobenzene derivative for the measurement of nitrate in dried foods by GC with thermal energy analyser, and reported a detection limit of 100 µg/kg.

9.2.5 Other methods

In recent years capillary electrophoretic (CE) methods have been developed for the simultaneous detection of nitrite and nitrate in foodstuffs (Marshall and Trenerry, 1996; Öztekin *et al.*, 2002). Anions are separated in a capillary coated with polyethyleneimine. Good recovery and reproducibility can be achieved, and the detection limit of the method is sufficiently low for the analysis of meat products and vegetables.

Damiani and Burini (1986) have described a fluorimetric assay for nitrite based on its reaction with 2,3-diaminonaphtalene. Application of the method to milk samples gave results that were approximately 10% lower than a colorimetric assay.

Nitrate-selective electrodes have found little application in the analysis of foodstuffs because of their potential interference from several commonly occurring anions such as chloride, sulfate and bicarbonate. Pentchuk *et al.* (1986) has reported that nitrate electrodes may give rise to positive interference when compared to other methods in the analysis of vegetables. Other methods that have been reported for the analysis of nitrate and nitrite include ion-pair extraction/atomic absorption spectrophotometry (Silva *et al.*, 1986), amperometry (Bertotti and Pletcher, 1997), polarography (Ximenes *et al.*, 2000) and stripping voltammetry (Van den Berg and Li, 1988).

9.3   INCIDENCE AND LEVELS OF OCCURRENCE

9.3.1   Vegetables

Vegetables constitute a major source of human exposure to nitrates contributing approximately 40–92% of the average daily intake (Penttilä, 1995; Dich *et al.*, 1996; Ximenes *et al.*, 2000; Eichholzer and Gutzwiller, 2003). Nitrate is present as a natural constituent in plants and may accumulate in different tissues. Each plant species have their own unique path of biological photosynthesis in leaves as well the transport mechanism for getting water and nutrients by roots. Those biological mechanisms are most essential factors influencing the nitrate level in plants.

The content of nitrate in different vegetables varies to large extent and according to that the vegetables can be divided into three groups (Tamme *et al.*, 2006):

1   Plants with nitrate content higher than 1000 mg/kg – lettuce, spinach, herbs, beetroot, rhubarb, turnip etc.
2   Plants with average content of nitrate (50–1000 mg/kg) – carrot, green beans, cauliflower, onion, pumpkin, eggplant, potato etc.
3   Plants with nitrate content lower than 50 mg/kg – berries, fruits, cereals, pod vegetables.

De Martin and Restani (2003) showed that leafy green vegetables accumulate the highest amounts of nitrates, concentrations reaching up to 6000 mg/kg. According to the data reported in Table 9.3, it can be concluded that the highest mean values of nitrates have been detected in spinach, lettuce and dill. Among root vegetables, nitrate concentration in beetroot is the highest. The lowest mean values of nitrates were detected in tomato, cucumber and onion. Nitrate concentrations in the vegetables measured in different countries are in good agreement with some exceptions. According to the data given in Table 9.3, the concentration of nitrates in potatoes differs a lot depending on the region of cultivation. The nitrate content in potatoes in Northern Europe remained below 100 mg/kg, while in the Far East the concentrations reached over 700 mg/kg (Penttilä, 1995; Chung *et al.*, 2003; Tamme *et al.*, 2006).

Nitrate concentrations in some salad crops of different varieties during the summer and winter seasons were screened by Escobar-Gutierrez *et al.* (2002): great variability between the cultivars and also varieties within one cultivar were detected. The exceeding of maximum

Table 9.3 Mean contents of nitrates in vegetables in different countries (Dejonckheere et al., 1994; Penttilä, 1995; Belitz and Grosch, 1999; Petersen and Stoltze, 1999; Ysart et al., 1999; Chung et al., 2003; Sušin et al., 2006; Tamme et al., 2006).

Vegetable commodity				Mean NO_3^- concentration (mg/kg)					
	Finland	Great Britain	Belgium	Denmark	Slovenia	Korea	Japan	Germany	Estonia
Potato	82	155	154	182	158	452	713	93	94
Carrot	264	97	278		264	316	193	232	148
Lettuce	1835	1051	2782	2631	1074	2430		1489	2167
Chinese cabbage	1057			1058		1740	1040		1243
Cabbage	607	338		333	881	725		451	437
Turnip	908	118							307
Tomato	170	17	36		<6			27	41
Sweet pepper	140		93			76	99		
Cucumber	240		344		93	212	384		160
Onion	140	48	59			23			55
Beetroot	1800	1211		1505				1630	1446
Green beans	455		585		298				
Radish			2136			1878	1060	2030	1309
Spinach		1631	2297	1783	965	4259	3560	965	2508
Leek			841	284		56			
Celery leaves			3135						565
Squash pumpkin						639			174
Spring onion						436	145		477
Garlic						124	455		174
Dill/fennel								1541	2936
Rhubarb								986	201
Parsley			2690						966
Endive			1246						
Cauliflower		86							
Canned vegetables		18							287

Table 9.4 Nitrite contents in vegetables in different countries.

| Country | NO$_2^-$ concentration (mg/kg) | | Reference |
	Range	Mean	
Finland	ND–1.5	1.0	Penttilä (1995)
Germany	<0.1–19.6	0.1	Belitz and Grosch (1999)
England	0.3–3.8	1.1	Ysart et al. (1999)
Japan	ND–7.0	2.1	Chung et al. (2003)
Korea	0.3–1.1	0.57	Chung et al. (2003)
Denmark	0.15–11	2.5	Petersen and Stoltze (1999)
Slovenia	<0.5–1.2		Sušin et al. (2006)
Poland	0.2–0.83		Nabrzyski and Gajewska (1994)

limit concentrations was more frequent in the summer season than in winter. This may be explained by lower maximum limit values valid for summer period. Tamme *et al.* (2006) found in their study that the content of nitrates in lettuce was lower in summer time (average 1952 mg/kg) than in winter (average 3024 mg/kg), and exceedings of limit concentrations were not detected. Szymczak and Prescha (1999) and Järvan (1993) reported that nitrate concentration in the greenhouse vegetables, lettuce, cucumber and radish, was greater than in the field-grown analogues.

Concerning organically farmed vegetables, controversial results have been achieved by different authors. Significantly higher nitrate content was found in Italian organically grown green salad and rocket compared with the same conventionally produced products (De Martin and Restani, 2003). On the contrary, US investigation (Worthington, 2001) stated that organic crops contained significantly less nitrates than conventionally grown analogues. The review of literature conducted by Heaton (2001) found 14 studies showing 50% lower nitrate content in organically grown crops and two studies showing insignificant differences.

For nitrite, the main source of exogenous human exposure is also food. The nitrite content of most fresh, frozen or canned vegetables is relatively low and usually of the order of 0–2 mg/kg (Siciliano *et al.*, 1975; Corré and Breimer, 1979). Comparison of data regarding the nitrite contents of vegetables obtained by different authors is shown in Table 9.4.

The studies have proved that ascorbic acid is very efficient at preventing the conversion of nitrate to nitrite in plant tissue and within the human body. Fresh vegetables that are rich in ascorbic acid, such as kale, green pepper and broccoli, may contain enough vitamin C to avoid significant nitrate reduction to nitrite and subsequent formation of nitrosoamines (Mackerness *et al.*, 1989; Kolb and Haug, 1997; Naidu, 2003).

9.3.1.1 *Factors affecting nitrate and nitrite levels in plants*

Nitrate concentrations in vegetables can vary from 1 to 10 000 mg/kg depending on biological properties of cultivars, light intensity, soil composition, air temperature, growth density, moisture, maturity of plant, duration of growth period, harvesting time, size of the vegetable, storage time, edible plant portion and nitrogen source (Tivo and Saskevic, 1990; Walker, 1990; WHO, 1995; Fytianos and Zarogiannis, 1999; Laslo *et al.*, 2000).

Plant portion. Accumulation of nitrate in plants differs largely between the parts of the crop (Fytianos and Zarogiannis, 1999). The root of the cabbage contains higher levels of nitrates compared to the leaves. On the contrary, root vegetables such as carrots and beetroots contain lower levels of nitrates than do their leaves. Experiments have shown that some potato breeds accumulate lower amounts of nitrates than others. Studies with carrots have indicated that the core of the carrot contains higher levels of nitrates than the outer layers do (Järvan, 1993). In the outer layers of cucumbers and radish, the nitrate concentration is 2–3 times higher than in the pulp (Golaszewska and Zalewski, 2001; Mozolewski and Smoczynski, 2004). The nitrate content is decreased by 40% by removing the stem and midrib of the spinach before stewing as the nitrate content in these plant portions was 2–3 times higher than in leaf tissue (Sokolov, 1987).

Lighting. Via the disturbance of nitrate reductase the decreased lighting conditions cause disorders in the formation of organic compounds and the concentrations of nitrates in plant stay high. Extended periods of cloudy weather increase nitrate content and dangerously high levels can occur when wet days follow a severe drought (Vulsteke and Biston, 1978).

Soil composition. In similar fertilization and growth conditions, the lowest nitrate concentrations are detected in vegetables grown in light sandy soils. Higher accumulation of nitrates is reported in clay- and humus-rich soils and the highest concentrations from low-lying swamps. At the same time, soils with higher concentrations of organic compounds are able to supply plants with nitrogen more equally compared with sandier soils, which have higher water filtration ability (Rückauf *et al.*, 2004).

Temperature. It is an important factor influencing the residual nitrate content in vegetables. Low temperatures in spring or autumn slow down photosynthesis and favour nitrate accumulation. Too high temperatures reduce nitrate reductase activity resulting in higher nitrate concentrations in plants. In optimal growth temperature conditions the stress in plants is avoided and no temperature-related nitrate accumulation is observed (Tivo and Saskevic, 1990; Järvan, 1993).

Moisture. Low moisture conditions favour accumulation of nitrates into the plants. Rückauf *et al.* (2004) reported that higher soil moisture resulted in more efficient plant nitrogen uptake. Moderate soil moisture conditions conduce the reasonable plant nitrogen nutrition while wet conditions increase the nitrate concentrations in plants.

Growth density. The connection between nitrate content and growth density has been described from two aspects. Firstly, the growth density influences the lighting conditions on leaves. Higher density will shadow the plants and causes decreased growth via the enzyme inhibitions. The same type of reductase inhibition has been reported as well for weedy fields. Secondly, there is a mutual relationship between growth density, soil fertility and final nitrate content of the plants. In the conditions of soil rich in plant nutrients and low growth density, overconsumption of nitrogen by plants may occur (Vulsteke and Biston, 1978; Järvan, 1993).

Plant maturity: Growth period. Nitrate content generally is highest in early stages of plant growth and decreases with maturity. Stems of vegetables contain higher amounts of nitrates than leaves. The growth period is generally species-specific, but sometimes it may be shortened due to leaf damages caused by night frost, hail, plant diseases, pests, herbicide drift etc. Longer and lighter growth periods favour the reduction of nitrates in plants for the time of harvesting (Järvan, 1993; Sheehy *et al.*, 2004).

Harvesting time. All season growing cultivars have more nitrates in early crops. Some vegetable species are harvested in different growing stage, which results in variable nitrate concentrations in plants harvested from the same field. According to the data by Järvan (1993), the nitrate content in radishes harvested in early growth stage nitrate concentration was 1.4-fold higher compared with the mature roots, for carrots the nitrate content in early plants was 2-fold higher.

Fertilization. Plants grown without excessive nitrogen fertilizer contain far less nitrate. Nitrate fertilizer applied shortly before harvest causes the greatest increase in nitrate levels and should be avoided. Use of slower nitrogen-releasing natural fertilizers such as animal and green manures enables vegetables to be produced with significantly lower nitrates.Low availability of phosphorus and potassium from soil can contribute to nitrate accumulation. Plant species, stress factors and plant growing conditions have been reported to have more influence to the nitrate levels in plants than amount of nitrogen fertilizer applied (Järvan, 1993; Hlusek *et al.*, 2000).

Storage conditions. Studies have shown that during storage, the nitrate content in vegetables decreases 15–20% (Sokolov, 1989). This is related with transformation of nitrates to nitrites. Nitrate concentration decreases and nitrite concentration increases. In optimal storage conditions, optimal temperature and moisture, the concentration of nitrates decreases slowly in all vegetables. Nitrite concentrations in vegetables may increase to elevated levels due to bacterial nitrification of nitrate to nitrite when vegetables are stored in rooms with high humidity and poor sanitation (Sokolov, 1987). Vegetable cuts, salads and raw juices have to be prepared preferably shortly prior to the consumption. Storage at room temperatures increases the nitrite concentrations to potentially hazardous level (Sokolov, 1987). Chung *et al.* (2004) found that during storage of leafy vegetables at ambient temperature, nitrate levels in the vegetables dropped significantly from the third day while nitrite levels increased dramatically from the fourth day of storage. Over 7 days, refrigerated storage did not lead to changes in nitrate and nitrite levels in the vegetables.

Food handling. High nitrate concentrations initially present in vegetables can be decreased during the treatment of food by utilization of the ability of nitrates and nitrites to dissolve in water (Buck and Osweiler, 1973). Dejonckheere *et al.* (1994) measured the nitrate loss in a number of vegetables after normal culinary practice such as washing, peeling, cooking and stewing. Washing of leafy vegetables with tap water reduced the nitrate concentration with 10–15% and for lettuce the elimination of the thick midrib resulted in a decrease of the nitrate content of 30–35%. Peeling and washing of vegetables can decrease the nitrate content of 20–30% (Laslo *et al.*, 2000). Studies have shown 25–30% decrease in nitrate content of potato, carrot, beetroot, turnip and cabbage after at least 1 hour of soaking (Mozolewski and Smoczynski, 2004). A slightly lower decrease of 20% was achieved for spinach, celery, dill and spring onion. Longer soaking reduces nitrate concentrations even more but the loss of beneficial food components is higher as well (Sokolov, 1987). Boiling of vegetables can decrease the nitrate concentrations by almost 80%. The reduction is related to dissolution of nitrates in the boiling water. Nitrate concentration decreases more when an abundant amount of boiling water is used and after cooking the vegetables are drained carefully. Adding sodium chloride is recommended at the end of boiling, as when this is done too early, nitrate solubility in water is reduced (Sokolov, 1987).

9.3.2 Fruits

The levels of nitrates in fruit are low compared with the vegetables. White (1976) reported the nitrate contents in fruits to be 10 mg/kg. Another earlier study by Herrmann (1972) showed that strawberries may contain nitrates over 100 mg/kg, grapes reached the level of 17 mg/kg. Nitrate concentrations in fruits and fruit products reported by different authors are represented

Table 9.5 Nitrate contents (mg/kg) in fruits and fruit products.

Fruits and products	Mean NO$_3^-$ concentration (mg/kg)	Reference
Apples	11	Dejonckheere et al., 1994
Apples	6	Sušin et al., 2006
Apples	19	Belitz and Grosch, 1999
Oranges	13	Dejonckheere et al., 1994
Bananas	402	Dejonckheere et al., 1994
Grapes	8.2	Belitz and Grosch, 1999
Grapes	46	Dejonckheere et al., 1994
Fresh fruits	27	Ysart et al., 1999
Fruit products	13	Ysart et al., 1999
Melons	375	Dejonckheere et al., 1994
Pears	14	Dejonckheere et al., 1994
Pears	1.0	Sušin et al., 2006
Peaches	10	Dejonckheere et al., 1994
Strawberries	156	Dejonckheere et al., 1994
Strawberries	94	Sušin et al., 2006
Strawberries	139	Belitz and Grosch, 1999
Strawberries	34	Nabrzyski and Gajewska, 1994

in Table 9.5. In a recent Slovenian study (Sušin et al., 2006), nitrate and nitrite contents were generally low: less than 6 mg/kg of nitrate was found in grapes, peaches, apples and pears. The highest average nitrate content, 94 mg/kg, was found in strawberries. In apples and pears, the average nitrite contents were 1.5 and 1.0 mg/kg, respectively. The content of nitrites did not exceed 0.5 mg/kg in other fruits. Nabrzyski and Gajewska (1994) determined concentrations of nitrates and nitrites in Polish fruit and berries during 1989–1992. The highest levels of nitrates were detected also in strawberry samples (maximum to 322 mg KNO$_3$/kg), mean level was found to be 59 mg KNO$_3$/kg. Other berries, such as currants, gooseberries, raspberries and cherries contained from 'not detected' to 36 mg KNO$_3$/kg. Very low level of nitrates was found in seven species of apples (from 1.3 to 9.7 mg KNO$_3$/kg). The concentration of nitrites in all samples remained from 'not detected' to fractions of mg/kg (Nabrzyski and Gajewska, 1994). The results are in good agreement with the earlier work by Gajewska et al. (1989) who reported that content of nitrates in frozen fruits (strawberries, black and red currants and plums) ranged from 2.5 to 57 mg KNO$_3$/kg, with the highest concentrations detected in garden strawberries. In cherry, strawberry, black and red currant jams, the concentrations were detected from 6.3 to 97 mg KNO$_3$/kg. The nitrite content in all these products was low, not exceeding 1 mg NaNO$_2$/kg, with the exception of plum jam where the maximal value of 1.6 mg NaNO$_2$/kg was found. The ranges of nitrate and nitrite concentrations in fruit juices have been reported to be 9.7–21 mg/L and 3.1–9.7 mg/L, respectively (Okafor and Ogbonna, 2003).

9.3.3 Milk and dairy products

Increased nitrate concentrations in milk are not only dangerous to human health as milk is used for production of baby and infant food, but may also cause many technological problems in milk processing (Baranova et al., 1993). Contamination of milk with nitrate may occur during and/or after secretion. Because of a very low carry-over rate of nitrate from forage to milk, the main route of contamination is post-secretory. Contamination of milk and dairy

Table 9.6 Nitrate and nitrite contents in milk and dairy products.

Product	Mean NO$_3$– concentration (mg/kg)	Mean NO$_2$– concentration (mg/kg)	Reference
Milk	4.8	–	Wright and Davison, 1964
Milk	8	–	Statens Leventsmiddelinstitut, 1981
Milk	5.3	–	Ysart et al., 1999
Milk	0.31	0.006	Luf and Brandl, 1986
Milk	1.4	–	Belitz and Grosch, 1999
Dairy products	27	–	Ysart et al., 1999
Soft cheese	1.5	0.17	Luf and Brandl, 1986
Edam cheeses	7.5	0.4	MAFF, 1987
Cheese	–	0.3	Belitz and Grosch, 1999
Cheese	16	0.9	Penttilä, 1995
Cheese	4	0.5	Nikolas et al., 1997
Cottage cheese	12.1	0.008	Luf and Brandl, 1986
English-type cheeses	2.8	0.4	MAFF, 1987
Danish cheeses	10	0.2	Statens Leventsmiddelinstitut, 1981
Milk powder	5.4	0.011	Luf and Brandl, 1986
Yoghurt	0.72	0.002	Luf and Brandl, 1986

products with nitrites during and/or after secretion is lower compared with nitrates (Blüthgen et al., 1997). Residues of nitric acid used as a cleaning reagent combined with inadequate rinsing, addition of water with high nitrate content or the use of nitrate as a food additive in cheese manufacturing are the reasons for increased nitrate contents in dairy products (Luf, 2002).

The investigation of Kammerer et al. (1989) showed that drinking water has no significant effect on milk nitrate content, which remains very low and does not constitute a health risk to consumers. A study of transport of nitrates and nitrites into the milk of dairy cows through the digestive system (Baranova et al., 1993) showed that following the per oral application of KNO$_3$ to dairy cows, a marked increase in nitrate content in milk appeared being dependent on applied KNO$_3$. Increased levels of residual nitrate in milk were found also after 38 hours of KNO$_3$ administration. In a study by Bouchard et al. (1999), the effect of endotoxin-induced mastitis leading to increased nitrite and nitrate concentrations in milk was reported. Results of Bouchard et al. (1999) suggest that nitric oxide production during endotoxin-induced mastitis resulted from the activity of the inducible form of nitric oxide synthase.

The nitrate/nitrite content in milk and dairy products is generally lower than in other foods such as vegetables, cured meat and drinking water. Nitrate and nitrite contents in milk and dairy products are presented in Table 9.6.

The content of nitrates and nitrites was determined in raw milk in a study by Przybylowski et al. (1989). The results of this investigation showed that most samples contained nitrates in amounts not exceeding 2 mg/kg, while nitrites were present in trace amounts or were not detected at all. Studies in Denmark estimated average nitrate concentration of 8 mg/L in milk. Danish cheeses contain comparable levels of nitrate and nitrite, approximately 10 and 0.2 mg/kg, respectively, whether or not nitrate had been used in processing (Statens Leventsmiddelinstitut, 1981). According to the data of Estonian Health Protection Inspectorate in 2004, the nitrate content of different Estonian cheeses on retail sale was within the range <7–49 mg/kg and nitrite content in all samples remained below 5 mg/kg. Edam cheeses on retail sale in the UK in 1980–1981 had nitrite contents within the range of 3.1–20 mg/kg whilst 15 samples of each of three English-type cheeses (Cheddar, Cheshire and Leicester) were relatively low in nitrate levels, 1.0–6.1 mg/kg (MAFF, 1987).

A survey of nitrates and nitrites in French dairy products was carried out by Amariglio and Imbert (1980). It was shown that 93% of dried milk samples contained nitrates at less than 30 mg/kg, and 82% of cheese samples at less than 5 mg/kg. The nitrate/nitrite content in milk and dairy products from Austrian dairies was determined in 1986. The mean value of nitrate and nitrite content in pasteurized milk was 0.31 and 0.006 mg/kg, respectively. In fresh cheese, the mean nitrate and nitrite concentrations were 4.58 and 0.013 mg/kg, respectively (Luf, 2002). In Greece where the use of nitrates in cheese production is prohibited, contamination of cheese products was related to using of nitrate containing fertilizers, animal feeds and drinking water at primary production level (Nikolas *et al.*, 1997).

The contribution of milk and dairy products to overall nitrate/nitrite ingestion is very low (Blüthgen *et al.*, 1997). In Austria an intake from milk and dairy products was estimated to be 0.096% of ADI for nitrate and 0.069% of ADI for nitrite (Luf, 2002).

9.3.4   Cereals and bread

Only few data are available on the content of nitrate and nitrite in cereals, bread, flour and various bakery products. This is probably related to the fact that the nitrate and nitrite contents in cereals and various bakery products are generally low. Early studies by Wu and McDonald (1976) reported that nitrate content in grains to be far lower than in stems or leaves of the plants. The nitrate content of winter wheat seeds depends on growth conditions and varies in the range of 0.4–11 mg/kg (McNamara *et al.*, 1971). Wu and McDonald (1976) reported that the nitrate concentration in white flour was 4–14 mg/kg. The data by Nabrzyski *et al.* (1990) showed that the content of nitrates in various bakery products varied from 0.96 in wheat rolls and baguettes to 44 mg KNO_3/kg in pumpernickel bread. The mean content of nitrites in bread varieties was 1.8 mg $NaNO_2$/kg. The same study reported that in white wheat flours the content of nitrates stayed in the range of 1.1–19 mg KNO_3/kg, and in the dishes produced from them under household conditions ranged from 0.5 to 16 mg KNO_3/kg. The content of nitrites in flour was found to be from 'not detected' (ND) to 4.2 mg $NaNO_2$/kg, and in corresponding bakery products from ND to 1.6 mg $NaNO_2$/kg. Eleven types of popular biscuits, wafers, gingerbread and hard cakes were tested in which the content of nitrates was found to be 3.7–17 KNO_3mg/kg, and that of nitrites from ND to 8.8 mg $NaNO_2$/kg (Nabrzyski *et al.*, 1990). Belitz and Grosch (1999) reported the nitrate and nitrite contents in cereals within the range of 0.3–19 and 0.3–1.0 mg/kg, respectively. Ysart *et al.* (1999) detected the nitrate concentrations in bread and miscellaneous cereals in the range of <4–20 mg/kg.

9.3.5   Fresh meat

Mostly the levels of naturally occurring nitrate determined in meat are low and only few data are reported regarding the nitrate and nitrite contents in fresh meat and fresh meat products. This chapter does not deal with cured meat products, to which nitrite and/or nitrate are added as food additives. Wright and Davison (1964) reported nitrate content of 0.9 mg/kg in fresh meat. Usher and Telling (1975) concluded in a series of studies that the nitrate concentration in fresh meat ranged from ND to 49 mg/kg. Ysart *et al.* (1999) reported the nitrate concentration in carcass meat and offal as a mean of 5.1 and 5.3 mg/kg, respectively.

Dry-cured hams treated only with sodium chloride and sugar contained nitrite 5 mg/kg in average (Kemp *et al.*, 1975). Fresh meat products may contain <2.7–9.5 mg NO_3^-/kg and <0.2–1.7 mg NO_2^-/kg (ECETOC, 1988).

9.3.6 Drinking water

Drinking water is regarded to be the second-largest source of nitrate in the diet after vegetables (Belitz and Grosch, 1999; Fytianos and Zarogiannis, 1999; Knobeloch *et al.*, 2000; Caballero Mesa and Rubio Armendáriz, 2003). According to the results of several studies, 20% of the total nitrate intake comes from the consumption of drinking water (White, 1983). Nitrate and nitrite can occur in drinking water mainly as a result of intensive agricultural activities. Contamination of soil with nitrogen-containing fertilizers, including anhydrous ammonia as well as animal or human natural organic wastes can raise the concentration of nitrate in water. Nitrate-containing compounds present in the soil are generally soluble and readily migrate into groundwater. As nitrite is easily oxidized to nitrate, nitrite levels in water are usually low, and nitrate is the compound predominantly found in groundwater and surface waters. Water in highly polluted wells may also contain nitrites at elevated levels. To guarantee drinking water safety, maximum allowable concentrations have been established for nitrate and nitrite, being 50 and 0.5 mg/L, respectively (EC, 1998). Contamination of drinking water with nitrates is a global problem. Studies have showed that in China, Botswana, Turkey, Senegal and Mexico, private well water nitrate levels exceeded the WHO guideline value of 50 mg/L, in some cases the levels of nitrate–nitrogen were over 68 mg/L (WHO, 2004). At the same time, all bottled water samples complied with legislative requirements, being therefore usable for the preparation of infant foods. The problem of high nitrates in water can be solved by using artesian wells as a substitute for draw wells as in artesian wells the water is taken from deep water layers in which the contamination is generally low.

A Finnish study in 1984 reported that only 0.3% of analysed samples exceeded nitrate levels of 30 mg/L (Lahermo, 2000). In Denmark and in Great Britain the reported concentrations of nitrates in drinking water were 13 and 14 mg/L, respectively (Kampmann, 1983; Kinght *et al.*, 1987).

Short-term exposure to drinking water with a nitrate level at or just above the health standard of 10 mg/L nitrate–nitrogen is a potential health problem primarily for infants. Infants and small children consume large quantities of water relative to their body weight, especially if water is used to mix powdered or concentrated milk formulae or juices. Also, the immature digestive system of infants is more vulnerable to the reduction of nitrate to nitrite (Spalding and Exner, 1993). During many years, studies in different countries have reported thousands of cases of children with nitrate–nitric methaemoglobinemia and more than hundred children have died (Hayes, 2001). Mild toxicoses were reported when the nitrate concentration was 80–100 mg/L in water used for infant food preparation.

Nitrate can be removed from drinking water by three methods: distillation, reverse osmosis and ion exchange. Heating or boiling is not applicable for reducing nitrate contents, and the concentration of nitrate in water even increases during boiling due to evaporation of 15–25% of water. Mechanical filters or chemical disinfections do not remove nitrate from water. The distillation process involves heating the water to boiling temperature and following collecting and condensing the steam by means of a metal coil. Nearly 100% of the nitrate–nitrogen can be removed in this process. In the reverse osmosis process, pressure is applied to water to force it through a semi-permeable membrane. The membrane filters out most of the impurities as the water passes through. It is known that 85–95% of the nitrate can be removed with reverse osmosis. Actual removal rates may vary, depending on the initial quality of the water, the system pressure and water temperature. For the nitrate removal process, special anion exchange resins are used that exchange chloride ions for nitrate and sulfate ions in the water as it passes through the resin. Since most anion exchange resins have a higher selectivity for sulfate than nitrate, the level of sulfate in the water is an important factor in the efficiency of an ion exchange system for removing nitrates (Jasa *et al.*, 2006).

9.4   TOXICITY OF NITRATE AND NITRITE

The health aspects can be divided into acute toxicity and the effects of chronic exposure.

The nitrate ion has a low level of acute toxicity, but when transformed into nitrite in food or human organism, it may constitute a health problem. The reduction of nitrate to nitrite may take place in the presence of bacteria or enzyme nitrate reductase, and in contact with metals. Nitrite is unstable at acidic pH values at which it can disproportionate into nitrate and nitrogen oxide or react with food components including amines, phenols and thiols (Hill, 1996). It has been estimated that 5–8% of the nitrate from the diet may be reduced to nitrite by the microflora in the oral cavity (Mensinga *et al.*, 2003). It has only recently been discovered that nitrate is manufactured endogenously in mammals by the oxidation of nitric oxide and that the nitrate formed has the potential for disinfecting the food we eat (Benjamin, 2000; Archer, 2002).

Nitrite has higher acute toxicity than nitrate. As an unstable ion it undergoes series of reactions when added to food. In an acidic environment, nitrite is converted into nitrous acid, which decomposes to nitric oxide. Nitric oxide, being an important product from the standpoint of colour fixation in cured meat, reacts with myoglobin to produce a red pigment – nitrosomyoglobin. The intake of nitrite is normally low compared to the dose that is acutely toxic, but nitrite in food is considered primarily causing health problems because its presence both in food and in the body may lead to the formation of carcinogenic nitrosoamines (JECFA, 1996; Vermeer *et al.*, 1998) and the clinical symptom of methemoglobinemia (WHO, 1995; Sanchez-Echaniz and Benito-Fernández, 2001).

The principal mechanism of nitrite acute toxicity is the oxidation of ferrous II ion (Fe^{2+}) in oxyhaemoglobin (Hb) to ferric III ion (Fe^{3+}) to produce methaemoglobin (Met-Hb). Methaemoglobin is unable to reversibly bind and transport oxygen. Over 10% Met-Hb of the total Hb causes cyanosis – the lips and skin become bluish-grey and the blood is chocolate brown in colour. The generally accepted lethal level of Met-Hb is 60% of the total Hb. Infants fed on infant formula, which is mixed with high-nitrate well water are particularly susceptible to methaemoglobinemia because of their high fluid intake per kg of body weight. In infant organism the upper gastrointestinal tract is heavily colonized by bacteria able to reduce nitrate to nitrite due to the lack of gastric acidity (Hill, 1996). Consumption of vegetables containing high level of nitrates and incorrect storage of homemade vegetable purees have also been found to be potential causes of infant methaemoglobinemia (Sanchez-Echaniz and Benito-Fernández, 2001).

N-nitroso compounds have been shown to be carcinogenic to multiple organs in several animal species, including higher primates (Eichholzer and Gutzwiller, 2003). Although carcinogenicity of *N*-nitrosoamines in humans cannot be tested, epidemiological studies have suggested a possible link to the incidence of various cancers in humans (Knekt *et al.*, 1999; Pegg and Shahidi, 2000). On the basis of the available data, a relationship between dietary *N*-nitroso compounds, nitrite and nitrate cannot be concluded or excluded. It is possible that other factors such as intake of vegetables, fruit and nitrosation inhibitors, or some other constituent of cured meat and salted fish could partly be responsible for the observed associations (Eichholzer and Gutzwiller, 2003).

The European Commission's Scientific Committee for Food (SCF) considered the possible influence of nitrate and nitrite on human health and set acceptable daily intake (ADI) values for nitrate and nitrite in 1990. The ADIs were reviewed in 1995, which resulted in present value for nitrate of 0–3.7 mg nitrate per kg of body weight and for nitrite 0–0.06 mg nitrite per kg of body weight established in 1995 (EU Scientific Committee for Food, 1995).

9.5 DIETARY EXPOSURE

Many assessments of nitrate and nitrite intakes have been reported in the literature, but most of them are difficult to interpret or compare, as all details of how they were conducted are not available. Ideally, all sources of nitrate and nitrite should be included in an intake assessment; however, in many cases only food and drinking water, known to be the major contributors to the overall exposure, have been included.

Dietary exposure to nitrate is very variable between individuals, regions and countries. The intakes of nitrate and nitrite from food were calculated at a global level on the basis of mean food consumption data and the mean concentrations in foods by WHO (2003). Intake from drinking water was added to the exposure obtained from food.

The estimated intakes of nitrate and nitrite from sources other than food additives are below their respective ADIs. The intakes of nitrates range from 70% of the ADI for the European diet and between 10 and 25% for other diets. The nitrite intakes represented 50% of the ADI for the Middle Eastern and Far Eastern diets and 40% of the ADI for the African, Latin American and European diets. Vegetables, including potatoes, were the main contributors to nitrate intake, accounting for 30–90% of total estimated value. Drinking water was the second highest contributor to the exposure of nitrates making up 5–40% of the intake (WHO, 2003).

The data for nitrate exposure calculated by WHO are in good agreement with the works by other authors. The mean total intake of nitrate per person estimated by different authors in Europe ranges between 50 and 140 mg/day and in the USA about 40–100 mg/day (Ysart et al., 1999; Mensinga et al., 2003). According to British intake estimations dietary exposure to nitrates for 1997 was 52 mg/day compared with 68 mg/day in 1994. The decrease can be partly explained by lower nitrate concentrations in green vegetables (Harrison, 2005). The estimated intakes of nitrate show that vegetables are the major contributors to total dietary intake, followed by water.

Several authors have tried to estimate nitrate intake by children, as infants are more susceptible to health implications possibly caused by nitrites and nitrates. The daily nitrate intake by Finnish adolescents and children was 48 mg (Penttilä, 1995). In a Polish study (Wawrzyniak et al., 2003), nitrates intake for 1- to 6-years-old children exceeded ADI twice. Nitrate intake by Estonian 1- to 6-years-old children was found to be 28 mg/day (46% of ADI), the mean nitrate intake from infant food by children aged 6–12 months was 7.8 mg/day (22% of ADI) (Tamme et al., 2006).

According to the WHO data (2003) the major contributors to nitrite intake are also sources other than food additives, including cereals, beverages and water. Cereals were the main contributor to nitrite intake, accounting for 35–60% of the estimated intake. Drinking water was the second highest contributor to the estimated intakes of 20–40% of nitrite (WHO,

2003). The mean nitrite exposure for whole population in 1997 was 1.3 mg/day, as compared with 1.7 mg/day in 1994 in UK (Harrison, 2005). Estimation of the impact of consumption of meat products, including cured meat, to overall nitrite exposure varies a lot among different authors. In the countries where consumption of cured meat is high, large part of nitrite intake has been reported to come from nitrite as food additive. In Polish study meat products supplied 98% of dietary nitrites, nitrite intake being less than 88% of ADI for 1- to 6-years-old children (Wawrzyniak et al., 1999; Wawrzyniak et al., 2003). According to Japanese studies (Murata et al., 2001; Murata et al., 2002) meat products provided 98% of nitrite intake. A Finnish study reported 5.3 mg/day for nitrite (150% of the ADI), and 95% of nitrite was derived from meat products (Dich et al., 1996). Especially for children ADI can easily be exceeded in result of frequent consumption of nitrite-treated sausages (Reinik et al., 2005).

The results of studies of the intake of nitrate and nitrite from all dietary sources showed mean consumptions of both nitrate and nitrite below the ADIs, although some consumers at high percentiles exceeded the ADI for both compounds (WHO, 2003).

9.6   REGULATIONS

In order to protect human health and taking into account the possible association of nitrates and nitrites in food with the formation of carcinogenic N-nitrosoamines, the level of these compounds should be reduced to as low as reasonably achievable (ALARA principle). At present time, regulatory limits for nitrates in food have been established in the EU only for spinach, lettuce and baby foods. The maximum level of nitrates in baby foods and processed cereal-based baby foods for infants and young children should not exceed 200 mg/kg. The content of nitrates in spinach is limited to 2000–3000 mg/kg, lettuce 2500–4500 mg/kg and 'iceberg' type lettuce 2000–2500 mg/kg (European Commission, 2006). The regulatory limit depends on the harvesting season and place of growth of the vegetables – highest concentrations are permitted in plants grown in winter period and/or greenhouse conditions.

The nitrate content of drinking water is limited to 50 mg/L by the current regulatory standard on the quality of water intended for human consumption 98/83/EEC (European Commission, 1998). The EU standard is based on the World Health Organization's guideline value for drinking water, which is also 50 mg/L; limit value for nitrite is set to 0.5 mg/L.

Conclusions

Nitrate and nitrite can be found in food as naturally occurring compounds, vegetables and drinking water being substantial sources of nitrate intake. They are also used as food additives in the processing of meat products due to their ability to inhibit the growth of Clostridium species and to give meat characteristic pink colour, texture and flavour. Although the permitted levels of added nitrates and nitrites have been decreased during last years, the other factors such as growing emissions or nitrogen oxides from fuel combustion, increased sewage recycling and use of nitrate-based fertilizers have led to net increase in exposure to nitrate in several countries.

When nitrate is transformed into nitrite in food or human organism, it may constitute a health problem because the presence of nitrite both in food and in the body may lead to the formation of carcinogenic nitrosoamines. Ingestion of high quantities of nitrates or nitrites by babies may result in methaemoglobinemia.

Nitrate levels, present in vegetables naturally via the nitrogen cycle, are affected by factors such as plant species, climatic and light conditions, soil characteristics and fertilization regime. The concentrations of nitrates in vegetables can vary enormously ranging from below 10 and up to 10 000 mg/kg. It has been estimated that vegetables constitute a major source of human exposure to nitrates contributing up to 92% of the average daily intake. The naturally occurring nitrate concentrations in other food commodities, such as milk and milk products, cereal products, fresh meat and fruits are generally much lower. Drinking water may be an essential source of nitrates for some consumers. Natural levels of nitrites in food are low, usually remaining under the limit of detection. ADI values established for nitrates and nitrites are not exceeded for the majority of consumers. Exceedings of ADI may occur more easily among the risk groups – small children and vegetarians in the case of nitrates and people consuming large quantities of products containing added nitrites.

References

Alonso, A., Etxaniz, B. and Martinez, M.D. (1992) The determination of nitrate in cured meat products. A comparison of the HPLC UV/VIS and Cd/spectrophotometric methods. *Food Additives and Contaminants*, **9**, 111–117.

Amariglio, S. and Imbert, A. (1980) Survey on the nitrate–nitrite content of various dairy products. *Annales de la nutrition et de l'alimentation*, **34**(5–6), 1053–1060.

Andrade, R., Viana, C.O., Guadagnn, S.G., Reyes, F.G.R. and Rath, S. (2003) A flow-injection spectrophotometric method for nitrate and nitrite determination through nitric oxide generation. *Food Chemistry*, **80**(4), 597–602.

AOAC (2000) *AOAC Official Methods of Analysis*, 17th edn. (ed. W. Horwitz). AOAC International, Gaithersburg, MD.

Archer, D.L. (2002) Evidence that ingested nitrate and nitrite are beneficial to health. *Journal of Food Protection*, **65**(5), 872–875.

Baranova, M., Mal'a, P. and Burdova, O. (1993) Transport of nitrates and nitrites into the milk of dairy cows through the digestive system. *Veterinarni Medicina*, **38**(10), 581–588.

Belitz, H.-D. and Grosch, W. (1999) Food contaminants, in *Food Chemistry*, 2nd edn. Springer, Berlin, pp. 1–992.

Benjamin, N. (2000) Nitrates in the human diet—good or bad? *Annales de Zootechnie*, **49**, 207–216.

Bertotti, M. and Pletcher, D. (1997) Amperometric determination of nitrite via reaction with iodide using microelectrodes. *Analytica Chimica Acta*, 337(1), 49–55.

Blüthgen, A., Burt, S. and Heeschen, W.H. (1997) Nitrate, nitrite, In nitrosoamines, *Monograph on Residues and Contaminants in Milk and Milk Products–Special Issue 9701* (ed. E. Hopkin), International Dairy Federation, Brussels, pp. 74–78.

Bouchard, L., Blais, S., Desrosiers, C., Zhao, X. and Lacasse, P. (1999) Nitric oxide production during endotoxin-induced mastitis in the cow. *Journal of Dairy Science*, **82**(12), 2574–2581.

Buck, W. and Osweiler, D. (1973) *Clinical and Diagnostic Veterinary Toxicology*. Kendall/Hunt Publishing Company, Dubuque, Iowa.

Butt, S.B., Riaz, M. and Iqbal, M.Z. (2001) Simultaneous determination of nitrite and nitrate by normal phase ion-pair liquid chromatography. *Talanta*, **55**(4), 789–797.

Cabello, P., Roldán, D.M. and Moreno-Vivián, C. (2004) Nitrate reduction and the nitrogen cycle in archaea. *Microbiology*, **150**, 3527–3546.

Caballero Mesa, J.M. and Rubio Armendáriz, C. (2003) Nitrate intake from drinking water on Tenerife island (Spain). *The Science of the Total Environment*, **302**, 85–92.

Casanova, J.A., Gross, L.K., McMullen, S.E. and Schenk, F.J. (2006) Use of Griess reagent containing vanadium(III) for post-column derivatization and simultaneous determination of nitrite and nitrate in baby food. *Journal of AOAC International*, **89**(2), 447–451.

Cheng, C.F. and Tsang, C.W. (1998). Simultaneous determination of nitrite, nitrate and ascorbic acid in canned vegetable juices by reverse-phase ion-interaction HPLC. *Food Additives and Contaminants*, **15**, 753–758.

Chou, S.S., Chung, J.C. and Hwang, D.F. (2003) A high performance liquid chromatography method for determining nitrate and nitrite levels in vegetables. *Journal of Food and Drug Analysis*, **11**(3), 233–238.

Chung, S.Y., Kim, J.S., Kim, M., Hong, M.K., Lee, J.O., Kim, C.M. and Song, I.S. (2003) Survey of nitrate and nitrite contents of vegetables grown in Korea. *Food Additives and Contaminants*, **20**(7), 621–628.

Chung, J.C., Chou, S.S. and Hwang, D.F. (2004) Changes in nitrate and nitrite content of four vegetables during storage at refrigerated and ambient temperature. *Food Additives and Contaminants*, **21**(4), 317–322.

Codispoti, L.A., Brandes, J.A., Christensen, J.P., Devol, A.H., Naqvi, S.W.A. and Paerl, H.W. (2001) The oceanic fixed nitrogen and nitrous oxide budgets: Moving targets as we enter the anthropocene? *Scientia Marina*, **65**(suppl 2), 85–105.

Corré, W.J. and Breimer, T. (1979) *Nitrate and Nitrite in Vegetables*, Literature Survey No. 39. Centre for Agricultural Publishing Documentation, Wageningen.

Cruz, C. and Martins Loução, M.A. (2002) Comparison of methodologies for nitrate determination in plants and soils. *Journal of Plant Nutrition*, **25**(6), 1185–1211.

Damiani, P. and Burini, G. (1986) Fluorimetric determination of nitrite. *Talanta*, **33**(8), 649–652.

Dejonckheere, W., Streubaut, W., Drieghe, S., Verstraeten, R. and Braeckman, H. (1994) Nitrate in food commodities of vegetable origin and the total diet in Belgium, 1992–1993. *Microbiologie-Aliments-Nutrition*, **12**, 359–370.

De Martin, S. and Restani, P. (2003) Determination of nitrates by a novel ion chromatographic method: Occurrence in leafy vegetables (organic and conventional) and exposure assessment for Italian consumers. *Food Additives and Contaminants*, **20**(9), 787–792.

Dennis, M.J., Key, P.E., Papworth, T., Pointer, M. and Massey, R.C. (1990) The determination of nitrate and nitrite in cured meats by HPLC/UV. *Food Additives and Contaminants*, **7**(4), 455–461.

Dich, J., Järvinen, R., Knekt, P. and Penttilä, P.-L. (1996) Dietary intakes of nitrate, nitrite and NDMA in the Finnish mobile clinic health examination survey. *Food Additives Contaminants*, **13**(5), 541–552.

Dugo, G., Pellicanò, T.M., La Pera, L., Lo Turco, V., Tamborrino, A. and Clodoveo, M.L. (2007) Determination of inorganic anions in commercial seed oils and in virgin olive oils produced from de-stoned olives and traditional extraction methods, using suppressed ion exchange chromatography (IEC). *Food Chemistry*, **102**(3), 599–605.

ECETOC (1988) *Nitrate and Drinking Water*. European Chemical Industry Ecology and Toxicology Centre, Brussels. Technical Report No. 27. Available from: http://www.who.int/water_sanitation_health/dwq/chemicals/en/nitratesfull.pdf

Eggers, N.J. and Cattle, D.L. (1986) High performance liquid chromatographic method for the determination of nitrate and nitrite in cured meat. *Journal of Chromatography*, **354**, 490–494.

Eichholzer, M. and Gutzwiller, F. (2003) Dietary nitrates, nitrites and N-nitroso compounds and cancer risk with special emphasis on the epidemiological evidence, in *Food Safety: Contaminants and Toxins* (ed. J.P.F. D'Mello). CABI Publishing, Wallingford, UK, pp. 217–234.

Elliott, W. and Elliott, D. (2002) *Biochemistry and Molecular Biology*. Oxford University Press, Oxford.

Ensafi, A.A. and Bagherian-Dehaghei, G. (1999) Ultra-trace analysis of nitrite in food samples by flow-injection with spectrophotometric detection. *Fresenius, Journal of Analytical Chemistry*, **363**(1), 131–133.

Ensafi, A.A., Rezaei, B. and Nouroozi, S. (2004) Simultaneous spectrophotometric determination of nitrite and nitrate by flow injection analysis. *Analytical Sciences*, **20**(12), 1749–1753.

Escobar-Gutierrez, A.J., Burns, I.G., Lee, A. and Edmondson, R.N. (2002) Screening lettuce cultivars for low nitrate content during summer and winter production. *The Journal of Horticultural Science and Biotechnology*, **7**(2), 232–237.

Estonian Health Protection Inspectorate (2004) Nitrates and Nitrites in the *Estonian National Monitoring Programme of Food Additives and Contaminants*. Available from: http://www.tklabor.ee/static/artiklid/1.lisaained%20kokku.pdf

EU Scientific Committee for Food (1995) *Opinion on Nitrate and Nitrite*. Expressed on 22 September 1995, annex 4 to document III/56/95, CS/CNTM/NO3/20-FINAL. European Commission DG III, Brussels.

European Commission (EC) (1998) Council Directive 98/83/EEC of 3rd November 1998 on the quality of water intended for human consumption. *Official Journal of the European Union*, **L330**, 32–54.

European Commission (EC) (2006) Commission Regulation No. 1881/2006 of 19 December 2006 setting maximum levels for certain contaminants in foodstuffs. *Official Journal of the European Communities*, **L364**, 5–24.

European Committee for Standardization (CEN) (1997) *EN 12014-2:1997 Foodstuffs – Determination of nitrate and/or nitrite content – Part 2: HPLC/IC method for the determination of nitrate content of vegetables and vegetable products*, Brussels, Belgium.

European Committee for Standardization (CEN) (1998) *EN 12014-7:1998 Foodstuffs – Determination of nitrate and/or nitrite content – Part 7: Continuous flow method for the determination of nitrate content of vegetables and vegetable products after Cadmium reduction*, Brussels, Belgium.

European Committee for Standardization (CEN) (2005a) *EN 12014-3:2005 Foodstuffs – Determination of nitrate and/or nitrite content – Part 3: Spectrometric determination of nitrate and nitrite content of meat products after enzymatic reduction of nitrate to nitrite*, Brussels, Belgium.

European Committee for Standardization (CEN) (2005b) *EN 12014-4:2005 Foodstuffs – Determination of nitrate and/or nitrite content – Part 4: Ion-exchange chromatographic (IC) method for the determination of nitrate and nitrite content of meat products*, Brussels, Belgium.

Farrington, D., Damant, A.P., Powell, K. Ridsdale, J., Walker, M. and Wood, R. (2006) A comparison of the extraction methods used in the UK nitrate residues monitoring programme. *Journal of the Association of Public Analysts (Online)*, **34**, 1–11.

Fox, J.B., Doerr, R.C. and Lakritz, L. (1982) Interaction between sample preparation techniques and three methods of nitrite determination. *Journal of Association of Analytical Chemistry*, **65**, 690–695.

Frolich, D.H. (1987) HPLC of anions on coated reversed phase columns using indirect UV-detection and eluants containing nitrophthalic acids. *Journal of High Resolution Chromatography and Chromatographic Communication*, **10**, 12–16.

Funazo, K., Tanaka, M. and Shono, T. (1980) Determination of nitrite at parts-per-billion levels by derivatization end electron capture gas chromatography. *Analytical Chemistry*, **52**, 1222–1224.

Fytianos, K. and Zarogiannis, P. (1999) Nitrate and nitrite accumulation in fresh vegetables from Greece. *Bulletin of Environmental Contamination and Toxicology*, **62**, 187–192.

Gajewska, R., Nabrzyski, M. and Szajek, L. (1989) Occurrence of nitrates and nitrites in certain frozen fruits, jams, stewed fruit and fruit-vegetable juices for children and in certain types of bee honey. *Roczniki Państwowego Zakładu Higieny*, **40**(4–6), 266–273.

Gal, C., Frenzel, W. and Möller, J. (2004) Re-examination of the cadmium reduction method and optimization of conditions for the determination of nitrate by flow injection analysis. *Microchimica Acta*, **146**(2), 155–164.

Gapper, L.W., Fong, B.Y., Otter, D.E., Indyk, H.E. and Woollard, D.C. (2004) Determination of nitrite and nitrate and dairy products by ion exchange LC with spectrophotometric detection. *International Dairy Journal*, **14**, 881–887.

Golaszewska, B. and Zalewski, S. (2001) Optimalisation of potato quality in culinary process. *Polish Journal of Food and Nutrition Sciences*, **10**(51), 59–63.

González, P.J., Correia, C., Moura, I., Brondino, C.D. and Moura, J.J.G. (2006) Bacterial nitrate reductases: Molecular and biological aspects of nitrate reduction. *Journal of Inorganic Biochemistry*, **100**, 1015–1023.

Hardisson, A. and Gonzáez Padrón, A. (1996) The evaluation of the content of nitrates and nitrites in food products for infants. *Journal of Food Composition and Analysis*, **9**, 13–17.

Harrison, N. (2005) Inorganic contaminants in food, in *Food Chemical Safety. Vol. 1: Contaminants* (ed. D.H. Watson). Woodhead Publishing Ltd, Cambridge, England, pp. 148–168.

Hayes, A.W. (2001) *Principles and Methods of Toxicology*, 4th edn. Raven Press, New York, pp. 59–60.

Heaton, S. (2001) *Organic Farming, Food Quality and Human Health*. Soil Association Report, Bristol, UK.

Hermann, K. (1972) Über den Nitrat und Nitritgehalt des Gemüses, Obstes und Wassers und deren Bedentung für die Ernährung. *Ernsehr. Umschau*, **11**, 398–402.

Hill, M.J. (1996) Nitrates and nitrites from food and water in relation to human disease nitrates and nitrites in food and water, in *Nitrates and Nitrites in Food and Water*, 2nd edn. (ed. M. Hill). Woodhead Publishing Ltd, Cambridge, pp. 163–187.

Hlusek, J., Zrust, J. and Juzl, J. (2000) Nitrate concentration in tubers of early potatoes. *Rostlinna Vyroba*, **46**, 17–21.

Hunt, J. and Seymour, D.J. (1985) Method for measuring nitrate–nitrogen in vegetables using anion-exchange high-performance liquid chromatography. *Analyst,* **10**, 131–133.

Jasa, P., Skipton, S., Varner, D. and Hay, D.L. (2006) *Drinking Water: Nitrate–Nitrogen*. University of Nebraska. Available from: http://www.ianrpubs.unl.edu/epublic/pages/publicationD.jsp?publicationId=477

Järvan, M. (1993) *Köögiviljade nitraatidesisaldust mõjutavad tegurid*. Dissertatsioon, Saku, Estonia.

JECFA (Joint FAO/WHO Expert Committee on Food Additives) (1996) Nitrate, toxicological evaluation of certain food additives and contaminants in food. *WHO Food Additives Series*, **35**, 325–360.

Kammerer, M., Pinault, L. and Pouliquen, H. (1989) Content of nitrate in milk. Relationship with its concentration in the water supply for livestock. *Annals of Veterinary Research*, **23**(2), 131–138.

Kampmann, J. (1983) Nitrat I drikkevand og grundvand og grundvand i Danmark. Redegørelse fra miljøstyrelsen (Kobenhagen) (in Danish).

Kazemzadeh, A. and Ensafi, A.A. (2001) Simultaneous determination of nitrate and nitrite in various samples using flow-injection spectrophotometric detection. *Microchemical Journal*, **69**(2), 61–68.

Kemp, J.D., Langlois, B.E., Fox, J.D. and Varney, W.Y. (1975) Effects of curing ingredients and holding times and temperatures on organoleptic and microbiological properties of dry-cured sliced ham. *Journal of Food Science*, **40**, 634–636.

Kirk, R.S. and Sawyer, R. (1991) *Pearson's Composition and Analysis of Foods*, 9th edn. Longman Scientific & Technical, Essex.

Kissner, R. and Koppenaol, W.H. (2005) Qualitative and quantitative determination of nitrite and nitrate with ion chromatography. *Methods in Enzymology*, **396**, 61–68.

Knekt, P., Jarvinen, R., Dich, J. and Hakulinen, T. (1999) Risk of colorectal and other gastro-intestinal cancers after exposure to nitrate, nitrite and N-nitroso compounds: A follow-up study. *International Journal of Cancer*, **80**(6), 852–856.

Knight, T.M., Forman, D., Al-Darbagh, S.A. and Doll, R. (1987) Estimation of dietary intake of nitrate in Great Britain. *Food and Chemical Toxicology*, **25**(4), 277–285.

Knobeloch, L., Salna, B., Hogan, A., Postle, J. and Anderson, H. (2000) Blue babies and nitrate-contaminated well water. *Environmental Health Perspectives*, **108**, 675–678.

Kolb, E. and Haug, M. (1997) Potential nitrosamine formation and its prevention during biological denitrification of red beet juice.*Food and Chemical Toxicology*, **35**, 219–224.

Lahermo, P.W. (2000) 1988 Pohjavedet kartoitettu, Studies on Finnish ground water (in Finnish). *Tiede 2000*, **8**, 53–57.

Laslo, C., Preda, N. and Bara, V. (2000) Relations between the administration of some nitrate fertilisers and the incidence of nitrates and nitrites in the fond products. *The University of Agricultural Sciences and Veterinary Medicine Cluj-Napoca*, **28**, 1–6.

Lucinski, R., Polcyn, W. and Ratajczak, L. (2002) Nitrate reduction and nitrogen fixation in symbiotic association Rhizobium-legumes. *Acta Biochimica Polonica*, **49**(2), 537–546.

Luf, W. (2002) Nitrates and nitrites in dairy products, in *Encyclopedia of Dairy Sciences* (eds. H. Roginski, J.W. Fuquau and P.F. Fox). Academic Press, Amsterdam, Boston, London, pp. 2097–2099.

Luf, W. and Brandl, E. (1986) Die aufnahme von nitrat und nitrit bei konsum von milch und milchprodukten. *Ernährung*, **10**, 683–688.

Mackerness, C.W., Leach, S.A., Thompson, M.H. and Hill, M.J. (1989) The inhibition of bacterially mediated *N*-nitrosation by vitamin C: Relevance to the inhibition of endogenous *N*-nitrosation in the achlorhydric stomach, Oxford Journals, Life Sciences. *Carcinogenesis*, **10**(2), 397–399.

MAFF (1987) *Nitrate, Nitrite and N-nitroso Compounds in Foods*. 20th Report of the Steering Group on Food Surveillance, Food Surveillance Paper No. 20. HMSO, London.

Marshall, P.A. and Trenerry, V.C. (1996) The determination of nitrite and nitrate in foods by capillary ion electrophoresis. *Food Chemistry*, **57**(2), 339–345.

Massey, R.C. (1996) Methods for the analysis of nitrate and nitrite in food and water, in *Nitrates and Nitrites in Food and Water*, 2nd edn. (ed. M. Hill). Woodhead Publishing Ltd, Cambridge, pp. 13–32.

Massey, R.C., Dennis, M.J., Pointer, M. and Key, P.E. (1990) An investigation of the levels of N-nitrosodimethylamine, apparent total N-nitroso compounds and nitrate in beer. *Food Additives and Contaminants*, **7**(5), 605–615.

Masson, P., de Raemaeker, F. and Bon, F. (2005) Quality control procedures for chloride and nitrate ions analysis in plant samples by ion chromatography. *Accreditation and Quality Assurance: Journal for Quality, Comparability and Reliability in Chemical Measurement*, **10**(8), 439–443.

McMullen, S.E., Casanova, J.A., Gross, L.K. and Schenck, F.J. (2005) Ion chromatographic determination of nitrate and nitrite in vegetable and fruit baby foods. *Journal of AOAC International*, **88**(6), 1793–1796.

McNamara, A.S., Klepper, L.A. and Hageman, R.H. (1971) Nitrate content of seeds of certain crop plants, vegetables and weeds. *Journal of Agricultural Food Chemistry*, **19**, 540–542.

Mensinga, T.T., Speijers, G.J.A. and Meulenbelt, J. (2003) Health implications of exposure to environmental nitrogenous compounds. *Toxicological Reviews*, **22**(1), 41–51.

Merino, L., Edberg, U., Fuchs, G. and Åman, P. (2000) Liquid chromatographic determination of residual nitrite/nitrate in foods: NMKL Collaborative Study. *Journal of AOAC International*, **83**, 365–375.

Monser, L., Sadok, S., Greenway, G.M., Shah, I. and Uglow, R.F. (2002) A simple simultaneous flow injection method based on phosphomolybdenum chemistry for nitrate and nitrite determinations in water and fish samples. *Talanta*, **57**(3), 511–518.

Moorcroft, M.J., Davis, J. and Compton, R.G. (2001) Detection and determination of nitrate and nitrite: A review. *Talanta*, **54**(5), 785–803.

Mozolewski, W. and Smoczynski, S. (2004) Effect of culinary processes on the content of nitrates and nitrites in potato. *Pakistan Journal of Nutrition*, **3**(6), 375–361.

Mullins, F.G.P. (1987) Determination of inorganic anions by non-suppressed ion chromatography with indirect ultraviolet absorption detection. *Analyst*, **112**, 665–671.

Murata, M. and Ishinaga, M. (2001) Daily intakes of nitrate and nitrite in middle-aged men by the duplicate portion method. *Shokuhin Eiseigaku Zasshi*, **42**, 215–219.

Murata, M., Kishida, N. and Ishinaga, M. (2002) Survey of the daily intake of nitrate and nitrite in school children by the duplicate portion method. *Journal of the Food Hygienic Society of Japan*, **43**, 57–61.

Nabrzyski, M., Gajewska, R. and Ganowiak, Z. (1990) Presence of nitrates and nitrites in baker's products and in certain other flour products. *Roczniki Państwowego Zakładu Higieny*, **41**(3–4), 187–193.

Naidu, K.A. (2003) Vitamin C in human health and disease is still a mystery? An overview. *Nutrition Journal*, **2**, 7.

Nabrzyski, M. and Gajewska, R. (1994) The content of nitrates and nitrites in fruits, vegetables and other foodstuffs. *Roczniki Państwowego Zakładu Higieny*, **45**(3), 167–180.

Nelson, D.L. and Cox, M.M. (2000) *Lehninger Principles of Biochemistry*, 3rd edn. Worth Publishers, New York.

Nikolas, B.K., Konstantina, T.-G. and Elvira, T.-B. (1997) Nitrate and nitrite content of Greek cheeses. *Journal of Food Composition and Analysis*, **10**, 343–349.

Norwitz, G. and Keliher, P.N. (1985) Study of interferences of the spectrophotometric determination of nitrite using composite diazotisation-coupling reagents. *Analyst*, **110**, 689–694.

Oenema, O., Kros, H. and de Vries, W. (2003) Approaches and uncertainties in nutrient budgets: Implications for nutrient management and environmental policies. *European Journal of Agronomy*, **20**, 3–16.

Okafor, P.N. and Ogbonna, U.I. (2003) Nitrate and nitrite contamination of water sources and fruit juices marketed in South-Eastern Nigeria. *Journal of Food Composition and Analysis*, **16**(2), 213–218.

Oliveira, S.M., Lopes, T.I.M.S. and Rangel, A.O.S.S. (2004) Spectrophotometric determination of nitrite and nitrate in cured meat by sequential injection analysis. *Journal of Food Science*, **69**(9), C690.

Öztekin, N., Nutku, M.S. and Erim, F.B. (2002) Simultaneous determination of nitrite and nitrate in meat products and vegetables by capillary electrophoresis. *Food Chemistry*, **76**(1), 103–106.

Pegg, R.B. and Shahidi, F. (2000) *Nitrite Curing of Meat. The N-Nitrosoamine Problem and Nitrite Alternatives*. Food & Nutrition Press, Trumbull, CT.

Pentchuk, J., Haldna, U. and Ilmoja, K. (1986) Determination of nitrate and chloride ions in food by single-column ion chromatography. *Journal of Chromatography*, **364**, 189–192.

Penttilä, P.L. (1995) *Estimation of Food Additive and Pesticide Intakes by Means of Stepwise Method*. Doctoral thesis, University of Turku, Finland.

Petersen, A. and Stoltze, S. (1999) Nitrate and nitrite in vegetables on the Danish market: Content and intake. *Food Additives and Contaminants*, **16**(7), 291–299.

Przybylowski, P., Kisza, J., Janicka, B. and Sajko, W. (1989) Presence of nitrates and nitrites in raw milk as subject to the system of milk purchasing. *Roczniki Panstwowego Zakladu Higieny*, **40**(1), 6–15.

Reece, P. and Hird, H. (2000) Modification of the ion exchange HPLC procedure for the detection of nitrate and nitrite in dairy products. *Food Additives and Contaminants*, **17**(3), 219–222.

Reinik, M., Tamme, T., Roasto, M., Juhkam, K., Jurtsenko, S., Tenno, T. and Kiis, A. (2005) Nitrites, nitrates and N-nitrosoamines in Estonian cured meat products: Intake by Estonian children and adolescents. *Food Additives and Contaminants*, **22**(11), 1098–1105.

Ross, N.D. and Hotchkiss, J.H. (1985) Determination of nitrate in dried foods by gas chromatography-thermal energy analyzer. *Journal of Association of Official Analytical Chemists*, **68**, 41–43.

Ruiter, A. and Bergwerff, A.A. (2005) Analysis of chemical preservatives in foods, in *Methods of Analysis of Food Components and Additives*, 1st edn. (ed. S. Ötles). CRC Press, Taylor & Francis Group, Boca Raton, FL, pp. 379–399.

Rückauf, U., Augustin, J., Russow, R. and Merbach, W. (2004) Nitrate removal from drained and reflooded fen soils affected by soil N transformation processes and plant uptake. *Soil Biology and Biochemistry*, **36**, 77–90.

Saccani, G., Tanzi, E., Cavalli, S. and Rohrer, J. (2006) Determination of nitrite, nitrate and glucose-6-phosphate in muscle tissues and cured meat by IC/MS. *Journal of AOAC International*, **89**(3), 712–719.

Sanchez-Echaniz, J. and Benito-Fernández, J. (2001) Methemoglobinemia and consumption of vegetables in infants. *Pediatrics*, **107**, 1024–1028.

Schroeder, D.C. (1987) The analysis of nitrate in environmental samples by reverse-phase HPLC. *Journal of Chromatographic Science*, **25**, 405–408.

Schuster, B.E. and Lee, K. (1987) Nitrate and nitrite methods of analysis and levels in raw carrots, processed carrots and in selected vegetables and grain products. *Journal of Food Science*, **52**, 1632–1641.

Sheehy, J.E., Mnzava, M., Cassman, K.G., Mitchell, P.L, Pablico, P., Robles, R.P., Samonte, H.P., Lales, J.S. and Ferrer, A.B. (2004) Temporal origin of nitrogen in the grain of irrigated rice in the dry season: The outcome of uptake, cycling, senescence and competition studied using a 15N-point placement technique. *Field Crops Research*, **89**, 337–348.

Siciliano, J., Krulick, S., Heisler, E.G., Schwartz, J.H. and White, J.W., Jr. (1975) Nitrate and nitrite of some fresh and processed market vegetables. *Journal of Agricultural and Food Chemistry*, **23**, 461–464.

Silva, M., Gallego, M. and Valcarel, M. (1986) Sequential atomic absorption spectrometric determination of nitrate and nitrite in meats by liquid–liquid extraction in a flow-injection system. *Analytica Chimica Acta*, **179**, 341–349.

Siu, D.C. and Henshall, A. (1998) Ion chromatographic determination of nitrate and nitrite in meat products. *Journal of Chromatography A*, **804**(1–2), 157–160.

Slack, P.T. (1987) *Analytical Methods Manual*, 2nd edn. Leatherhead Food Research Association, Leatherhead, Surrey, England.

Sokolov, O.A. (1987) Osobennosti raspredelenija nitratov I nitritov v ovoščah. *Kartofel'i i obošči*, **6**.

Sokolov, O.A. (1989) Hranenie i kulinarnaja obrabotka ovoščei. *Nauka i žizn*, **2**, 152–153.

Spalding, R.F. and Exner, M.E. (1993) Occurrence of nitrate in ground-water—a review. *Journal of Environmental Quality*, **22**, 392–402.

Stalikas, C.D., Konidari, C.N. and Nanos, C.G. (2003) Ion chromatographic method for the simultaneous determination of nitrite and nitrate by post-column indirect fluorescence detection. *Journal of Chromatography A*, **1002**(1–2), 237–241.

Statens Leventsmiddelinstitut (1981) Nitrat og Nitrit i Kdvarer. Rapport fra in *Intern Arbejdsgrappe*, Soborg, Denmark, pp. 63.

Szymczak, J. and Prescha, A. (1999) Content of nitrates and nitrites in market vegetables in Wroclaw in the years 1996–1997. *Roczniki Państwowego Zakładu Higieny*, **50**(1), 17–23.

Sušin, J., Kmecl, V. and Gregorčič, A. (2006) A survey of nitrate and nitrite content of fruit and vegetables grown in Slovenia during 1996–2002. *Food Additives and Contaminants*, **23**(4), 385–390.

Tamme, T., Reinik, M., Roasto, M., Juhkam, K., Tenno, T. and Kiis, A. (2006) Nitrates and nitrites in vegetables and vegetable-based products and their intakes by the Estonian population. *Food Additives and Contaminants*, **23**(4), 355–361.

Tivo, P.F. and Saskevic, L.A. (1990) Nitratõ sluhi i realnost, Minsk; Byelorussia.

Usher, C.D. and Telling, G.M. (1975) Analysis of nitrate and nitrite in foodstuffs – critical review. *Journal of the Science of Food and Agriculture*, **26**, 1793–1805.

Vaessen, H.A.M.G. and Schothorst, R.C. (1999) The oral nitrite intake in The Netherlands: Evaluation of the results obtained by HPIC analysis of duplicate 24-hour diet samples collected in 1994. *Food Additives and Contaminants*, **16**(5), 181–188.

Van den Berg, C.M.G. and Li, H. (1988) The determination of nanomolecular levels for nitrite in fresh and sea water by cathodic stripping voltammetry. *Analytica Chimica Acta*, **212**, 31–41.

Vermeer, I.T., Pachen, D.M.F.A., Dallinga, J.W., Kleinjans, J.C.S. and van Maanen, J.M.S. (1998) Volatile N-nitrosoamine formation after intake of nitrate at the ADI level in combination with an amine-rich diet. *Environmental Health Perspectives*, **106**(8), 459–463.

Vulsteke, G. and Biston, R. (1978) Factors affecting nitrate content in field-grown vegetables. *Plant Foods for Human Nutrition*, **28**(1), 71–78.

Walker, R. (1990) Nitrates and N-nitroso compounds: A review of the occurrence in food and diet and the toxicological implications. *Food Additives and Contaminants*, **7**, 717–768.

Wawrzyniak, A., Gronowska-Seneger, A. and Gorecka, K. (1999) The evaluation of nitrates and nitrites food intake among Polish households in 1991–1995. *Roczniki Panstwowego Zakladu Higieny*, **50**, 269–287.

Wawrzyniak, A., Hamulka, J. and Skibinska, E. (2003) The evaluation of nitrate, nitrite and antioxidant vitamin intake in daily food ration of children aged 1–6 year of age. *Roczniki Panstwowego Zakladu Higieny*, **54**, 65–72.

White, J.W. (1976) Correction: Relative significance of dietary sources of nitrate and nitrite. *Journal of Agricultural and Food Chemistry*, **24**, 202.

White, R.J. (1983) Nitrate in British Waters. *Aqua*, **2**, 51–57.

WHO (2004) *Nitrates and Nitrites in Drinking-Water*. WHO/SDE/WSH/04.08/56. Rolling revision of the WHO guidelines for drinking-water quality. Draft for review and comments. World Health Organization, Geneva. Available from: http://www.who.int/water_sanitation_health/dwq/chemicals/en/nitratesfull.pdf

WHO Food Additives Series: 50 (2003) Safety evaluation of certain food additives 1059. *Nitrite and Nitrate: Intake assessment Fifty-ninth Meeting of the Joint FAO/WHO Expert Committee on Food Additives (JECFA)*. World Health Organization, Geneva. Available from: http://www.inchem.org/documents/jecfa/jecmono/v50je07.htm

WHO (1995) Evaluation of certain food additives and contaminants, Joint FAO/WHO Expert Committee on Food Additives. *WHO Technical Report*, **859**, 29–35.

Worthington, V. (2001) Nutritional quality of organic versus conventional fruits, vegetables and grains. *Journal of Alternative and Complementary Medicine*, **7**(2), 161–173.

Wright, M.J. and Davison, K.L. (1964) Nitrate accumulation on crops and nitrate poisoning in animals. *Advances in Agronomy*, **16**, 197–247.

Wu, K.Y. and McDonald, C.E. (1976) Effect of nitrogen fertiliser on nitrogen fractions of wheat and flour. *Cereal Chemistry*, **53**, 242–249.

Wu, H.L., Chen, S.H., Funazo, K., Tanaka, M. and Shono, T. (1984) Electron capture gas chromatographic determination of nitrite as the pentafluorobenzyl derivative. *Journal of Chromatography*, **291**, 409–415.

Ysart, G., Miller, P., Barret, G., Farrington, D., Lawrance, P. and Harrison, N. (1999) Dietary exposures to nitrate in the UK. *Food Additives and Contaminants*, **16**(12), 521–532.

Ximenes, M.I.N., Rath, S. and Reyes, F.G.R. (2000) Polargraphic determination of nitrate in vegetables. *Talanta*, **51**(1), 49–56.

Zanardi, E., Dazzi, G., Madarena, G. and Chizzolini, R. (2002) Comparative study on nitrite and nitrate ions determination. *Annali della Facoltà di Medicina Veterinaria Di Parma*, **22**, 79–86.

10 Acrylamide in Heated Foods

Vural Gökmen and Hamide Z. Şenyuva

Summary

Upon the discovery of acrylamide in foods, international authorities and experts concluded that the presence of acrylamide was a major concern and additional research on mechanisms of formation and toxicity was seriously needed. However, it is still not really known the extent to which acrylamide consumption poses a health risk. Meanwhile, an earlier feeding study in rats has demonstrated a relationship between acrylamide in animal foods and specific haemoglobin adducts. Long-term exposure to acrylamide may cause damage to the nervous system both in humans and animals to a certain extent, and acrylamide is also considered as a potential genetic and reproductive toxin exhibiting mutagenic and carcinogenic properties in experimental mammalians both in vitro and in vivo.

Numerous analytical methods for acrylamide in foods have been published based on either GC or LC techniques coupled with MS, of which some produce satisfactory results for the determination of acrylamide in difficult matrices. Development of a generic method for sample preparation (extraction and clean-up) for an accurate, precise and reliable determination of acrylamide in all foods is nevertheless still challenging.

Although a favoured mechanism for the formation of acryamide is the Maillard reaction, involving the condensation of the amino acid asparagine and a carbonyl source (e.g. sugars), fundamental studies have very recently revealed new avenues, identifying acrylic acid as a potential precursor of acrylamide in model systems.

10.1 INTRODUCTION

Acrylamide is a white crystalline solid, is odourless and has high solubility in water (2155 g/L water). It is a reactive chemical, which is used as a monomer starting material in the synthesis of polyacrylamides used, e.g. in purification of water, and in the formulation of grouting agents. Acrylamide is a known component in tobacco smoke. Acrylamide is primarily reactive through its ethylenic double bond (Fig. 10.1) and polymerization of acrylamide occurs through radical reactions with the double bond. Acrylamide can also react as an electrophile by 1,4-addition to nucleophiles, e.g. SH- or NH_2-groups in biomolecules ((McCollister et al., 1964; Croll and Simkins, 1972; Cutie and Kallos, 1986).

In April 2002, the Swedish National Food Administration and the University of Stockholm jointly announced that certain foods, which are processed/cooked at high temperatures, contain relatively high amounts of acrylamide (Swedish National Food Administration, 2002), following on an earlier rat feeding study that demonstrated a link of acrylamide in fried

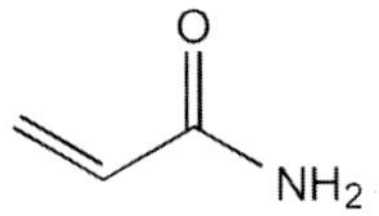

Fig. 10.1   Chemical structure of acrylamide.

animal food to specific haemoglobin adducts (Tareke *et al.*, 2000). Exposure to acrylamide causes damage to the nervous system in humans and animals (Tilson, 1981; Lopachin and Lehning, 1994) and acrylamide is also considered a reproductive toxin (Dearfield *et al.*, 1988; Costa *et al.*, 1992), with mutagenic and carcinogenic properties in experimental mammalian in vitro and in vivo systems (Dearfield *et al.*, 1995). Acrylamide is classified as a probable human carcinogen by the International Agency for Research on Cancer (IARC, 1994).

Because flocculants made from acrylamide are sometimes used in the treatment of drinking water, the Environmental Protection Agency (EPA) and the World Health Organization (WHO) require water suppliers to limit acrylamide to 0.5 µg/L or less. The European Union (EU) guideline for acrylamide in drinking water is 0.1 µg/L. Although the potential health risk of acrylamide in food has been considered by a number of government agencies and national authorities, no maximum permitted concentration has yet been established for acrylamide in processed foods.

10.2   MECHANISM OF FORMATION IN FOODS

10.2.1   Mechanistic pathways

A number of theoretical mechanisms have been proposed for the formation of acrylamide in heated foods. The major mechanistic pathway for the formation of acrylamide in foods so far established is via the Maillard reaction (Becalski *et al.*, 2002; Mottram *et al.*, 2002; Sanders *et al.*, 2002; Stadler *et al.*, 2002, 2004). Studies to date clearly show that the amino acid asparagine is mainly responsible for acrylamide formation in cooked foods after condensation with reducing sugars or a carbonyl source (Fig. 10.2). Moreover, the sugar-asparagine adduct, *N*-glycosylasparagine, generates high amounts of acrylamide, suggesting the early Maillard reaction as a major source of acrylamide (Stadler *et al.*, 2002). In addition, decarboxylated asparagine (3-aminopropionamide), when heated, can generate acrylamide in the absence of reducing sugars (Zyzak *et al.*, 2003). A recent study revealed that, besides acrylamide, 3-aminopropionamide, which may be a transient intermediate in acrylamide formation, was also formed during heating when asparagine was reacted in the presence of glucose (Granvogl and Scieberle, 2006).

Other possible routes involve the Strecker reaction of asparagine with the Strecker aldehyde as the direct intermediate (Mottram *et al.*, 2002), or a mechanism via acrylic acid (Gertz and Klostermann, 2002; Lingnert *et al.*, 2002; Becalski *et al.*, 2003; Stadler, 2003). The oxidation and/or thermal degradation of lipids in fried foods have been discussed as a possible mechanistic route contributing to acrylamide formation of via an acrylic acid intermediate (Yasuhara *et al.*, 2003).

Good evidence supporting the early Maillard reaction as the main reaction pathway involving early decarboxylation of the Schiff base, rearrangement to the resulting Amadori product, and subsequent β-elimination to release acrylamide has been presented (Yaylayan *et al.*, 2003).

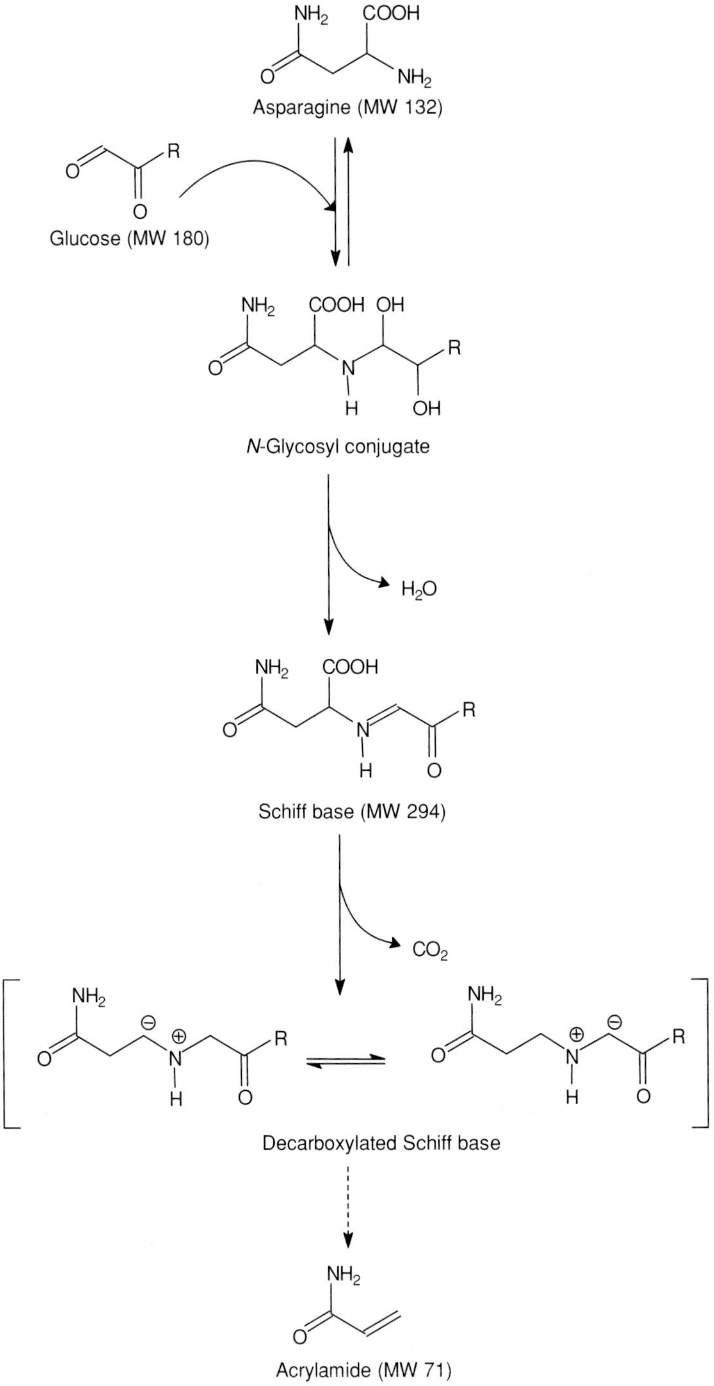

Fig. 10.2 Formation of acrylamide during the pyrolysis of asparagine with glucose. (Adapted from Gökmen, V. and Şenyuva, H.Z. Effects of some cations on the formation of acrylamide and furfurals in glucose–asparagine model system. *European Food Research and Technology*, **225**, 815–820. Copyright 2007b with kind permission from Springer Science and Business Media.)

10.2.2   Formation in foods during thermal processing

The resulting acrylamide concentration in foods ultimately depends on the concentrations of the precursors (asparagine and reducing sugars), as well as the processing conditions. However, other factors such pH and water activity have been shown to influence the levels of acrylamide in heated foods (Friedman, 2003). Concerning reducing sugars, fructose has been found more effective than glucose in forming acrylamide during heating. However, both the chemical reactivity of the sugars and their physical state play an important role in acrylamide formation. The melting points of fructose and glucose are 126 and 157°C, respectively (Robert *et al.*, 2004). This explains why fructose is more reactive than glucose on acrylamide formation during thermal processing. Concerning frying and baking processes, they are characterized as open processes in which heat and mass transfer occur simultaneously. This influences the physical states of reaction precursors, especially sugars. As the moisture reduces due to evaporation, sugars initially dissolved in water begin to form a saturated solution and then crystallize. After crystallization, melting is required to change their state to liquid, so to make them chemically reactive. In this respect, reducing sugar having a lower melting point is expected to form acrylamide earlier during heating.

The influence of temperature on acrylamide formation has been repeatedly demonstrated (Mottram *et al.*, 2002; Tareke *et al.*, 2002; Becalski *et al.*, 2003; Biedermann and Grob, 2003; Rydberg *et al.*, 2003). In general, the rate of acrylamide formation increases as the temperature increases, but the rate of its elimination also increases (Gökmen and Şenyuva, 2006a). For shorter heating times as in the frying operation of potato chips or strips, lowering the frying temperature may significantly reduce the amount of acrylamide formed. The same may not be true for longer heating periods as in the roasting of coffee beans where extending the operation may result in a decrease in the amount of acrylamide persisted in the final product.

The fact that acrylamide is not formed during boiling indicates that higher temperatures and/or low moisture conditions are needed for its formation. During heating under atmospheric conditions, higher temperatures can be reached only if simultaneous drying takes place, which is the case in frying, baking and roasting. Temperature and time have been shown to be significant factors affecting the amount of acrylamide formed in potatoes during frying (Matthäus *et al.*, 2004; Pedreschi, *et al.*, 2005; Gökmen *et al.*, 2006a,b,c, 2007).

The widespread presence of acrylamide has been largely reported in fried, baked and roasted foods. Thermally processed foods based on potato, cereals and coffee are known to contain acrylamide in a range between a few micrograms to milligrams per kilogram. In addition to the parameters related to the composition, the conditions of thermal processing are important on acrylamide formation in a particular food. Therefore, understanding the effects of frying, baking and roasting in depth is crucial to optimize the conditions of thermal processing for minimal formation.

10.2.2.1   *Frying of potatoes*

Fried potatoes are in the food category with probably the highest concentrations of acrylamide recorded so far (Friedman, 2003), and the category for which the most work has been done to control or reduce levels of acrylamide. Experimental trials that have been conducted on acrylamide in potatoes have shown that the major determinants of acrylamide formation are reducing sugars, as well as the free asparagine. The glucose content has a positive and significant effect on acrylamide formation in the finished product (Biedermann *et al.*, 2002a;

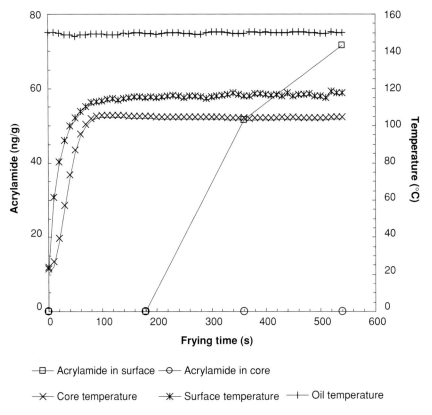

Fig. 10.3 Surface and core temperature profiles with corresponding acrylamide concentrations of French fries during frying at 150°C. (Reprinted from Gökmen, V., Palazoğlu, T.K., Şenyuva, H.Z. Relation between the acrylamide formation and time–temperature history of surface and core regions of French fries. *Journal of Food Engineering*, **77**, 972–976. Copyright 2006a with permission from Elsevier.)

Becalski *et al.*, 2003; Haase *et al.*, 2003). The content of sugar in the raw potato shows a strong correlation with the amount of acrylamide formed upon heating (Biedermann *et al.*, 2002b; Gökmen *et al.*, 2007).

Widely varying concentrations of asparagine, glucose, fructose and sucrose have been found in potatoes (Amrein *et al.*, 2003). This variability may be one important explanation for the difference in the amounts of acrylamide that may be formed in the products during processing. The frying conditions in terms of temperature and time are also considered as important determinants of acrylamide formation.

Temperature measurements during frying have indicated that the surface temperature of a strip of potato does not exceed 120°C during 9 minutes of frying at 150°C (Fig. 10.3). Even though this is the case, the fact that some acrylamide still forms suggests that the temperature need be no higher than 120°C for acrylamide to form (Gökmen *et al.*, 2006a).

During frying at any temperature, all the heat transferred from the hot oil is utilized to increase the internal energy of the strip until the surface reaches slightly above the boiling point of water (103–104°C). After this point, moisture evaporation starts extracting a large amount of the incoming energy. When the frying oil temperature is 150°C, the energy input to the potato strip is limited. This prevents the surface temperature from reaching temperatures

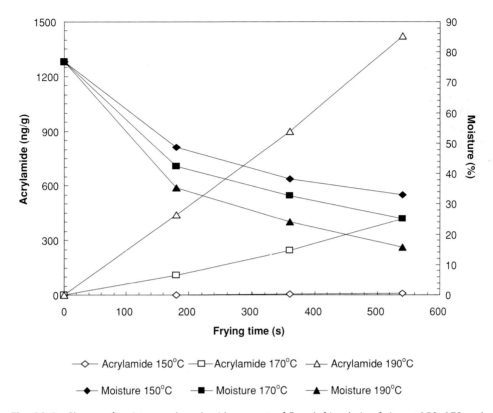

Fig. 10.4 Change of moisture and acrylamide contents of French fries during frying at 150, 170 and 190°C. (Reprinted from Gökmen, V., Palazoğlu, T.K. and Şenyuva, H.Z. Relation between the acrylamide formation and time–temperature history of surface and core regions of French fries. *Journal of Food Engineering*, **77**, 972–976. Copyright 2006a with permission from Elsevier.)

above 120°C within 9 minutes. However, when the oil temperature is high enough (170°C), the energy input is sufficient for both moisture evaporation and temperature increase to take place in the same duration, which favours the formation of acrylamide (Fig. 10.4).

Moisture evaporation is an important barrier to internal energy increase, suggesting that acrylamide reduction methods that involve pre-drying may actually result in increased acrylamide contents. Frying rapidly forms a crust, enveloping the potato strip like a skin. The formation of acrylamide takes place mainly at the surface and in the near-surface regions, because during the process of frying, the conditions in this part of the potato strip become favourable for acrylamide formation as a result of simultaneous drying. As a consequence, any treatment like washing of the cut-surface of potato strips may decrease concentrations of precursors on the surface where the chemical reactions responsible for the acrylamide formation take place.

In addition to other factors such as colour or sensory impression, the moisture remaining in the product characterizes the quality and taste of fried potatoes. According to the industry, residual moisture for an ideal product should be in a range between 38 and 45% (Matthäus *et al.*, 2004).

Recent studies using stable isotope labelled compounds have been conducted on the elimination of acrylamide in potatoes, showing that acrylamide declines fairly rapidly within the

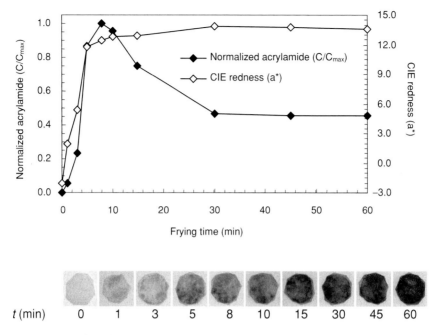

Fig. 10.5 Change of acrylamide concentration and CIE redness parameter a* in potato chips during frying at 170°C. For a color version of this figure, please see Plate 2 of the color plate section that falls between pages 224 and 225. (Reprinted from Gökmen, V., Şenyuva, H.Z., Dülek, B. and Çetin, A.E. Computer vision-based image analysis for the estimation of acrylamide concentrations of potato chips and french fries. *Food Chemistry*, **101**, 791–798. Copyright 2006b with permission from Elsevier.)

first 10–15 minutes at temperatures of approximately 120°C (Biedermann *et al.*, 2002a). It has been further confirmed that acrylamide concentration of potato chips tended to decrease after reaching to an apparent maximum within 10 minutes of frying at 170°C (Gökmen *et al.*, 2006a).

There are characteristic colour transitions observed in potato chips during frying (Fig. 10.5). The CIE a* which is the indication of redness has been shown to be correlated with acrylamide level for early frying times only (Gökmen *et al.*, 2006b).

10.2.2.2 Baking of cereals

Breakfast cereals are made by a variety of distinct processes that yield many forms of flake, puffs, extrudates and biscuit-like pieces. Common to all of the processes is that grains are cooked in water, in most cases, with added sugars to a mass that can be formed into individual cereal pieces. The individual cereal pieces are then dried in air, generally in some form of oven that develops toasted flavours. The Maillard reaction develops flavours and colour in both the cooking and the toasting steps of the process, but acrylamide is formed in the toasting step (Taeymans *et al.*, 2004).

In certain bakery products acrylamide contents up to 1000 mg/kg have been observed (Croft *et al.*, 2004). The highest contents have been found in products prepared with the baking agent ammonium hydrogen carbonate such as gingerbread products (Konings *et al.*, 2003; Amrein *et al.*, 2004). Model experiments showed that ammonium hydrogen carbonate strongly promotes acrylamide formation in sweet bakery (Biedermann and Grob, 2003;

Weisshaar, 2004). Replacing this baking agent by sodium hydrogen carbonate presents a very effective way to limit the acrylamide content of bakery goods (Amrein *et al.*, 2004; Vass *et al.*, 2004). Besides the baking agent, the content of reducing sugars and free asparagine as well as the process conditions influence the acrylamide formation in bakery goods (Springer *et al.*, 2003; Amrein *et al.*, 2004; Surdyk *et al.*, 2004; Vass *et al.*, 2004).

Some reports in the literature would lead one to expect little or no acrylamide to form at the low temperature at the centre of the biscuit where the temperature does not exceed 120°C due to evaporative cooling. However, it has been shown that acrylamide is present in both zones of the biscuit, but apparently lower amounts in the centre of biscuit (Taeymans *et al.*, 2004). Similar to biscuits, the crust and crumb of bread contain significantly different amounts of acrylamide (Surdyk *et al.*, 2004). The crust layers have been shown to contain up to 718 μg/kg of acrylamide while the crumb was free of acrylamide (Şenyuva and Gökmen, 2005a). These results suggest that acrylamide formation in baked cereals mostly occurs by a surface phenomenon. Even though the baking temperature is high enough to produce acrylamide in the crust; the total baking time is not long enough to increase the centre temperature above 120°C at which point acrylamide begins to form.

When biscuits are toasted to a near burnt state, the acrylamide concentration is decreased by up to 50%. Similar results have been shown for several other forms of cereal. Acrylamide levels in biscuits are in contrast with those for the potato, but are consistent with the suggestion made elsewhere that the acrylamide content results from a balance between formation and elimination, with the latter being more rapid at higher temperature (Taeymans *et al.*, 2004).

Ingredients play an important role in acrylamide formation, as different ingredients have various amounts of free asparagine and reducing sugar precursors. Sugars seem to be the most important ingredient in the dough formula from the viewpoint of acrylamide formation, because the free asparagine level of wheat flour is relatively low. Noti *et al.* (2003) reported levels of 150–400 mg/kg of asparagine in 10 samples of wheat flour. Surdyk *et al.* (2004) measured asparagine levels of 170 mg/kg in white wheat flour.

Empirical reformulation trials have been carried out on commercial products to investigate which ingredients may influence acrylamide formation. The addition of whole wheat flour and bran to biscuit formulas tended to increase acrylamide compared with their plain counterparts. Reducing the amount of the raising agent, ammonium bicarbonate, in formulas lowered acrylamide in plain flour matrices. The addition of lactic acid also lowered acrylamide content in plain flour matrices (Taeymans *et al.*, 2004).

Baking temperature and time are closely related with acrylamide formation in the baking process. There is no acrylamide present in uncooked dough, but the acrylamide level rises with time. In cookies containing sucrose, acrylamide concentrations showed a rapid increase after an initial lower rate period, reaching to a plateau within a baking time of 15–20 minutes at 180°C or higher temperatures. When sucrose was replaced with glucose or fructose, initial lower rate period disappears and acrylamide concentration of cookies increases rapidly onset of baking, attaining the plateau values earlier (Summa *et al.*, 2006).

10.2.2.3   *Roasting of coffee beans*

Compared to the many other fried, roasted, and baked food products, roasted and ground coffee has been reported to contain relatively low concentrations (<400 μg/kg) of acrylamide (Friedman, 2003). Furthermore, the dietary contribution of coffee to the total acrylamide intake varies (Konings *et al.*, 2003; Svensson *et al.*, 2003). In addition, more uncertainty has

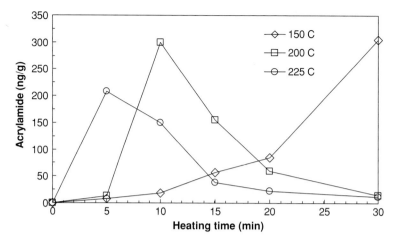

Fig. 10.6 Change of acrylamide concentration in coffee during heating at 150, 200 and 225°C. [Reprinted from Şenyuva, H.Z. and Gökmen, V. Study of acrylamide in coffee using an improved liquid chromatography mass spectrometry method: Investigation of colour changes and acrylamide formation in coffee during roasting. *Food Additives and Contaminants,* **22**, 214–220. Copyright 2005b with permission from Taylor & Francis Ltd, http://www.tandf.co.uk/journals (http://www.informaworld.com).]

been linked to the analytical methods established so far for coffee, due to the complexity of the matrix and difficulty to achieve reliable analytical data.

Similar to frying of potatoes, the amount of acrylamide measured in coffee increases exponentially at the onset of roasting, reaching an apparent maximum, and then decreasing rapidly as the rate of degradation exceeds the rate of formation (Taeymans *et al.*, 2004; Şenyuva and Gökmen, 2005b). However, this effect is more pronounced in coffee as illustrated in Fig. 10.6, because the roasting temperatures of coffee beans are much higher than the frying temperatures of potatoes. Thus, light roasted coffees may be expected to contain relatively higher amounts of acrylamide than very dark roasted beans (Şenyuva and Gökmen, 2005b).

Prolonged heating has been repeatedly shown to reduce the amount of acrylamide in several other systems (Stadler *et al.*, 2002; Tareke *et al.*, 2002; Becalski *et al.*, 2003; Grob *et al.*, 2003; Rydberg *et al.*, 2003; Yasuhara *et al.*, 2003; Bråthen and Knutsen, 2005). This clearly indicates that acrylamide reacts further and/or is eliminated through evaporation.

10.2.3 Kinetic modelling

An understanding of kinetics is of great importance to our ability to control acrylamide formation, ideally over the range of temperatures involved in food processing. Acrylamide formation in foods is under kinetic control, i.e., the amount which is measured in thermally processed food depends on the conditions under which the food has been heated and the time of heating. With respect to acrylamide, so far only limited data with kinetic analysis has been reported in the literature (Claeys *et al.*, 2005; Knol *et al.*, 2005; Wedzicha *et al.*, 2005; Gökmen and Şenyuva, 2006a).

Kinetic models tend to be much simpler than mechanistic models because they describe the rates of formation of substances in terms of the rate limiting processes. The kinetic approach may be used to determine how the control points in acrylamide formation are affected by

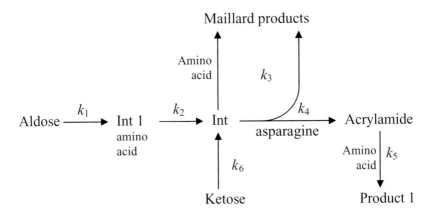

Fig. 10.7 Kinetic model for the formation of acrylamide as a result of the reaction of asparagine with key intermediates in the Maillard reaction. (Reprinted from Wedzicha, B.L., Mottram, D.S., Elmore, J.S., Koutsidis, G. and Dodson, A.T. Kinetic models as a route to control acrylamide formation in food. *Advances in Experimental Medicine and Biology*, **561**, 235–253. Copyright 2005 with kind permission from Springer Science and Business Media.)

reaction conditions. Wedzicha *et al.* (2005) have proposed a multi-response kinetic model taking into account the formation of acrylamide as a result of the reaction of asparagine with key intermediates in the Maillard reaction (Fig. 10.7).

Based on the results reported in literature, formation and degradation of acrylamide occurs at the same time during heating at elevated temperatures. The kinetics of acrylamide loss has only been determined for the reactions which take place after long heating times, i.e. after all the amino acids has been exhausted (Wedzicha *et al.*, 2005).

It is possible to consider the concentration of acrylamide measured in food at any time as a net result from two consecutive reactions at the temperature studied. Then the overall reaction may be written in a more simplified form given below:

$$\text{Carbonyl compound (A)} + \text{Asparagine (B)} \xrightarrow{k_1} \text{Acrylamide (C)} \xrightarrow{k_2} \text{Degradation product (D)}$$

at which, the rate of acrylamide formation is expressed as

$$\frac{dC_C}{dt} = k_1 C_A^a C_B^b - k_2 C_C^c \tag{10.1}$$

where t is time; C_A, C_B, C_C are concentrations of carbonyl compound, asparagine, acrylamide, and a, b, c are the rate orders with respect to carbonyl compound, asparagine, acrylamide, respectively. The differential equation is solved to determine the apparent rate constants for formation (k_1) and degradation (k_2) once the rate orders are known. Although there are many reports of the levels of acrylamide formation in model systems composed of different ratios of asparagine and sugars or in real food systems, these reports do not give useful information for a satisfactory kinetic analysis.

Figure 10.8 depicts the typical kinetic pattern of acrylamide formation in a fructose–asparagine model system at 150°C. It has been shown that approximately 0.05 μ moles of acrylamide forms in the reaction mixture composed of 5 μ moles of fructose and asparagine.

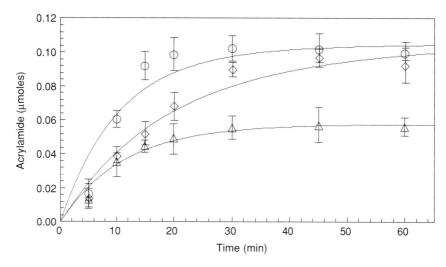

Fig. 10.8 Formation of acrylamide in fructose-asparagine model system during heating at 150°C. The amounts of fructose:asparagine in µmoles are; –△– 5 : 5 , –○– 10 : 5 , –◇– 5 : 10. [Reproduced from Gökmen, V. and Şenyuva, H.Z. A simplified approach for the kinetic characterization of acrylamide formation in fructose-asparagine model system. *Food Additives and Contaminants*, **23**, 348–354. Copyright 2006a with permission from Taylor & Francis Ltd, http://www.tandf.co.uk/journals (http://www.informaworld.com).]

The amount of acrylamide doubles when the concentration of asparagine is doubled under the same conditions.

The initial rate of acrylamide formation is also as important as it determines the maximum amount of acrylamide attained during the reaction. To determine the initial rate, the overall reaction may be considered to proceed as follows:

$$\text{Carbonyl compound (A)} + \text{Asparagine (B)} \xrightarrow{k_1} \text{Acrylamide (C)}$$

by neglecting the concurrent thermal degradation. For the experimental data shown in Fig. 10.8, this approach reveals that the initial rate is approximately 0.0051 µmol/min, and it increases to 0.0098 µmol/min by increasing fructose concentration from 5 to 10 µmoles. However, increasing asparagine from 5 to 10 µmoles has no significant effect on the initial rate (0.0053 µmol/min) while it significantly affects the maximum concentration of acrylamide formed in the reaction (Gökmen and Şenyuva, 2006a). These observations suggest that acrylamide formation is first-order with respect to fructose ($\alpha = 1$) and zero-order with respect to asparagine ($\beta = 0$).

In order to obtain a solution for the differential equation (Equation (10.1)), which describes the kinetic constants involved in the reaction responsible for acrylamide formation and degradation; it is necessary to know the rate order for the degradation of acrylamide. It has been shown, by using deuterium labelled acrylamide, the degradation follows first-order kinetic pattern ($c = 1$). So, the rate equation may be written as follows:

$$\frac{dC_C}{dt} = k_1' C_A - k_2 C_C \tag{10.2}$$

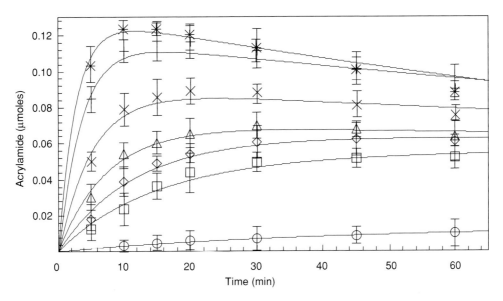

Fig. 10.9 Formation of acrylamide during Maillard reaction in fructose–asparagine model system at different temperatures (–◯– 120°C –□– 140°C, –◇– 150°C, –△– 160°C, –×– 170°C, – + – 180°C, –✳– 200°C). [Reproduced from Gökmen, V. and Şenyuva, H.Z. A simplified approach for the kinetic characterization of acrylamide formation in fructose-asparagine model system. *Food Additives and Contaminants*, **23**, 348–354. Copyright 2006a with permission from Taylor & Francis Ltd, http://www.tandf.co.uk/journals (http://www.informaworld.com).]

where, k_1' is the efficient apparent rate constant for acrylamide formation and expressed simply as

$$k_1' = k_1 C_B \tag{10.3}$$

A convenient way for integrating the rate equation, which is a system of first-order ordinary differential equation, is to use the Laplace transform.

$$C_C = \frac{C_{A_0}.k_1'}{k_2 - k_1'} [\exp(-k_1't) - \exp(-k_2t)] \tag{10.4}$$

Figure 10.9 shows kinetic patterns of acrylamide formation in a fructose–asparagine model system in the temperature range between 120 and 200°C with model fit of experimental data. The decline of the curves is most likely due to the polymerization. The simplified model fits well to the experimental data giving useful estimations of apparent rate constants for formation and concurrent degradation of acrylamide at different temperatures (Gökmen and Şenyuva, 2006a)

The temperature plays a major role in the Maillard reaction. The temperature dependence of acrylamide formation and degradation obeys the Arrhenius law. The activation energies have been determined to be 52.1 and 72.9 kJ/mol for the formation and concurrent degradation of acrylamide, respectively (Gökmen and Şenyuva, 2006a). Others reported activation energies varying between 40.1 and 94.4 kJ/mol for steps involved in a reaction network (Knol *et al.*, 2005).

Considering the various possible formation mechanisms of acrylamide, it seems to be an almost impossible task to elucidate the kinetics of all the pathways involved. Artificial neural network (ANN) modelling, on the other hand, may be a viable alternative to the kinetic models. While the use of phenomenological models requires a simultaneous solution of a large number of non-linear algebraic equations that need long and exhaustive iterative processes, the solution based on an ANN approach is simpler and quicker, and it consists of a solution of a system of linear algebraic equations (Gonçalves *et al.*, 2005).

ANNs are non-linear mathematical models that learn from the examples through iterations. They are made of a large number of nodes or artificial neurons, which are disposed in a parallel structure. Each ANN has one input layer containing one node for each independent variable, one or more hidden layers, where the data are processed, and one output layer, containing one node for each dependent variable. The data from the input layer are propagated through the hidden layer and then to all the network, which are associated with a scalar weight. Neurons in the hidden and output layers calculate their inputs by performing a weighted summation of all the outputs they receive from the layer before. Their outputs, on the other hand, are calculated by transforming their inputs using a non-linear transfer function. Then, the network output is compared with the actual output provided by the user. The difference is used by the optimization technique to train the network. Thus, the training process requires a forward pass to calculate an output and a backward pass to update the weights in feed-forward back-propagation networks (Bishop, 1994; Gonçalves *et al.*, 2005).

A great advantage of ANN models is that, they do not require prior knowledge of the relationship between the input and output variables, and instead of that, they figure out these relationships through training. Therefore, complex processes can be optimized to produce the desired outputs using successfully trained ANN models. In order to reduce acrylamide levels in food, a systematic evaluation of potential strategies is required for processing conditions. A successfully trained ANN model for the prediction of acrylamide concentration is thought to give useful information for the optimization of thermal processes of foods (Serpen and Gökmen, 2006).

An important use of a predictive model is to conduct 'what if' experiments, whereby the response to an imposed change in reaction conditions (e.g. concentration, temperature, time) can be determined without actually doing the experiment. For complicated reactions like the Maillard reaction (Fig. 10.10), the main shortcoming of kinetic models is the use of simplified reaction networks on the basis of several assumptions to derive the governing equations. Nevertheless, a pragmatic approach of ANN modelling that considers only the input and output variables rather than the complex chemistry can already be sufficient for obtaining information of direct applicability to food processing (Serpen and Gökmen, 2006).

Figure 10.11 shows an example of ANN model to predict acrylamide and browning in potato chips during frying (Serpen and Gökmen, 2006). Obtaining experimental data, which reflects the common variations in the input variables (glucose, asparagine, temperature and time), is obtained by an appropriate design, the ANN model is built by training and testing sequentially. A good correlation between the predicted and measured acrylamide concentrations is usually obtained, but the correlation coefficient itself is not enough to indicate the success of an ANN model. The model should also be capable of predicting its output variables, i.e. the acrylamide concentration in a time-dependent manner. Using successfully trained ANN model, it is also possible to observe the combined effects of various input variables on acrylamide concentration of the resulting product (Serpen and Gökmen, 2006).

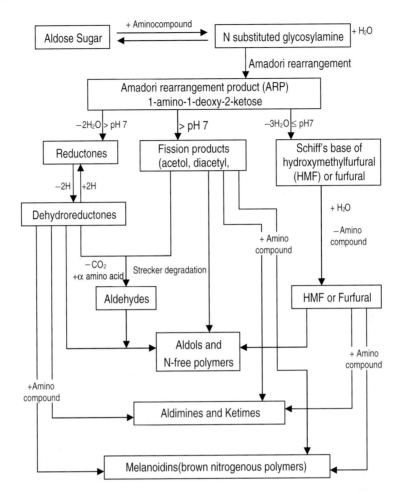

Fig. 10.10 The major steps in the Maillard reaction.

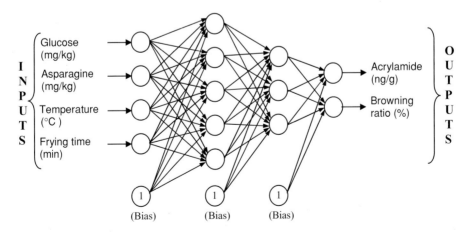

Fig. 10.11 Structure of an ANN model to predict acrylamide and browning in potato chips during frying (Serpen and Gökmen, 2006).

Contrary to kinetic models, ANN modelling is more realistic since it depends only on experimental observations without assumption. It seems to be a feasible approach for the prediction of acrylamide concentrations in potato chips based on the parameters related to both potato composition and frying conditions. The prediction capability of ANN models offers an advantage for the optimization of processing conditions to minimize undesirable changes in foods.

10.3 METHODS OF ANALYSIS

Detection of acrylamide levels in processed foods has been an intensive area of research shortly after the discovery of acrylamide in heated foods by the Swedish researchers in April 2002. Numerous methods have been developed in the past years to determine the acrylamide, especially in water, biological fluids and uncooked foods (sugar, field crops, mushrooms), and the majority are classical methods based on liquid chromatography (LC) or gas chromatographic (GC) techniques (Tekel et al., 1989; Castle, 1993; Environmental Protection Agency, 1996; Bologna et al., 1999). However, because of the complexity of food matrices, these methods do not suffice for the analysis of acrylamide in heat-treated foods at trace levels. Particularly, they lack selectivity and the additional degree of analyte certainty required to confirm the presence of acrylamide in a complex food matrix.

Rosén and Hellenäs (2002) firstly reported the analysis of acrylamide in different heat-treated foods using the isotope dilution LC-MS technique. They developed a mass spectrometry method for direct detection of acrylamide, which would unequivocally verify the presence of acrylamide in a range of heat-treated foods. The choice of LC-MS was due to the hydrophilic properties of acrylamide, and MS/MS was used for a high degree of verification if several transitions could be found. Since the initial Swedish announcement, other laboratories have worked intensively on the development of LC-MS-based methods to determine acrylamide in processed and cooked foods (Riediker and Stadler, 2003; Roach et al., 2003; Wenzl et al., 2003; Zyzak et al., 2003). These methods are based mainly on MS as the determinative technique for better identification of acrylamide in processed foods, coupled with a chromatographic step either by LC (Ahn et al., 2002; Höfler et al., 2002; Nemoto et al., 2002; Tareke et al., 2002; Becalski et al., 2003; Gutsche et al., 2002; Murkovic, 2004; Şenyuva and Gökmen, 2005a; Gökmen and Şenyuva, 2006b) or GC, the latter either after derivatization of the analyte (Gertz and Klostermann, 2002; Ono et al., 2003; Pittet et al., 2003), or in a few cases analysis of the compound directly (Biedermann et al., 2002c; Tateo and Bononi, 2003). The advantage of the LC-MS-based methods is that acrylamide can be analysed without prior derivatization (e.g., bromination), which considerably simplifies the analysis.

Hitherto, it can be summarized from recent methodological studies that GC-MS and LC-MS or LC-MS/MS appeared to be acknowledged as the most useful and authoritative method for acrylamide determination. A recent European inter-laboratory study conducted to validate two analytical procedures (LC-MS and GC-MS) for the determination of acrylamide in bakery and potato products has shown that an LC-MS-based method was undoubtedly fit-for-purpose (Wenzl et al., 2006).

However, many of these methods do not perform well in difficult matrixes such as cocoa and coffee. As previously highlighted by some researchers, the selected mass transitions cannot always provide baseline separation in certain profiles during LC-MS/MS analysis. The significant loss of the analyte throughout the sample preparation steps and ion-suppression effects leading to a low response of acrylamide are common problems encountered after a two-step extract clean-up (Andrzejewski *et al.*, 2004).

The following sections mainly focus on the parameters for sample extraction and clean-up, chromatographic separation and detection procedures for the determination of acrylamide in heat-treated foods. The schematic representation of a generic method for the determination of acrylamide in wide variety of thermally processed foods is given in Fig. 10.12.

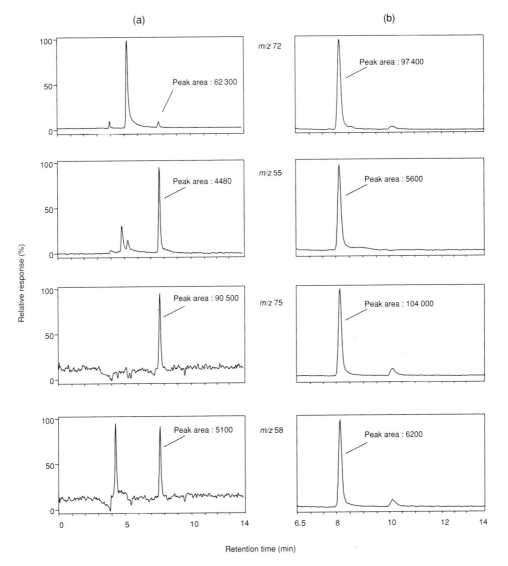

Fig. 10.12   LC-MS chromatograms of potato chips containing 1020 ng/g of acrylamide and 1000 ng/g $^{13}C_3$-labelled acrylamide (a) no delay time (b) delay time is 6.5 minutes. Chromatographic conditions; column: Inertsil ODS-3 (1), mobile phase: 0.01 mM acetic acid in 0.2% aqueous solution of formic acid and 0.2% acetic acid in acetonitrile (98:2, v/v). (Reprinted from Şenyuva, H.Z. and Gökmen, V. Interference-free determination of acrylamide in potato and cereal-based foods by a laboratory validated liquid chromatography–mass spectrometry method. *Food Chemistry*, **97**, 539–545. Copyright 2006 with permission from Elsevier.)

10.3.1 Sample preparation

An incomplete extraction or consequent loss of analyte during the removal of fat or the evaporation of extracting solvent are most likely cause of erroneous results during the analysis of acrylamide. A common approach for the extraction of acrylamide from foods entails an extraction with water followed by sample clean-up with solid phase extraction (SPE) prior to LC-MS analysis (Ahn *et al.*, 2002; Rosén and Hellenäs, 2002; Tareke *et al.*, 2002; Becalski *et al.*, 2003). Disintegration of the samples to small particles prior to extraction makes additional homogenization, thus avoiding the need for vigorous mixing by Ultra Turrax to increase the extraction yield (Petersson *et al.*, 2006). Potato- and cereal-based food samples are usually composed of high amounts of colloids (starch and proteins) and fat which should be separated after the extraction with water. Some laboratories have included a defatting step before or in combination with the extraction step. This has been carried out by extraction with hexane, toluene or cyclohexane. However, it has been recently demonstrated that the removal of fat by an organic solvent gives a negative effect on the extraction yield (Petersson *et al.*, 2006). This may be due to a loss of acrylamide during defatting. Another approach has been the use of an accelerated solvent extraction (ASE) device (Cavalli *et al.*, 2003; Höfler *et al.*, 2002). However, it has been shown by others that an extract free from the colloids and fat can be obtained by Carrez clarification and cold centrifugation after extracting the sample with acidified water (Şenyuva and Gökmen, 2005a, 2006).

A great diversity exists among the SPE clean-up procedures described in literature. One approach has been to combine Oasis HBL (Waters, Milford, MA, USA) and Bond Elut-Accucat (mixed mode: C8, SAX and SCX) (Varian, Palo Alto, CA, USA) cartridges. Becalski *et al.* (2003) used a combination of three different cartridges: Oasis MAX (mixed-mode anion exchange) (Waters), Oasis MCX (mixed-mode cation exchange) and ENVI-Carb (graphitized carbon) (Supelco, Bellefonte, PA, USA). A similar combination of SPE cartridges consisting of Bond Elut C18, Bond Elut Jr-PSA (anion exchange) and Bond Elut Accucat (all Varian) were chosen for clean-up of samples, which were measured by LC-MS with column switching (Takatsuki *et al.*, 2003). Another group of laboratories have used Isolute Mixed Mode (C18, SAX and SCX) cartridges (International Sorbent Technology, Hengoed, UK) combined with filtration and/or ultracentrifugation to avoid blockage of the chromatographic system.

The small molecules such as amino acids are difficult to avoid during extraction and clean-up in the analysis of acrylamide. It has been shown that amino acid valine, which yields a compound-specific product ion having *m/z* of 72 prevents an accurate quantitation of acrylamide (Şenyuva and Gökmen, 2006). The adverse ion suppression effect of this interfering co-extractive can be eliminated by instrumentally adjusted delay time for ionization (Fig. 10.13) or by SPE clean-up using an appropriate cation exchanger sorbent, i.e. Oasis MCX (Fig. 10.14). Since valine (pI ∼6.0) present in the final extract is positively charged, it is easily retained by passing the extract through the SPE cartridge packed with Oasis MCX.

Co-detection of acrylamide and interferences in coffee extract has been confirmed by analysing the purity of co-eluted peaks in scan mode. As illustrated in Fig. 10.15, the ion having *m/z* of 71 appears as the major interference during the MS detection of acrylamide in SIM mode (Şenyuva and Gökmen, 2005b).

Using a single SPE clean-up in different modes does not remove the compounds, which interfere with the detection of acrylamide by mass spectrometry, present in the aqueous extract of coffee. Some researchers have used sequential SPE cartridge clean-up using Oasis HLB cartridge and then a Bond Elut-Accucat (cation and anion exchange sorbent) cartridge. However, a number of peaks have been observed both before and after the acrylamide peak

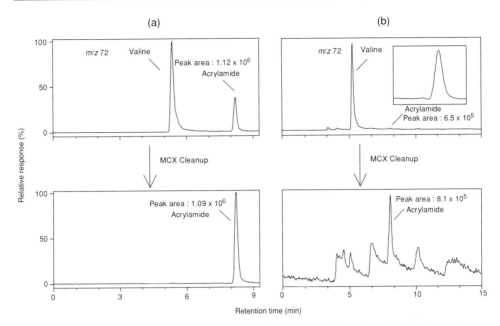

Fig. 10.13   LC-MS chromatograms of (a) a standard mixture containing 5 µg/mL of valine and acrylamide and (b) potato crisp sample (FAPAS T3007) before and after Oasis MCX clean-up. Chromatographic conditions are same as given in Fig. 10.15. (Reprinted from Şenyuva, H.Z. and Gökmen, V. Interference-free determination of acrylamide in potato and cereal-based foods by a laboratory validated liquid chromatography–mass spectrometry method. *Food Chemistry*, **97**, 539–545. Copyright 2006 with permission from Elsevier.)

in the ion profiles monitored for coffee extracts during LC-MS/MS analysis, despite two SPE cartridge clean-up steps (Andrzejewski *et al.*, 2004).

Considering its solubility (155.0 g/100 mL), methanol can be considered as an alternative for extraction. It has been previously shown that methanol as the extraction solvent can be successfully applied for potato chips and crisps with subsequent Carrez clarification and SPE

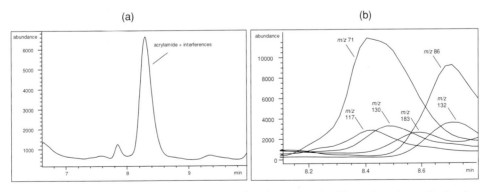

Fig. 10.14   (a) LC-MS chromatogram of instant coffee showing acrylamide peak overlapped by interference present in coffee. (b) Interfering ions detected for the peaks co-eluted at 8.335 minutes. Chromatographic conditions are same as given in Fig. 10.15. [Reprinted from Şenyuva, H.Z. and Gökmen, V. Study of acrylamide in coffee using an improved liquid chromatography mass spectrometry method: Investigation of colour changes and acrylamide formation in coffee during roasting. *Food Additives and Contaminants*, **22**, 214–220. Copyright 2005b with permission from Taylor & Francis Ltd, http://www.tandf.co.uk/journals (http://www.informaworld.com).]

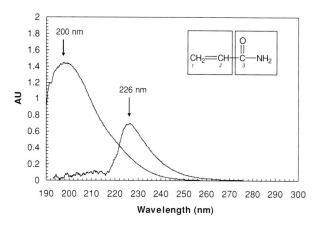

Fig. 10.15   UV spectrum of acrylamide in water and acid. (Reprinted from Gökmen, V., Şenyuva, H.Z., Acar, J. and Sarıoğlu, K. Determination of acrylamide in potato chips and crisps by high-performance liquid chromatography. *Journal of Chromatography A*, **1088**, 193–199. Copyright 2005 with permission from Elsevier.)

clean-up prior to LC coupled to UV detection at 226 nm (Gökmen *et al.*, 2005). Since methanol does not extract the colloids (starch, proteins, etc.) as much as water does, it usually yields clearer extract than water even without subsequent centrifugation. In addition, it can be easily evaporated under a gentle stream of nitrogen to improve the limit of quantitation. It has been successfully shown that this approach, which entails the extraction with methanol, followed by Carrez clarification, evaporation and reconstitution in water, and SPE clean-up using Oasis HLB, works also well for coffee during the LC-MS analysis of acrylamide (Şenyuva and Gökmen, 2005b). Carrez clarification not only purifies the extract by precipitation of dissolved colloids, but also prevents the loss of acrylamide during the evaporation of methanol under nitrogen. Following evaporation to complete dryness, the residue is redissolved in water. By changing the solvent, lipophilic and some of brown coloured co-extractives are excluded leaving them as a residue on the wall of glass flask, but acrylamide is completely transferred into aqueous phase. The extract is further cleaned up by using Oasis HLB cartridge in a pass through strategy. Doing so, a colourless or slightly coloured final extract is obtained prior to LC-MS analysis.

10.3.2   Chromatographic separation

Since co-extractives cannot be completely removed during extraction and clean-up, analytical methods may fail in the detection step as a result of poor chromatographic separation. Acrylamide is very polar molecule with poor retention ($k' < 2.0$) in conventional LC-reversed phase sorbents. A wide variety of reversed-phase columns have been used for the chromatographic separation of acrylamide (Ahn *et al.*, 2002; Rosén and Hellenäs, 2002; Tareke *et al.*, 2002; Becalski *et al.*, 2003; Ono *et al.*, 2003). Ion-exchange columns have also been used to increase the capacity factor value compared with reversed-phase columns, resulting in good separation of acrylamide from matrix compounds even for untreated sample extracts.

It has been shown that reversed phase columns greatly differ in their ability to retain acrylamide. Atlantis HILIC and Inertsil ODS-3 packing materials have a somewhat higher retention capacity for acrylamide (Table 10.1).

Since the columns packed with Inertsil ODS-3 have a reasonable capacity factor, these may be the optimum choice for the chromatographic separation of acrylamide. The aqueous

Table 10.1 Capacity factors and plate numbers calculated for the separation of acrylamide on different LC columns[a].

Column	Dimensions	t_R	k'	N
Atlantis HILIC	250 × 4.6 mm, 5μ	8.9	3.67	17 373
Zorbax SIL	250 × 4.6 mm, 5μ	8.4	1.80	19 944
Inertsil ODS-3 (1)	250 × 4.0 mm, 5μ	10.0	3.11	11 400
Inertsil ODS-3 (2)	250 × 4.6 mm, 5μ	9.42	3.09	10 500
Atlantis dC$_{18}$	250 × 4.6 mm, 5μ	13.8	2.54	19 944
Zorbax StableBond C$_{18}$	250 × 4.6 mm, 5μ	11.2	2.39	19 250
HiChrom 5C$_{18}$	300 × 4.6 mm, 5μ	10.9	1.79	18 233
Luna C$_{18}$	250 × 4.6 mm, 5μ	12.6	1.93	5232
Synergi MAX-RP	250 × 4.6 mm, 5μ	12.6	1.74	8076

[a] Mobile phase: 0.5 mL/min of water at 25°C.
Reprinted from Gökmen, V., Şenyuva, H.Z., Acar, J. and Sarıoğlu, K. Determination of acrylamide in potato chips and crisps by high-performance liquid chromatography. *Journal of Chromatography A*, **1088**, 193–199. Copyright 2005 with permission from Elsevier.

mobile phase which is used in HPLC should be modified with organic acid and acetonitrile or methanol to increase ionization yield and improve repeatability. It has been shown that the retention of acrylamide can be improved by both hydrophilic and hydrophobic interaction chromatography by avoiding the organic modifiers in the aqueous mobile phase (Gökmen et al., 2005). This will result in a better resolution of acrylamide from the interfering compounds commonly present in roasted coffee samples (Gökmen and Şenyuva, 2006b).

10.3.3   Detection

For the detection of acrylamide after LC separation, tandem mass spectrometry is most often the method of choice. There are just a few exceptions in which UV and single quadrupole MS (in SIM mode) were used (Höfler et al., 2002). Since acrylamide is a derivative of carboxylic acid, it has a maximum absorption within the wavelength range of 195–205 nm. However, all co-extractives from the food matrix also absorb well in this wavelength range adversely affecting the specificity of detection. On the other hand, acrylamide has also a characteristic absorption at 226 nm due to double bond between C1 and C2 (Fig. 10.16). Comparing to the absorbance at 200 nm, the absorbance of acrylamide is almost twice as low at 226 nm (Gökmen et al., 2005).

UV detection can be a choice for the analysis of acrylamide in potato chips and crisps when coupled to successful LC separation conditions and sample preparation procedures as shown in Fig. 10.17 (Gökmen et al., 2005). However, the lack of selectivity would hamper the determination of acrylamide in more complex matrices. Since an additional degree of analyte certainty is required to confirm the presence of acrylamide in the complex food matrix, MS becomes the choice for the detection step coupled to LC. Although MS has higher selectivity, the mass of acrylamide itself or its fragment ions are not specific due to presence of co-extractives that yield the same magnitude of m/z with acrylamide in the sample matrix. Compounds other than acrylamide were found to be present in the chromatograms of certain type of foods despite the use of MS/MS.

Many studies have shown that tandem MS-MS with positive electrospray ionization (ESI) is a powerful tool for the detection of acrylamide in foods at low levels (Rosén and Hellenäs, 2002; Tareke et al., 2002; Riediker and Stadler, 2003). Under the positive ion ESI conditions, frequent cleaning of the detector inlet is required due to heavy contamination with salts. In

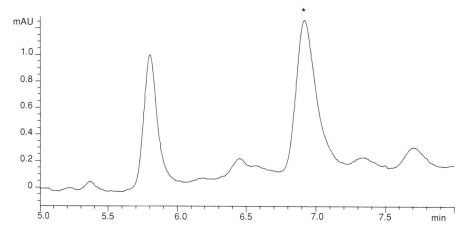

Fig. 10.16 LC-UV chromatogram of acrylamide in potato chips (acrylamide concentration: 1020 ng/g). Chromatographic conditions; column: Atlantis dC18, mobile phase: water at 1.0 mL/min. (Reprinted from Gökmen, V., Şenyuva, H.Z., Acar, J. and Sarıoğlu, K. Determination of acrylamide in potato chips and crisps by high-performance liquid chromatography. *Journal of Chromatography A*, **1088**, 193–199. Copyright 2005 with permission from Elsevier.)

contrast, LC-MS with atmospheric pressure chemical ionization (APCI) allows the determination of acrylamide sensitively and precisely without problems. During MS detection of acrylamide, $^{13}C_3$-labelled acrylamide is usually added to sample to confirm the native acrylamide. [M+H]+ ions (*m/z* 72 and 75 for acrylamide and $^{13}C_3$-labelled acrylamide) with compound-specific product ions due to loss of NH_3 from the protonated molecule can be sensitively detected under the positive APCI conditions (Fig. 10.18). The ratio of corresponding ions is used to confirm the identification of acrylamide in the sample.

10.3.3.1 *Rapid methods: challenges ahead*

Although the analytical methods based on MS detection coupled either LC or GC perform well for quality control purposes in a food analysis laboratory, they are laborious, costly and cannot be adopted easily for process control purposes by the food industry. To satisfy the increased awareness, and greater expectation of consumers, as well as demands by the regulatory authorities, it is necessary to improve quality evaluation of food products. The following sections describe the use of colour spectrophotometer, computer vision-based image analysis and near infrared spectroscopy for rapid detection of acrylamide in certain thermally processed foods.

10.3.4 Colour spectrophotometer

It is a fact that both brown coloured products and acrylamide are formed during the Maillard reaction at high temperatures. Previous findings have suggested that surface colour may be correlated with acrylamide in thermally processed foods (Pedreschi *et al.*, 2004; Surdyk *et al.*, 2004; Şenyuva and Gökmen, 2005b). Earlier attempts have been directed to measure colour in CIE Lab space units, which is an international standard for colour measurements, adopted by the Commission Internationale d'Eclairage (CIE) in 1976.

A significant correlation has been found between the level of acrylamide and colour measured as CIE a* value in roasted coffee. This correlation has been fitted to a non-linear

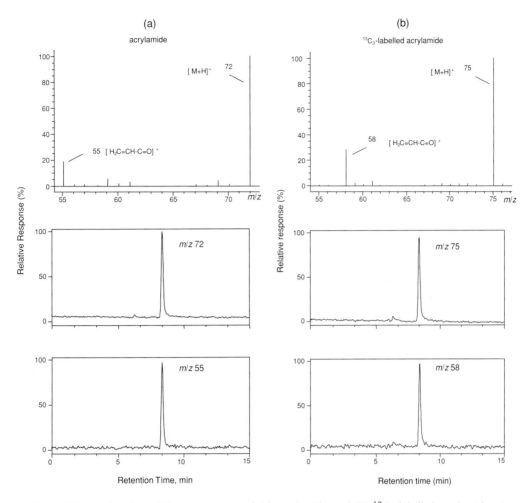

Fig. 10.17 Positive ion APCI mass spectra of (a) acrylamide and (b) $^{13}C_3$-labelled acrylamide at 100 ng/mL. Chromatographic conditions are same as given in Fig. 10.15. [Reproduced from Şenyuva, H.Z. and Gökmen, V. Survey of acrylamide in Turkish foods by an in-house validated LC-MS method. *Food Additives and Contaminants*, **22**, 204–209. Copyright 2005a with permission from Taylor & Francis Ltd, http://www.tandf.co.uk/journals (http://www.informaworld.com).]

logarithmic function (Fig. 10.19), which has been used to predict acrylamide concentration from the measured CIE a* value in a variety of commercial coffee samples. It has been concluded that the level of acrylamide in roasted coffee may be approximately estimated from the CIE a* value (Şenyuva and Gökmen, 2005b).

10.3.5 Computer vision-based image analysis

Although the CIE redness parameter has been shown to a certain extent to be correlated with the level of acrylamide, it may not be a reliable predictor of acrylamide concentration in thermally processed foods, i.e. potato chips, where the surface colour is not homogenous. In this case, an examination of colour in depth may be a choice by means of computer vision-based image analysis.

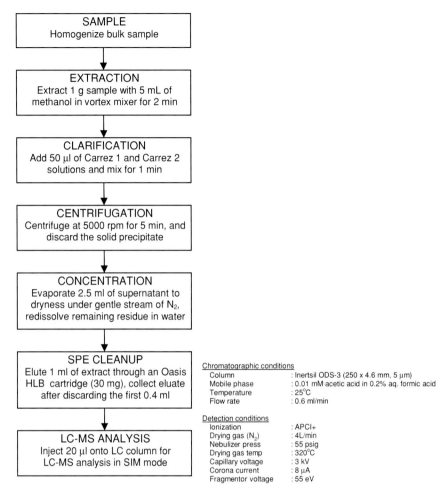

Fig. 10.18 The generic method for the determination of acrylamide in foods. (Reprinted from Gökmen, V. and Şenyuva, H.Z. A generic method for the determination of acrylamide in thermally processed foods. *Journal of Chromatography A*, **1120**, 194–198. Copyright 2006b with permission from Elsevier.)

Being an objective, rapid and non-invasive tool, computer vision can be a powerful technique for inspection and evaluation purposes by means of a rapid prediction of acrylamide level in food products.

Computer vision is the construction of explicit and meaningful descriptions of physical objects from images (Ballard and Brown, 1982). Image analysis is the core of computer vision with numerous algorithms and methods available to achieve the required classification and measurements. From the viewpoint of acrylamide, surface colour is an important food attribute that can be used to predict the acrylamide level.

A typical image captured by a digital camera consists of an array of vectors called pixels. Each pixel has red, green and blue colour values:

$$x[n, m] = \begin{bmatrix} x_r(n, m) \\ x_g(n, m) \\ x_b(n, m) \end{bmatrix}$$

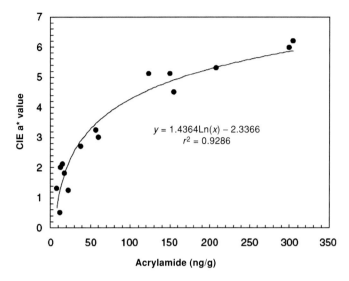

Fig. 10.19 Non-linear function for the correlation between acrylamide concentration and CIE a* value for roasted coffee. [Reprinted from Şenyuva, H.Z. and Gökmen, V. Study of acrylamide in coffee using an improved liquid chromatography mass spectrometry method: Investigation of colour changes and acrylamide formation in coffee during roasting. *Food Additives and Contaminants*, **22**, 214–220. Copyright 2005b with permission from Taylor & Francis Ltd, http://www.tandf.co.uk/journals (http://www.informaworld.com).]

where $x_r(n, m)$, $x_g(n, m)$ and $x_b(n, m)$ are values of the red, green and blue components of the (m, n)th pixel $x[n, m]$, respectively. In digital images, x_r, x_g and x_b colour components are represented in eight bits, i.e. they are allowed to take integer values between 0 and 255 $(= 2^8 - 1)$ (Gonzales and Woods, 2002).

In fried potato chip images, there are three regions which typically have bright yellow, yellowish brown and dark brown colours. Digital image pixel values can be used to estimate the acrylamide level of a potato chip after a useful feature is extracted from the digital image. After the representative mean red, green and blue values of the pixels of three regions are determined, pixels of the fried potato images may be classified into three sets based on their Euclidian distances to the representative mean values. This process is also called vector quantization (Cetin and Weerackody, 1988; Rabiner and Juang, 1993).

Since yellowish brown coloured potato chips have been found to contain higher levels of acrylamide, a new parameter called the normalized area of Set-II pixels (NA2) has been defined to correlate with the level of acrylamide. The NA2 ratio is computed from the segmented images (Gökmen *et al.*, 2006b). Figure 10.20 shows the original and segmented images of a potato chip sample using semi-automatic segmentation algorithm. There is a strong linear correlation between NA2 values and measured acrylamide levels of potato chips as shown in Fig. 10.21.

Acrylamide levels of number of commercial and laboratory-made fried potato chips ($n = 60$) have been predicted by means of this correlation data. The percentage error of estimates has averaged 29% with a standard deviation of 21% for the test samples. From the viewpoint of food safety evaluation and inspection, it is important to define a threshold level for acrylamide in potato chips. However, no maximum permitted concentration has been established for acrylamide in thermally processed foods by the legal authorities. Once a critical level of acrylamide is adopted by the food industry, the semi-automatic segmentation algorithm may

(a) (b)

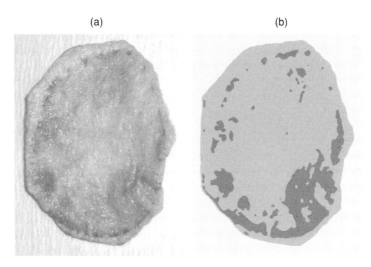

Fig. 10.20 (a) Original fried potato chip image, (b) result of semi-automatic segmentation (NA2 = 0.1996). (Reprinted from Gökmen, V., Şenyuva, H.Z., Dülek, B. and Çetin, A.E. Computer vision-based image analysis for the estimation of acrylamide concentrations of potato chips and french fries. *Food Chemistry*, **101**, 791–798. Copyright 2006b with permission from Elsevier.)

be used as a tool for safety evaluation and inspection purposes in order to sort out poor quality chips. To exemplify the inspection capability of computer vision approach using the test samples, a provisional maximum permitted concentration of acrylamide has been assumed as 1000 ng/g in the finished product. Potato chips exceeding this threshold limit could be accurately recognized by means of the image segmentation algorithm with one exception (Gökmen *et al.*, 2006c).

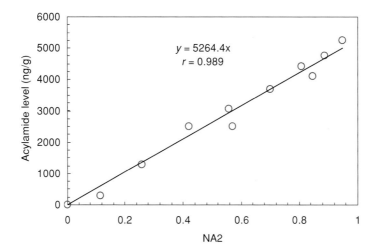

Fig. 10.21 Correlation between defined parameter NA2 values and measured acrylamide levels for potato chips. (Reprinted from Gökmen, V., Şenyuva, H.Z., Dülek, B. and Çetin, A.E. Computer vision-based image analysis for the estimation of acrylamide concentrations of potato chips and french fries. *Food Chemistry*, **101**, 791–798. Copyright 2006b with permission from Elsevier.)

In the food industry, quality evaluation still heavily depends on manual inspection, which is tedious, laborious, and costly, and is easily influenced by physiological factors, inducing subjective and inconsistent evaluation results (Brosnan and Sun, 2004). With the advantages of speed and accuracy, computer vision is considered to be a powerful tool for safety evaluation and inspection purposes for the food industry. In such a system, a digital camera can be installed in the packaging line and fried potato images can be analysed in real-time and those products with NA2 values exceeding the pre-defined threshold limit. Present technology in terms of the speed of computers allows this kind of data processing to be performed in reasonable times.

10.3.6   Near-infrared spectrophotometer

Near-infrared (NIR) spectroscopy has been established as a technique for the analysis of the proximate composition of complex foods, mainly due to its speed and ability to perform contact-free measurements. NIR spectroscopy is a non-specific technique that utilizes several wavelengths in order to predict one or more chemical components. The wavelengths are strongly co-linear, which requires regression techniques that can handle co-linearity without over-fitting the data. Principal component regression (PCR) and partial least squares regression (PLSR) are probably the most used regression techniques for multivariate calibration of rapid spectroscopic instruments (Segtnan et al., 2006).

Most NIR applications focus on bulk components like fat, water, carbohydrate and protein (Osborne et al., 1997), but several studies have shown that the technique is also capable of separating and quantifying different species and states within these main classes. In particular, water, which is the strongest NIR absorber in food systems, has shown interesting features that can reflect small changes in other molecules that get attached to the water molecules through hydrogen bonding.

Recently, NIR spectra acquired from the processed samples have been found to be correlated with the acrylamide content. The best prediction model has been found using the whole spectral region (400–2500 nm), using extended multiplicative signal correction (EMSC) transformation, standardization and jack knifing for removal of non-significant variables. An average prediction error of the model has been reported as 247 µg/kg ($r = 0.952$).

10.4   OCCURRENCE IN FOODS AND EXPOSURE ASSESSMENTS

Extensive amounts of data on the levels of acrylamide have been collected by European authorities, the Confederation of the Food and Drink Industries in the EU (CIAA) and the US Food and Drug Administration (FDA). A website has been created to allow the rapid compilation of data by EU Joint Research Centre (JRC). It is apparent from the acrylamide monitoring database that fried potato products are among the foods which contain the highest amount of acrylamide, followed by bakery products and roasted coffee (Table 10.2).

Estimating consumer exposure to acrylamide is a high priority for governments and industry. After the initial finding of considerable acrylamide levels in various foods, several independent studies have estimated overall exposure for the average consumer (Table 10.3).

The usual way to assess exposure to acrylamide from food is to use 'simplified exposure equation' which is widely used by the FDA to estimate the exposure of a wide range of

Table 10.2 Acrylamide levels in food as collected by the EU Joint Research Center (updated June 2006).

Food commodity	n	Min	Median	Max
Potato chips	1377	5	186	4653
Potato crisps	831	5	528	4215
Potato fritter	121	10	352	3072
Fine bakery ware	1056	4	145	3324
Gingerbread	1003	5	303	7834
Crispbread	411	5	244	2838
Infant food	157	5	79	910
Diabetics cakes + biscuits	402	5	230	3044
Breakfast cereals	333	5	60	1649
Coffee roasted	221	79	286	975
Coffee substitutes	108	116	773	2955

Data were taken from the monitoring database on acrylamide levels in food (http://www.irmm.jrc.be/) maintained by the IRMM (Institute for Reference Materials and Measurements, together with Health and Consumer Protection DG (DG SANCO).

substances in food.

$$\mathrm{EDI}_x = \sum_{f=1}^{F} \frac{\mathrm{Freq}_f \mathrm{Port}_f \mathrm{Con}_{xf}}{N}$$

where EDI_x is the estimated daily intake of substance x; F is the total number of foods in which x can be found; Freq_f is the number of eating occasions for food f over N survey days; Port_f is the average portion of size for food f; Conc_{xf} is the concentration of the substance x in the food f; and N is the number of survey days.

Freq_f and Port_f are estimated from the relevant survey while Conc_{xf} is determined experimentally based on exploratory data on acrylamide in foods. These concentration values can be updated as more data become available.

The foods that contribute most to acrylamide exposure vary depending upon the population's eating habits and the way the foods are processed and prepared. French fries and other potato products are consumed in relatively high amounts in the United States (35% of the average daily acrylamide intake), whereas coffee and bread contribute relatively little to the average daily US acrylamide intake (7% for coffee, 11% for toast and soft bread). The contribution of potato products is even higher in the Netherlands, with French fries and chips taken together contributing up to 50% (Konings et al., 2003). On the other hand, the contribution of coffee and bread or crisp bread is much higher in European countries. According to Dybing and Sanner (2003), potato products contributed ~30% to the daily acrylamide intake in Norway, whereas coffee and bread contributed 28 and 20%, respectively. Similar results for Sweden were reported by Svensson et al. (2003) (coffee 39%, potato products 26%, bread and crisp bread 17%). Generally, the most important categories of food appear to be: fried potato products such as French fries and chips, ready-to-eat breakfast cereals, baked goods such as cookies, pies and cakes, brewed coffee and breads (Dybing et al., 2005).

10.5 BIOLOGICAL EFFECTS

Acrylamide is metabolized in the body to glycidamide, a reactive compound formed through epoxidation of the double bond (Lopachin and Lehning, 1994). The major route of metabolism

Table 10.3 Exposure estimates from 2002 to 2004[a].

Exposure assessment	Daily intake μg/kg bw/day		Source
	Mean (age group)	95th percentile; [a]90th percentile	
FAO/WHO (2000)	0.3–0.8		http://www.who.int/foodsafety/publications/chem/en/acrylamide_full.pdf
EU, SCF (2002)	0.2–0.4		http://europa.eu.int/comm/food/fs/sc/scf/out131_en.pdf
BfR, Germany (2002)	1.1 (15–18)	3.4	http://www.bfr.bund.de/cm/208/abschaetzung_der_acrylamid_aufnahme_durch_hochbelastete_nahrungsmittel_in_deutschland_studie.pdf
BAG, Switzerland (2002)	0.28 (16–57)		http://www.bag.admin.ch/verbrau/aktuell/d/DDS%20acrylamide%20preliminary%20communication.pdf
AFSSA, France (2002)	0.5 (>15) 1.4 (2–14)	1.1 2.9	http://www.afssa.fr/ftp/afssa/basedoc/acrylpoint2sansannex.pdf
FDA (2002)	0.7		http://www.jifsan.umd.edu/presentations/acry2004/acry_2004_dinovihoward_files/frame.htm
FDA (2003)	0.37 (>2) 1.00 (2–5)	0.81[a] 2.15[a]	http://www.jifsan.umd.edu/presentations/acry2004/acry_2004_dinovihoward_files/frame.htm
FDA (2004)	0.43 (>2) 1.06 (2–5)	0.92[a] 2.31[a]	http://www.jifsan.umd.edu/presentations/acry2004/acry_2004_dinovihoward_files/frame.htm
SNFA, Sweden (2002)	0.45 (18–74)	1.03	Svensson et al. (2003)
NFCS, Netherlands	0.48 (1–97) 1.04 (1–6) 0.71 (7–18)	0.6 1.1 0.9	Konings et al. (2003)
SNT, Norway (2003)	0.49 (males) 0.46 (females) 0.36 (9, boys) 0.32 (9, girls) 0.52 (13, boys) 0.49 (13, girls)	1.01[a] 0.86[a] 0.72[a] 0.61[a] 1.35[a] 1.2[a]	Dybing and Sanner (2003)

[a] Reprinted from Dybing, E., Parmer, P.B., Andersen, M., Fennel, T.R., Lalljie, S.P.D., Müller, D.J.G., Olin, S., Petersen, B.J., Schlatter, J., Scholz, G., Scimeca, J.A., Slimani, N., Törnqvist, M., Tuijelaars, S. and Verger, P. Human exposure and internal dose assessments of acrylamide in food. *Food and Chemical Toxicology*, **43**, 365–410. Copyright 2005 with permission from Elsevier.

is the conjugation of acrylamide and glycidamide to glutathione by Michael-type reactions leading to water-soluble thioethers. Acrylamide is also considered a reproductive toxin, with mutagenic and carcinogenic properties in experimental mammalian in vitro and in vivo systems (Dearfield et al., 1995). Specific adducts formed with haemoglobin are used as biomarkers of exposure, and a rat feeding study has recently demonstrated in fried animal foods a link between acrylamide levels and such haemoglobin adducts (Tareke et al., 2000).

A joint FAO/WHO consultation on the health implications of acrylamide in food has estimated an average daily food intake of acrylamide of the general population in the range of 0.3–0.8 µg/kg bw/day. For the non-cancer endpoint, a no-observed-adverse-effect level (NOAEL) of 0.2 mg/kg/day has been derived for nerve damage (sub-chronic rat study). Considering an average human daily exposure of up to 0.8 µg/kg bw/day, a margin of at least 250 can be calculated. Therefore, no neurotoxic effects are to be expected from the levels of acrylamide found in food (Dybing and Sanner, 2003.). Since cancer induced by genotoxic compounds is generally considered a non-threshold effect, no acceptable daily intake has been allocated for acrylamide.

The weight of evidence for the carcinogenicity of acrylamide in humans is based primarily on animal data. With a few exceptions, epidemiology studies have not found a statistically significant association between acrylamide exposure and risk of cancer. Three separate prospective cohort studies examining two different cohorts have attempted to address the relationship between acrylamide exposure and various types of cancer. In 1986, Sobel et al. published a study that examined the relationship between acrylamide exposure and cancers of the central nervous system, thyroid gland, other endocrine glands and mesothelium in a cohort of 371 employees who were involved with acrylamide monomer and polymerization operations. The authors concluded that acrylamide exposure is not related to overall mortality, total malignant neoplasms, or any specific cancers. In 1989, Collins et al. published their observations of the relationship between all-cause and cause-specific mortality rates and acrylamide exposure in a cohort of 8854 men, 2293 of which were exposed to acrylamide. The cohort consisted of four factory populations in the United States and the Netherlands. No relationship between acrylamide exposure and the risk of mortality from several cancer sites was observed. Mucci et al. (2003) examined a population-based Swedish case–control study that included cases with large bowel, bladder, or kidney cancer matched to healthy controls. When acrylamide exposure from food has been determined by linking extensive food frequency data with acrylamide levels in certain food items recorded by the Swedish National Food Administration, they found no association between acrylamide exposure from food and these types of cancers. Pelucchi et al. (2003) found no association between fried/baked potato consumption and cancer risk in a series of hospital-based case–control studies conducted in Italy and Switzerland. The cancer sites examined included oral cavity/pharynx, oesophagus, larynx, large bowel, breast and ovary. However, a study by Bosetti et al. (2002) did find a significant and positive association between laryngeal cancer and consumption of several types of fried foods.

Even though epidemiology studies do not provide strong, conclusive evidence for an association between acrylamide exposure and cancer risk, this does not necessarily indicate that no relationship exists. Existing epidemiological studies do not have the statistical power to detect cancer risk from acrylamide exposure at the levels suggested by toxicology studies. Epidemiology studies are also subject to many confounding factors. Some of the potential confounding factors for cohort studies examining the relationship between acrylamide exposure in the workplace and cancer risk include inexact exposure estimates due to changes in the workplace environment over time, failure to control for smoking or other sources of acrylamide, and reporting errors in the cause of death. Potential confounding effects also exist for cohort and case–control studies examining the relationship between acrylamide exposure from food and cancer risk. If the range of acrylamide exposure from food in the study population is very narrow, then statistical significance cannot be determined. Also, hospital-based studies may be subject to selection bias.

Future epidemiology studies investigating the relationship between acrylamide exposure and cancer risk should take precautions to minimize selection bias by focusing primarily on cohort, nested case–control, and population-based case–control studies. Controlling for known confounding factors, such as smoking and other dietary components (e.g. heterocyclic amines, polycyclic aromatic hydrocarbons) is essential. Efforts to understand how biomarker data might be utilized to validate dietary assessment tools or to improve exposure assessment should also be examined.

10.6   METHODS TO REDUCE FORMATION IN FOODS

Minimization of acrylamide formed during thermal processing of foods is of great importance from the viewpoint of food safety. Since the thermal processing of foods is rather complex due to the wealth of components in the reaction pool, the results of a single treatment would be diverse. So, any approach put forward to prevent any adverse effect like acrylamide formation during thermal processing should also consider other possible changes likely to be made to the resulting food product.

To date, several approaches have been found to lower the levels of acrylamide formed in foods. Studies have generally applied to fried and baked potatoes and baked cereal products.

10.6.1   Potato products

Numerous possible avenues for the reduction of acrylamide in potato products have been highlighted in several reports. These entail controlling the temperature of storage of the raw potato (Grob *et al.*, 2003; Noti *et al.*, 2003), selection of certain varieties (Biedermann *et al.*, 2002c; Haase *et al.*, 2003), modifying frying conditions (Biedermann *et al.*, 2002a,b,c; Grob *et al.*, 2003; Haase *et al.*, 2003; Jung *et al.*, 2003), and assessing the impact of protein coating (Vattem and Shetty, 2003).

It is feasible to select potato varieties with low levels of reducing sugars. These varieties can be used for cut potato products intended for frying and baking to lower the levels of acrylamide formed. Storage of potatoes below 8°C is known to cause increased levels of reducing sugars. Storage at 8°C or above will therefore reduce the potential for the formation of acrylamide in the potato product upon baking or frying. Avoiding cold conditions can be readily achieved for short-term storage. There is a practical problem for long-term agricultural storage. To maintain supplies of potatoes throughout the year it is necessary for producers to store potatoes for periods of several months. This can only be achieved successfully if the storage conditions prevent the potatoes from sprouting. Cool temperatures are often used to prevent sprouting. Chemical treatments of potatoes with sprout suppressing agents can be used, although the use of such chemicals is not always desired by the consumer or may not be permitted. Alternatively, low-dose irradiation of potato tubers can also be used as an effective post-harvest treatment for sprout suppression during long-term storage at elevated temperatures (Gökmen *et al.*, 2007).

Levels of reducing sugars can be lowered by pre-blanching the cut potatoes in warm or hot water or soaking them in water at room temperature before they are fried or baked. Soaking cut potatoes in water also reduces the level of asparagine. It has been shown that acrylamide formed on frying potatoes can be reduced up to 25% by soaking into water or blanching. If the soaking or blanching solution contains organic acids (citric, acetic), the level of acrylamide can be reduced up to 90% (Jung *et al.*, 2003; Kita *et al.*, 2004). However, this approach can cause souring of flavour if a precise procedure is not followed and also the frying oil can become rancid. It has also been shown that the addition of citric acid limits the generation of volatiles during thermal processing (Low *et al.*, 2006).

Asparagine is an important amino acid component of potatoes. Since it is necessary to form acrylamide in the presence of reducing sugars, one possible approach may be lowering asparagine concentration prior to frying or baking. The use of asparaginase is a possible approach to interrupt the interaction of asparagine with reducing sugars. Another possibility is to use other amino acids to compete with asparagine during the Maillard reaction. It has been shown that the addition of amino acids such as glycine or glutamine during soaking or blanching can reduce acrylamide by up to 30% (Bråthen *et al.*, 2005). However, this treatment may also affect the sensory properties of thermally processed food by changing the volatile profile (Low *et al.*, 2006).

Previous research suggests that acrylamide formation can easily be limited by reducing the frying time. In deep fat frying, conduction within the food material is the rate-controlling heat transfer mechanism, which implies that the frying time can be reduced if the potato strips are cooked before going into the fryer. This would eliminate the need to rely on conduction of heat during frying to render the interior cooked. In this case frying is performed just to bring about the desirable surface characteristics (crust and colour) of the already-cooked strips. Microwave pre-cooking differs from blanching in that the potato strip is rendered cooked volumetrically in a very short time without losing the reducing sugars and asparagine, which are also responsible for the product's desirable surface characteristics. It has been successfully shown that microwave pre-cooking is effective in reducing acrylamide levels in the surface region of French fries, where most of the acrylamide formation takes place. The reduction is a consequence of the combined effect of reduced frying time and surface temperature (Demirkol *et al.*, 2006). Microwave pre-treatment would also fit favourably into the continuous process line for industrial production of French fries (Erdoğdu *et al.*, 2007).

In order to reduce acrylamide formed upon heating in foods, a promising approach seems the use of cations (Na^+, Ca^{2+}) to mitigate acrylamide in thermally processed foods (Lindsay and Jang, 2005; Kolek *et al.*, 2006). The results of model studies have confirmed the inhibitory effects of many cations on acrylamide formation. In addition, the soaking of potato strips in sodium chloride or calcium chloride has been shown to reduce the level of acrylamide formed upon frying up to 50 or 95%, respectively (Gökmen and Şenyuva, 2007a). The addition of cations may also be considered beneficial in certain foods prior to thermal processing not only to mitigate acrylamide, but also to enrich calcium in the finished product. However, any potential side effects of this treatment need to be understood in depth before implementing it as a mitigation strategy. Recent findings have shown that the addition of cations results in a change in the reaction path leading to excessive formation of hydroxymethyl furfural in the model system (Gökmen and Şenyuva, 2007b).

10.6.2 Bakery products

The two major approaches for reduction of acrylamide levels in cookies are replacing NH_4HCO_3 by $NaHCO_3$, and the use of a sucrose solution instead of inverted sugar syrup (Graf *et al.*, 2006). In crackers, a combination of the first two approaches has been even more effective (Vass *et al.*, 2004) which could also apply to other products. The measure focusing on the replacement of reducing sugars by sucrose is limited to products where browning is not of primary importance. The addition of glycine during dough making has been shown to reduce acrylamide in flat breads and bread crusts up to 90% (Bråthen *et al.*, 2005).

A moderate addition of organic acids may also be considered for mitigation of acrylamide in cookies. However, care should be taken particularly when cookies are formulated with sucrose instead of inverted syrups. Lowering pH may result in excessive hydrolysis of sucrose, which will make the composition more favourable to form acrylamide during baking.

For many bakery products perhaps the most straightforward way to lower the levels of acrylamide formed is to reduce the baking time to avoid excess browning. Reducing the baking temperatures can also help to achieve reduction. For example, lighter baking of sweet biscuits has resulted in significantly lowering levels of acrylamide in some products. However, for some bakery products excess baking can have the same effect and also lead to lower levels of acrylamide.

References

Ahn, J.S., Castle, L., Clarke, D.B., Lloyd, A.S., Philo, M.R. and Speck, D.R. (2002) Verification of the findings of acrylamide in heated foods. *Food Additives and Contaminants*, **19**, 1116–1124.

Amrein, T.M., Bachmann, S., Noti, A., Biedermann, M., Barbosa, M.F., Biedermann-Brem, S., Grob, K., Keiser, A., Realini, P., Escher, F. and Amadó, R. (2003) Potential of acrylamide formation, sugars, and free asparagine in potatoes: A comparison of cultivars and farming systems. *Journal of Agriculture and Food Chemistry*, **51**, 5556–5560.

Amrein, T.M., Schönbächler, B., Escher, F. and Amadó, R. (2004) Acrylamide in gingerbread: Critical factors for formation and possible ways for reduction. *Journal of Agricultural and Food Chemistry*, **52**, 4282–4288.

Andrzejewski, D., Roach, J.A., Gay, M.L. and Musser, S.M. (2004) Analysis of coffee for the presence of acrylamide by LC–MS/MS. *Journal of Agricultural and Food Chemistry*, **52**, 1996–2002.

Ballard, D.A. and Brown, C.M. (1982) *Computer Vision*. Prentice-Hall, Eaglewoood Cliffs, NeJ.

Becalski, A., Lau, B.P.-Y., Lewis, D. and Seaman, S. (2002) *Acrylamide in Foods; Occurrence and Sources*. AOAC Annual Meeting, Los Angeles, CA, 22–26 September 2002.

Becalski, A., Lau, B.P.-Y., Lewis, D. and Seaman, S. (2003) Acrylamide in foods: Occurrence, sources and modeling. *Journal of Agricultural and Food Chemistry*, **51**, 802–808.

Biedermann, M., Biedermann-Brem, S., Noti, A. and Grob, K. (2002a) Methods for determining the potential of acrylamide formation and its elimination in raw materials for food preparation, such as potatoes. *Mitteilungen aus Lebensmitteluntersuchung und Hygiene*, **93**, 653–667.

Biedermann, M., Noti, A., Biedermann-Brem, S., Mozzetti, V. and Grob, K. (2002b) Experiments on acrylamide formation and possibilities to decrease the potential of acrylamide formation in potatoes. *Mitteilungen aus Lebensmitteluntersuchung und Hygiene*, **93**, 668–687.

Biedermann, M., Biedermann-Brem, S., Noti, A., Grob, K., Egli, P. and Mändli, H. (2002c) Two GC-MS methods for the analysis of acrylamide in foods. *Mitteilungen aus Lebensmitteluntersuchung und Hygiene*, **93**, 638–652.

Biedermann, M. and Grob, K. (2003) Model studies on acrylamide formation in potato, wheat flour and corn starch; ways to reduce acrylamide contents in bakery ware. *Mitteilungen aus Lebensmitteluntersuchung und Hygiene*, **94**, 406–422.

Bishop, M.C. (1994) Neural network and their applications. *Review in Scientific Instruments*, **64**, 1803–1831.

Bologna, L.S., Andrawes, F.F., Barvenik, F.W., Lentz, R.D. and Sojka, R.E.J. (1999) Analysis of residual acrylamide in field crops. *Chromatographic Science*, **37**, 240–244.

Bosetti, C., Talamini, R., Levi, F., Negri, E., Franceschi, S., Airoldi, L. and La Vecchia, C. (2002) Fried foods: A risk factor for laryngeal cancer? *British Journal of Cancer*, **87**, 1230–1233.

Bråthen, E., Kita, A., Knutsen, S.H. and Wicklund, T. (2005) Addition of glycine reduces the content of acrylamide in cereal and potato products. *Journal of Agricultural and Food Chemistry*, **53**, 3259–3264.

Bråthen, E. and Knutsen, S.H. (2005) Effect of temperature and time on the formation of acrylamide in starch-based and cereal model systems, flat breads and bread. *Food Chemistry*, **92**(4), 693–700.

Brosnan, T. and Sun, D.-W. (2004) Improving quality inspection of food products by computer vision – a review. *Journal of Food Engineering*, **61**, 3–16.

Castle, L. (1993) Determination of acrylamide monomer in mushrooms grown on polyacrylamide gel. *Journal of Agriculture and Food Chemistry*, **41**, 1261–1263.

Cavalli, S., Maurer, R. and Hofler, F. (2003) Fast determination of acrylamide in food samples using accelerated solvent extraction followed by ion chromatography with UV or MS detection. *LC GC Europe*, (suppl S), 9–11.

Cetin, A.E. and Weerackody, V. (1988) Design of vector quantizers using simulatedannealing. *IEEE Transactions on Circuits Systems*, **35**, 1550.

Claeys, W.L., Vleeschouwer, K.D. and Hendrickx, M.E. (2005) Kinetics of Acrylamide formation and elimination during heating of an asparagine–sugar model system. *Journal of Agricultural and Food Chemistry*, **53**, 9999–10005.

Collins, J., Swaen, G., Marsh, G., Utidjian, H., Caporossi, J. and Lucas, L. (1989) Mortality patterns among workers exposed to acrylamide. *Journal of Occupational Medicine*, **31**, 614–617.

Comaniciu, D. and Meer, P. (2002) Mean shift: A robus approach toward feature space analysis. *IEEE Transactions on Pattern Analysis and Machine Intelligence*, **24**, 603–619.

Costa, L.G., Deng, H., Greggotti, C., Manzo, L., Faustman, E.M., Bergmark, E. and Calleman, C.J. (1992) Comparative studies on the neuro and reproductive toxicity of acrylamide and its epoxide metabolite glycidamide in the rat. *Neurotoxicology*, **13**, 219–224.

Croft, M., Tong, P., Fuentes, D. and Hambridge, T. (2004) Australian survey of acrylamide in carbohydrate-based foods. *Food Additives and Contaminants*, **21**, 721–736.

Croll, B.T. and Simkins, G.M. (1972) Determination of acrylamide in water by using electron-capture gas chromatography. *Analyst*, **97**(1153), 281.

Cutie, S.S. and Kallos, G.J. (1986) Determination of acrylamide in sugar by thermospray liquid chromatography mass-spectrometry. *Analytical Chemistry*, **58**(12), 2425–2428.

Dearfield, K.L., Abernathy, C.O., Ottley, M.S., Brantner, J.H. and Hayes, P.F. (1988) Acrylamide: Its metabolism, developmental and reproductive effects, genotoxicity, and carcinogenicity. *Mutation Research*, **195**, 45–77.

Dearfield, K.L., Douglas, G.R., Ehling, U.H., Moore, M.M., Sega, G.A. and Brusick, D.J. (1995) Acrylamide: A review of its genotoxicity and an assessment of heritable genetic risk. *Mutation Research*, **330**, 71–99.

Demirkol, E., Erdogdu, F. and Palazoglu, T.K. (2006) Experimental determination of mass transfer coefficient: Moisture content and humidity ratio driving force approaches during baking. *Journal of Food Process Engineering*, **29**(2), 188–201.

Dybing, E. and Sanner, T. (2003) Risk assessment of acrylamide in foods. *Toxicological Sciences*, **75**, 7–15.

Dybing, E., Parmer, P.B., Andersen, M., Fennel, T.R., Lalljie, S.P.D., Müller, D.J.G., Olin, S., Petersen, B.J., Schlatter, J., Scholz, G., Scimeca, J.A., Slimani, N., Törnqvist, M., Tuijtelaars, S. and Verger, P. (2005) Human exposure and internal dose assessments of acrylamide in food. *Food and Chemical Toxicology*, **43**, 365–410.

Ehling, S. and Shibamoto, T. (2005) Correlation of acrylamide generation in thermally processed model systems of asparagine and glucose with color formation, amounts of pyrazines formed, and antioxidative properties of extracts. *Journal of Agricultural and Food Chemistry*, **53**, 4813–4819.

Elmore, J.S., Koutsidis, G., Dodson, A.T., Mottram, D.S. and Wedzicha, B.L. (2005) Measurement of acrylamide and its precursors in potato, wheat, and rye model systems. *Journal of Agricultural and Food Chemistry*, **53**, 1286–1293.

EPA (1996) SW 846, Method 8032A, U.S. Environmental Protection Agency. Washington, DC.

Erdoğdu, S.B., Palazoğlu, T.K., Gökmen, V., Şenyuva, H.Z. and Ekiz, İ. (2007) Reduction of acrylamide formation in French fries by microwave pre-cooking of potato strips. *Journal of the Science of Food and Agriculture*, **87**(1), 133–137.

Friedman, M. (2003) Chemistry, biochemistry, and safety of acrylamide. *A Review Journal of Agricultural and Food Chemistry*, **51**, 4504–4526.

Gertz, C. and Klostermann, S. (2002) Analysis of acrylamide and mechanisms of its formation in deep-fried products. *European Journal of Lipid Science Technology*, **104**, 762–771.

Gökmen, V., Şenyuva, H.Z., Acar, J. and Sarıoğlu, K. (2005) Determination of acrylamide in potato chips and crisps by high-performance liquid chromatography. *Journal of Chromatography A*, **1088**, 193–199.

Gökmen, V., Palazoğlu, T.K. and Şenyuva, H.Z. (2006a) Relation between the acrylamide formation and time–temperature history of surface and core regions of French fries. *Journal of Food Engineering*, **77**, 972–976.

Gökmen, V., Şenyuva, H.Z., Dülek, B. and Çetin, A.E. (2006b) Computer vision-based image analysis for the estimation of acrylamide concentrations of potato chips and french fries. *Food Chemistry*, **101**, 791–798.

Gökmen, V., Şenyuva, H.Z., Dülek, B. and Çetin, A.E. (2006c) Computer vision based analysis of potato chips: A tool for rapid detection of acrylamide level. *Molecular Nutrition and Food Research*, **50**, 805–810.

Gökmen, V., Akbudak, B., Serpen, A., Acar, J., Turan, Z.M. and Eriş, A. (2007) Effects of controlled atmosphere storage and low-dose irradiation on potato tuber components affecting acrylamide and color formations upon frying. *European Food Research and Technology*, **224**(6), 681–687.

Gökmen, V. and Şenyuva, H.Z. (2006a) A simplified approach for the kinetic characterization of acrylamide formation in fructose–asparagine model system. *Food Additives and Contaminants*, **23**, 348–354.

Gökmen, V. and Şenyuva, H.Z. (2006b) A generic method for the determination of acrylamide in thermally processed foods. *Journal of Chromatography A*, **1120**, 194–198.

Gökmen, V. and Şenyuva, H.Z. (2007a) Acrylamide formation is prevented by divalent cations during the Maillard reaction. *Food Chemistry*, **103**(1), 196–203.

Gökmen, V. and Şenyuva, H.Z. (2007b) Effects of some cations on the formation of acrylamide and furfurals in glucose–asparagine model system. *European Food Research and Technology*, **225**(5–6), 815–820.

Gonçalves, E.C., Minim, L.A., Coimbra, J.S.R. and Minim, V.P.R. (2005) Modeling sterilization process of canned foods using artificial neural networks. *Chemical Engineering Processing*, **44**, 1269–1276.

Gonzales, R.C. and Woods, R.E. (2002) *Digital Image Processing*. Prentice-Hall, Eaglewoood Cliffs, NJ.

Graf, M., Amrein, T.M., Graf, S., Szalay, R., Escher, F. and Amadò, R. (2006) Reducing theacrylamide content of a semi-finished biscuit on industrial scale. *LWT*, **39**, 724–728.

Granvogl, M. and Scieberle, P. (2006) Thermally generated 3-aminopropionamide as a transient intermediate in the formation of acrylamide. *Journal of Agriculture and Food Chemistry*, **54**, 5933–5938.

Grob, K., Biedermann, M., Biedermann-Brem, S., Noti, A., Imhof, D., Amrein, T., Pfefferle, A. and Bazzocco, D. (2003) French fries with less than 100 µg/kg acrylamide. A collaboration between cooks and analyst. *European Food Research and Technology*, **271**, 185–194.

Gutsche, B., Weisshaar, R. and Buhlert, J. (2002) Acrylamid in Lebensmitteln—Ergebnisse aus der amtlichen Lebensmittelüberwachung Baden-Württembergs. *Deutsche Lebensmittel Rundschau*, **98**, 437–443.

Haase, N.U., Matthaeus, B. and Vosmann, K. (2003) Acrylamide formation in foodstuffs-minimizing strategies for potato crisps. *Deutsche Lebensmittel Rundschau*, **99**, 87–90.

Höfler, F., Maurer, R. and Cavalli, S. (2002) Schnelle analyse von Acrylamid in Lebensmitteln mit ASE und LC/MS. *GIT Labor-Fachzeitschrift*, **48**, 986–970.

International Agency for Research on Cancer (IARC) (1993) Monographs. Acrylamide, in *International Agency for Research on Cancer*.

International Agency for Research on Cancer (IARC) (1994) Acrylamide chemicals, in *IARC Monographs on the Evaluation of the Carcinogenic Risk of Chemicals to Humans*, Vol. 60, pp. 389–433. IARC, Lyon, France.

Johnson, K.A., Gorzinski, S.J., Bodner, K.M., Campbell, R.A., Wolf, C.H., Friedman, M.A., *et al.* (1986) Chronic toxicity and oncogenicity studyon acrylamide incorporated in the drinking water of Fischer 344 rats. *Toxicology and Applied Pharmacology*, **85**, 154–168.

Jung, M.Y., Choi, D.S. and Ju, J.W. (2003) A novel technique for limitation of acrylamide formation in fried and baked corn chips and in French fries. *Journal of Food Science*, **68**, 1287–1290.

Kita, A., Bråthen, E., Knutsen, S.H. and Wicklund, T. (2004) Effective ways of decreasing acrylamide content in potato crisps during processing. *Journal of Agricultural and Food Chemistry*, **52**, 7011–7016.

Knol, J.K., van Loon, W.A.M., Linssen, J.P.H. Ruck, A.-L., van Boekel, M.A.J.S. and Voragen, A.G.J. (2005) Toward a kinetic model for acrylamide formation in a glucose–asparagine reaction system. *Journal of Agricultural and Food Chemistry*, **53**, 6133–6139.

Kolek, E., Simko, P. and Simon, P. (2006) Inhibition of acrylamide formation in asparagine/D-glucose model system by NaCl addition. *European Food Research Technology*, **224**(2), 283–284.

Konings, E.J.M., Baars, A.J., van Klaveren, J.D., Spanjer, M.C., Rensen, P.M., Hiemstra, M., van Kooij, J.A. and Peters, P.W.J. (2003) Acrylamide exposure from foods of the Dutch population and an assessment of the consequent risk. *Food and Chemical Toxicology*, **41**, 1569–1579.

Lindsay, R.C. and Jang, S. (2005) Model systems for evaluating factors affecting acrylamide formation in deep fried foods. *Advances in Experimental Medicine and Biology*, **561**, 329–341.

Lingnert, H., Grivas, S., Jägerstad, M., Skog, K., Törnqvist, M. and Åman, P. (2002) Acrylamide in food – mechanisms of formation and influencing factors during heating of foods. *Scandinavian Journal of Nutrition*, **46**, 159–172.

Lopachin, R.M. and Lehning, E.J. (1994) Acrylamide induced distal axon degeneration. A proposed mechanism of action. *Neurotoxicology*, **15**, 247–260.

Low, M.Y., Koutsidis, G., Parker, J.K., Elmore, J.S., Dodson, A.T. and Mottram, D.S. (2006) Effect of citric acid and glycine addition on acrylamide and flavor in a potato model system. *Journal of Agricultural and Food Chemistry*, **54**, 5976–5983.

Matthäus, B., Haase, N.U. and Vosmann, K. (2004) Factors affecting the concentration of acrylamide during deep-fat frying of potatoes. *European Journal of Lipid Science and Technology*, **106**, 793–801.

McCollister, D.D., Rowe, V.K. and Sadek, S.E. (1964) Toxicologic studies of acrylamide – acrylic acid resins. *Toxicology and Applied Pharmacology*, **6**(3), 353.

Mottram, D.S., Wedzicha, B.I. and Dodson, A.T. (2002) Acrylamide is formed in the Maillard reaction. *Nature*, **419**, 448.

Mucci, L., Dickman, P., Steineck, G., Adami, H. and Augustsson, K. (2003) Dietary acrylamide and cancer of the large bowel, kidney, and bladder: Absence of an association in a population-based study in Sweden. *British Journal of Cancer*, **88**, 84–89.

Mucci, L., Lindblad, P., Steineck, G. and Adami, H. (2004) Dietary acrylamide and risk of renal cell cancer. *International Journal of Cancer*, **109**, 774–776.

Murkovic, M. (2004) Acrylamide in Austrian foods. *Journal of Biochemical and Bioophysical Methods*, **61**, 161–167.

Nemoto, S., Takatsuki, S., Sasaki, K. and Maitani, T. (2002) Determination of acrylamide in food by GC/MS using 13C-labelled acrylamide as internal standard. *Journal of Food Hygiene Society Japan*, **43**, 371–376.

Noti, A., Biedermann-Brem, S., Biedermann, M., Grob, K., Albisser, P. and Realini, P. (2003) Storage of potatoes at low temperature should be avoided to prevent increased acrylamide during frying or roasting. *Mitteilungen aus Lebensmitteluntersuchung und Hygiene*, **94**, 167–180.

Ono, H., Chuda, Y., Ohnishi-Kameyama, M., Yada, H., Ishizaka, M., Kobayashi, H. and Yoshida, M. (2003) Analysis of acrylamide by LC-MS/MS and GC-MS in processed Japanese foods. *Food Additives and Contaminants*, **20**, 215–220.

Osborne, S.D., Jordan, R.B. and Kunnemeyer, R. (1997) Method of wavelength selection for partial least squares. *The Analyst*, **122**(12), 1531–1537.

Pedreschi, F., Mery, D., Mendoza, F. and Aguilera, J.M. (2004) Classification of potato chips using pattern recognition. *Journal of Food Science*, **69**(6), E264–E270.

Pedreschi, F., Moyano, P., Kaack, K. and Granby, K. (2005) Color changes and acrylamide formation in fried potato slices. *Food Research International*, **38**, 1–9.

Pelucchi, C., Franceschi, S., Levi, F., Trichopoulos, D., Bosetti, C., Negri, E. and La Vecchia, C. (2003) Fried potatoes and human cancer. *International Journal of Cancer*, **105**, 558–560.

Petersson, E.V., Rosen, J., Turner, C., Danielsson, R. and Hellenas, K.E. (2006) Critical factors and pitfalls affecting the extraction of acrylamide from foods: An optimisation study. *Analytica Chimica Acta*, **557**, 287–295.

Pittet, A., Périsset, A. and Oberson, J.-M. (2003) Trace level detection of acrylamide in cereal-based foods by gas chromatography mass spectrometry. *Journal of Chromatography A*, **1035**, 123–130.

Rabiner, L. and Juang, B.H. (1993) *Fundamentals of Speech Recognition*. Prentice-Hall, Eaglewoood Cliffs, NJ.

Riediker, S. and Stadler, R.H. (2003) Analysis of acrylamide in food using isotope-dilution liquid chromatography coupled with electrospray ionization tandem mass spectrometry. *Journal of Chromatography A*, **1020**, 121–130.

Roach, J.A.G., Andrzejewski, D., Gay, M.L., Nortrup, D. and Musser, S.M. (2003) Rugged LC-MS/MS survey analysis for acrylamide in foods. *Journal of Agriculture and Food Chemistry*, **51**, 7547–7554.

Robert, F., Vuataz, G., Pollien, P., Saucy, F., Alonso, M.-I., Bauwens, I. and Blank, I. (2004) Acrylamide formation from asparagine under low-moisture Maillard reaction conditions. 1. Physical and chemical aspects in crystalline model systems. *Journal of Agricultural and Food Chemistry*, **52**, 6837–6842.

Rosén, J. and Hellenäs, K.-E. (2002) Analysis of acrylamide in cooked foods by liquid chromatography tandem mass spectrometry. *The Analyst*, **127**, 880–882.

Rydberg, P., Eriksson, S., Tareke, E., Karlsson, P., Ehrenberg, L. and Törnqvist, M. (2003) Investigations of factors that influence the acrylamide content of heated foodstuffs. *Journal of Agricultural and Food Chemistry*, **51**, 7012–7018.

Sanders, R.A., Zyzak, D.V., Stojanovic, M., Tallmadge, D.H., Eberhart, B.L. and Ewald, D.K. (2002) An LC/MS acrylamide method and its use in investigating the role of Asparagine. Presentation at the *Annual AOAC International Meeting*, Los Angeles, CA, 22–26 September 2002.

Schulz, M., Hertz-Picciotto, I., van Wijngaarden, E., Hernandez, J. and Ball, L. (2001) Dose-response relation between acrylamide and pancreatic cancer. *Occupational and Environmental Medicine*, **58**, 609.

Segtnan, V.H., Kita, A., Mielnik, M., Jorgensen, K. and Knutsen, S.H. (2006) Screening of acrylamide contents in potato crisps using process variable settings and near-infrared spectroscopy. *Molecular Nutrition and Food Research*, **50**, 811–817.

Şenyuva, H.Z. and Gökmen, V. (2005a) Survey of acrylamide in Turkish foods by an in-house validated LC-MS method. *Food Additives and Contaminants*, **22**, 204–209.

Şenyuva, H.Z. and Gökmen, V. (2005b) Study of acrylamide in coffee using an improved liquid chromatography mass spectrometry method: Investigation of colour changes and acrylamide formation in coffee during roasting. *Food Additives and Contaminants*, **22**, 214–220.

Şenyuva, H.Z. and Gökmen, V. (2006) Interference-free determination of acrylamide in potato and cereal-based foods by a laboratory validated liquid chromatography–mass spectrometry method. *Food Chemistry*, **97**, 539–545.

Serpen, A. and Gökmen, V. (2006) Modeling acrylamide formation using artificial neural network. *COST – IMARS Joint Workshop*, Naples, Italy, 24–27 May 2006.

Sobel, W., Bond, G., Parsons, T. and Brenner, F. (1986) Acrylamide cohort mortality study. *British Journal of Industrial Medicine*, **43**, 785–788.

Stadler, R.H. (2003) Understanding the formation of acrylamide and other Maillard-derived vinylogous compounds in foods. *European Journal of Lipid Science and Technology*, **105**, 199–200.

Stadler, R.H., Blank, I., Varga, N., Robert, F., Hau, J., Guy, P.A., Robert, M.-C. and Riediker, S. (2002) Acrylamide from Maillard reaction products. *Nature*, **419**, 449.

Stadler, R.H., Robert, F., Riediker, S., Varga, N., Davidek, T., Devaud, S., Goldmann, T., Hau, J. and Blank, I. (2004) In-depth mechanistic study on the formation of acrylamide and other vinylogous compounds by the Maillard reaction. *Journal of Agricultural and Food Chemistry*, **52**, 5550–5558.

Summa, C., Wenzl, T., Brohee, M., De La Calle, B. and Anklam, E. (2006) Investigation of the correlation of the acrylamide content and the antioxidant activity of model cookies. *Journal of Agriculture and Food Chemistry*, **54**, 853–859.

Surdyk, N., Rosén, J., Andersson, R. and Åman, P. (2004) Effects of asparagine, fructose, and baking conditions on acrylamide content in yeast-leavened wheat bread. *Journal of Agricultural and Food Chemistry*, **52**, 2047–2051.

Svensson, K., Abramsson, L., Becker, W., Glynn, A., Hellenäs, K.-E., Lind, Y. and Rosén, J. (2003) Dietary intake of acrylamide in Sweden. *Food and Chemical Toxicology*, **41**, 1581–1586.

Swedish National Food Administration (2002) Information about Acrylamide in Food. 24 April 2002. Available from: www.slv.se

Taeymans, D., Wood, J., Ashby, P., Blank, I., Studer, A., Stadler, R.H., Gonde, P., Van Eijck, P., Lalljie, S., Lingnert, H., Lindblom, M., Matissek, R., Müller, D., Tallmadge, D., O'Brien, J., Thompson, S., Silvani, D. and Whitmore, T. (2004) A review of acrylamide: An industry perspective on research, analysis, formation and control. *Critical Reviews in Food Science and Nutrition*, **44**, 323–347.

Takatsuki, S., Nemoto, S., Sasaki, K. and Maitani, T. (2003) Determination of acrylamide in processed foods by LC/MS using column switching. *Journal of Food Hygienic Society of Japan*, **44**, 89–95.

Tareke, E., Rydberg, P., Karlsson, P., Eriksson, S. and Törnqvist, M. (2000) Acrylamide: A cooking carcinogen? *Chemical Research in Toxicology*, **13**, 517–522.

Tareke, E., Rydberg, P., Karlsson, P., Eriksson, S. and Törnqvist, M. (2002) Analysis of acrylamide, a carcinogen formed in heated foodstuffs. *Journal of Agricultural and Food Chemistry*, **50**, 4998–5006.

Tateo, F. and Bononi, M. (2003) A GC/MS method for the routine determination of acrylamide in food. *Italian Journal of Food Science*, **15**, 149–151.

Taubert, D., Harlfinger, S., Henkes, L., Berkels, R. and Schömig, E. (2004) Influence of processing parameters on acrylamide formation during frying of potatoes. *Journal of Agricultural and Food Chemistry*, **52**, 2735–2739.

Tekel, J., Farkas, P. and Kovác, M. (1989) Determination of acrylamide in sugar by capillary GLC with alkali flame-ionization detection. *Food Additives and Contaminants*, **6**, 377–381.

Tilson, H.A. (1981) The neurotoxicity of acrylamide: An overview. *Neurobehavioral Toxicology and Teratology*, **3**, 445–461.

Vass, M., Amrein, T.M., Schönbächler, B., Escher, F. and Amadó, R. (2004) Ways to reduce the acrylamide formation in cracker products. *Czech Journal of Food Science*, **22**, 19–21.

Wedzicha, B.L., Mottram, D.S., Elmore, J.S., Koutsidis, G. and Dodson, A.T. (2005) Kinetic models as a route to control acrylamide formation in food. *Advances in Experimental Medicine and Biology*, **561**, 235–253.

Weisshaar, R. (2004) Acrylamid in Backwaren-Ergebnisse von Modellversuchen (in German). *Deutsche Lebensmittel-Rundschau*, **100**, 92–97.

Wenzl, T., Karasek, L., Rosen, J., Hellenaes, K.E., Crews, C., Castle, L. and Anklam, E. (2006) Collaborative trial validation study of two methods, one based on high performance liquid chromatography–tandem mass spectrometry and on gas chromatography–mass spectrometry for the determination of acrylamide in bakery and potato products. *Journal of Chromatography A*, **1132**, 211–218.

Wenzl, T., de la Calle, B. and Anklam, E. (2003) Analytical methods for the determination of acrylamide in food products: A review. *Food Additives and Contaminants*, **20**, 885–902.

Yasuhara, A., Tanaka, Y., Hengel, M. and Shibamoto, T. (2003) Gas chromatographic investigation of acrylanmide formed in browning model systems. *Journal of Agricultural and Food Chemistry*, **51**, 3999–4003.

Yaylayan, V.-A., Wnorowski, A. and Perez-Locas, C. (2003) Why asparagine needs carbohydrates to generate acrylamide. *Journal of Agricultural and Food Chemistry*, **51**, 1753–1757.

Zhou, J., Sun, G., Han, Z., Liao, G., Wu, R., Zhang, Q., et al., (2003) Static light scattering study of the formation of partially hydrolyzed poly(acrylamide)/calcium (II) complexes. *Polymer Preprints*, **44**, 907–908.

Zyzak, D., Sanders, R.A., Stojanovic. M., Tallmadge, D., Eberhart, B.L., Ewald, D.K., Gruber, D.C., Morsch, T.R., Strothers, M.A., Rizzi, G.P. and Villagran, M.D. (2003) Acrylamide formation mechanism in heated foods. *Journal of Agricultural and Food Chemistry*, **51**, 4782–4787.

11 Furan in Processed Foods

Imre Blank

Summary

The US Food and Drug Administration (FDA) published in May 2004 a survey of furan in canned and jarred foods that undergo heat treatment. Since then, there have been a number of contributions that are summarized in this review with focus on formation, occurrence and analysis. Headspace (HS) gas chromatography coupled to mass spectrometry and the use of a deuterated internal standard is currently the analytical method of choice to obtain reliable quantitative data on furan in foods. Furan occurs in a variety of foods such as coffee, canned and jarred products including baby food containing meat, and various vegetables. There are multiple routes of furan formation. Recent data indicate polyunsaturated lipids and ascorbic acid to be the major sources of furan, followed by carotenoids, carbohydrates and certain amino acids. Furan formation is mainly associated with lipid oxidation, but it can also be generated by non-enzymatic browning reactions. Furan can be formed from an intact carbon chain or by condensation reactions of carbonyls obtained from different sources. Furan comprises an intact C_4-unit of ascorbic acid (mainly C-3 to C-6) generated by liberating two C_1-units, i.e. carbon dioxide and formic acid, with 2-deoxyaldoteroses and 2-furoic acid as possible intermediates. Furan mitigation in food is a challenging task due to the great number of precursors occurring in food. A major challenge remains the development of concepts leading to a reduction of furan upon industrial and domestic food processing while maintaining the overall food quality.

11.1 INTRODUCTION

Furan is a five-member ring colourless liquid (C_4H_4O, CAS-No. 110-00-9; Fig. 11.1) which can induce tumours and liver toxicity in experimental animals and is classified as '*possibly carcinogenic to humans*' (group 2B) by the International Agency for Research on Cancer (IARC, 1995). The announcement of the US Food and Drug Administration (FDA) on the occurrence of furan levels in foods amounting to up to 125 µg/kg (FDA, 2004a) initiated a number of studies to evaluate a potential safety concern upon food consumption. In the chemical industry, furan serves as an intermediate in the synthesis and preparation of numerous linear polymers (NTP, 1993).

Furan has been known for long time as food constituent (review by Maga, 1979). It has probably been first reported in coffee (Johnston and Frey, 1938). Furan was also found in cooked chicken (Grey and Shrimpton, 1967), white bread (Mulders *et al.*, 1972), canned beef (Persson and von Sydow, 1973), *Maillard*-type systems containing reducing sugars and

Fig. 11.1 Chemical structure of furan.

amino acids or proteins (Walter and Fagerson, 1968; Yaylayan *et al.*, 1994), glucose caramel (Sugisawa, 1966), lactose/casein (Ferretti *et al.*, 1970), birch syrup (Kallio *et al.*, 1989) and heated proteins such as soy, casein and fish (Qvist and von Sydow, 1974). Usually, the concentrations have not been determined due to lack of reliable quantification methods.

The data published in the last 3 years mainly deal with the analytical methodology, occurrence of furan in food and its manifold formation pathways upon heat treatment. Therefore, the present review paper focuses on recent furan research, in particular on formation, occurrence and analysis. Toxicological effects, exposure assessment and regulations will not be discussed in detail as for furan there is very limited recent information to review on these aspects. A concise summary of toxicological data on furan has recently been published by Crews and Castle (2007).

11.2 FORMATION AND MECHANISMS

There is only limited information on the mechanisms of furan formation under conditions simulating industrial food processing or domestic cooking. Most published data are based on model studies aiming at identifying potential precursors and elucidating the various formation mechanisms, in some cases by using labelled precursors or intermediates. In the following, a brief summary is given on furan formation including concepts to reduce the levels of this food-borne process contaminant.

11.2.1 Model studies

Based on literature data and a systematic study involving ^{13}C-labelled precursors, Perez and Yaylayan (2004) identified key building blocks of furan generation, such as aldotetrose derivatives, acetaldehyde, glycolaldehyde and 4-hydoxy-2-butenal. These key intermediates may originate from different precursors such as sugars, certain amino acids, ascorbic acid and polyunsaturated lipids. The authors found ascorbic acid derivatives to be very efficient in generating furan under pyrolysis conditions (250–350°C, 20 s). Becalski and Seaman (2005) confirmed the role of ascorbic acid derivatives under pressure cooking conditions (118°C, 30 min). In addition, they mentioned polyunsaturated fatty acids (PUFAs), the corresponding triacylglycerides and carotenoids as important precursors of furan. Model experiments suggested oxidative reactions play a vital role in furan formation from lipids. The general scheme, shown in Fig. 11.2, summarizes the various sources of furan and the corresponding major precursor classes.

In their study on the formation of furan under roasting conditions (220°C), Märk *et al.* (2006) found ascorbic acid and polyunsaturated lipids as the most effective precursors generating up to 10 and 5 mmol/mol furan, respectively (Fig. 11.3). In contrast, dehydroascorbic acid showed low efficiency, most likely due to its melting point (~230°C) that was above the roasting temperature. A similar trend was also observed by Perez and Yaylayan (2004). On the other side, Becalski and Seaman (2005) reported about 10-times higher furan amounts

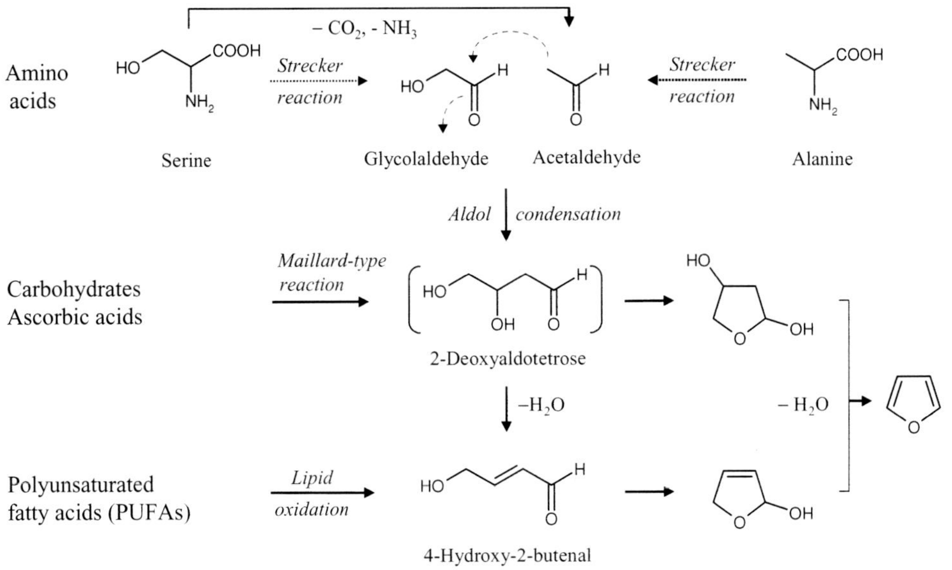

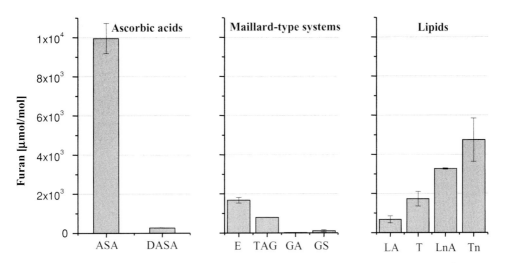

Fig. 11.2   Different precursors leading to parent furan upon thermal treatment. (Adapted from Perez and Yaylayan, 2004.)

formed from dehydroascorbic acid under pressure cooking conditions. These data could not be confirmed by Limacher *et al.* (2007) who found (Fig. 11.4) that dehydroascorbic acid formed only slightly more furan at pH 7 than ascorbic acid, but significantly less at pH 4, thus indicating the effect of the pH on furan formation under aqueous conditions. It seems that the efficiency of the ascorbic acid derivatives to form furan depends very much on the reaction system.

Fig. 11.3   Formation of furan from various precursors heated at 220°C and on-line monitored by PTR-MS. ASA, ascorbic acid; DASA, dehydroascorbic acid; E, erythrose; TAG, threonine/alanine/glucose; GA, glycolaldehyde/alanine; GS, glycolaldehyde/serine; LA, linoleic acid; T, trilinoleate; LnA, linolenic acid; Tn, trilinolenate. (Adapted from Märk et al., 2006.)

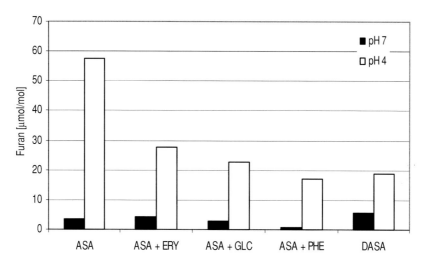

Fig. 11.4   Furan contents (μmol/mol) obtained in aqueous model systems heated at pH 4 and pH 7 (25 min, 121°C). ASA, ascorbic acid; ERY, erythrose; GLC, glucose; PHE, phenylalanine; DASA, dehydroascorbic acid. [Reproduced from Limacher, A., Kerler, J., Conde-Petit, B. and Blank, I. Formation of furan and methylfuran from ascorbic acid in model systems and food. *Food Additives and Contaminants*, **24**(S1), 122–135. Copyright 2007 with permission from Taylor & Francis Ltd, http://www.tandf.co.uk/journals (http://www.informaworld.com).]

Interestingly, the highest furan levels were obtained in pure ascorbic acid systems. The furan amounts drop drastically in the presence of additional compounds, such as sugars, amino acids and lipids. This effect was observed in both dry and aqueous reaction systems. Moreover, even combinations of potential furan precursors lead to reduced furan amounts (Fig. 11.5).

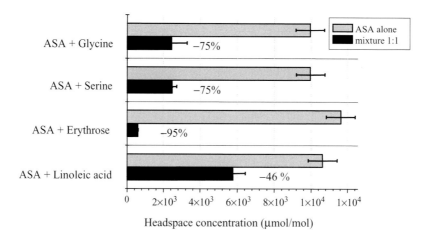

Fig. 11.5   Comparison of furan formation from ascorbic acid (ASA) alone (grey bars) and binary mixtures, i.e. in the presence of equimolar amounts of additional compounds (black bars). (Reprinted with permission from Märk, J., Pollien, Ph., Lindinger, Ch., Blank, I. and Märk, T., Quantitation of furan and methylfuran formed in different precursor systems by proton transfer reaction mass spectrometry, *Journal of Agricultural and Food Chemistry*, **54**, 2786–2793, copyright 2006, American Chemical Society.)

Mixtures of ascorbic acid and erythrose or linoleic acid lead to a reduction of 95 and 45%, respectively. This phenomenon suggests competing reaction pathways that increase in more complex systems. Therefore, results obtained from model systems should be taken with much care and general conclusions must not be drawn.

11.2.2 Formation mechanisms

The first systematic study on furan formation (Perez and Yaylayan, 2004) already involved ^{13}C-labelled precursors, such as differently labelled glucose and serine isotopomers. Limacher *et al.* (2007) used ^{13}C-labelled ascorbic acid to elucidate the formation mechanisms of furan. Basically, furan can be formed (i) from an intact C4 skeleton or (ii) by recombination of fragments. In both cases, there is a multitude of possible sources which are active depending on the reaction conditions.

11.2.2.1 Amino acids

The mechanism of furan formation from amino acids was studied by pyrolysing serine independently labelled at C-1, C-2 and C-3. The label incorporation in the parent furan indicated that two of the four carbon atoms of furan originated from C-2 of serine, and the remaining two carbon atoms originated from C-3. No incorporation of the C-1 atom of serine was detected. These observations are consistent with the proposed aldol condensation mechanism (Fig. 11.2) between an acetaldehyde and a glycolaldehyde moiety, both originating from serine.

These studies suggest that those amino acids capable of forming acetaldehyde and glycolaldehyde can generate furan through aldol condensation followed by cyclization and dehydration (Fig. 11.2). Serine and cysteine are such amino acids. Other amino acids such as aspartic acid, alanine and threonine are able to generate only acetaldehyde and therefore they need the presence of other components that can furnish glycolaldehyde (e.g. reducing sugars, serine, cysteine). Acetaldehyde might be formed from serine by decarboxylation and subsequent loss of ammonia from the intermediary ethanolamine. Glycolaldehyde can be generated through *Strecker*-type reactions in the presence or absence of reducing sugars (Yaylayan, 2003). Threonine has been shown to produce acetaldehyde (Yaylayan and Wnorowski, 2001) and hence can generate furan in the presence of sugars.

11.2.2.2 Carbohydrates

Based on ^{13}C-labelling experiments, Perez and Yaylayan (2004) have compared the relative efficiency of different sugars to generate furan. The order of reactivity under pyrolysis conditions was D-erythrose > D-ribose > D-sucrose > D-glucose = D-fructose. The four carbon sugar D-erythrose generated eightfold excess of furan compared to either fructose or glucose. Labelling studies have indicated that four carbon sugar modules are also involved in the generation of furan from glucose. Labelling studies based on glucose and serine have indicated that the major pathway (50%) of glucose degradation (pathway A in Fig. 11.6) leading to furan incorporated C3–C4–C5–C6 carbon atoms of glucose through formation of an aldotetrose moiety (e.g. erythrose), 10% incorporated C1–C2–C3–C4 carbon atoms of glucose (pathway B) through formation of a 2-deoxy-3-keto-aldotetrose moiety and another 10% incorporated C2–C3–C4–C5 carbon atoms of glucose (pathway C) through formation of a 2-deoxy-aldotetrose moiety. The remaining 30% of furan was formed from serine degradation alone.

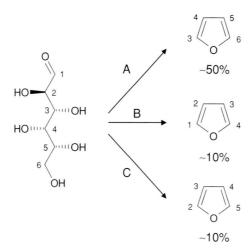

Fig. 11.6 Origin of carbon atoms incorporated into the furan ring from D-glucose in glucose/serine model system where 30% of furan originated from serine. A, aldotetrose pathway; B, 2-doxy-3-keto-aldotetrose pathway; C, 2-deoxy-aldotetrose pathway. (Reprinted with permission from Perez and Yaylayan in *Journal of Agricultural and Food Chemistry*, **42**, 6830–6836. Copyright 2004 with permission from the American Chemical Society.)

11.2.2.3 Polyunsaturated lipids

Model systems (118°C, 30 min) have indicated that only PUFAs such as linoleic and linolenic acids can generate furan upon heating (Becalski and Seaman, 2005). Linolenic acid generated about four times more furan than linoleic acid and catalytic amounts of ferric chloride increased furan formation by several folds. Triglycerides of linoleic and linolenic acids also generated comparable amounts of furan. However, in the presence of ferric chloride the triglycerides generated less furan compared to free fatty acids.

These findings suggest lipid oxidation as general mechanism for the formation of furan. As shown in Fig. 11.2, furan could be formed from 4-hydroxy-2-butenal through cyclization and subsequent dehydration. Similar reactions are known in lipid chemistry. As an example, the origin of the furan derivative 5-pentylfuran used as a chemical marker for rancidity is linked to the formation of 4-hydroxy-2-nonenal (Sayre *et al.*, 1993), a higher homologue of 4-hydroxy-2-butenal. In general, oxidative degradation of PUFAs may lead to lipid hydroperoxides non-enzymatically by reactive oxygen species or enzymatically by lipoxygenases. Subsequent homolytic cleavages of PUFA hydroperoxides, catalysed by transition metal ions may result in the formation of 4-hydroxy-2-alkenals as reactive intermediates.

11.2.2.4 Ascorbic acid

Sodium salts of the acids generated less furan compared to their free acid counterparts, suggesting decarboxylation to take part in ascorbic acid degradation. Due to the ease of hydrolysis and oxidation of ascorbic acid in food (Liao and Seib, 1987) resulting in 2,3-diketogulonic acid, Perez and Yaylayan (2004) proposed the formation of C4 precursors such as aldotetrose and 2-deoxyaldotetrose that could be converted into furan (Fig. 11.7). Pathway B is, however, less probable as both an oxidative and a reductive step is required to yield furan, whereas pathway A directly leads to furan by decarboxylation, cyclization and dehydration.

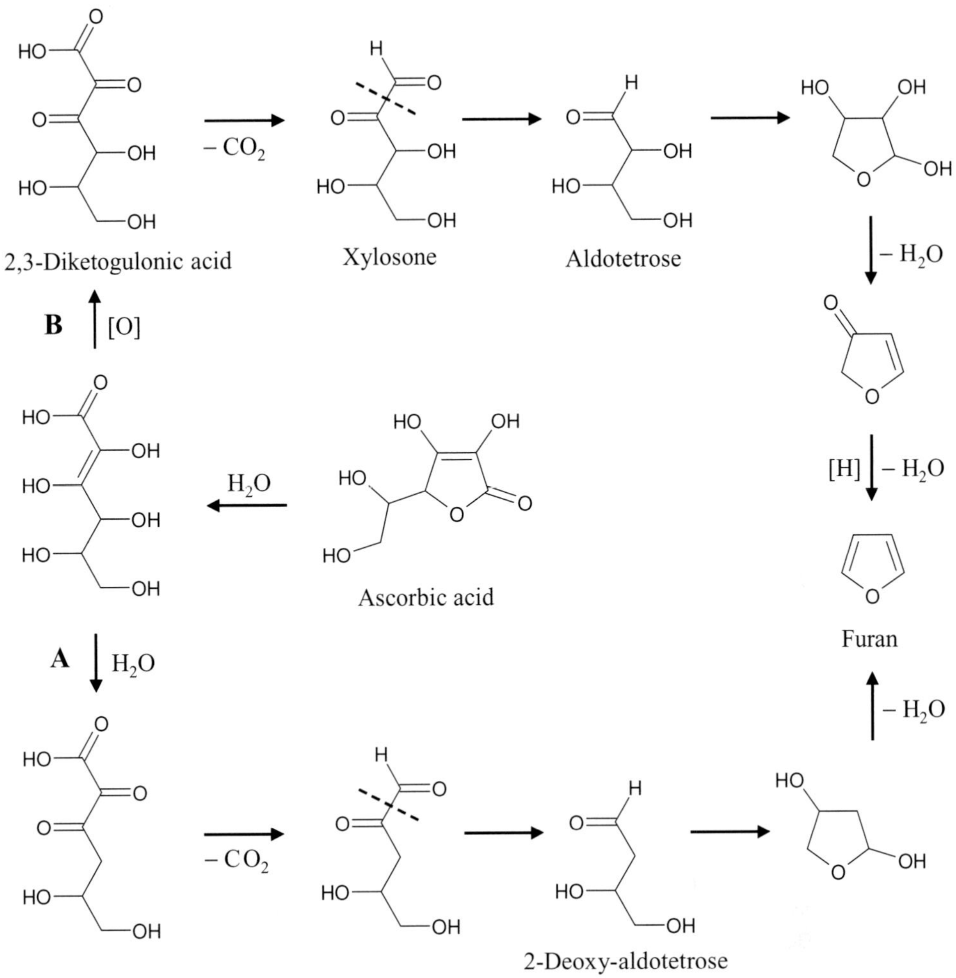

Fig. 11.7 Proposed thermal decomposition mechanism of ascorbic acid to produce furan. Dotted lines indicate dicarbonyl cleavage; [O], oxidation; [H], reduction. (Reprinted with permission from Perez and Yaylayan in *Journal of Agricultural and Food Chemistry*, **42**, 6830–6830. Copyright 2004 with permission from the American Chemical Society.)

Limacher *et al.* (2007) have shown using [1-^{13}C]- and [2-^{13}C]-labelled ascorbic acid isotopomers that there was no incorporation of C-1 and C-2 into furan whereas [6-^{13}C]-ascorbic acid only led to mono-labelled furan (Table 11.1). These data suggest that furan is exclusively formed from the intact ascorbic acid skeleton. Furthermore, the labelling pattern was independent of the reaction conditions (dry and aqueous at pH 4 or 7). Modified CAMOLA experiments (Carbon Module Labelling, Schieberle, 2005) based on equimolar mixtures of unlabelled ascorbic acid and fully ^{13}C-labelled glucose ([U-^{13}C$_6$]-GLC) did not indicate any recombination of fragments as only unlabelled or fully labelled furan were obtained. Under dry-heating conditions, the model system ASA/[U-^{13}C$_6$]-GLC resulted in 73% unlabelled furan and 27% fully labelled (^{13}C$_4$) furan, indicating that they were formed either from ascorbic acid or from [U-^{13}C$_6$]-glucose. In aqueous systems at pH 7, higher relative amounts were obtained from ascorbic acid (84%) that further increased to 99% at pH 4. This is in

Table 11.1   Per cent labelling distribution of furan generated from different ascorbic acid (ASA) and glucose (GLC) isotopomers.

Model system	$[M]^+$ (m/z 68)	$[M+1]^+$ (m/z 69)	$[M+2]^+$ (m/z 70)	$[M+3]^+$ (m/z 71)	$[M+4]^+$ (m/z 72)
L-[1-^{13}C]-ASA[a,b,c]	100	0	0	0	0
L-[2-^{13}C]-ASA[a,b,c]	100	0	0	0	0
L-[6-^{13}C]-ASA[a,b,c]	0	100	0	0	0
L-ASA + D-[U-^{13}C$_6$]-GLC (1:1)[a]	73	0	0	0	27
L-ASA + D-[U-^{13}C$_6$]-GLC (1:1)[b]	84	0	0	0	16
L-ASA + D-[U-^{13}C$_6$]-GLC (1:1)[c]	99	0	0	0	1

[a] Results from dry-heating systems (200°C, 10 min).
[b] Results from aqueous solutions at pH 7 (121°C, 25 min).
[c] Results from aqueous solutions at pH 4 (121°C, 25 min).

agreement with the observation that furan was preferably generated from ascorbic acid at lower pH (Fig. 11.4).

These data are in agreement with pathway A in Fig. 11.8: cyclization and subsequent dehydration of 2-deoxyaldotetrose, which is composed of the C3–C4–C5–C6 carbon atoms of ascorbic acid, give rise to parent furan. Alternatively, 3-deoxypentosulose may dehydrate

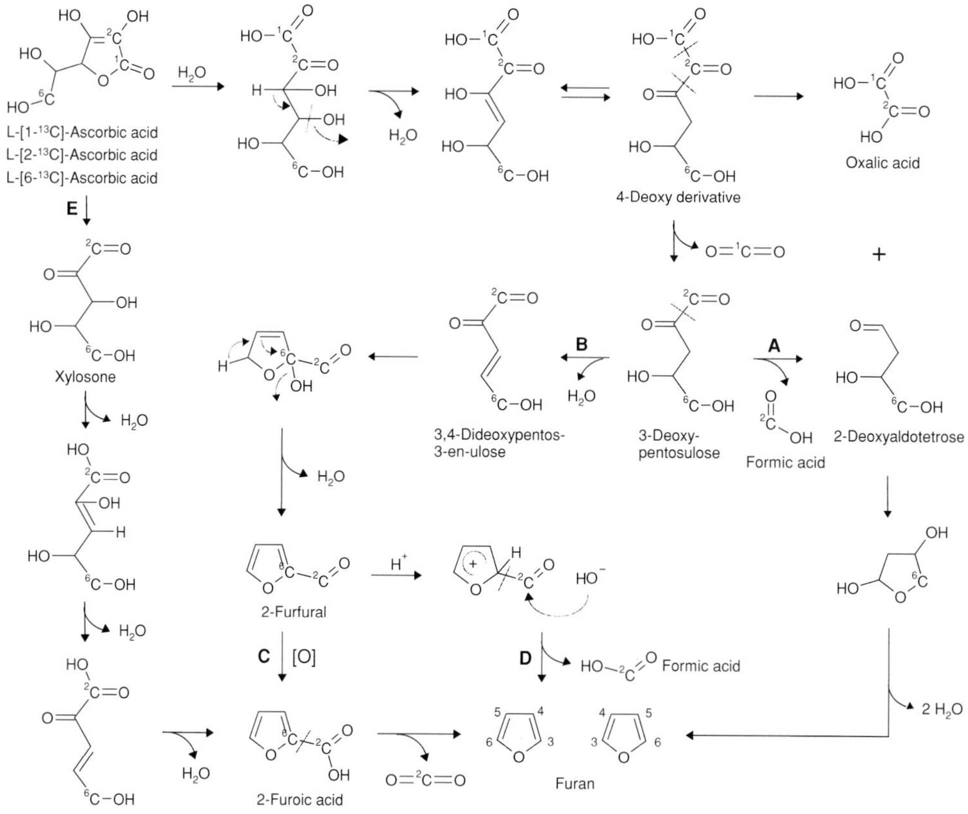

Fig. 11.8   Proposed thermal decomposition mechanism of ascorbic acid to produce furan. [O], oxidation. See text for more information. (Adapted from Perez and Yaylayan, 2004; Shinoda et al., 2005; Limacher et al., 2007.)

Table 11.2 Amounts of furan (μmol/mol) generated from intermediates in model systems simulating food process conditions.

Model system	Roasting[a]	Pressure cooking[b] (pH 7)	Pressure cooking[b] (pH 4)
2-Furoic acid	1656 ± 114	31.4 ± 3.0	326 ± 24
2-Furfural	265 ± 43	97.7 ± 0.6	70.4 ± 7.6

[a] Roasting: 200°C, 10 minutes.
[b] Pressure cooking: 121°C, 25 minutes.

at C-4 (= C-5 of ascorbic acid) via pathway B leading to 2-furfural by cyclization and dehydration. Further oxidation leads to 2-furoic acid (pathway C) with concomitant release of CO_2. Alternatively, 2-furoic acid may be formed through the xylosone pathway E (Shinoda et al., 2005). Furan may also be generated directly from 2-furfural by an electrophilic aromatic substitution-type reaction via pathway D with formic acid as by-product. Model experiments have shown (Table 11.2) that both 2-furfural and 2-furoic acid are fairly good precursors of furan (Limacher et al., 2007). The intermediacy of 2-furoic acid generating furan by decarboxylation has first been proposed by Becalski and Seaman (2005).

Unequivocal evidence for the decarboxylation step was obtained by monitoring the CO_2 by SmartNose. Relatively more labelled $^{13}CO_2$ was found in the model system based on [1-^{13}C]-ascorbic acid as indicated by the ratio 1:1.8 of unlabelled and labelled CO_2 (Table 11.3). On the other side, unlabelled CO_2 dominated in the presence of [2-^{13}C]- and [6-^{13}C]-ascorbic acid. However, traces of $^{13}CO_2$ were also found, suggesting that decarboxylation may partially take place at C-2 (via 2-furoic acid as intermediate) and C-6 as well. Furthermore, formic acid was identified as a typical decomposition product of ascorbic acid. The highest percentage of labelled formic acid (32%) was found from [2-^{13}C]-ascorbic acid (Table 11.3). However, also unlabelled formic acid was detected in this sample, thus demonstrating that formic acid is mainly derived from C-2 of ascorbic acid, but also other C-atoms can be transformed into formic acid.

11.2.3 Food products

As indicated in model experiments in which the furan levels decreased with increasing complexity of the system, it is deemed important to check the potential of precursors to form furan in real food samples. Limacher et al. (2007) extended their study to food products containing fruits and vegetables heated under conditions that simulate the sterilization process (123°C, 22 min). The products were based on vegetable purée (pumpkin) and vegetable and fruit juice (carrot and orange). Spiking experiments were conducted with unlabelled and [6-^{13}C]-labelled ascorbic acid to estimate the validity of the reaction mechanisms in a real food environment.

The concentrations of ascorbic acid in the model products ranged from less than 0.1 mg/100 g to about 50 mg/100 g. The amounts added prior to heat treatment were 56–58 mg/100 g. As shown in Table 11.4, the highest increase in furan formation (+124%)

Table 11.3 Carbon dioxide and formic acid generated by thermal decomposition of ascorbic acid (ASA) isotopomers.

Model system	CO_2 to $^{13}CO_2$ ratio	HCOOH (m/z = 46)	H^{13}COOH (m/z = 47)
L-[1-^{13}C]-ASA	1:1.8	80%	20%
L-[2-^{13}C]-ASA	1:0.3	68%	32%
L-[6-^{13}C]-ASA	1:0.2	84%	16%

Table 11.4 Spiking experiments with ascorbic acid (ASA) and [6-^{13}C]-ASA in food products.

Product	Vegetable purée (pumpkin)	Vegetable juice (carrot)	Fruit juice (orange)
Natural ASA concentration (mg/100 g)	<0.1	7.4	48.3
Total ASA content after spiking (mg/100 g)	57.1	63.2	106.4
Added ASA content (%)[a]	100	88	55
Furan content without spiking (µg/kg)	48.6 ± 1.6	30.2 ± 5.8	20.7 ± 0.4
Furan content with [6-^{13}C]-ASA spiking (µg/kg)	108.9 ± 13.5	41.7 ± 0.3	17.8 ± 0.2
Deviation in furan level (%)[b]	+124	+38	−16
Unlabeled furan content (spiked sample) (%)[c]	100	96	97
[1-^{13}C]-Furan content (spiked sample) (%)[c]	0	4	3
pH before spiking	6.26	6.47	3.90
pH after spiking	5.76	6.30	3.81

[a] Added ASA content (%) relative to total ASA amount.
[b] '+' means additional formation of furan, '−' means overall loss of furan.
[c] Spiked amount of [6-^{13}C]-ASA in mg/100 g: 53.57 (pumpkin vegetable), 54.56 (carrot juice), 56.47 (orange juice).

was observed in the pumpkin vegetable purée caused by the addition of 57.1 mg/100 g ascorbic acid followed by the carrot juice (+38%) to which 55.8 mg/100 g ascorbic acid was added. Interestingly, the addition of 58.1 mg ascorbic acid to 100 g orange juice resulted in less total furan levels (−16%). These results clearly miss a direct correlation between the concentration of ascorbic acid present in the samples and the furan amounts generated.

According to the formation pathway of furan from ascorbic acid, C-6 of ascorbic acid is always incorporated into furan (Fig. 11.8). Surprisingly, the content of mono-labelled furan ([1-^{13}C]-furan) was very low (<4 %) in all food samples spiked with [6-^{13}C]-ascorbic acid (Table 11.4). For example, labelled furan was hardly detectable in the pumpkin sample though about 124% more furan was found after spiking of ascorbic acid. This clearly demonstrates that, under the experimental conditions chosen, ascorbic acid is only a minor precursor of furan in the food products studied. The data also suggest that other compounds must function as precursors, the transformation of which into furan is accelerated in the presence of ascorbic acid.

One explanation for the increase in furan contents in pumpkin purée and carrot juices could be the pro-oxidant effect of ascorbic acid at high concentrations. Especially in the pumpkin purée in which the oxidative degradation of lipids is of importance, a pro-oxidative ingredient may favour furan formation. Another explanation for such a trend could be the pH decrease from 6.26 to 5.73 after addition of the ascorbic acid, which favours lipid oxidation in general. This points to lipids as potential furan precursors that have been suggested in previous studies (Becalski and Seaman, 2005; Märk et al., 2006). Indeed, spiking of pumpkin purée with increasing amounts of linoleic acid resulted in an almost linear increase of furan after the sterilization process (Limacher et al., unpublished data). The results, shown in Fig. 11.9, suggest PUFAs as major furan precursors in pumpkin-based purées.

In fruit juices, a fairly good correlation was found between the amount of furan formed upon autoclaving and the decomposition of ascorbic acid (Fig. 11.10). Heating orange juice to boiling for 5 minutes produced 1.4 ng/mL furan; however much higher furan levels were produced on autoclaving for 25 minutes at 121°C (Fan, 2005a). Very little furan was produced on warming apple juice but on autoclaving, far more furan was produced than when autoclaving orange juice. Subsequently, it was shown that furan formation by heating or irradiation was affected by the pH and concentration of sugars and ascorbic acid in solution (Fan, 2005b).

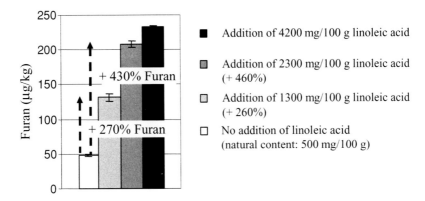

Fig. 11.9   Furan levels in pumpkin purée spiked with increasing amounts of linoleic acid. Dotted arrows indicate the increase in furan levels after the thermal treatment. (*Source:* Limacher *et al.*, unpublished results.)

Şenyuva and Gökmen (2007) found temperature as the most important processing param-eter related to furan formation in hazelnuts, in particular when exceeding 120°C (Fig. 11.11). Potential precursors of furan such as PUFAs (linoleic acid), amino acids (threonine, alanine) and sugars (glucose) were present in hazelnuts at amounts sufficient to generate furan upon

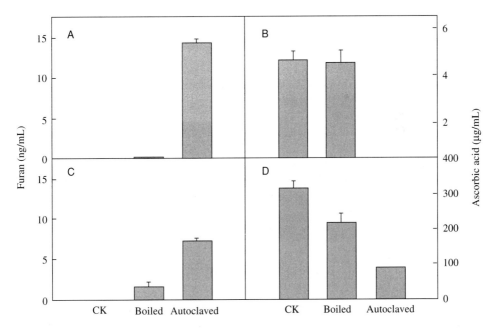

Fig. 11.10   Effect of thermal treatment on the formation of furan (A, B) and decomposition of ascorbic acid (C, D) in apple (A, C) and orange (B, D) juices. Fresh juices were either non-treated (CK), boiled (100°C, 5 min), or autoclaved (120°C, 25 min). A, furan in apple juice; B, furan in orange juice; C, ascorbic acid in apple juice; D, ascorbic acid in orange juice. (Reproduced from Fan, X. Impact of ionizing radiation and thermal treatments on furan levels in fruit juice. *Journal of Food Science*, **70**, E409–E414. Copyright 2005a with permission of Blackwell Publishing.)

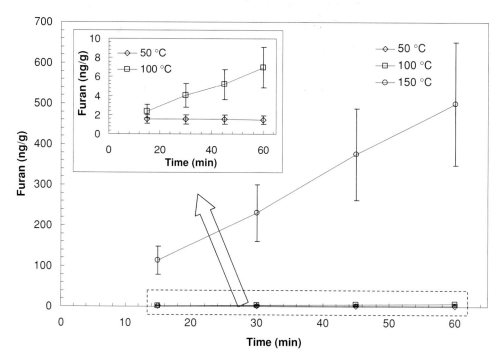

Fig. 11.11 Kinetics of furan formation in hazelnuts during heating at different temperatures. [Reproduced from Şenyuva, H.Z. and Gökmen, V. Potential of furan formation in hazelnuts during heat treatment. *Food Additives and Contaminants*, **24**(S1), 136–142. Copyright 2007 with permission from Taylor & Francis Ltd, http://www.tandf.co.uk/journals (http://www.informaworld.com).]

heating. The composition of lipid fraction in terms of relative percentages of fatty acids was relatively stable during heating at 150°C for 30 minutes, but the concentrations of amino acids and sugars decreased significantly at the end of heating period. Therefore, the authors concluded that the *Maillard* reaction is possibly the primary mechanism responsible for the formation of furan in hazelnuts during heating. Spiking experiments would provide critical information to check the validity of this hypothesis.

11.2.4 Mitigation concepts

Since the carcinogenicity of furan is probably attributable to a genotoxic mechanism (EFSA, 2004a) levels in food should be kept as low as reasonably achievable. With the broad range of furan precursors that belong to many chemical classes, furan can easily be generated by many reactions including the two major and ubiquitous reactions in food chemistry, i.e. lipid oxidation and *Maillard* reactions. Therefore, reduction of furan in foods is likely to be more challenging compared to other process contaminants. The obvious way to reduce furan as a heat-induced chemical in food is by changing the heating regime or by reducing the content of potential precursors.

There are only a few studies on the efficiency (molar yield) of furan formed from precursors in model systems. Becalski and Seaman (2005) found highest furan levels generated from PUFAs under pressure cooking conditions. Märk *et al.* (2006) have reported ascorbic acid to be very efficient under roasting conditions. However, the molar yields remain below 1%

(Limacher *et al.*, 2007). Recent studies performed in our laboratory indicate that PUFAs have to be considered as the major source of furan, in particular in aqueous food systems (Limacher *et al.*, unpublished results).

11.2.4.1   Thermal food processing

There is often little room of freedom to lower heating times and temperatures because the processes of pasteurization and sterilization are for the microbiological safety of foods. Changing the heating regime may significantly modify the product characteristics. In addition, milder heating conditions may lead to higher levels of acrylamide, another food-born process contaminant. It is known that only a small fraction of the total acrylamide formed during coffee roasting remains in the final product as it is degraded at higher temperatures (Guenther *et al.*, 2006). Thus, reducing the heat load during roasting will lead to increased acrylamide levels in coffee.

Furthermore, furan can be formed from a wide range of precursors, with ascorbic acid and PUFAs showing the highest potential, followed by carotenoids, sugars and amino acids, all of them being intrinsic food constituents. It should also be noted that ascorbic acid, PUFAs and carotenoids are regarded as desirable food components because of their health benefits. Therefore, the point of fortification in the process may be crucial. As an example, our studies indicated that the addition of ascorbic acid prior to thermal treatment of food products that contain other furan precursors leads to higher overall furan levels (Table 11.4). Assuming that these furan precursors belong to the chemical class of polyunsaturated lipids (e.g. PUFAs), it is not recommended to fortify such food with vitamin C prior to thermal treatment. This may in particular be important for canned and jarred products such as baby food.

Another mitigation measure that could be explored is to make use of the volatility of furan, e.g. de-aeration prior to thermal processing. However, again this may be of limited applicability because for microbiological reasons canned and jarred foods have to be sealed hermetically. Aseptic filling might be considered in this context, but it requires sterilized functional ingredients for fortification after the heat treatment step. For coffee it would be technically difficult to selectively purge coffee of furan whilst retaining all the flavour and aroma substances required for a balanced coffee note. The structure (i.e. pore size, inner surface) may affect the release behaviour of furan (and possibly of aroma constituents).

11.2.4.2   Competing reactions

The best approaches appear so far to involve intervention in the reaction mechanisms. For example, formation of 2-furfural from ascorbic acid in model orange juice was repressed by the presence of ethanol and mannitol acting as free radical scavengers (Shinoda *et al.*, 2005). Reduction of atmospheric oxygen slows down the autoxidation of unsaturated fatty acids and also reduces furan formation from several precursors, notably ascorbic acid (Fig. 11.12). Therefore, modification of the atmospheres within heating systems might be effective in reducing furan in foods (Märk *et al.*, 2006). The authors found an even more pronounced effect on furan reduction (-70%) in the presence of sulphite, which most likely interferes by trapping reactive intermediary carbonyls required for furan formation. On the other side, the antioxidant BHT showed only a minor effect (-25%). Therefore, given the complicated and competing reaction pathways available in autoxidation processes, limiting autoxidation may not always lead to a corresponding reduction in furan formation.

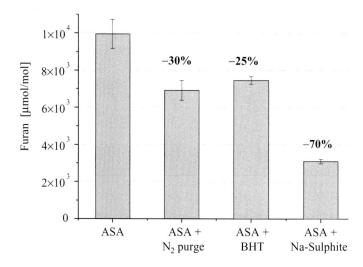

Fig. 11.12 Formation of furan from ascorbic acid (ASA) in model systems as affected by selected additives. BHT, 2,6-bis(1,1-dimethylethyl)-4-methylphenol; N_2, gaseous nitrogen. Adapted from Mörk et al. (2006). (Reproduced from Fan, X.T. and Sommers, C.H. Effect of gamma radiation on furan formation in ready-to-eat products and their ingredients. *Journal of Food Science*, **71**, C407–C412. Copyright 2006 with permission of Blackwell Publishing.)

Furan formation is quite sensitive towards changes of the reaction conditions and precursor compositions indicating complex reaction pathways. Furan amounts can be reduced to a great extent by favouring competing reactions and/or intervening at the redox system level. Therefore, the furan levels are definitely much lower in more complex systems such as foods than one would expect from the data obtained with pure precursors.

11.2.4.3 Ionizing radiation

In general, γ-radiation has been reported to initiate furan formation in fruit juices (Fan, 2005a), honey and corn syrup (Fan and Sommers, 2006), and model systems based on ascorbic acid or sugars (Fan, 2005b). However, irradiation-induced furan formation was less pronounced in ready-to-eat meat and poultry products containing those critical ingredients. In certain cases, γ-radiation may even be a suitable means to achieve some furan reduction (Fan and Sommers, 2006). As an example, irradiation at doses up to 4.5 kGy induced formation of furan in aqueous solutions of Na-ascorbate, Na-erythorbate, glucose, honey and corn syrup. Addition of Na-nitrite into these solutions prior to irradiation completely eliminated or significantly reduced furan formation (Fig. 11.13).

The possibility of using ionizing radiation to reduce the levels of thermally induced furan and acrylamide in water and selected foods was investigated by Fan and Mastovska (2006). Irradiation at a low dose (1.0 kGy) destroyed almost all furan and acrylamide in water. However, in real foods such as sausages and infant sweet potatoes, the reduction of furan was less effective than in water, whereas the reduction in acrylamide was minimal. Irradiation at 2.5–3.5 kGy, doses that can inactivate 5-log of most common pathogens, reduced furan levels in the food samples by 25–40%. Even at a dose of 10 kGy, irradiation had a very limited effect on acrylamide levels in oil and in potato chips.

In a real food, whether irradiation results in a decrease or an increase in furan levels will depend on many factors such as composition, with water content being one of the most

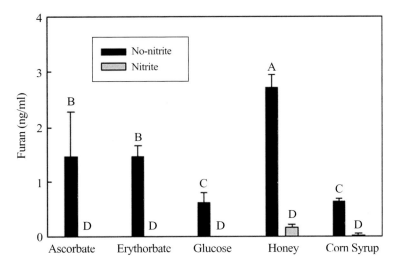

Fig. 11.13 Effects of sodium nitrite on the formation of furan from Na-ascorbate, Na-erythorbate, glucose, honey and corn syrup solutions. All solutions were irradiated with 4.5 kGy gamma rays at 5°C. Vertical bars represent standard deviation of means (n = 4). Means with the same letter are not significantly (P > 0.05) different. (*Source*: Fan and Sommers, 2006.)

crucial factors, which may affect the formation and degradation of furan. Antioxidants and other competing food components may reduce formation and degradation of furan. The reactivity of food components and proximity to the primary free radicals may also influence furan accumulation. It appears that furan was relatively sensitive to irradiation, and the rate of furan formation was relatively lower in food products. However, because of the limited effectiveness in most foods, other factors such as the possibility of nutrient loss and off-odour compound formation, and economical aspects, irradiation is unlikely to be used for the sole purpose of reducing furan (or acrylamide).

11.3   LEVELS OF OCCURRENCE AND EXPOSURE

11.3.1   Data collection

In 2004, the US FDA reported the first results for furan in 334 selected foods, mainly heat-processed foods sold in jars and cans (FDA, 2004a). The majority were baby foods, followed by canned vegetables, fruit, meat and fish, pasta sauces, nutrition drinks, fruit preserves, beers and coffees. The FDA found that many heat-treated foods contained detectable furan (limit of quantification: 2–5 μg/kg), in particular baby foods sold in jars or in cans. The highest levels were for vegetables, particularly beans, squash and sweet potatoes.

The European Food Safety Authority (EFSA) reported provisional findings (EFSA, 2004a,b) for some European samples (quantification limits of 1–2 μg/kg). The EFSA data were summarized with those of the FDA and expressed as ranges over 11 food categories. Table 11.5 covers an enlarged range of food categories and provides average furan levels on the basis of data reported by EFSA and FDA (FDA, 2004a; EFSA, 2004b, updated November 2005). Furan levels of over 100 μg/kg were found principally in three categories of major foods: coffee, baby food, and sauces and soups. Furan was detectable in 262 of the 273 baby food samples reported (96%) with an average level of 28 μg/kg.

Table 11.5 Levels of furan in foods (mostly canned and jarred) reported by EFSA and FDA.[a]

Sample type	Number		Furan (µg/kg)		Average of positives[b] (µg/kg)
	Total	Negative	Minimum	Maximum	
Baby foods	273	11	1	112	28
Infant foods	71	1	1.3	87.3	27
Infant formulae[c]	42	14	2.5	27	12
Coffee beans/powder	19	0	239	5050	1500
Coffee brewed	38	4	3	125	50
Beers	14	4	0.8	1	6
Bread	13	12	30	30	30
Bread, toasted	6	5	18	18	18
Crisp breads/crackers	4	0	4.2	18.6	12
Candy and chocolate	20	8	0.5	10.3	3
Desserts/puddings	12	1	1.5	13	4
Fish	9	3	1.5	8	6
Fruit	28	2	0.9	11	4
Fruit juices	51	27	0.5	31	5
Fruit preserves	47	2	0.9	37	7
Gravies	8	0	13.3	174	48
Malt	6	1	16	195	91
Convenience meals	32	0	3	94	35
Meats	15	9	1.7	39	18
Miscellaneous	47	9	1	420[d]	28
Nutrition drinks/shakes	22	2	1.1	174	29
Sauces	41	4	3.3	46	11
Soups	58	7	3	125	32
Soy sauces	12	0	17.2	91	50
Maple syrup	3	0	8.6	88.3	45
Tortilla and potato chips	6	1	4.4	39.1	16
Vegetable juices	8	0	3.2	40	10
Vegetables (in cans and jars)	50	9	0.8	48	9
Baked beans	26	0	23.3	122	59

[a] Data taken from EFSA (2004a,b) and FDA (2004a).
[b] Some values are approximate (estimated from ranges).
[c] Includes ready to eat samples and concentrates prepared as for consumption but with cold water.
[d] Two samples (caramels) contained 192 and 400 **g**/kg with all remaining samples below 100 **g**/kg.

However, based on this limited set of data, it is not possible to draw firm conclusions concerning the tendency of different foods to form furan, because exact details on food composition and the heating conditions of time and temperature were not known. Both the US and European surveys were restricted to specific products thought likely to contain furan generated upon heating in sealed containers. Therefore, EFSA has called for further concentration data for furan in food to estimate consumers' exposure using data that better represent the actual distribution of furan in foods (EFSA, 2006c).

11.3.2 Coffee

Furan and furan derivatives have long been known as intrinsic components of coffee flavour. Some of them play a key role for the characteristic aroma of coffee, such as 2-furfurylthiol perceived as coffee-like at high dilution (Mayer et al., 2000). Green coffee beans contain only traces of furan. High levels of furan (~4 mg/kg) are found in roasted coffee beans

Table 11.6 Furan in coffee beans and its transfer into brewed coffee.[a]

Coffee type	Brewing method	Furan in bean (μg/kg)	Furan in brew (μg/kg)
Whole beans	Home machine	3600–6100	9–33
Whole beans	Manual machine	3600–6100	17–24
Whole beans	Automatic machine	3600–6100	57–115
Whole beans	French press	3600–6100	33–66
Regular powder	In cup	800–3400	9–66
Decaffeinated powder		1000–2800	8–31
Espresso		1500–2000	28–60
Instant 1		200–700	3–15
Instant 2		2000–2200	14–25

[a] Adapted from Crews and Castle (2007), data summarized from Kuballa et al. (2005).

(Table 11.6), probably due to the roasting process where the high temperatures exceed most other food processing procedures. Roast and ground coffee contains about 2 mg/kg furan with no effect of the decaffeination step (Kuballa et al., 2005). On the other side, relatively low amounts of furan (~0.9 mg/kg) were found in instant coffee. Even coffee alternatives contain furan with a mean level of about 1.2 mg/kg.

However, of outmost importance for estimation of the daily exposure are the furan levels in the final ready-to-drink coffee beverage. As shown in Table 11.6, losses are substantial during coffee brewing by all of the usual methods such as expression or filtration, and also in the preparation of instant coffee (Kuballa et al., 2005). Automatic coffee machines produced brews with the highest levels of furan, because a higher ratio of coffee powder to water is used giving a lower dilution factor, and because of the closed system favouring retention of furan. Much lower levels were produced by standard home coffee-making machines and by manual brewing. These data show the substantial reduction on brewing coffee from coffee powder.

11.3.3 Baby food and other dietary sources

The FDA survey included 31 infant formula products (FDA, 2004a). Furan was found in 11 of 13 samples with added iron and in 6 of 18 samples without iron, typically at 8–10 μg/kg. The data reported in the annex to the EFSA report of the CONTAM panel on provisional findings on furan in food is shown in Fig. 11.14 for baby food and infant formulae (EFSA, 2004a). The furan amounts (median) found in baby food and infant formulae are in the range of 5–20 μg/kg. Similar levels were reported in baby food purchased on the Swiss market, i.e. 12 μg/kg (median, n = 102) including extreme values of up to 153 μg/kg (Reinhard et al., 2004). Unfortunately, the representativeness of the values reported in the literature is not always satisfactory.

The following foods were shown by EFSA and the FDA to contain relatively high levels of furan: malts (16–195 μg/kg), gravies (13–174 μg/kg), caramels (220–400 μg/kg) and soy sauce (17–90 μg/kg). A recent survey of the US FDA in 300 processed foods indicated furan levels ranging from non-detectable to about 100 μg/kg (Morehouse et al., 2007). In pre-packaged, processed adult foods, vegetable beef soup and baked beans resulted in the highest values (85–90 μg/kg) followed by soy sauce and coffee brew (ca. 50 μg/kg). In baby food, highest furan amounts were found in sweet potatoes and garden vegetables (ca. 80 μg/kg).

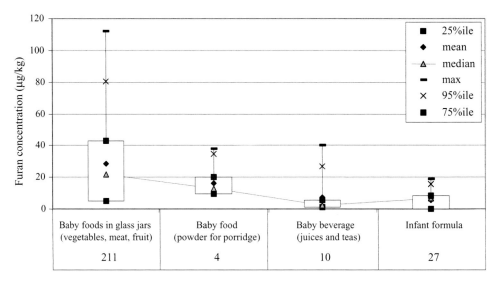

Fig. 11.14   Summary of data on baby food and infant formulae collected by the EFSA. (*Data source:* Annex to EFSA report of the CONTAM Panel on provisional findings on furan in food, dated 22 December 2004. http://www.efsa.eu.int/science/contam/contam_scientific_documents/catindex_en.html.)

Foods in which no furan was found include milk, margarine, mayonnaise, yogurt, sour cream, cottage cheese and pasteurized eggs.

Substantial levels of furan (20–200 µg/kg) have also been reported in foods not cooked in closed containers, such as potato crisps, crackers and crisp breads (Hoenicke *et al.*, 2004) and toasted bread (Hasnip *et al.*, 2006). These findings are surprising and worth further study, in particular with regard to the levels of furan present in these products as consumed.

11.3.4   Domestic cooking

Most of the results in the literature to date are for heat-processed retail foods. There have been a few studies to investigate if furan persists in food during normal preparation processes. It was thought that such a volatile substance may simply evaporate from the food, e.g. when cans or jars were opened. However, furan was found to be rather persistent in foods. As shown in Table 11.7, furan formed in foods following heating in sealed containers during industrial processing is not lost to a very significant extent when the foods are warmed prior to consumption (Hasnip *et al.*, 2006). The exceptions are vigorous boiling or cooking where furan can be evaporated or swept with large volumes of steam. On the other hand, warming the food even in lightly lidded containers can increase furan levels, thus indicating the presence of additional precursors with the potential of in situ furan generation (Hasnip *et al.*, 2006).

In general, furan levels did not decrease as much when foods were cooked in a microwave when compared to the same foods cooked in a saucepan. Furan levels decreased in most canned and jarred foods after heating in a saucepan. Low levels of furan in soups in cartons were not changed by any procedure. Furan decreased slightly in foods on standing before consumption but did so more rapidly on stirring, i.e. the furan amount was decreased from about 20 µg/kg by a factor of ~10 upon stirring for 10 minutes (Fig. 11.15) (Roberts

Table 11.7   Effects of heating in a closed vessel on furan concentration in food.[a]

Food type	As purchased/ heated	Open/ closed	Range[b] (ng/vial)	Mean (ng/vial)	Mean (ng/g food)	MU[c] (ng/vial)
Dried vegetables	As purchased		<1	<1	<1	–
	Heated	Closed	46–50	48	96	9
	Heated	Open	1–2	2	4	8
Rice pudding (jar)	As purchased		141	141	27	16
	Heated	Closed	255–289	272	53	27
Rice pudding (can)	As purchased		86	86	19	12
	Heated	Closed	105–109	107	21	13
Fruit purée (foil-sealed plastic pot)	As purchased		3	3	<1	8
	Heated	Closed	99–106	103	19	13
Baby food (jar)	As purchased		237	237	46	24
	Heated	Closed	534–547	541	103	52
Baby food (can)	As purchased		353	353	70	35
	Heated	Closed	300–323	312	61	31

[a] *Source*: Hasnip et al. (2006).
[b] Distilled water control (blank) showed a background of 8–10 ng/vial, which was subtracted from the results.
[c] Analytical measurement uncertainty (MU) only since heating experiments were carried out in the same batch and inclusion of between-batch heating variation is not appropriate.
Reproduced from Hasnip, S., Crews, C. and Castle, L. Some factors affecting the formation of furan in heated foods. *Food Additives and Contaminants*, **23**, 219–227. Copyright 2006 with permission from Taylor & Francis Ltd, http://www.tandf.co.uk/journals (http://www.informaworld.com).

et al., 2007). The levels also decreased slightly when foods were left to stand on plates. This observation can be attributed to the volatility of furan and is similar to the findings reported by Goldmann *et al.* (2005) showing evidence for the decrease of furan concentrations over time upon heating in open jars, thus leading to 85% loss compared with 50% loss in non-heated samples.

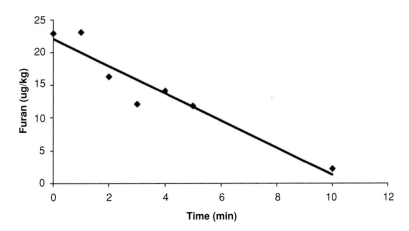

Fig. 11.15   Decrease in furan concentration in a hot water heated canned carrot and lamb baby food (on standing) with stirring. [Reproduced from Roberts, D., Crews, C., Grundy, H., Mills, G. and Matthews, W. Effect of consumer cooking on furan in convenience foods. *Food Additives and Contaminants*, (2008) **25**, 25–31. Copyright 2008, with permission from Taylor & Francis Ltd, http://www.tandf.co.uk/journals (http://www.informaworld.com).]

Table 11.8   Estimate of exposure to furan from various adult food types.[a]

Food type	Exposure estimate (μg/kg-bw/day)[b]
Brewed coffee	0.15
Chili	0.04
Soups containing meat	0.01
Pork and beans	0.004
Canned pasta	0.004
Canned string beans	0.004
Instant coffee	0.001
Spaghetti sauces	0.001
Juices	0.001
Canned corn	0.0003
Canned tuna (packed in water)	0.00008
Broths	0.000005

[a] According to Morehouse et al. (2007).
[b] Micrograms of furan per kilogram body weight per day.
Reproduced from Morehouse, K.M., Nyman, P.J., McNeal, T.P., DiNovi, M.J. and Perfetti, G.A. Survey of furan in heat processed foods by headspace gas chromatography/mass spectrometry and an estimate of adult exposure. *Food Additives and Contaminants*, DOI: 10.1080/02652030701552949. Copyright 2007 with permission from Taylor & Francis Ltd, http://www.tandf.co.uk/journals (http://www.informaworld.com).

11.3.5   Dietary exposure

In general, there are still limited data available to perform a reliable assessment of the dietary exposure. However, based on this limited data set, EFSA estimated exposure to furan and found baby food of particular interest as a high proportion of samples sold in jars and cans contained furan and such foods may form the sole diet of many babies (EFSA, 2004a). The estimated intake based on consumption of baby food from glass jars was <0.2 to 26 μg furan per day or <0.03 to 3.5 μg/kg bw per day for a 6-month-old baby weighing 7.5 kg. The daily intake for adults from canned or jarred vegetables (35 samples) was estimated to be 1.1–23 μg/person, and from beer (12 samples) it was 1.3–50 μg/person. The daily intake from coffee based on data from 45 samples was 2.4–116 μg/person, making coffee the major dietary source for adults.

On the basis of recent results obtained by analysing 300 processed foods, Morehouse *et al.* (2007) estimated the exposure to furan from the average diet. Although these data do not include all foods, they do represent a broad spectrum of foods found in the average diet. For the average adult diet, brewed coffee is the major source of furan (Table 11.8). The estimate of the mean exposure to furan for the average US consumer is about 0.2 μg/kg-bw/day (Table 11.9) and it does not vary among the examined sub-populations. This exposure estimate

Table 11.9   Estimates of mean exposure to furan for different subpopulations.[a]

Subpopulation	Mean exposure (μg/kg-bw/day)[b]
Males and females age 2–5 years	0.23
Females age 15–45 years	0.24
Males and females age 2 years and older	0.25

[a] According to Morehouse et al. (2007).
[b] Micrograms of furan per kilogram body weight per day.
Reproduced from Morehouse, K.M., Nyman, P.J., McNeal, T.P., DiNovi, M.J. and Perfetti, G.A. Survey of furan in heat processed foods by headspace gas chromatography/mass spectrometry and an estimate of adult exposure. *Food Additives and Contaminants*, DOI: 10.1080/02652030701552949. Copyright 2007 with permission from Taylor & Francis Ltd, http://www.tandf.co.uk/journals (http://www.informaworld.com).

does not necessarily express the actual consumption of furan by the general public, since other factors, such as the inclusion of foods not yet analysed and/or further processing of the food product prior to consumption, could change the estimate. Indeed, the authors stress that furan intakes from some food products analysed for the US survey may be over-estimated since the can/jar contents are sometimes warmed and stirred prior to use, which could lead to lower furan levels in the foods that are consumed due to volatilization.

11.4   METHODS OF ANALYSIS

Headspace (HS) sampling is the most appropriate method for the analysis of furan due to its volatility. Two approaches were developed for the quantitative analysis of furan in food samples, i.e. (i) static HS gas chromatography/mass spectrometry GC/MS and (ii) solid phase micro-extraction (SPME)-GC/MS. Both methods use the isotope dilution assay (IDA) methodology with $[^2H_4]$-furan ($= d_4$-furan) as internal standard. The use of the labelled analogue allows for accurate measurement of trace amounts of the analyte in complex matrices (e.g. food) by compensating for losses during sample preparation (Pickup and McPherson, 1976; De Bièvre, 1990; Milo and Blank, 1998).

The first quantitative method for furan in food was published by the FDA in June 2004 (FDA, 2004b) and revised later by Nyman *et al.* (2006). Basically, it comprises sample preparation (5–10 g) with d_4-furan fortification in sealed HS vials under cold conditions and static HS sampling following incubation at 80°C. Furan and d_4-furan are detected by GC/MS in the full scan mode. Chromatographic separation uses a porous layer open tubular (PLOT) column with a bonded polystyrene-divinylbenzene based phase (PLOTQ) which separates small volatile molecules effectively. Furan is quantified by using a standard additions curve requiring seven extractions per sample, where the concentration of furan in the fortified test portions is plotted versus the furan/d_4-furan response factors using ions at m/z 68 and m/z 72 for furan and d_4-furan, respectively. Since then the interest in furan analysis grew and several variations of this procedure were tested and introduced by other research groups summarized in Table 11.10. Various critical steps that may affect furan quantification are discussed in the following.

11.4.1   Sample preparation

Due to its volatility (b.p. 31.4°C), furan should be analysed in refrigerated samples as quickly as possible after their opening and/or preparation. Effective homogenization of the sample is recommended using a chilled blender with the sample immersed in an ice bath. Depending on the physical state of the sample, homogenization should be carried out under chilled or cryogenic conditions with an adapted mixer, e.g. freezing the sample portion with liquid nitrogen. Salt is added to improve the equilibrium distribution of furan towards the HS gas. The internal standard is added into the sample extract, not simply dispensed on its surface.

A simplified version of the initial FDA HS method was reported by Health Canada where the actual experiments were conducted in the autosampler vials to minimize possible losses of the analyte by reducing the number of transfer steps (Becalski *et al.*, 2005; Becalski and Seaman, 2005). This method involves HS sampling from an autosampler vial (2 mL) containing less sample (1 g). The samples are spiked with d_4-furan, homogenized with water in a blender at 0°C and transferred to vials containing sodium sulphate. After equilibration at 30°C, an aliquot of HS (50 µL) is injected and analysed by GC/MS in the single ion monitoring (SIM) mode.

Table 11.10 Quality criteria of various headspace methods for the quantification of furan in food.[a]

Method	LOD[b] (µg/kg)	LOQ[c] (µg/kg)	Linearity (µg/kg)	Precision[d] CV (%)	Recovery (%)	Application	Reference
1. Static headspace							
GC/ITMS, split, PorarisQ, 80°C	n.d.[e]	0.5–2	n.d.[e]	6 (n = 6) / 8 (n = 6)	n.d.[e]	Baby food / Coffee	Reinhard et al., 2004
GC/MS (SIM), splitless, CP-PoraBOND Q	0.1	n.d.[e]	0.4–1000 ($R^2 > 0.999$)	1.6	92–122 (25–250 µg/kg)	Coffee brew, rice soup	Becalski et al., 2005
GC/MS, split, HP-PLOT Q	0.2 / 0.9	0.6 / 2.9	0–44 ($R^2 = 0.9915$)	4.0–8.6 / 11.7–17.3	98–102 / 97–101	Apple juice / Peanut butter	Nyman et al., 2006
2. SPME							
GC/MS, split, CP-PoraBOND U	0.3	0.8	3.2–177.7 ($R^2 = 0.9953$)	4.6	92–102 (38–152 µg/kg)	Coffee brew	Ho et al., 2005
GC/MS (SIM), splitless, HP-PLOT Q	0.034[f]	0.086[g]	0.2–5 ($R^2 \sim 0.999$)	7–16 (coffee)	87–93 (2–10 µg/kg)	Coffee brew / Orange juice	Goldmann et al., 2005
GC/MS (SIM), splitless, HP-InnoWAX	0.03	0.04	1–100 ($R^2 = 0.998$)	1.3–6.2	92–96	Baby food	Bianchi et al., 2006
GC/IT-MS, splitless, BPX-volatiles	0.008–0.07	0.03–0.25	0.2–5 ($R^2 > 0.999$)	6–10	n.d.[e]	Coffee, baby food, etc.	Altaki et al., 2007
GC/MS (SIM), split, ZebronWAX	0.03	0.09	0–1133[h] ($R^2 > 0.999$)	5–8 (model systems, n = 6)	93–104 (2–10 µg/kg)	Model systems, pap, juice	limacher et al., 2007

[a] All methods listed in the table are based on Isotope Dilution Assay using d_4-furan as internal standard.
[b] LOD, limit of detection.
[c] LOQ, limit of quantification.
[d] CV, coefficient of variation.
[e] n.d., not determined.
[f] Detection capability (CCβ
[g] Decision limit (CCβ
[h] The value corresponds to mol/mol precursor.

For the analysis of complex food samples (Goldmann *et al.*, 2005), powdered beverages are reconstituted as proposed on the product label and an aliquot ($\sim$0.5 g) is transferred to an HS vial (10 mL) containing NaCl (0.2 g). Similarly, a refrigerated and homogenized wet food sample (e.g. fruit purée) is transferred to an HS vial containing NaCl and water. Dry food samples are refrigerated and freshly homogenized and then mixed with NaCl and water in an HS vial. The internal standard solution (d_4-furan, 1 μL, 0.1 ng/mL in water) is added and the vial immediately sealed. In case of higher furan content, the sample is diluted with water (9 + 1, w/w). When using a sample dilution step the internal standard is added in the HS vial and not directly to the sample prior to dilution.

11.4.2   Sampling methods

In direct HS analysis, a portion of the HS gas is taken and injected directly into the GC/MS. In contrast, in SPME a needle coated with a polymeric material is first exposed to the HS vapours to absorb volatiles, and is then desorbed thermally in the injection port of the GC. The food samples need to be chilled or frozen before handling to avoid losses. Partitioning of furan from the food sample into the HS in the vial is affected by time, temperature and the mobility of the sample. Effective partitioning is ensured by prolonging the equilibration incubation time (Becalski *et al.*, 2005), or improving the efficiency of automated shaking by adding glass beads to the HS vial (Hasnip *et al.*, 2006).

Normal HS procedures involve heating the sample to promote volatilization of analyte into the HS gas phase. Excessive heating is not recommended for furan analysis. Becalski *et al.* (2005) showed that increasing the HS incubation temperature from 30 to 50°C caused only a 50% increase in the furan peak area of an aqueous standard. Addition of salt more than doubled the furan signal. In view of the danger of forming extra furan even at quite low temperatures, most workers employ an incubation temperature of 50°C or below and this gives adequate sensitivity of about 1 μg/kg. Şenyuva and Gökmen (2005) demonstrated furan formation in green coffee (4 μg/kg) and in tomato and orange juices on incubation at 40°C for 30 minutes (Fig. 11.16). Therefore, it is necessary to check the analytical procedure for any new sample type, especially the HS incubation temperature used, to avoid the formation of extra furan.

In SPME the fibre is first exposed to the HS to absorb volatiles (e.g. for 10–60 min) followed by thermal desorption (1–5 min, 90–300°C) in the GC injection port. The first SPME-GC/MS approach was developed by Ho *et al.* (2005) in which the sample (1 g) is transferred into a vial (10 mL), and then water (5 mL) and salt (4 g) is added. The vial is sealed and the internal standard solution is added through the septum using a microlitre syringe. The vials are vortexed (30 s) prior to analysis. The method employs a SPME fibre coated with CarboxenTM/polydimethylsiloxane (film thickness 85 μm). The vials are agitated (20 min, 30°C) using an automated system and desorbed (90°C) onto a CP-Pora Bond U capillary column.

In general, SPME allows sample concentration and affords higher sensitivity. Limits of detection in the low ng/kg range can be achieved for furan as shown in Table 11.10. However, in practice the effective detection limit can be constrained by background levels of furan in the blank samples. Different SMPE fibres were compared including polyacrylate, Carbowax®/divinylbenzene, polydimethylsiloxane/divinylbenzene, and CarboxenTM/polydimethylsiloxane. In terms of sensitivity, CarboxenTM/polydimethylsiloxane showed best results (Goldmann *et al.*, 2005; Fan and Sommers, 2006). The use of this fibre (film thickness 75 μm) combined with cryofocusing technique using HP-PLOT Q column

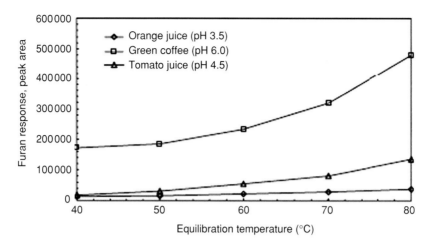

Fig. 11.16 Change of furan peak response during equilibration of green coffee, tomato juice and orange juice at temperatures between 40 and 80°C for a fixed time of 30 minutes. [Reproduced from Senyuva, H.Z. and Gökmen, V. Analysis of furan in foods. Is headspace sampling a fit-for-purpose technique? *Food Additives and Contaminants*, **22**, 1198–1202. Copyright 2005 with permission from Taylor & Francis Ltd, http://www.tandf.co.uk/journals (http://www.informaworld.com).]

generated a very sensitive method for furan detection. Several authors have reported SPME methods for furan quantification (Table 11.10), e.g. applied to coffee (Ho *et al*., 2005), orange juice (Fan, 2005a), foods in general (Goldmann *et al*., 2005; Bianchi *et al*., 2006; Fan and Sommers, 2006).

11.4.3 GC-MS analysis

The identification of furan is assured by checking for the correct GC retention time along with the correct ratio of the molecular ion at m/z 68 compared to its fragment ion at m/z 39. According to Crews and Castle (2007), the deviation between analyte and standard furan should not exceed ±2% for the retention time and ±10% for the mass ratio. However, the low abundance of the m/z 39 ion constrains the detection and quantification limits.

Quantification has been based on standard additions or external calibration graphs, both incorporating a deuterium-labelled internal standard (d_4-furan). High levels of internal standard must be avoided to prevent a contribution to the m/z 68 analyte signal from the [M-d_2]$^+$-fragment ion (Crews and Castle, 2007). Analytical recovery for the major matrices, brewed coffee and baby food, is consistently better than 90% with limits of detection of less than 1 μg/kg. A number of method performance characteristics have been reported in detail (Table 11.10) and show that the current HS methods perform very well.

A typical MS (SIM) chromatogram for dry pet food that represents a complex matrix is illustrated in Fig. 11.17. Among the five ion chromatograms depicted, the m/z 39 trace shows interference, and the baseline for m/z 44 is rather poor. However, reliable quantification is possible using the ion traces at m/z 68 and at m/z 72. It is mandatory to define qualifiers and quantifiers for both furan and d_4-furan. Recently developed methods (Table 11.10) do not require tedious standard addition curves for each analysis. The use of the furan/d_4-furan ratio is sufficient and no advantage in precision was found when applying the standard addition curve (Reinhard *et al*., 2004). A comparison of both approaches did not reveal a

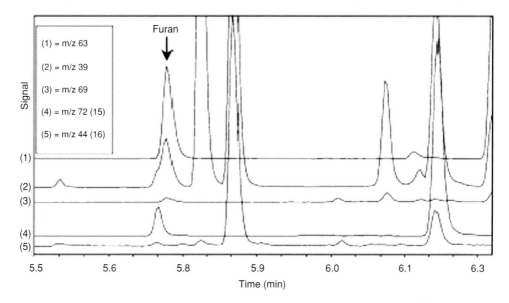

Fig. 11.17 MS (SIM) chromatogram of a dry pet food sample showing interference in the m/z 39 trace and a poor baseline for m/z 44. Reliable quantification is possible using the ion traces m/z 68 and m/z 72. (Reproduced from Goldmann, T., Perisset, A., Scanlan, F. and Stadler, R.H. Rapid determination of furan in heated foodstuffs by isotope dilution solid phase microextraction–gas chromatography–mass spectrometry (SPME-GC-MS). *Analyst*, **130**, 878–883. Copyright 2005 by permission of the Royal Society of Chemistry.)

significant difference in terms of method precision and accuracy (Altaki *et al.*, 2007). In general, the recent sample preparation procedures allow high throughput analyses (e.g. 48 GC runs within 24 h, equates to 1 calibration curve and 20 samples in duplicate) in many different food matrices.

11.4.4 Method validation

It is of outmost importance to collect reliable data for estimating the daily exposure through the normal diet, which requires thorough method validation. As shown in Table 11.10, low limits of detection (LOD) and quantification (LOQ) were achieved in the recently developed methods by sensitive GC-MS techniques. However, these limits may very much depend on the complexity of the food matrix. Coffee, chocolate and fat-rich foods require particular care to achieve acceptable repeatability. Becalski *et al.* (2005) reported good agreement for within-run duplicates. Each homogenized sample was sub-sampled twice and analysed. Results for duplicate testing for the foods analysed are shown in Fig. 11.18, where results of one determination of a particular sample is referenced using x-axis while duplicate of that sample is referenced using y-axis. All data points are essentially on the $y = x$ line indicating a good agreement for within-run duplicates.

Characteristic mass units are used as qualifier and quantifier to sort out interferences. The US FDA method uses two ions for furan, i.e. m/z 39 as qualifier and m/z 68 as quantifier as well as the characteristic ratio of m/z 39/68, and one ion for d_4-furan (m/z 72 as quantifier, no qualifier ion and thus no ion ratio determined). Goldmann *et al.* (2005) suggest the use of three mass transitions for the identification of furan (1 quantifier and 2 qualifier ions)

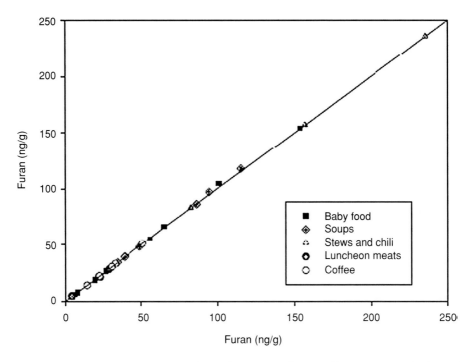

Fig. 11.18 Comparison of within-run duplicates for the quantification of furan in various food samples. [Reproduced from Becalski, A., Forsyth, D., Casey, V., Lau, B.P.-Y., Pepper, K. and Seaman, S. Development and validation of a headspace method for determination of furan in food. *Food Additives and Contaminants*, **22**, 535–540. Copyright 2005 with permission from Taylor & Francis Ltd, http://www.tandf.co.uk/journals (http://www.informaworld.com).]

(Table 11.11) following EC guidelines (2002). Even if the MS acquisition proposed by the US FDA is recorded in the full scan mode, the method proposes no identification criteria based on the mass spectrum.

The variation in furan analysis obtained by two analysts on different days with different instruments was studied by Nyman *et al.* (2006). In general, the results showed good within-laboratory precision (Table 11.12). A minimum of three replicates of different sub-samples from the same lot were analysed by each analyst. The furan levels ranged from none detected to 122 µg/kg. For most of the foods, the difference between the amounts of furan found was ≤18%. Apple juice and one of the infant formulas showed the largest differences, but

Table 11.11 Qualifiers and quantifiers for the analysis of furan.[a]

Compound	Mass (m/z)	Function
Furan	68 ([M]$^+$)	Quantifier
	39 ([M-CHO]$^+$)	Qualifier
	69 ([M+1]$^+$)	Qualifier
d_4-Furan	72 ([M+4]$^+$)	Quantifier
	42 ([M+4-C^{2}HO]$^+$)	Qualifier

[a] Adapted from Goldmann *et al.* (2005) and Limacher *et al.* (2007).

Table 11.12   Amounts of furan found in sub-samples of the same lot analysed by two analysts on different days with different instruments.[a]

Product	Analyst A		Analyst B		Difference (%)
	Furan (µg/kg)	RSD (%)[b]	Furan (µg/kg)	RSD (%)[b]	
Infant formula with iron, concentrate, brand 1	8.5	2.4	8.6	1.5	1.2
Sweet potatoes, brand 1, baby food	93.1	0.3	91.0	3.1	2.3
Spaghetti sauce, brand 1	6.1	2.1	5.9	1.0	3.3
Baked beans, brand 1	122	0.5	117	4.1	4.2
Canned luncheon meat loaf	1.9	3.0	1.8	3.3	5.4
Sweet potatoes, brand 2, baby food	81.1	0.4	75.7	1.3	6.9
Pork and beans, brand 2	78.7	1.4	85.6	2.9	8.4
Chicken broth, brand 1	8.2		9.1	1.8	10
Infant formula with iron, concentrate, brand 2	18.8	1.5	16.8	1.1	11
Sweet potatoes, brand 3, baby food	82.9	1.0	73.8	6.4	12
Spaghetti sauce, brand 2	3.1	3.1	3.5	2.9	12
Peanut butter	6.1	5.8	7.1	6.4	15
Chicken broth, brand 2	15.2	1.0	18.2	4.1	18
Cut green beans	5.9	1.9	8.2	2.3	28
Chicken dinner, baby food	39.7	1.1	29.4	2.3	30
Chicken and stars, baby food	23.6	0.9	15.9	9.3	39
Apple juice	1.2	10.3	2.2	5.3	59

[a] Adapted from Nyman et al. (2006).
[b] Per cent relative standard deviations (RSD) determined with a minimum of three replicates.

they also contained very low furan levels. This result may reflect a difference in sensitivity between the two instruments. The chicken baby foods and green beans also showed a larger difference, which may reflect a difference in the manner in which these foods were homogenized.

Laboratory proficiency testing has been performed by FAPAS® (2005) to allow an interlaboratory comparison of furan analysis in baby food. Satisfactory results were obtained for 85% of the 13 participating laboratories from several countries. The assigned value was calculated as 25.0 µg/kg with a median absolute deviation of 7.41 µg/kg and a target standard deviation of 5.50 µg/kg. The z-scores varied from −2.2 to +4.1, with two values being outside of the satisfactory range of $|z| > 2$ (see Fig. 11.19). It should be mentioned that no recommendations were given as to the analytical method to be applied, and indeed the conditions varied much along the analytical procedure. Standardization of the analytical methodology may lead to further improvement.

Despite the use of d_4-furan as internal standard and the isotope dilution assay methodology, the quantitative analysis of furan remains a challenging task due to its volatility and because it can readily be formed in situ at relatively low temperatures (Şenyuva and Gökmen, 2005). An important factor to consider in determining the furan content in food is the time dependence of the concentration of furan in the samples due to its extreme volatility. Furan concentrations tend to decrease over time in open jars: up to 85% loss of analyte over a period of 5.5 hours in heated samples was observed compared with 50% loss in non-heated samples (Goldmann et al., 2005).

Moreover, the structure of the food product and the degree of grinding may affect the furan results, possibly leading to further complication. Furan trapped in pores of solid materials may be lost by evaporation during the sample preparation step by destroying the porous

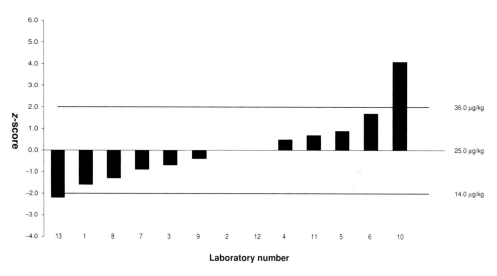

Fig. 11.19 FAPAS® Z-scores for 13 participants taking part in proficiency testing for furan in baby food. Assigned value for furan = 25.0 µg/kg. Z-scores between −2 and +2 are satisfactory.

structure. On the other hand, materials with very fine pore size may more readily release furan to be analysed. Indeed, roast and ground coffee of various degree of freeness may result in up to 50% difference, the finer ground material yielding higher furan levels (J. Kerler *et al.*, personal communication). However, furan seems to be persistent under certain conditions, which may be explained by π–π stacking-type interactions of furan with other food ingredients, such as polyphenolic components. Such interactions leading to relatively stable complexes have been reported for caffeine with chlorogenic acid (Horman and Viani, 1971) and theaflavin (Charlton *et al.*, 2000).

11.4.5 On-line monitoring

A rapid method based on proton transfer reaction mass spectrometry (PTR-MS) has been developed as an on-line monitoring tool (Märk *et al.*, 2006). This PTR-MS procedure can be employed for rapid screening purposes and direct comparison of effluent gases during processing. It results in semi-quantitative data obtained in a few minutes and could therefore be used for on-line quality control. So far it has been applied to model studies based on simple precursor systems. As shown in Fig. 11.20, a large number of volatile organic compounds was produced upon thermal treatment of ascorbic acid indicated by selected mass traces. The major signals were represented by the ions at m/z 69, 47 and 97, followed by many other ion traces. The ion trace at m/z 69 corresponds to furan, which was unambiguously identified by PTR-MS/GC-MS coupling. This is achieved by trapping aliquots of the HS on Tenax® cartridges and analysing off-line the volatile composition by GC-MS, as already described for acrylamide (Pollien *et al.*, 2003).

Concluding remarks

Reduction of furan in foods seems to be more challenging compared to other process contaminants such as acrylamide. This is due to (i) the limited flexibility to lower heating times and temperatures because of microbiological safety achieved by pasteurization and sterilization,

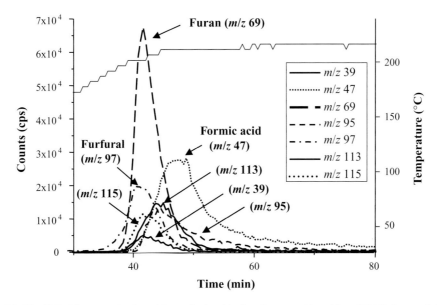

Fig. 11.20   PTR-MS traces of selected ions obtained by heating ascorbic acid at 200°C. Some of these traces were identified by coupling with GC-MS after trapping the headspace on Tenax® tubes. For a color version of this figure, please see Plate 3 of the color plate section that falls between pages 224 and 225. (Reprinted with permission from Märk, J., Pollien, Ph., Lindinger, Ch., Blank, I. and Märk, T., Quantitation of furan and methylfuran formed in different precursor systems by proton transfer reaction mass spectrometry, *Journal of Agricultrual and Food Chemistry*, **54**, 2786–2793, copyright 2006, American Chemical Society.)

(ii) the ease of furan formation by lipid oxidation and *Maillard*-type reactions from (iii) a wide range of precursors with ascorbic acid and PUFAs showing the highest potential followed by carotenoids, sugars and amino acids, which are intrinsic food constituents occurring naturally in raw materials.

Furan formation is mainly associated with lipid oxidation while acrylamide is in first instance a *Maillard* reaction issue. However, both reaction cascades may lead to either of the processing contaminant. Therefore, a major challenge for academia and industry remains the development of concepts leading to a reduction of heat-induced risky compounds such as furan upon industrial and also domestic food processing without changing the desired organoleptic properties of the food. Other potential process contaminants should be considered as well as recently done by Fan and Mastovska (2006).

Due to the complexity of furan formation in general and the differences observed in model systems as compared to food, conclusions should be drawn with much caution, avoiding data extrapolation from oversimplified model systems based on single precursors to complex food products. In fact, any study using model systems (preferably simulating food processing conditions) should be followed up by a validation step in the corresponding food environment.

Further contributions are expected on furan research in food as indicated by the abstracts of the ACS symposium in San Francisco in 2006 (Granvogl and Schieberle; Becalski; Morehouse *et al.*, 2007). We have also generated results on furan formation from sugars, *Maillard* systems and lipids, including PUFAs and carotenes, which will be published soon.

Additional work is required to get a better control of heat-induced formation of furan and other food-born process contaminants. Such a study should be combined with monitoring of positive food quality attributes, i.e. aroma, taste, colour, texture. This may help to understand the formation mechanisms of furan and other targeted compounds from various types of precursors under food processing conditions with the aim of reducing their formation while keeping the desired food quality. Finally, a special effort is expected for generating and gathering reliable analytical data on furan levels in food. This is the basis for a reliable assessment of the dietary mean exposure for the average consumer to furan.

References

Altaki, M.S., Santos, F.J. and Galceran, M.T. (2007) Analysis of furan in foods by headspace solid-phase microextraction–gas chromatography–ion trap mass spectrometry. *Journal of Chromatography A*, **1146**, 103–109.

Becalski, A., Forsyth, D., Casey, V., Lau, B.P.-Y., Pepper, K. and Seaman, S. (2005) Development and validation of a headspace method for determination of furan in food. *Food Additives and Contaminants*, **22**, 535–540.

Becalski, A. and Seaman, S. (2005) Furan precursors in food: A model study and development of a simple headspace method for determination of furan. *Journal of AOAC International*, **88**, 102–106.

Bianchi, F., Careri, M., Mangia, A. and Musci, M. (2006) Development and validation of a solid phase micro-extraction–gas chromatography–mass spectrometry method for the determination of furan in baby-food. *Journal of Chromatography A*, **1102**, 268–272.

Charlton, A.J., Davis, A.L., Jones, D.J., Lewis, J.R., Davies, A.P., Haslam, E. and Williamson, M.P. (2000) The self association of the black tea polyphenol theaflavin and its complexation with caffeine. *Journal of the Chemical Society: Perkin Transactions*, **2**, 317–322.

Crews, C. and Castle, L. (2007) A review of the occurrence, formation and analysis of furan in heat-processed foods. *Trends in Food Science and Technology*, **18**, 365–372.

De Bièvre, P. (1990) Isotope dilution mass spectrometry: What can it contribute to accuracy in trace analysis? *Fresenius Journal of Analytical Chemistry*, **337**, 766–771.

EC guidelines, Commission Decision 2002/657/EC (12 August 2002) Implementing Council Directive 96/23/EC concerning the performance of analytical methods and the interpretation of results. *Official Journal of European Communities*, **L221**, 8–36.

EFSA (2004a) Report of the Scientific Panel on Contaminants in the Food Chain on provisional findings of furan in food. *EFSA Journal*, **137**, 1–20. Available from: http://www.efsa.eu.int/science/contam/ contam_documents/760/contam_furan_report7-11-051.pdf

EFSA (2004b) Report of the CONTAM Panel on provisional findings on furan in food. Annexe corrigendum. Available from: http://www.efsa.europa.eu/etc/medialib/efsa/science/contam/ contam_documents/760.Par.0002.File.dat/furan_annex1.pdf

EFSA (2006c) Invitation to submit data on furan in food and beverages. Available from: http://www. efsa.europa.eu/en/science/data_collection/furan.html

Fan, X. (2005a) Impact of ionizing radiation and thermal treatments on furan levels in fruit juice. *Journal of Food Science*, **70**, E409–E414.

Fan, X. (2005b) Formation of furan from carbohydrates and ascorbic acid following exposure to ionizing radiation and thermal processing. *Journal of Agricultural and Food Chemistry*, **53**, 7826–7831.

Fan, X.T. and Mastovska, K. (2006) Effectiveness of ionizing radiation in reducing furan and acrylamide levels in foods. *Journal of Agricultural and Food Chemistry*, **54**, 8266–8270.

Fan, X.T. and Sommers, C.H. (2006) Effect of gamma radiation on furan formation in ready-to-eat products and their ingredients. *Journal of Food Science*, **71**, C407–C412.

FAPAS (2005) Acrylamide and other food processing derived contaminants. Proficiency test 3012, May 2005–June 2005.

FDA (2004a) Exploratory data on furan in food. Available from: http://www.cfsan.fda.gov/wdms/furandat. htm

FDA (2004b) Determination of furan in foods. Available from: http://www.cfsan.fda.gov/wdms/furan.html

Ferretti, A., Flanagan, V.P. and Ruth, J.M. (1970) Nonenzymatic browning in a lactose-casein model system. *Journal of Agricultural and Food Chemistry*, **18**, 13–18.

Goldmann, T., Perisset, A., Scanlan, F. and Stadler, R.H. (2005) Rapid determination of furan in heated food-stuffs by isotope dilution solid phase microextraction–gas chromatography–mass spectrometry (SPME-GC-MS). *Analyst*, **130**, 878–883.

Grey, T. and Shrimpton, D.H. (1967) Volatile components in the breast muscle of chicken of different ages. *British Poultry Science*, **8**, 35–41.

Guenther, H., Lantz, I., Ternite, R., Wilkens, J., Hoenicke, K. and van der Stegen, G. (2006) Acrylamide in roast coffee-investigations of influencing factors. *Colloque Scientifique International sur le Cafe*, **21**, 582–589.

Hasnip, S., Crews, C. and Castle, L. (2006) Some factors affecting the formation of furan in heated foods. *Food Additives and Contaminants*, **23**, 219–227.

Ho, I.P., Yoo, S.J. and Tefera, S. (2005) Determination of furan levels in coffee using automated solid-phase microextraction and gas chromatography/mass spectrometry. *Journal of AOAC International*, **88**, 574–576.

Hoenicke, K., Fritz, H., Gatermann, R. and Weidemann, S. (2004) Analysis of furan in different foodstuffs using gas chromatography mass spectrometry. *Czech Journal of Food Science*, **22**, 357–358.

Horman, I. and Viani, R. (1971) The caffeine-chlorogenate complex of coffee. An NMR study, in *Proceeding of ASIC, 5e colloque*, Lisbonne, pp. 102–111.

IARC (International Agency for Research on Cancer) (1995) *Monographs on the Evaluation of Carcinogenic Risks to Humans*, Vol. 63. Dry Cleaning, Some Chlorinated Solvents and Other Industrial Chemicals. IARC Lyon, France, pp. 3194–3407.

Johnston, W.R. and Frey, C.N. (1938) The volatile constituents of roasted coffee. *Journal of the America Chemical Society*, **60**, 1624–1627.

Kallio, H., Leino, M. and Salorinne, L. (1989) Analysis of the headspace of foodstuffs near room temperature. *Journal of High Resolution Chromatography*, **12**, 174–177.

Kuballa, T., Stier, S. and Strichow, N. (2005) Furan in Kaffee und Kaffeegetränken. *Deutsche Lebensmittel-Rundschau*, **101**, 229–235.

Liao, M.L. and Seib, P.A. (1987) Selected reactions of L-ascorbic acid related to foods. *Food Technology*, **41**(11), 104–107, 111.

Limacher, A., Kerler, J., Conde-Petit, B. and Blank, I. (2007) Formation of furan and methylfuran from ascorbic acid in model systems and food. *Food Additives and Contaminants*, **24**(S1), 122–135.

Maga, J.A. (1979) Furans in foods. *CRC Critical Review in Food Science and Nutrition*, **11**, 355–400.

Märk, J., Pollien, Ph., Lindinger, Ch., Blank, I. and Märk, T. (2006) Quantitation of furan and methylfuran formed in different precursor systems by proton transfer reaction mass spectrometry. *Journal of Agricultural and Food Chemistry*, **54**, 2786–2793.

Mayer, F., Czerny, M. and Grosch, W. (2000) Sensory study of the character impact aroma compounds of a coffee beverage. *European Food Research and Technology*, **211**, 272–276.

Milo, C. and Blank, I. (1998) Quantification of impact odorants in food by isotope dilution assay – strength and limitations, in *Flavor Analysis – Developments in Isolation and Characterizing*; ACS Symposium Series 705 (eds. C.J. Mussinan and M.J. Morello). American Chemical Society, Washington, DC, pp. 250–259.

Morehouse, K.M., Nyman, P.J., McNeal, T.P., DiNovi, M.J. and Perfetti, G.A. (2007) A survey of furan in heat processed foods by headspace gas chromatography/mass spectrometry and an estimate of adult exposure. *Food Additives and Contaminants*, DOI: 10.1080/0265203070/15252949.

Mulders, E.J., Maarse, H. and Weurman, C. (1972) The odour of white bread. *Zeitschrift für Lebensmittel Untersuchung und Forschung*, **150**, 68–74.

NTP (National Toxicology Program) (1993) Toxicology and carcinogenesis studies of furan (CAS No. 110-00-9) in F344/N rats and B6C3Fl mice (gavage studies). *NTP Technical Report No. 402*, U.S. Department of Health and Human Services, Public Health Service, National Institutes of Health, Research Triangle Park, NC.

Nyman, P.J., Morehouse, K.M., McNeal, T.P., Perfetti, G.A. and Diachenko, G.W. (2006) Single-laboratory validation of a method for the determination of furan in foods by using static headspace sampling and gas chromatography/mass spectrometry. *Journal of AOAC International*, **89**, 1417–1424.

Perez, L.C. and Yaylayan, V.A. (2004) Origin and mechanistic pathways of formation of the parent furan – A food toxicant. *Journal of Agricultural and Food Chemistry*, **52**, 6830–6836.

Persson, T. and von Sydow, E. (1973) Aroma of canned beef: Gas chromatographic and mass spectrometric analysis of the volatiles. *Journal of Food Science*, **38**, 377–385.

Pickup, J.F. and McPherson, K. (1976) Theoretical considerations in stable isotope dilution mass spectrometry for organic analysis. *Analytical Chemistry*, **48**, 1885–1890.

Pollien, P., Lindinger, C., Yeretzian, C. and Blank, I. (2003) Proton transfer reaction mass spectrometry, a tool for on-line monitoring of acrylamide in food and Maillard systems. *Analytical Chemistry*, **75**, 5488–5494.

Qvist, I.H. and von Sydow, E.C.F. (1974) Unconventional proteins as aroma precursors. Chemical analysis of the volatile compounds in heated soy, casein, and fish protein model systems. *Journal of Agricultural and Food Chemistry*, **22**, 1077–1084.

Reinhard, H., Sager, F., Zimmermann, H. and Zoller, O. (2004) Furan in foods on the Swiss market – Method and results. *Mitteilungen in Lebensmittel Hygiene*, **95**, 532–535.

Roberts, D., Crews, C., Grundy, H., Mills, G. and Matthews, W. (2008) Effect of consumer cooking on furan in convenience foods. *Food Additives and Contaminants*, **25**, 25–31.

Sayre, L.M., Arora, P.K., Iyer, R.S. and Salomon, R.G. (1993) Pyrrole formation from 4-hydroxynonenal and primary amines. *Chemical Research in Toxicology*, **6**, 19–22.

Schieberle, P. (2005) The carbon module labeling (CAMOLA) technique: A useful tool for identifying transient intermediates in the formation of Maillard-type target molecules. *Annals of the New York Academy of Sciences*, **1043**, 236–248.

Şenyuva, H.Z. and Gökmen, V. (2005) Analysis of furan in foods. Is headspace sampling a fit-for-purpose technique? *Food Additives and Contaminants*, **22**, 1198–1202.

Şenyuva, H.Z. and Gökmen, V. (2007) Potential of furan formation in hazelnuts during heat treatment. *Food Additives and Contaminants*, **24**(S1), 136–142.

Shinoda, Y., Komura, H., Homma, S. and Murata, M. (2005) Browning of model orange juice solution: Factors affecting the formation of decomposition products. *Bioscience Biotechnology and Biochemistry*, **69**, 2129–2137.

Sugisawa, H. (1966) Thermal degradation of sugars. II. Volatile decomposition products of glucose caramel. *Journal of Food Science*, **31**, 381–385.

Walter, R.H. and Fagerson, I.S. (1968) Volatile compounds from heated glucose. *Journal of Food Science*, **33**, 294–297.

Yaylayan, V. (2003) Recent advances in the chemistry of Strecker degradation and Amadori rearrangement: Implications to aroma and color formation. *Food Science and Technology Research*, **9**, 1–6.

Yaylayan, V.A., Forage, N.G. and Mandeville, S. (1994) Microwave and thermally induced Maillard reactions, in *Thermally Generated Flavors* (eds. T.H. Parliment, M.J. Morello and R.J. MacGorrin). American Chemical Society, Washington, DC, pp. 449–456.

Yaylayan, V. and Wnorowski, A. (2001) The role of L-serine and L-threonine in the generation of sugar-specific reactive intermediates during Maillard reaction, in *Food Flavors and Chemistry – Advances of the New Millennium* (eds. A. Spanier, F. Shahidi, T. Parliment, C. Mussinan, C.-T. Ho and E. Contis). Royal Society of Chemistry, Cambridge, UK, pp. 313–317.

12 Chloropropanols and their Fatty Acid Esters

Colin G. Hamlet

Summary

Although the occurrence of chloropropanols and their fatty acid esters (chloroesters) has long since been associated with acid-hydrolysed vegetable protein, an ingredient used in processed savoury food products, the occurrence of these contaminants in other foodstuffs is relatively recent and has prompted much new research. This chapter brings together for the first time information on chloropropanols and chloroesters arising from a wide range of processes used in the manufacture of foods, including food contact materials and foods prepared in the home. A brief historical perspective introduces the reader to the topic while developments in toxicology, state-of-the-art methods of analysis, occurrence, mechanisms of formation and mitigation efforts of the food industry are reviewed, along with estimates of exposure and controlling legislation. Several new lines of evidence indicate that chloroesters are widespread in processed foods and that levels of these contaminants are higher than the corresponding chloropropanols. This is of concern because model studies have shown that chloroesters can be hydrolysed to known toxic chloropropanols such as 3-chloropropane-1,2-diol (3-MCPD) by enzymes present in foodstuffs. The toxicological significance of chloroesters and their contribution to dietary intakes of chloropropanols is not yet known.

12.1 INTRODUCTION

Chloropropanols and chloroesters are contaminants that are formed during the processing and manufacture of certain foods and ingredients. The presence of these compounds in foods is of concern because toxicological studies have shown that they could endanger human health (Lynch *et al.*, 1998).

12.1.1 Historical perspective: 1979–2000

Velíšek (Velíšek *et al.*, 1978, 1979, 1980, 1982; Velíšek and Davídek, 1985) and Davídek (Davídek *et al.*, 1980, 1982) were the first to demonstrate that chloropropanols and chloroesters could be formed in hydrolysed vegetable proteins (HVP) produced by hydrochloric acid hydrolysis of proteinaceous by-products from edible oil extraction such as soybean meal, rapeseed meal and maize gluten. It was shown that hydrochloric acid could react with residual glycerol and lipids associated with the proteinaceous materials to yield a range of chloropropanols. The main chloropropanol found in HVP was 3-chloropropane-1,2-diol (3-MCPD) together with lesser amounts of 2-chloropropane-1,3-diol (2-MCPD),

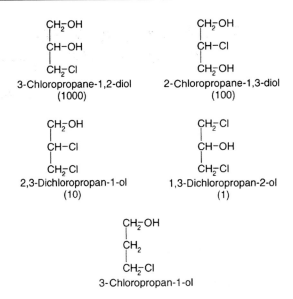

Fig. 12.1   Chloropropanols found in hydrolysed vegetable proteins and their relative amounts (in parentheses).

1,3-dichloropropanol (1,3-DCP), 2,3-dichloropropanol (2,3-DCP) and 3-chloropropan-1-ol (see Fig. 12.1). Although fatty acid esters of 1,3-DCP and 3-MCPD (mono- and diesters) were formed in acid HVP (Velíšek *et al.*, 1980; Velíšek, 1989; Velíšek and Ledahudcová, 1993), it was shown that only traces remained in the final commercial product since the majority were effectively removed during filtration of the hydrolysate. Collier *et al.* (1991) later showed that chloroesters were intermediates in the formation of dichloropropanols (DCPs) and monochloropropanediols (MCPDs) from lipids and the relative amounts of each chloropropanol formed in acid HVP were dependent on the composition of the lipid in the raw materials used.

The first reported occurrence of chloroesters in a natural unprocessed food was by Cerbulis *et al.* (1984) who identified a small but significant quantity (<1% of total neutral lipids) of diesters of 3-MCPD in raw milk from several herds of goats. The chloroesters were composed of molecular species containing C10–C18 fatty acids and corresponded closely in carbon number to sn-1,2-diacylglycerols (sn = stereospecific numbering system) moieties of the goat milk triacylglycerols (Kuksis *et al.*, 1986). Stereospecific analysis (Myher *et al.*, 1986) showed that the 3-MCPD diesters were racemic mixtures of both enantiomers (see Fig. 12.2). These results appeared to exclude their biosynthesis in vivo; however, it was not clear whether these compounds could have formed from dietary substances, e.g. 3-MCPD or its esters, passed through the organism or from anthropogenic chlorine containing compounds (such as chlorine-based sanitisers used in dairy operations) post-secretion.

In 1994 and 1997, the EU Scientific Committee for Food concluded that 3-MCPD should be regarded as a genotoxic carcinogen and since a safe threshold dose could not be determined, residues in foods should be undetectable by the most sensitive analytical method (European Commission, 1997). This spurred intense analytical (Vanbergen *et al.*, 1992; Hamlet and Sutton, 1997; Hamlet, 1998; Meierhans *et al.*, 1998) monitoring (MAFF, 1999a,b; Wu and Zhang, 1999; Macarthur *et al.*, 2000) and mitigation activities in many countries. In the case

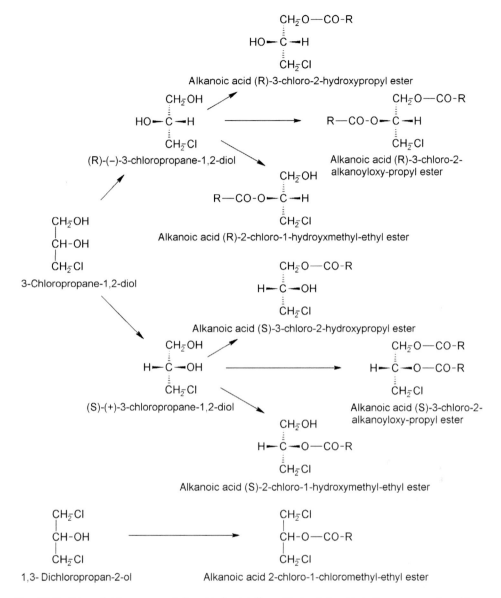

Fig. 12.2 Esters of chloropropanols found in foodstuffs and their relationship to the corresponding chloropropanol isomers.

of HVP, progress made by European manufacturers in implementing initial chloropropanol reduction processes (Faesi *et al.*, 1987; Brown *et al.*, 1989; De Rooij *et al.*, 1989; Hirsbrunner and Weymuth, 1989; Payne, 1989) have significantly reduced levels of both MCPDs and DCPs in these products.

12.1.2 Scope of chapter

The subsequent discovery of chloropropanols in food contact materials treated with wet strength resins (Boden *et al.*, 1997), drinking water, and a range of processed foods that do not use acid-HVP as an ingredient (FAC, 2000), has renewed interest in these contaminants. This chapter brings together the information available (since approximately 2000) on toxicology, methods of analysis, occurrence, mechanisms and precursors, mitigation options, exposure and regulatory status for chloropropanols and chloroesters in foodstuffs.

12.2 TOXICOLOGY

12.2.1 3-MCPD

The chloropropanol 3-MCPD has been investigated in short- and long-term toxicity studies and the most recent toxicology, mutagenicity and carcinogenicity data have been summarised previously by the Joint FAO/WHO Expert Committee on Food Additives (JECFA) at its meeting in 2001 (Schlatter *et al.*, 2002a). In rats and mice the kidney was the main target organ for toxicity with effects also observed on male fertility. Studies have demonstrated that 3-MCPD has mutagenic activity in vitro (Lynch *et al.*, 1998; Robjohns *et al.*, 2003) although negative results reported from a bone marrow micronucleus assay in rats and a rat liver unscheduled DNA synthesis (UDS) assay (Fellows, 2000; Marshall, 2000) have provided reassurance that the mutagenic activity seen in vitro was not expressed in vivo (COM, 2000). No epidemiological or clinical studies in humans have been reported.

In June 2001, JECFA (2002) assigned a provisional maximum tolerable daily intake (PMTDI) for 3-MCPD of 2 μg kg^{-1} body weight on the basis of the lowest observed effect level (LOEL) and a safety factor of 500. The safety margin included a factor of five for extrapolation from an LOEL to a no observed effect level (NOEL) and was considered adequate to account for the effects on male fertility and for inadequacies in the reproductive toxicity data.

12.2.2 Dichloropropanols

The available toxicology, mutagenicity and carcinogenicity data for 1,3-DCP have been summarised previously by the 57th session of JECFA in 2001 (Schlatter *et al.*, 2002b). JECFA concluded that 1,3-DCP was hepatotoxic, induced a variety of tumours in various organs in the rat, and was genotoxic in vitro. The UK Committees on Mutagenicity (COM) and Carcinogenicity (COC) of Chemicals in Food, Consumer Products and the Environment considered 1,3-DCP in 2003 and 2004, respectively, following the publication of results from in vivo rat bone marrow micronucleus and rat liver UDS tests (Fellows, 2000; Marshall, 2000). The COM (2003) concluded that 1,3-DCP is not genotoxic in vivo in the tested tissues. However, the COC (2004) concluded that 1,3-DCP should be regarded as a genotoxic carcinogen as it was not possible to exclude a genotoxic mechanism for the induction of tumours of rat tongue observed in the 2-year carcinogenicity study. The Committee also recommended that further investigation regarding the mechanisms of 1,3-DCP carcinogenicity in vivo is needed. The most recent assessment of 1,3-DCP (JECFA, 2006) concluded that the critical effect of 1,3-dichloro-2-propanol was carcinogenicity and that a genotoxic mode of action could not be excluded. On the basis of these findings a tolerable daily intake for 1,3-DCP has not been set.

There are very few data on the absorption, distribution and excretion of 2,3-DCP. Theoretically, 2,3-DCP could be metabolised to produce epichlorohydrin (and subsequently glycidol) and therefore there are structural alerts for genotoxicity and carcinogenicity (COM, 2004).

Limited in vitro mutagenicity data indicate that 2,3-DCP is genotoxic with and without metabolic activation in bacterial and mammalian cells (COM, 2004). Recently published in vivo rat bone marrow micronucleus and rat liver UDS assays were negative (Fellows, 2000; Marshall, 2000). No appropriate carcinogenicity studies of 2,3-DCP are available. The UK COM (2004) considered that 2,3-DCP has no significant genotoxic potential in vivo in the tissues evaluated (i.e. bone marrow and liver in the rat). The COC (2004) considered that no conclusions regarding carcinogenicity of 2,3-DCP could be reached.

12.2.3   Chloroesters

The recent discovery of chloroesters in foods other than acid-HVP has raised questions about the bioavailability of chloropropanols from the dietary intake of these substances. To date, there are insufficient data for expert bodies to evaluate dietary intake or the toxicological significance of chloroesters, and JECFA (2006) has recommended that studies be undertaken to address these questions.

12.3   ANALYTICAL METHODS

12.3.1   Chloropropanols

The measurement of chloropropanols at trace contaminant levels in foodstuffs presents a challenge for the analyst: the absence of suitable chromophores has made approaches based on high performance liquid chromatography (HPLC) with ultraviolet or fluorescence detection unsuitable; the low volatility and high polarity of, e.g. MCPDs give rise to unfavourable interactions with components of gas chromatography (GC) systems that result in poor peak shape and low sensitivity; while the low molecular weight of both MCPDs and DCPs make mass detection difficult since diagnostic ions cannot be reliably distinguished from background chemical noise. Many of these limitations have been overcome and methods based on the formation of stable GC volatile derivatives with mass-selective (MS) detection predominate. Restrictions on the commercial availability of some reference standards have necessitated custom syntheses of, e.g. 2-MCPD. The latter compound has been quantified to a first approximation by using the response factors from readily available 3-MCPD calibration standards (Hamlet *et al.*, 2002).

12.3.1.1   *Heptafluorobutyryl ester derivatives*

Procedures based on the GC/MS detection of heptafluorobutyryl (HFB) esters of chloropropanols have been universally applied to the analysis of a wide range of foodstuffs and have subsequently become the normative reference methods (Brereton *et al.*, 2001; British Standards Institution, 2004). Heptafluorobutyryl imidazole (HFBI) and heptafluorobutyric anhydride (HFBA) are the derivatisation reagents of choice and the formation of the corresponding HFB esters is selective for the hydroxyl function and hence the analysis of all chloropropanols (see Fig. 12.3). Since HFBI and HFBA are moisture sensitive, derivatisation must be carried out under anhydrous conditions.

Brereton *et al.* (2001) validated by collaborative trial the procedures developed by Hamlet and Sutton (1997) and Hamlet (1998) for the determination of 3-MCPD at the low μg kg^{-1} level in a wide range of foods and ingredients. The water-soluble 3-MCPD was extracted into saline solution and then partitioned into diethyl ether using a solid phase extraction technique based on diatomaceous earth. Dried and concentrated extracts were treated with

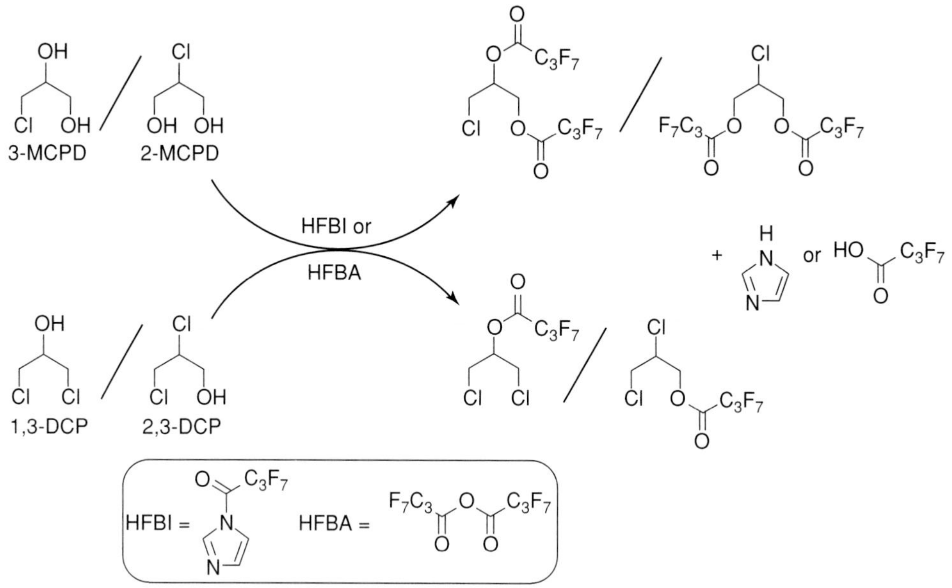

Fig. 12.3 Derivatisation reactions of chloropropanols with HFBI and HFBA.

heptafluorobutyrylimidazole (HFBI) to convert 3-MCPD to the corresponding HFB di-ester (see Fig. 12.3) prior to analysis by GC and mass spectrometry (MS). Quantification was by a stable isotope internal standard method using deuterium labelled 3-MCPD (3-MCPD-d_5) added to the sample prior to extraction. The method was suitable for the quantification of 3-MCPD at levels of ≥ 10 µg kg^{-1} and was adopted as an AOAC First Action Official Method. The three main advantages of this approach were (a) high sensitivity resulting from the formation of a GC volatile 3-MCPD derivative; (b) high specificity associated with mass spectrometric detection at higher mass; and (c) accurate quantification from the use of a stable isotope internal standard.

The procedure of Brereton *et al.* (2001) has subsequently been modified and extended to the analysis of MCPDs and/or DCPs. Nyman *et al.* (2003b) optimised a method for the analysis of 1,3-DCP in soy sauces using 10% diethyl ether/hexane to isolate the dichloropropanol at the solid phase extraction stage. The mean recovery of 1,3-DCP from spiked test samples was 100% with a relative standard deviation of 1.32%: the method LOQ was 0.185 µg kg^{-1}. Xu *et al.* (2006) added hexane at the initial aqueous extraction stage to remove fat from samples and also reported improved sensitivity and selectivity over electron ionisation (EI)-MS when using negative ion chemical ionisation MS. The authors also showed that the derivatisation of DCPs with HFBA was more effective when a catalytic amount of triethylamine was added. Chung *et al.* (2002) used silica gel columns with ethyl acetate as the elution solvent for the solid phase extraction step in the analysis of soy sauces. Abu-El-Haj *et al.* (2007) used dichloromethane as the chloropropanol elution solvent: aliquots of aqueous samples, e.g. soy sauce and soups, were pre-absorbed onto alumina columns while dry samples, e.g. crumbed cereal products, were eluted directly following a lipid removal step with hexane. Chung *et al.* (2002) and Abu-El-Haj *et al.* (2007) both used HFBA for the derivatisation step and the limits of quantification (LOQ) for these methods were 3–5 µg kg^{-1}.

The EI mass spectra of the HFB esters of 3-MCPD, 1,3- and 2,3-DCP and the characteristic ions for all chloropropanols are given in Fig. 12.4 and Table 12.1, respectively.

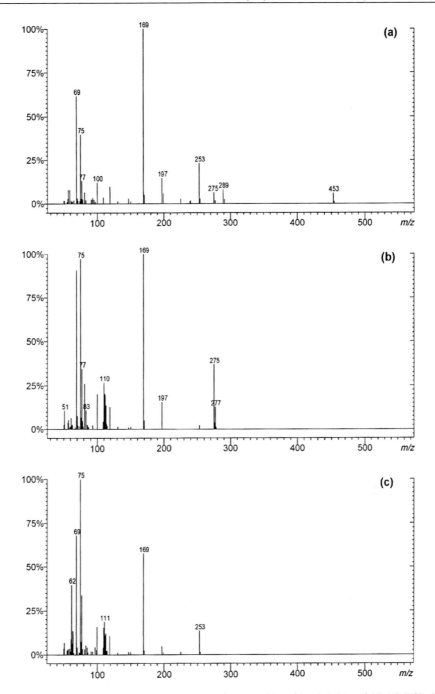

Fig. 12.4 EI mass spectra of the HFB derivatives of (a) 3-MCPD, (b) 1,3-DCP and (c) 2,3-DCP. (Data courtesy of RHM Technology, UK.)

Table 12.1 Characteristic ions (m/z) in the mass spectra of the HFB esters of chloropropanols.

(a) EI mass spectra

HFB ester of:	MW	[M-CH$_2$Cl]$^+$	[M-C$_3$F$_7$CO$_2$]$^+$	[M-C$_3$F$_7$CO$_2$CH$_2$]$^+$	[M-C$_3$F$_7$CO$_2$-HCl]$^+$	[M-C$_3$F$_7$CO$_2$-C$_3$F$_7$CO$_2$H]$^+$
3-MCPD	502	453	289/291[a]	275/277[a]	253	75/77[a]
3-MCPD-d$_5$	507	456	294/296[a]	278/280[a]	257	79/81[a]
2-MCPD	502	–	289/291[a]	–	253	75/77[a]
1,3-DCP	324	275/277[a]	111/113/115[a]	–	75/77[a]	–
1,3-DCP-d$_5$	329	278/280[a]	116/118/120[a]	–	79/81[a]	–
2,3-DCP	324	–	111/113/115[a]	–	75/77[a]	–

(b) NCI mass spectra

HFB ester of:	MW	[M-HF]$^-$	[M-HF-HCl]$^-$	[M-HF-HCl-HF]$^-$
3-MCPD	502	482/484[a]	446	426
3-MCPD-d$_5$	507	486/488[a]	449	428
2-MCPD	502	482/484[a]	446	426
1,3-DCP	324	304/306/308[a]	268/270[a]	248/250[a]
1,3-DCP-d$_5$	329	308/310/312[a]	271/273[a]	250/252[a]
2,3-DCP	324	304/306/308[a]	268/270[a]	248/250[a]

[a] Isotopic chlorine cluster ions.

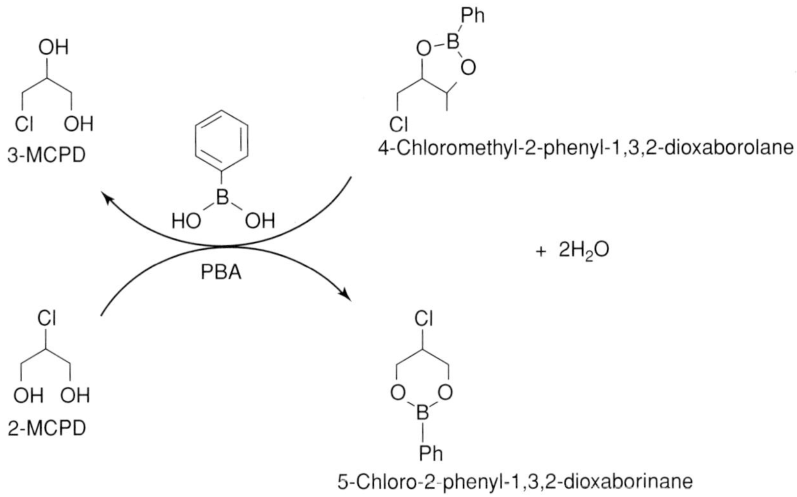

Fig. 12.5 Derivatisation reactions of MCPDs with PBA.

12.3.1.2 *Dioxaborolane/dioxaborinane derivatives*

Boronic acids react with 1,2- and 1,3-diols to give cyclic dioxaborolane/dioxaborinane derivatives. Rodman and Ross (1986) carried out derivatisation under anhydrous conditions while Pesselman and Feit (1988) showed that a boronic acid derivative of 3-MCPD could be prepared under aqueous conditions, despite the liberation of water from the formation reaction (see Fig. 12.5). Hence, this approach has been adopted for the analysis of liquid seasonings such as HVP and soy sauces (Plantinga *et al.*, 1991; Wu and Zhang, 1999). Since boronic acids react specifically with diol compounds other chloropropanols such as 1,2- and 1,3-DCP cannot be determined by this procedure.

Breitling-Utzmann *et al.* (2003) and Divinová *et al.* (2004) prepared boronic acid derivatives of MCPDs extracted from a wide variety of retail foodstuffs. Breitling-Utzmann *et al.* used saline solution for both extraction and derivatisation while Divinová *et al.* used hexane/acetone (1:1 v/v) for the initial extraction followed by a partition step into saline solution. Derivatisation was carried out in the saline solution using phenylboronic acid (PBA) and the reaction products were extracted into hexane. Quantification was by a stable isotope internal standard method using 3-MCPD-d_5 and GC/MS operating in selected ion monitoring (SIM) mode. Typical LOQs ranged from 10 µg kg^{-1} (e.g. soy sauce) to 50 µg kg^{-1} (e.g. toasted bread). Velíšek *et al.* (2002) used a Chiraldex G-TA fused silica capillary column to separate the optical isomers of 3-MCPD prior to quantification by MS. Kuballa and Ruge (2004) showed that lower LOQs could be achieved using GC combined with tandem mass spectrometry (MS/MS) on a triple quadrupole mass spectrometer operating in the selected reaction monitoring mode. In a recent publication Huang *et al.* (2005) demonstrated that the PBA derivative of 3-MCPD prepared in liquid HVPs could be extracted and quantified at the low µg kg^{-1} level using headspace solid-phase microextraction (SPME) combined with GC/MS detection.

The EI mass spectra of the PBA derivative of 3-MCPD and the characteristic ions for all MCPDs are given in Fig. 12.6 and Table 12.2, respectively.

Table 12.2 Characteristic ions (m/z) in the EI mass spectra of the PBA derivatives of MCPDs.

PBA derivative of:	MW	[M]+	[M-CH₂Cl]+	Other structurally significant ions
3-MCPD	196	196/198[a]	146/147[b]	103/104[b] [Ph-BO]+, 91 [C₇H₇]+
3-MCPD-d₅	201	201/203[a]	149/150[b]	103/104[b] [PhBO]+, 93 [C₇H₅D₂]+
2-MCPD	196	196/198[a]	–	103/104[b] [PhBO]+, 91 [C₇H₇]+

[a] Isotopic chlorine cluster ions.
[b] Isotopic boron cluster ions.

12.3.1.3 Dioxolane/dioxane derivatives

The reaction of 1,2- and 1,3-diols with aldehydes and ketones to give cyclic acetals and ketals is well known (March, 1983) and Meierhans et al. (1998) were the first to report a quantitative method for 2- and 3-MCPD using dry acetone as the derivatising reagent in the presence of toluene-4-sulphonic acid (TsOH) according to the reaction scheme given in Fig. 12.7. The resulting 1,3-dioxolane and 1,3-dioxane derivatives were characterised by GC/MS. MCPDs were partitioned from aqueous samples/extracts using solid phase extraction on diatomaceous earth and diethyl ether elution described previously. In keeping with the HFB reagents, anhydrous conditions are required for derivatisation.

Subsequent modifications to the procedure of Meierhans et al. (1998) include: the addition of MCPD-d₅ to samples prior to extraction for quantification (Jin et al., 2001; Dayritt and Ninonuevo, 2004; Rétho and Blanchard, 2005); the use of 3-pentanone or 4-heptanone/TsOH for derivatisation (Dayritt and Ninonuevo, 2004; Rétho and Blanchard, 2005); the use of ethyl acetate for the solid phase partition step and additional clean up post-derivatisation using basic aluminium oxide cartridges (Rétho and Blanchard, 2005). The LOQs for these methods are in the range 1–5 μg kg⁻¹.

Although the occurrence of 2-MCPD has been reported in some foodstuffs (Rétho and Blanchard, 2005) no mass spectral data have been given for the acetone derivative (dioxane)

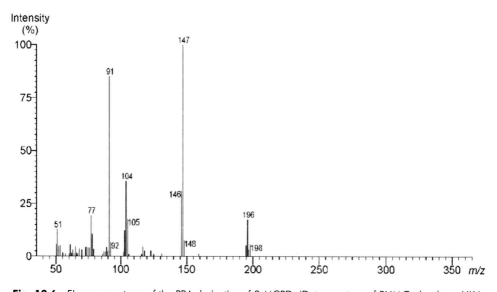

Fig. 12.6 EI mass spectrum of the PBA derivative of 3-MCPD. (Data courtesy of RHM Technology, UK.)

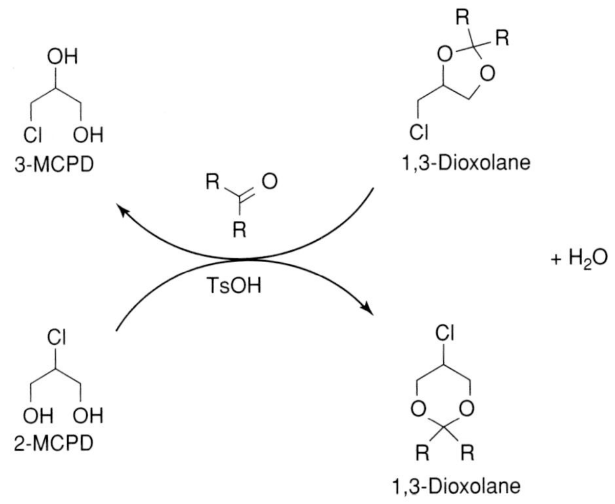

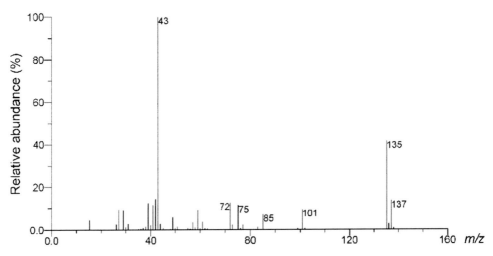

Fig. 12.7 Derivatisation reactions of MCPDs with ketones.

of this isomer. These methods have not been used for the determination of dichloropropanols since these compounds do not form cyclic derivatives with ketones. The EI mass spectra of the acetone derivative of 3-MCPD and the characteristic ions for all MCPDs are given in Fig. 12.8 and Table 12.3, respectively.

12.3.1.4   Other methods of interest

The higher volatility and lower polarity of dichloropropanols, compared to MCPDs, has permitted the direct determination of 1,3-DCP using GC/MS, i.e. without derivatisation (Wittmann, 1991). Schumacher *et al.* (2005) extracted 1,3-DCP from river water samples

Fig. 12.8   EI mass spectrum of the acetone derivative of 3-MCPD. (Data courtesy of RHM Technology, UK.)

Table 12.3   Characteristic ions (m/z) in the EI mass spectra of the dioxolane derivatives of MCPDs.

Chloropropanol	Dioxolane of:	MW	$[M-C_nH_{2n+1}]^+$	Other structurally significant ions
3-MCPD	acetone	150	135/137[a]	43 $[C_2H_3O]^+$
3-MCPD-d$_5$	acetone	155	140/142[a]	43 $[C_2H_3O]^+$
3-MCPD	3-pentanone	178	149/151[a]	57 $[C_3H_5O]^+$
3-MCPD-d$_5$	3-pentanone	183	154/156[a]	57 $[C_3H_5O]^+$
3-MCPD	4-heptanone	206	163/165[a]	71 $[C_4H_7O]^+$
3-MCPD-d$_5$	4-heptanone	211	168/170[a]	71 $[C_4H_7O]^+$

[a] Isotopic chlorine cluster ions.

using ethyl acetate while Crews *et al.* (2002) showed that 1,3-DCP was sufficiently volatile to be quantified in the headspace of soy sauces. Both groups used GC/MS to monitor ions at *m/z* 79 and *m/z* 81 for 1,3-DCP and *m/z* 82 for the stable isotope internal standard 1,3-DCP-d$_5$. The procedure of Crews *et al.* (2002) was subsequently validated by a collaborative trial in which participants used both static headspace and SPME, and the method was shown to be rapid, accurate and fit for purpose (Hasnip *et al.*, 2005) with a method LOD of $\leq$10 μg kg^{-1}.

Although the formation of trimethylsilyl ethers of chloropropanols has previously been used for the analysis of MCPDs and DCPs in HVPs (Wittmann, 1991) and resin-treated papers (Boden *et al.*, 1997), recent applications have been limited to the analysis of soy sauces (Mingxia *et al.*, 2003).

Leung *et al.* (2003) recently demonstrated the feasibility of using molecularly imprinted polymers (MIP) as a qualitative tool for the screening of food products for 3-MCPD. An MIP derived from 4-vinylphenylboronic acid was shown to act as a potentiometric chemosensor via the increase in Lewis acidity of the receptor sites upon reaction of the aryl boronic acid with 3-MCPD (see Fig. 12.5). A simple pH glass electrode was sufficient to monitor the analyte-specific binding and in water a linear response was obtained over 0–350 mg kg^{-1}.

Xing and Cao (2007) described a capillary electrophoresis technique with electrochemical detection for the rapid analysis of 3-MCPD in soy sauces. The linear range for of the method was 6.6–200 mg L^{-1} with an LOD of 0.13 mg L^{-1}.

12.3.2   Chloroesters

There are fewer methods for the analysis of chloroesters. Hamlet and Sadd (2004), Hamlet *et al.* (2004a) and Zelinková *et al.* (2006) developed methods for the direct analysis of esters of 3-MCPD in cereal products and edible oils based on an adaptation of the earlier procedure of Davídek *et al.* (1980). Fat containing the chloroesters was extracted from samples using either ethyl acetate or diethyl ether and separated by preparative TLC into individual fractions containing the diesters and monoesters of 3-MCPD. The individual fractions were analysed by GC/MS and the diesters and monoesters of MCPDs were identified by comparison with synthetic 3-MCPD-palmitate esters and reference mass spectral data (Kraft *et al.*, 1979; Davídek *et al.*, 1980). All MCPD-esters were quantified as 3-MCPD-dipalmitate using 5-α-cholestane as an internal standard. No recent methods for the analysis of monoesters of 1,3- and 2,3-DCP have been reported.

The isolation and measurement of all chloroesters by these methods is a lengthy process due to the many species arising from the different fatty acid combinations associated with

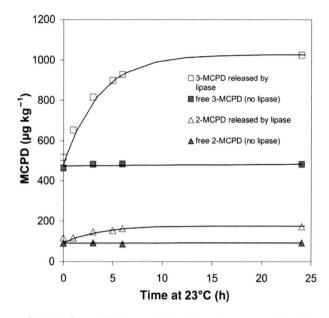

Fig. 12.9   Release of MCPDs from MCPD-esters in bread crust by enzyme hydrolysis.

each chloropropanol moiety. Chloroesters may be hydrolysed to the corresponding toxic chloropropanol, either during food processing or in vivo.Consequently, methods have been devised to measure the total amount of chloropropanol, i.e. free and ester derived.

Hamlet *et al.* (2004a) and Hamlet and Sadd (2004) measured total MCPDs in dried cereal products, i.e. free and ester derived. The MCPDs esterified with fatty acids were released by enzyme hydrolysis following incubation in 0.1 M phosphate buffer (pH 7.0) with a commercial lipase from *Aspergillus oryzae* for 24 hours (see Fig. 12.9). Total MCPDs were determined as the HFB-esters by GC/MS as described previously (see Section 12.3.1.1). The recovery of 3-MCPD from 3-MCPD-dipalmitate reference standard was 106% and 91% at 250 µg kg^{-1}and 750 µg kg^{-1}(sample basis), respectively. The repeatability of the method, expressed as a coefficient of variation, was 3.7% and the limit of quantification was <10 µg kg^{-1}. Divinová *et al.* (2004) developed a method to measure total 3-MCPD in a wide range of foodstuffs. Free and esterified 3-MCPD was isolated from samples as fat using diethyl ether as the extraction solvent. The solvent-free fat extract was then subjected to methanolysis using methanolic sulphuric acid and the 3-MCPD generated was quantified as the PBA derivative using GC/MS as described previously (see Section 12.3.1.1). The LOD and LOQ were determined to be 1.1 mg kg^{-1} and 3.3 mg kg^{-1} respectively on the extracted fat basis. Samples and extracts prepared by these methods were not analysed for the corresponding DCPs.

12.4   OCCURRENCE

12.4.1   Chloropropanols

12.4.1.1   *Soy sauces and related products*

Following reports in 1999 that unacceptably high levels of chororpopanols (Joint Food Safety and Standards Group, 1999; Macarthur *et al.*, 2000) had been found in some soy sauce

Table 12.4 Published data on the occurrence of 3-MCPD and 1,3-DCP in soy sauces and similar products.

Author	3-MCPD (mg kg^{-1})			1,3-DCP (mg kg^{-1})		
	Incidence[a]	Mean	Range	Incidence[a]	Mean	Range
Macarthur et al. (2000)	19/40	2.48[b]	<0.01–30.5	–	–	–
Jin et al. (2001)	15/30	–[c]	<0.01–>10	–	–	–
Fromberg (2001)	9/21	7.67[b]	<0.01–69	–	–	–
FSA (2001a,b)	31/100	3.73[b]	<0.01–93.1	17/100	0.023[b]	<0.005–0.345
FSA (2002)	14/273	0.253[b]	<0.001–23.7			
Nyman et al. (2003b)	33/55	44.07[b]	<0.025–876	14/39	0.650[b]	<0.025–9.84
Crews et al. (2003)	8/99	0.235[b]	<0.01–21.2	1/99	0.003[b]	0.017
FSANZ (2003)	18/39	15.4[b]	<0.01–148.2	9/39	0.070[b]	<0.01–0.60
European Commission (2004)	714/2035	9.164[b]	<0.01–1779	60/282	0.092[b]	0.010–1.37
Cheng et al. (2004)	104/214	–[c]	<0.01–6.63	–	–	–
Wong et al. (2006)	45/421	0.45[d]	<0.01–>3	–	–	–

[a] Number of samples above the limit of detection.
[b] Derived value from reported data (one half of the reported limit of detection value used to calculate the mean).
[c] Data not available.
[d] Upper bound mean level.

samples, 3-MCPD and 1,3-DCP have been monitored extensively in these products around the world (see Table 12.4).

Many of these products originated from Asia and via manufacturing processes that utilised acid hydrolysis of defatted soya beans, a process that without adequate controls can give rise to chloropropanol contamination (Velíšek et al., 1978). A breakdown of the occurrence of 3-MCPD in soy sauces by country of origin, taken from a major compilation of data produced by EU member states (European Commission, 2004), is shown in Fig. 12.10. It should be emphasised that these data do not necessarily reflect the current status of products from a particular country of origin since some of the data may originate from samples produced prior to the introduction of chloropropanol reduction methods. Furthermore, the data are likely to be skewed as a result of targeted analyses of products suspected of having high levels of chloropropanols.

Less data are available for other chloropropanols in soy sauces: in samples where quantifiable levels of 3- and 2-MCPD were measured (European Commission, 2004), the two isomers were correlated ($r^2 = 0.95$, $n = 25$) and the ratio of 3-MCPD:2-MCPD was about 8:1. All samples containing DCPs also contained 3-MCPD (Nyman et al., 2003a; European Commission, 2004). Nyman et al. (2003a) found that 3-MCPD and 1,3-DCP were loosely correlated ($r^2 = 0.73$) in samples containing quantifiable levels of both chloropropanols, although this trend was not apparent for all DCP isomers in data from other surveys (European Commission, 2004). However, it could be shown that samples containing in excess of 10 mg kg^{-1} 3-MCPD would be expected to contain 1,3-DCP in concentrations ranging from 0.25 kg^{-1} to 10 mg kg^{-1} (Nyman et al., 2003b).

12.4.1.2 Retail foods other than soy sauces

The chloropropanol 3-MCPD has been found in a wide range of retail ready-to-eat processed foods including cereal, coffee, cheese, liquorice, fish and meat suggesting that formation

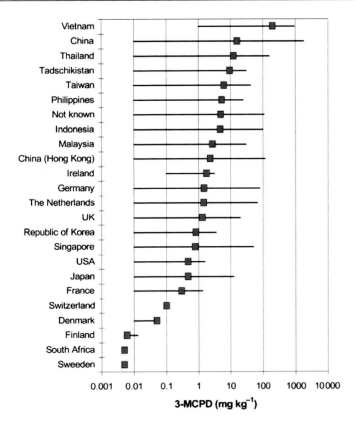

Fig. 12.10   Distribution of 3-MCPD in soy sauces (mean and range) by countries of origin.

may occur by a range of mechanisms/processes (see Table 12.5). Reassuringly, the levels of 3-MCPD are much lower than those reported for some soy sauces.

Thermally processed products, e.g. cereals, account for the greatest incidence of 3-MCPD with some of the highest levels found in products attaining high temperatures, e.g. bread crust and toasted bread. The mechanisms of formation of 3-MCPD in thermally processed cereal products are discussed in Section 12.6.

Low levels of 3-MCPD have also been found in products not subjected to high temperature treatments such as cheese, salami and cold smoked fish. Some possible occurrence routes for 3-MCPD in cheese and salami include enzymatic release of 3-MCPD from chloroesters and/or migration from epichlorohydrin containing resin-treated food contact materials (see Section 12.4.1.4). In the case of smoked products, the occurrence of 3-MCPD in smoke generated from the wood during the process and from liquid smoke ingredients may contribute to the levels found in these products (Kuntzer and Weisshaar, 2006). High levels of 3-MCPD, up to 1150 µg kg^{-1}, have been reported in uncooked convenience chicken products, e.g. chicken nuggets and crumb-dressed chicken breast (Kuntzer and Weisshaar, 2006). The authors offered the suggestion that these products may have been treated with highly contaminated protein hydrolysates (to increase water-binding capacity) although no further details were given.

Although information on other chloropropanols in retail foods is scarce at present, data from the recent EC SCOOP task (European Commission, 2004) indicate that the incidence and levels of 2-MCPD (8/115), 1,3-DCP (0/42) and 2,3-DCP (1/28) in retail foods is likely

Table 12.5 Retail, ready to eat foodstuffs with quantifiable levels of 3-MCPD.

Foodstuff	Incidence[a]	Mean[b] (μg kg^{-1})	Range (μg kg^{-1})	Reference
Cereal products				
Breads	14/27	12	<10–49	FSA (2001c)
Bread	6/9	23	<10—76	Breitling-Utzmann et al. (2003)
Crust	9/9	91	24–275	Breitling-Utzmann et al. (2003)
Toast	26/26	214	30–679	Breitling-Utzmann et al. (2003)
Toast	10/10	136	20–322	Crews et al. (2001)
Cake, fruit	8/10	78	<10–210	European Commission (2004)
Crackers/toasts	30/34	38	<10–134	FSA (2001c)
Doughnuts	5/5	18	11–24	FSA (2001c)
Rusks	10	21	<10–48	European Commission (2004)
Sweet biscuits	5/19	5	<10–32	FSA (2001c)
Cheese	4/30	8	<10–31	FSA (2001c)
Cheese	11/105	8	<10–95	European Commission (2004)
Cheese	4/9	12	<10–37	Crews et al. (2001)
Cheese	3/3	42	13–83	Svejkovská et al. (2004)
Coffee	11/15	12	<9–19	Doležal et al. (2005)
Fish				
Anchovies, in olive oil	2/2	48	15–81	FSA (2001c)
Crumbed	5/6	37	<5–83	FSANZ (2003)
Smoked fish	6/8	37	<10–191	European Commission (2004)
Liquorice	2/2	22	20–23	European Commission (2004)
Meats				
Bacon	3/6	11	<5–22	FSANZ (2003)
Beef burger/hamburger	5/7	25	<10–71	FSA (2001c)
Beef burger/hamburger	6/6	15	7–49	FSANZ (2003)
Salami	9/20	12	<10–69	FSA (2001c)
Salami	2/2	31	14–48	Svejkovská et al. (2004)
Sausages	3/6	16	<5–69	FSANZ (2003)
Smoked meats				
Bacon	2/10	11	<10–47	FSA (2001c)
Fermented sausages/ smoked ham	29/33	43	<5–74	Kuntzer and Weisshaar (2006)

[a] Number of samples above the limit of detection.
[b] Derived value from reported data (one half of the reported limit of detection value used to calculate the mean).

to be very low. However, results from a recent survey of food products carried out by Food Standards Australia New Zealand (FSANZ, 2003) revealed that, unlike soy sauces, 1,3-DCP may be present in food without the presence of 3-MCPD or at levels substantially higher than those of 3-MCPD. These foods were all raw meats (see Table 12.6) and the levels of 1,3-DCP decreased when the samples were cooked. A subsequent survey of 28 raw meats from retail outlets in the UK (FSA, 2004) did not find any quantifiable levels of 3-MCPD or 1,3-DCP and to date the occurrence route for 1,3-DCP in the raw meats from the 2003 FSANZ survey remains unknown.

12.4.1.3 Commercial food ingredients

Survey data on the occurrence of 3-MCPD in commercial food ingredients are given in Table 12.7. Occurrence data for other chloropropanols are very limited.

Table 12.6 Occurrence data for chloropropanols in raw meats.[a]

Sample	Number of samples	3-MCPD Incidence[b]	3-MCPD Range ($\mu g\ kg^{-1}$)	1,3-DCP Incidence	1,3-DCP Range ($\mu g\ kg^{-1}$)
Beef, minced	10	0/10	<5	9/10	<3–110
Beef, steak	5	0/5	<5	1/5	70
Sausages	10	3/10	<5–13	7/10	<3–69
Ham, leg	5	2/5	<5–27	2/5	<3–21
Lamb, chops	6	0/6	<5	1/6	91

[a] Adapted from FSANZ (2003).
[b] Number of samples above the limit of detection.

Processed garlic accounted for the highest incidence and concentration, with 20/21 samples containing between 5 and 690 $\mu g\ kg^{-1}$3-MCPD. The mechanism of formation of 3-MCPD in these products is discussed in Section 12.6.

Mean 3-MCPD concentrations in 72 acid-HVPs ranged from 61 to 78 $\mu g\ kg^{-1}$ and are consistent with the efforts made by industry to reduce levels in these products. Levels of 2-MCPD in these products were frequently higher than 3-MCPD as a consequence of the chloropropanol reduction process, discussed in Section 12.7. It is interesting to note that 3-MCPD was also found in HVPs produced by enzyme hydrolysis, a process developed to avoid chloropropanol contamination by acid hydrolysis. The mechanism of formation of 3-MCPD in these products is not known although formation from residual 3-MCPD esters present in the feedstock may be a possibility (see Section 12.6).

The presence of 3-MCPD in commercial smoke flavourings has recently been reported by Kuntzer and Weisshaar (2006) in the concentration range 200–760 $\mu g\ kg^{-1}$. Based on these findings the estimated contribution to, e.g. smoke-flavoured sausages was 9 $\mu g\ kg^{-1}$.

The occurrence of 3-MCPD was widespread in malt products and could be attributed to the additional heat treatments to which these malts had been subjected to produce the desired colours and flavours (Brereton *et al.*, 2005). Although processing details were not available for the modified starches, the two samples with quantifiable levels of 3-MCPD were both

Table 12.7 Commercial food ingredients containing quantifiable levels of 3-MCPD.

Food ingredient	Incidence[a]	Mean[b] ($\mu g\ kg^{-1}$)	Range ($\mu g\ kg^{-1}$)	Reference
Breadcrumbs	1/6	7	<10–14	Hamlet *et al.* (2002)
Garlic, processed	20/21	106	<10–690	European Commission (2004)
HVP				
Acid hydrolysed	21/50	78	<10–649	European Commission (2004)
Enzyme hydrolysed	1/3	26	78	Hamlet *et al.* (2002)
Flavouring ingredients	11/22	61	<10–550	Fromberg (2001)
Liquid smokes	6/6	–[c]	200–760	Kuntzer and Weisshaar (2006)
Malts	31/63	96	<10–850	European Commission (2004)
Meat extracts	1/5	7	<10–14	Hamlet *et al.* (2002)
Modified starches	2/9	59	<10–488	European Commission (2004)

[a] Number of samples above the limit of detection.
[b] Derived value from reported data (one half of the reported limit of detection value used to calculate the mean).
[c] Data not available.

maize yellow dextrins. These samples can be prepared by heat treatment in the presence of a mineral acid, a process that can produce chloropropanols by a mechanism analogous to that of acid-HVP (Hamlet *et al.*, 2002).

12.4.1.4   Other sources of chloropropanols in foodstuffs

Epichlorohydrin copolymers with polyamines and/or polyamides are used to provide wet strength to food contact papers. Typical applications include tea bag paper, coffee filters, absorbents packaged with meats, and cellulose casings (for ground meat products such as sausage). These resins can accumulate chloropropanol forming species during storage resulting in contamination of foodstuffs by migration of chloropropanols from treated papers (Masten, 2005). Levels of 3-MCPD ranging from 0.005 to 219 mg kg^{-1} (mean 13.9 mg kg^{-1}) were found in 7 of 16 edible sausage casings reported in the EC data compilation on chloropropanols in foods (EC, 2004).

Levels of 1,3-DCP up to 1000 mg kg^{-1} have also been reported in dimethylamine-epichlorohydrin copolymer (DEC) used at concentrations of up to 150 ng g^{-1} by weight of sugar solids in sugar refining (Masten, 2005). DEC is used as a flocculent or decolourising agent for sugar liquors. It is also used to immobilise glucose isomerase enzymes for production of high-fructose corn syrup (Masten, 2005).

Chloropropanols have been found in epichlorohydrin polyamine polyelectrolyes used in drinking water treatment chemicals as coagulation and flocculation products (Masten, 2005) and low concentrations of 3-MCPD have been found in finished water from flocculent use in the United Kingdom (CCFAC, March 2001). In the UK, the Drinking Water Inspectorate (DWI, 2003) concluded that limiting the dosing rate of the flocculent to no more than 2.5 mg/L drinking water would indirectly regulate concentrations of the chloropropanols 1,3-DCP, 2,3-DCP and 3-MCPD.

Kuntzer and Weisshaar (2006) isolated 3-MCPD from smoke generated by heating commercial food grade wood pellets; hence, foods in contact with smoke could become contaminated with 3-MCPD. Dichloropropanols were not detected in the generated smoke.

12.4.1.5   Other compounds of interest

Rétho and Blanchard (2005) identified 3-bromopropane-1,2-diol and 2-bromopropane-1,3-diol at levels of 35, 45 and 7 µg kg^{-1} in grape seed, rape seed and sesame oils (total of both isomers). In all cases it was a mixture (about 2.5:1.0) of the 3-bromo- and 2-bromo isomers.

12.4.2   Chloroesters

Data on chloroesters in foods, other than HVPs, are sparse at present and confined to the occurrence of esters of 3-MCPD. However, recent findings indicate that the formation of 3-MCPD-esters (monoesters and diesters with higher fatty acids) may be widespread in processed foods derived from cereals, potatoes, meat, fish, nuts and oils (Hamlet *et al.*, 2004a; Hamlet and Sadd, 2004; Svejkovská *et al.*, 2004; Doležal *et al.*, 2005; Zelinková *et al.*, 2006). These compounds represent a new class of food contaminants, which might release 3-MCPD into foods during processing and storage or possibly in vivo as a result of lipase-catalysed hydrolysis reactions. Table 12.8 shows the levels of 3-MCPD-esters, measured as 3-MCPD released by hydrolysis of the esters, in a wide range of foods. In many cases the amount of 3-MCPD-esters exceeded that of the free 3-MCPD.

Table 12.8 Levels of 3-MCPD-esters in foodstuffs quantified as 3-MCPD.

Foodstuff	Number of samples	Mean ($\mu g\ kg^{-1}$)	Range ($\mu g\ kg^{-1}$)	Reference
Bread	1	6.7	6.7	Hamlet et al. (2004a);
Crumb	1	4.9	4.9	Hamlet and Sadd (2004)
Crust	1	547	547	
Toast	7	86[a]	60–160	
Crispbread	1	420	420	Svejkovská et al. (2004)
Coffee	15	140[a]	<100–390	Doležal et al. (2005)
Cracker	1	140	140	Svejkovská et al. (2004)
DATEM	1	66	66	Hamlet et al. (2004a);
Doughnut	1	1210	1210	Hamlet and Sadd (2004)
French fries	1	6100	6100	Svejkovská et al. (2004)
Malt, dark	1	580	580	Svejkovská et al. (2004)
Nuts, roasted	3	1370	433–500	Svejkovská et al. (2004)
Oils				Zelinková et al. (2006)
Virgin seed oils	9	63	<100–<300	Zelinková et al. (2006)
Roasted	1	337	337	
Virgin olive oils	4	75	<100–<300	
Virgin germ oils	2	100	<100–<300	
Refined seed oils	5	524	<300–1234	
Refined olive oils	5	1464	<300–2462	
Pickled herring	1	280	280	Svejkovská et al. (2004)
Salami	1	1760	1760	Zelinková et al. (2006)
Wheat flour	1	<5	<5	Hamlet et al. (2004a); Hamlet and Sadd (2004)

[a] Derived value from reported data (one half of the reported limit of detection value used to calculate the mean).

Hamlet et al. (2004a) and Hamlet and Sadd (2004) measured 3-MCPD-esters in bread and toast. Highest levels were found in regions of the bread that attained the highest temperature, i.e. the crust, and levels increased from 60 to 160 $\mu g\ kg^{-1}$ when the bread was toasted over 40–120 seconds. The highest level of 3-MCPD-esters (6100 $\mu g\ kg^{-1}$) was found in a sample of French fries (Svejkovská et al., 2004). The level of 3-MCPD-esters in roast coffee was relatively low and varied between 6 $\mu g\ kg^{-1}$ (soluble coffees) and 390 $\mu g\ kg^{-1}$ (decaffeinated coffee) although it exceeded the free 3-MCPD level by a factor of 8–33 times (Doležal et al., 2005). The presence of 3-MCPD esters in bread crumb (Hamlet et al., 2004a; Hamlet and Sadd, 2004), pickled olives and herrings suggests that these compounds can also form at relatively low temperatures and even in acid media (Svejkovská et al., 2004).

Zelinková et al. (2006) analysed 25 retail virgin and refined edible oils for the content of free 3-MCPD and 3-MCPD-esters (as 3-MCPD released from its esters with higher fatty acids). The oils contained free 3-MCPD ranging from <3 $\mu g\ kg^{-1}$ (LOD) to 24 $\mu g\ kg^{-1}$. Surprisingly, the 3-MCPD-ester level was much higher and varied between <100 $\mu g\ kg^{-1}$ (LOD) and 2462 $\mu g\ kg^{-1}$. Generally, virgin oils had relatively low levels of 3-MCPD-esters ranging from <100 $\mu g\ kg^{-1}$ (LOD) to <300 $\mu g\ kg^{-1}$ (LOQ). Higher levels of 3-MCPD-esters were found in oils obtained from roasted oilseeds (337 $\mu g\ kg^{-1}$) and in the majority of refined oils (<300 $\mu g\ kg^{-1}$ to 2462 $\mu g\ kg^{-1}$), including refined olive oils. In general, it appeared that the formation of 3-MCPD-esters in oils was linked with the preliminary heat treatment of oilseeds and with the process of oil refining. The analysis of crude, degummed, bleached and deodorised rapeseed oil showed that the level of MCPD-esters decreased during the refining process. However, additional heating of seed oils for 30 minutes at temperatures ranging from 100 to 280°C, and heating at 230 and 260°C for up to 8 hours led to an increase of the level of 3-MCPD-esters. Conversely, heating olive oil resulted in a decrease in the 3-MCPD-ester level.

Analysis of fat isolated from a salami containing 1670 µg kg^{-1}of 3-MCPD-esters mainly consisted of 3-MCPD-diesters together wither lesser amounts of 3-MCPD-monoesters (Zelinková *et al.*, 2006). The major types of 3-MCPD-diesters (about 85%) were mixed-diesters of palmitic acid with C18 fatty acids (stearic, oleic, linoleic acids); 3-MCPD dis-tearate (11%); and 3-MCPD dipalmitate (4%). The level of 3-MCPD as the free compound in the fat extract was 31 µg kg^{-1}.

12.5   MECHANISMS AND PRECURSORS

12.5.1   From hydrochloric acid and glycerol/acylglycerols

In the laboratory, MCPDs and DCPs can be prepared by the action of concentrated hydrochloric acid or dry hydrogen chloride gas on glycerol alone, or in the presence of glacial acetic acid (Conant and Quayle, 1947a,b). These reactions require prolonged heating at temperatures of about 100°C and are applicable to the formation of chloropropanols in acid-HVP (Velíšek *et al.*, 1979; Collier *et al.*, 1991).

12.5.1.1   *Formation in acid-HVP*

In further studies of chloropropanols formed during the acid-hydrolysis of HVPs, Velíšek *et al.* (2002) showed that (R)-3-MCPD and (S)-3-MCPD were formed in equimolar concentrations. Model experiments with glycerol, triolein and soya lecithin heated with hydrochloric acid in solution confirmed that these materials were precursors of 2- and 3-MCPD and, as expected, yielded racemic 3-MCPD. The mechanisms of formation were in agreement with the earlier studies by Collier *et al.* (1991) and the yields of 3-MCPD decreased in the order triolein > lecithin > glycerol. A summary of the main 3-MCPD forming reactions in HVP given in Fig. 12.11 also illustrates how chloroesters are formed from acylglycerols.

12.5.2   From hypochlorous acid and allyl alcohol

Hypochlorous acid (HOCl, from chlorine and water) will add to the double bond of allyl alcohol (2-propenol) to give 2- and 3-MCPD in accordance with Markovnikov's rules governing carbenium ion stability (see Fig. 12.12). The reaction is reported to proceed rapidly at 50–60°C with an 88% yield of MCPDs (Myszkowski and Zielinski, 1965). Sources of allyl alcohol include: (S)-allyl-L-cysteine sulfoxide (alliin), a cysteine amino acid found in garlic and related species (Kubec *et al.*, 1997) and a soft cheese made from ewe's milk (Carbonell *et al.*, 2002). This mechanism is believed to account for the levels of 3-MCPD found in processed garlic (Hamlet *et al.*, 2002).

12.5.3   Formation in thermally processed foods and ingredients

12.5.3.1   *Model system studies with (sodium) chloride and glycerol/acylglycerols*

Hasnip *et al.* (2002), Doležal *et al.* (2003) and Calta *et al.* (2004) demonstrated that 3-MCPD could be formed by heating sodium chloride with emulsified (Tween 80) glycerol/acylglycerol

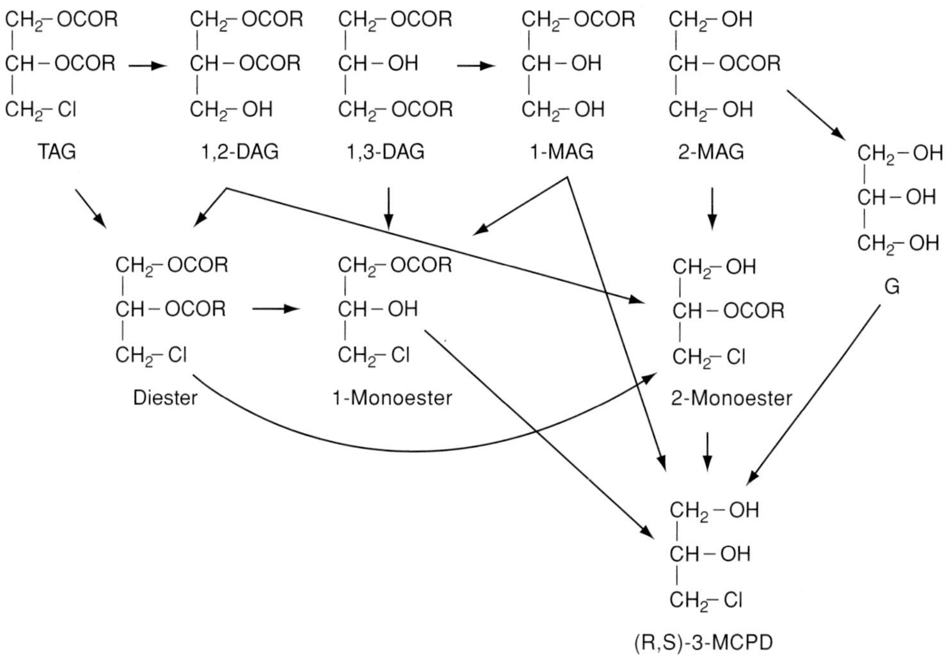

Fig. 12.11 Summary of main chloropropanol forming reactions from acylglycerols in HVPs according to Velíšek *et al.* (2002). TAG, triacylglycerol; DAG, diacylglycerol; MAG, monoacylglycerol; G, glycerol.

precursor mixtures in sealed vials. The generation of 3-MCPD increased with increasing sodium chloride concentration and maximum formation occurred at a water content of between 10 and 20%. The relative amounts of 3-MCPD (relative to monoacylglycerols, mol/mol) formed by heating sealed samples at 200°C for 30 minutes were monoacylglycerols (1.0) < triacylglycerols (1.1) < glycerol (1.6) < diacylglycerols (1.8) < lecithin (3.3).

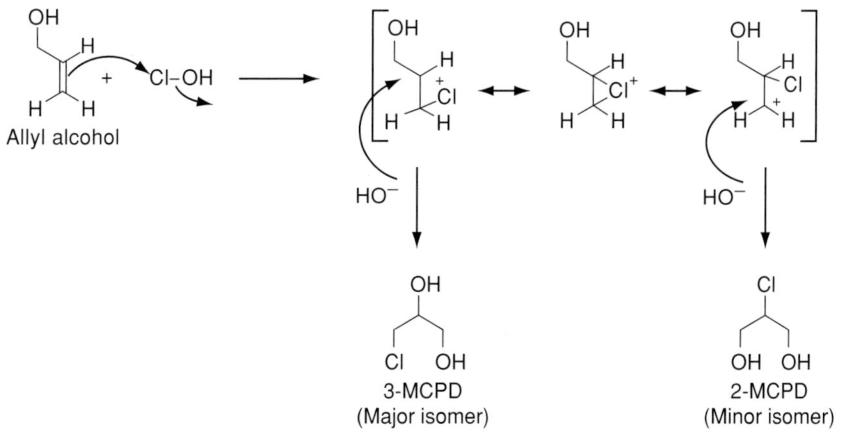

Fig. 12.12 Formation of MCPDs from allyl alcohol and hypochlorous acid.

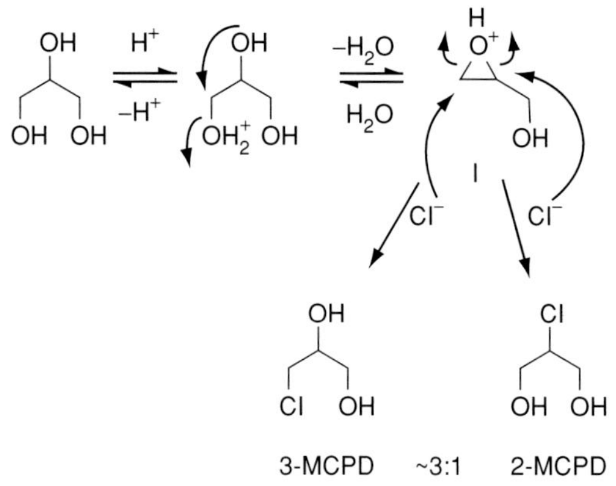

Fig. 12.13   Formation of MCPDs from glycerol via the intermediate epoxide, glycidol (I).

12.5.3.2   Formation in bakery products

Hamlet (2004) developed and validated a model system to determine the major precursors, intermediates and the mechanisms of formation of MCPDs in bread. Free glycerol was shown to be the major precursor of MCPDs in leavened dough (Hamlet *et al.*, 2004b). This glycerol was generated by yeast during proving and its subsequent reaction with added chloride during baking could account for approximately 70% of the MCPDs formed. The production of glycerol by yeast in dough was dependent upon time and temperature and limited by available sugar. The addition of glucose promoted MCPD generation in dough but not via increased glycerol production by yeast. The promoting effect of glucose was subsequently shown to be due to the removal of potential amino inhibitors, e.g. amino acids, via the Maillard reaction (Hamlet and Sadd, 2005). Breitling-Utzmann *et al.* (2005) found that the addition of glucose and sucrose to bread dough promoted 3-MCPD formation during toasting of bread slices, presumably by a similar mechanism. The generation of MCPDs in breads also showed strong moisture dependence: levels increased with decreasing dough moisture to a point where formation was limited by the solubility of chloride and competing reactions involving glycerol and the reaction intermediate, glycidol (Hamlet *et al.*, 2004b) shown in Fig. 12.13.

Minor precursors of MCPDs (together with chloride) in dough were found to be monoacylglycerols, lysophospholipids and phosphatidylglycerols, all present at significant levels in white flour and germ, together with DATEM (diacetyl tartaric acid esters of monoacyl glycerols), an emulsifier used in breadmaking (Hamlet *et al.*, 2004c). These compounds could account quantitatively for the remaining contribution (30%) to MCPD levels in bread. A characteristic of these precursors was the proximity of -OH with respect to either acyl- or phosphoryl-groups on the glycerol skeleton. It was concluded that (a) the subsequent reaction with chloride ions involved a neighbouring group mechanism, and (b) the removal of the acyl group from the resulting MCPD-ester intermediate was rate determining. These mechanistic features could explain the relatively non-reactive nature of the di- and triacylglycerols with respect to MCPD generation. The generation of MCPDs in bread was pH dependent and increased with decreasing pH (see Fig. 12.16). This effect was not due to reduced degradation of MCPDs at reduced pH (see Section 12.7) but attributed to an increased rate of hydrolysis of intermediate MCPD-esters (Hamlet and Sadd, 2005).

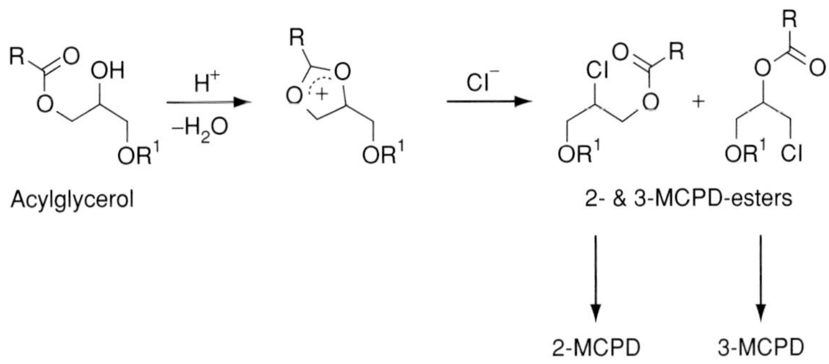

Fig. 12.14 Proposed mechanism of formation of MCPD-esters from mono- and di-acylglycerols (R = alkyl, R^1 = H or COR).

Hamlet *et al.* (2005a) studied the formation of 3-MCPD (and other process contaminants) in bread prepared in domestic bread machines in the UK. It is usual for such as the domestic machines to employ extended proof times (e.g. 2–3 h) compared to commercial bread production (e.g. 50 min). This extended proof leads to increased yeast activity and hence higher glycerol precursor levels in the dough (Hamlet *et al.*, 2004b). Despite these conditions, MCPD generation was found to be only slightly higher than that of commercial products. An explanation for this is the lower baking temperatures used in domestic bread machines.

Evidence for the formation of MCPD-ester intermediates in bread was provided by Hamlet *et al.* (2004a) and Hamlet and Sadd (2004) who showed that chloroester formation was correlated with MCPD generation and that levels of both species increased on heating. The presence of low levels of chloroesters in bread crumb, i.e. at temperatures <100°C, illustrates that these compounds may form readily from partial acylglycerols, presumably as a consequence of facile cyclic acyl oxonium ion formation and subsequent ring opening by chloride ion (see Fig. 12.14).

12.5.3.3 Formation in malts and roasted cereals

Studies on the formation of chloropropanols in malts (Brereton *et al.*, 2005) have shown that 3-MCPD can form when malted or unmalted barley is dry roasted (above 170°C), and that endogenous lipids and glycerol/acylglycerols in the grain are sufficient to promote this synthesis. The 3-MCPD produced during roasting was correlated with colour development while extended heating at temperatures of 200°C or greater gave a reduction in 3-MCPD levels, presumably as a result of degradation (Brereton *et al.*, 2005).

12.5.4 Mechanisms involving enzymes

The formation of MCPDs has been demonstrated in model systems comprising pepper, oil and sodium chloride and over a period of several days (Robert and Stadler, 2003). As spices are known to be a source of enzyme activity, this was subsequently shown to be a general reaction of a lipase in the presence of chloride and lipid. The highest yield of 3-MCPD was obtained in reaction mixtures containing lipase from *Rhizopus oryzae*, all the lipases studied exhibited a high hydrolytic activity towards triglycerides from palm and peanut oil (Robert *et al.*, 2004). The authors proposed that MCPDs were formed by a lipase-catalysed reaction between a triacylglycerol and chloride ion. However, it is more likely that MCPDs were formed by the hydrolysis of residual MCPD-esters that are now known to be present in the raw materials used (Zelinková *et al.*, 2006).

12.5.5   Other mechanisms of interest

In a study of the smoking process Kuntzer and Weisshaar (2006) proposed that 3-MCPD generated in smoke from burning wood chips could be derived from cellulose. A mechanism was proposed for the reaction of chloride ions with acetol (3-hydroxyacetone), a thermal degradation product of cellulose and wood smoke. It is interesting to note that 3-hydroxyacetone is isomeric with the known 3-MCPD intermediate glycidol.

12.6   REACTIONS OF CHLOROPROPANOLS

12.6.1   Stability of MCPDs

Hamlet *et al.* studied the degradation reactions of MCPDs in pure water (Hamlet and Sadd, 2002) and in bread (Hamlet *et al.*, 2003; Hamlet, 2004). Under aqueous conditions, 3-MCPD decayed to glycerol, via the intermediate epoxide glycidol, according to first-order kinetics. The stability of 3-MCPD was sensitive to both pH and temperature, particularly over a range applicable to baked cereal products. In bread dough, the decay of 2-MCPD and 3-MCPD was slower than in pure water: the decay reaction was inhibited by a decrease in moisture content and the pH drop seen in cooked bread dough at elevated temperatures. At high temperatures, i.e. $>100°C$, 2- and 3-MCPD exhibited similar stability: at lower temperatures, i.e. $<100°C$ this behaviour was reversed and 2-MCPD was more stable. It was shown that in bread dough, added 3-MCPD was converted into the isomeric compound 2-MCPD in a mechanism involving the known decay intermediate, glycidol (Hamlet, 2004).

12.6.2   Reactions of 3-MCPD

The structure of 3-MCPD indicates that the compound should undergo reactions characteristic of both alcohols and alkyl chlorides. For example, 3-MCPD is known to react readily with alcohols, aldehydes, amino compounds, ammonia, ketones, organic acids and thiols (Velíšek *et al.*, 1991a). These reactions are summarised in Fig. 12.15. Some of these reaction products, e.g. 3-MCPD derived amino alcohols, dihydroxypropylamines and amino acids, have been identified in HVPs (Velíšek *et al.*, 1991a,b, 1992). Model system studies (Velíšek *et al.*, 1991b, 1992) have shown that the reaction of 3-MCPD with either ammonia or amino acids occurs readily and at moderate temperatures (20–90°C). The reactions of cysteine and glutathione with 3-MCPD have also been reported (Velíšek *et al.*, 2003).

12.7   MITIGATION

Manufacturers of HVPs and soy sauces and producers of wet strength resins for food contact applications have made considerable efforts to reduce chloropropanols in, and contamination of foods from, their products. For example, in alkaline media both 2- and 3-MCPD are

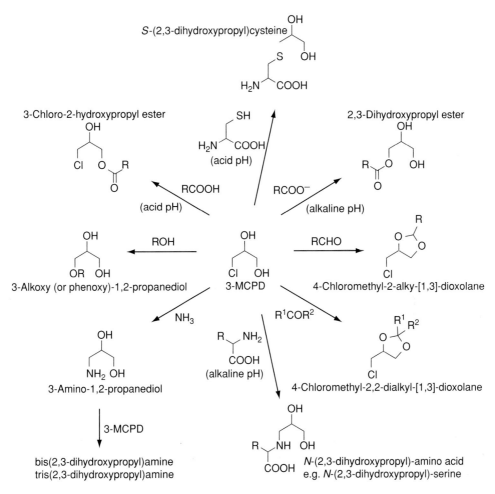

Fig. 12.15   Reactions of 3-MCPD. (Adapted from Velíšek *et al.*, 1991a, 2003).

decomposed to glycerol (Doležal and Velíšek, 1992, 1995) and hence alkalisation is a method that is used commercially to reduce the level of MCPDs in protein hydrolysates (Sim *et al.*, 2004) and also in wet strength resins (Riehle, 2006). However, strategies to reduce chloropropanols and chloroesters in other products have not yet been fully explored and may not be possible for all foodstuffs. These strategies need to consider whether interventions to reduce the risk of chloropropanols and chloroesters might increase the risk of other process contaminants, e.g. furan or acrylamide (Konings *et al.*, 2007). For example, Fig. 12.16 illustrates how increasing dough pH to reduce 3-MCPD formation in cereal products has the opposite effect on acrylamide generation. A summary of potential mitigation options together with literature references is given in Table 12.9.

12.8   EXPOSURE

The most recent exposure assessments have considered 1,3-DCP and 3-MCPD and were carried out by JECFA (2006). Unlike previous evaluations (JECFA, 2002), that mainly considered

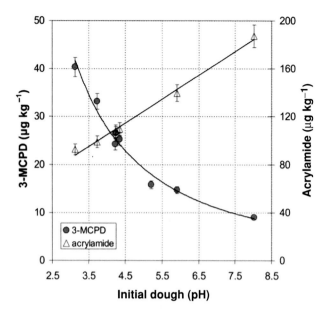

Fig. 12.16   Effect of pH on 3-MCPD and acrylamide formation in baked cereal products. (Adapted from Hamlet and Sadd, 2005 and Hamlet et al., 2005b.)

exposure from soy sauces, the most recent evaluation was based on contributions from all food groups in the diet using data from, e.g. the FSANZ survey (2003) and the EC compilation produced under tasks for scientific cooperation (EC, 2004). Particular consideration was given to groups that might have higher levels of exposure to chloropropanols.

JECFA also noted recent reports that fatty acid esters of 3-MCPD are present in foods, and recommended that studies should be undertaken to enable their intake or toxicological significance to be evaluated.

12.8.1   3-MCPD

Estimated exposures at the national level considered a wide range of foods, including soy sauce and related products, ranged from 1 to 35% of the PMTDI for average exposure in the general population. For the consumers at the high percentile (95th), the estimated intakes ranged from 3 to 85% and up to 115% of the PMTDI in young children. The estimates were based on concentrations of 3-MCPD derived before any remedial action had been taken by government or industry. JECFA noted that reduction in the concentration of 3-MCPD in soy sauce and related products made with acid-HVP could substantially reduce the intake of this contaminant by certain consumers of this condiment.

JECFA (2006) concluded that a representative mean intake of 1,3-DCP for the general population was of 0.051 µg kg^{-1} body weight per day and an estimated high-level intake (young children included) was 0.136 µg kg^{-1} body weight per day. Comparison of these mean and high-level intake values for consumers with the lowest calculated dose for incidence data on tumour-bearing animals indicated margins of exposure of approximately 65 000 and 24 000, respectively. Based on these margins of exposure, JECFA (2006) concluded that the estimated intakes of 1,3-dichloro-2-propanol were of low concern for human health.

Table 12.9 Summary of potential mitigation measures for chloropropanols and chloroesters in foods and food contact materials.

Product/application	Potential control measure	Chloropropanol(s)	Reference
Bread/toast	Time/temperature/minimum fermentation/ recipe management	3-MCPD	Hamlet (2004); Hamlet et al. (2004b,c, 2005b); Breitling-Utzmann et al. (2005)
HVP and soy sauces produced by acid-hydrolysis	Alkalisation/neutralisation	3-MCPD	Sim et al. (2004)
	Degradation by *Pseudomonas* sp. OS-K-29	Esters of dichloropropanols	Kasai et al. (2006)
Roasted cereals/malts	Time/temperature	3-MCPD	Brereton et al. (2005)
Savoury foodstuffs	Lipase inactivation of ingredients (temperature, pH, water activity)	3-MCPD released from chloroesters	–
Smoke	CaCO$_3$ pre-treatment of wood pellets	3-MCPD	Kuntzer and Weisshaar (2006)
Unspecified	Degradation by bacterial cultures	1,3-DCP & 3-MCPD 1,3-DCP	Mamma et al. (2006) Bastos et al. (2002); Yonetani et al. (2004)
	Degradation by *Saccharomyces cerevisiae*	3-MCPD	Bel-Rhlid et al. (2004)
	Reactive food additives, e.g. cysteine, disodium carbonate, glutathione, sodium bicarbonate	3-MCPD	Velišek et al. (2003)
Wet strength resins	Chemical (base)/enzyme/process treatment of epichlorohydrin copolymers	1,3-DCP	Laurent et al. (2002)
		MCPDs	Riehle et al. (2003, 2006); Yamamoto et al. (2001)

Table 12.10 International maximum limits/specifications for chloropropanols in foodstuffs.

Country/region	DCPs (mg/kg)	3-MCPD (mg/kg)	Scope	Reference
Australia/New Zealand	0.005	0.2	Soy/oyster sauces	FSANZ (2006)
Canada	–	1	Soy/oyster sauces	Health Canada (2005)
European Community	–	0.02	HVP and soy sauces (40% solids)	European Commission (2001)
Korea	–	0.3	Soy sauce containing acid-HVP	WTO (2001a)
	–	1	HVP	WTO (2001a)
Malaysia	–	0.02	Liquid foods with acid-HVP	WTO (2001b)
	–	1	Acid-HVP, industrial product	WTO (2001b)
Switzerland	0.05	0.2	Savoury sauces	DFI (2006)
Thailand	–	1	Hydrolysed soybean protein	WTO (2002)
USA	0.05	1	Acid-HVP	Committee on Food Chemicals Codex (1996)

12.9 REGULATORY STATUS

To date, maximum levels/specifications have been adopted for HVPs and soy sauces (see Table 12.10) since chloropropanol intake levels from other foods are very low. With appropriate manufacturing controls the levels of chloropropanols can be controlled in HVPs and soy sauces produced by acid hydrolysis. In instances where brand loyalty could result in regular consumption of, e.g. contaminated soy sauces, these controls could markedly reduce intake levels (JECFA, 2002). It has been demonstrated that 1,3-DCP is typically associated with high concentrations of 3-MCPD. Hence, regulatory control of 3-MCPD should negate the need for specific limits on 1,3-DCP, although some countries have imposed maximum levels (see Table 12.10).

Conclusions and future prospects

The recent observations of chloropropanols in a wide range of foodstuffs have led to a renewal of research worldwide on methods of analysis, occurrence and formation mechanisms of these contaminants. Analytical methods for MCPDs and DCPs have been updated significantly and robust procedures are available covering a wide range of matrices. Considerable progress has been made on occurrence routes, particularly in thermally treated cereal products, where the precursors and mechanisms of MCPD formation have been elucidated. Overall, the contribution from foodstuffs (excluding soy sauces) to the dietary intake of chloropropanols is relatively low and to date regulatory controls to minimise exposure have not been deemed necessary. However, higher levels of chloropropanols, e.g. 3-MPCD, found in certain soy sauces have necessitated maximum limits for these materials in some countries. With the appropriate manufacturing controls, contamination of soy sauces by chloropropanols can be eliminated.

There is growing evidence that chloroesters may be widespread in processed foods and that levels of these contaminants are higher than the corresponding chloropropanols. This is of concern because model studies have shown that chloroesters can be hydrolysed to, e.g. 3-MCPD by enzymes present in some foodstuffs. The extent to which the latter may occur in foods, the toxicological significance of chloroesters and their contribution to dietary intakes of chloropropanols will need to be assessed. Several potential mitigation measures for chloroesters in foodstuffs have already been identified. However, it is important that nutritionists, food chemists and toxicologists jointly consider the wider risks and benefits of these measures. This will ensure that interventions to reduce chloroesters do not have a negative impact on health and nutrition or increase the risk of other process contaminants.

References

Abu-El-Haj, S., Bogusz, M.J., Ibrahim, Z., Hassan, H. and Al Tufail, M. (2007) Rapid and simple determination of chloropropanols (3-MCPD and 1,3-DCP) in food products using isotope dilution GC–MS. *Food Control*, **18**(1), 81–90.

Bastos, F., Bessa, J., Pacheco, C.C., De Marco, P., Castro, P.M.L., Silva, M. and Jorge, R.F. (2002) Enrichment of microbial cultures able to degrade 1,3-dichloro-2-propanol: A comparison between batch and continuous methods. *Biodegradation*, **13**(3), 211–220.

Bel-Rhlid, R., Talmon, J.P., Fay, L.B. and Juillerat, M.A. (2004) Biodegradation of 3-chloro-1,2-propanediol with Saccharomyces cerevisiae. *Journal of Agricultural and Food Chemistry*, **52**(20), 6165–6169.

Boden, L., Lunddgren, M., Stensio, K. and Gorzynski, M. (1997) Determination of 1,3-dichloro-2-propanol in papers treated with polyamidoamine-epichlorohydrin wet-strength resins by gas chromatography–mass spectrometry using selective ion monitoring. *Journal of Chromatography A*, **788**, 195–203.

Breitling-Utzmann, C.M., Kobler, H., Herbolzheimer, D. and Maier, A. (2003) 3-MCPD – Occurrence in bread crust and various food groups as well as formation in toast. *Deutsche Lebensmittel-Rundschau*, **99**(7), 280–285.

Breitling-Utzmann, C.M., Hrenn, H., Haase, N.U. and Unbehend, G.M. (2005) Influence of dough ingredients on 3-chloropropane-1,2-diol (3-MCPD) formation in toast. *Food Additives and Contaminants*, **22**(2), 97–103.

Brereton, P., Kelly, J., Crews, C., Honour, S., Wood, R. and Davies, A. (2001) Determination of 3-chloro-1,2-propanediol in foods and food ingredients by gas chromatography with mass spectrometric detection: Collaborative study. *Journal of AOAC International*, **84**(2), 455–465.

Brereton, P., Crews, C., Hasnip, S., Reece, P., Velisek, J., Dolezal, M., Hamlet, C., Sadd, P., Baxter, D., Slaiding, I. and Muller, R. (2005) The origin and formation of 3-MCPD in foods and food ingredients, report No FD 04/12. A report prepared for the Food Standards Agency. Central Science Laboratory, York.

British Standards Institution (2004) BS EN 14573: 2004 Foodstuffs. Determination of 3-monochloropropane-1,2-diol by GC/MS.

Brown, D.A., Van Meeteren, H.W. and Simmons, J.D. (1989) Process for preparing improved hydrolysed protein. European patent EP03610595B1.

Calta, P., Velíšek, J., Doležal, M., Hasnip, S., Crews, C. and Réblová, Z. (2004) Formation of 3-chloropropane-1,2-diol in systems simulating processed foods. *European Food Research and Technology*, **218**(6), 501–506.

Carbonell, M., Nunez, M. and Fernandez-Garcia, E. (2002) Evolution of the volatile components of ewe raw milk La Serena cheese during ripening. Correlation with flavour characteristics. *Lait*, **82**(6), 683–698.

Cerbulis, J., Parks, O., Liu, R., Piotrowski, G. and Farrell, H. (1984) Occurrence of diesters of 3-chloro-1,2-propanediol in the neutral lipid fraction of goats' milk. *Journal of Agricultural and Food Chemistry*, **32**, 474–476.

Cheng, W.C., Chen, H.C., Lin, Y.P., Lee, H.F., Chang, P.C. and Chou, S.S. (2004) Survey on 3-monochloro-1,2-propandiol (3-MCPD) contents of soy sauce products during fiscal year 2002 in Taiwan. *Journal of Food and Drug Analysis*, **12**(4), 336–341.

Chung, W.C., Hui, K.Y. and Cheng, S.C. (2002) Sensitive method for the determination of 1,3-dichloropropan-2-ol and 3-chloropropane-1,2-diol in soy sauce by capillary gas chromatography with mass spectrometric detection. *Journal of Chromatography A*, **952**(1–2), 185–192.

Codex Committee on Food Additives and Contaminants (CCFAC) (2001) Position paper on chloropropanols, CX/FAC 01/31, thirty-third Session the Hague, the Netherlands, 12–16 March 2001. Available from: http://www.codexalimentarius.net/ccfac33/fa01_01e.htm [accessed November 2006].

Collier, P.D., Cromie, D.D.O. and Davies, A.P. (1991) Mechanism of formation of chloropropanols present in protein hydrolysates. *Journal of the American Oil Chemists Society*, **68**(10), 785–790.

Committee on Carcinogenicity of Chemicals in Food, Consumer Products and the Environment (COC) (2004) Carcinogenicity of 1,3-dichloropropan-2-ol (1,3-DCP) and 2,3-dichloropropan-1-ol (2,3-DCP), COC/04/S2 – June 2004. Available from: http://www.advisorybodies.doh.gov.uk/coc/1,3-2,3dcp04.htm [accessed November 2006].

Committee on Food Chemicals Codex (1996) *Food Chemicals Codex: First Supplement to Fourth Edition*. Institute of Medicine of the National Academies, Washington, DC. Available from: http://www.iom.edu/report.asp?id=4590 [accessed January 2007].

Committee on Mutagenicity of Chemicals in Food, Consumer Products and the Environment (COM) (2000) Mutagenicity of 3-monochloro propane 1,2-diol (3-MCPD), COM statement COM/00/S4 – October 2000. Available from: http://www.advisorybodies.doh.gov.uk/com/mcpd2.htm [accessed November 2006].

Committee on Mutagenicity of Chemicals in Food, Consumer Products and the Environment COM (2003) Statement on the mutagenicity of 1,3-dichloropropan-2-ol Com/03/S4 – October 2003. Available from: http://www.advisorybodies.doh.gov.uk/com/1,3-dcp.htm [accessed November 2006].

Committee on Mutagenicity of Chemicals in Food, Consumer Products and the Environment (COM) (2004) Statement on the mutagenicity of 2,3-dichloropropan-1-ol, Com/04/s1 – may 2004. Available from: http://www.advisorybodies.doh.gov.uk/com/2,3dcp04.htm [accessed November 2006].

Conant, J.B. and Quayle, O.R. (1947a) Glycerol, -dichlorohydrin. *Organic Syntheses Collective*, **1**, 292–294.

Conant, J.B. and Quayle, O.R. (1947b) Glycerol, -monochlorohydrin. *Organic Syntheses Collective*, **1**, 294–296.

Crews, C., Brereton, P. and Davies, A. (2001) The effects of domestic cooking on the levels of 3- monochloro-propanediol in foods. *Food Additives and Contaminants*, **18**(4), 271–280.

Crews, C., LeBrun, G. and Brereton, P.A. (2002) Determination of 1,3-dichloropropanol in soy sauces by automated headspace gas chromatography-mass spectrometry. *Food Additives and Contaminants*, **19**(4), 343–349.

Crews, C., Hasnip, S., Chapman, S., Hough, P., Potter, N., Todd, J., Brereton, P. and Matthews, W. (2003) Survey of chloropropanols in soy sauces and related products purchased in the UK in 2000 and 2002. *Food Additives and Contaminants*, **20**(10), 916–922.

Davídek, J., Velíšek, J., Kubelka, V., Janîeek, G. and Šimicová, Z. (1980) Glycerol chlorohydrins and their esters as products of the hydrolysis of tripalmitin, tristearin and triolein with hydrochloric acid. *Lebensmittel-Untersuchung und-Forschung*, **171**, 14–17.

Davídek, J., Velíšek, J., Kubelka, V. and Janíček, G. (1982) New chlorine containing compounds in protein hydrolysates, in *Recent Developments in Food Analysis, Proceedings of Euro Food Chem I*, Vienna, Austria, 17–20 February 1981 (eds. W. Baltes, P.B. Czedik-Eysenberg and W. Pfannhauser). Institute of Chemical Technology, Prague, Czech Republic, pp. 322–325.

Dayritt, F.M. and Ninonuevo, M.R. (2004) Development of an analytical method for 3-monochloropropane-1,2-diol in soy sauce using 4-heptanone as derivatizing agent. *Food Additives and Contaminants*, **21**(3), 204–209.

De Rooij, J.F.M., Ward, B.A. and Ward, M. (1989) Process for preparing improved hydrolysed protein. European patent EP0361597 (A1).

Département fédéral de l'intérieur (DFI) (2006) Ordonnance du DFI du 26 juin 1995 sur les substances étrangères et les composants dans les denrées alimentaires, RS 817.021.23 (DFI schedule on foreign substances and components in foodstuffs). DFI, Berne, Switzerland. Available from: http://www.admin.ch/ch/f/rs/c817_021_23.html [accessed January 2007].

Divinová, M., Svejkovská, B., Doležal, M. and Velíšek, J. (2004) Determination of free and bound 3-chloropropane-1,2-diol by gas chromatography with mass spectrometric detection using deuterated 3-chloropropane-1,2-diol as internal standard. *Czech Journal of Food Sciences*, **22**, 182–189.

Doležal, M. and Velíšek, J. (1992) Kinetics of 3-chloro-1,2-propanediol degradation in model systems, in *Proceedings of Chemical Reactions in Foods II*, Prague, Czech Republic, 24–26 September 1992, pp. 297–302.

Doležal, M. and Velíšek, J. (1995) Kinetics of 2-chloro-1,3-propanediol degradation in model systems and in protein hydrolysates. *Potravinaoske Vèdy*, **2**, 85–91.

Doležal, M., Calta, P., Velíšek, J. and Hasnip, S. (2003) Study on the formation of 3-chloropropane-1,2-diol from lipids in the presence of sodium chloride, in *Euro Food Chem XII Conference on 'Strategies for Safe Food: Analytical Industrial, and Legal Aspects; Challenges in Organisation and Communication'*,

24–26 September 2003 (eds. T. Eklund, H. De Brabender, E. Daeselerie, I. Dirinck and W. Ooghe). KVCV, Brugge, Belgium, pp. 251–254.

Doležal, M., Chaloupská, M., Divinová, M., Svejkovská, B. and Velíšek, J. (2005) Occurrence of 3-chloropropane-1,2-diol and its esters in coffee. *European Food Research and Technology*, **221**, 221–225.

Drinking Water Inspectorate (DWI) (2003) The Use of Polyamine Coagulants in Public Water Supplies. DWI, London. Available from: http://www.dwi.gov.uk/cpp/chloro.htm [accessed November 2006].

European Commission (1997) Opinion on 3-monochloro-propane-1,2-diol (3-MCPD), expressed on 16 December 1994, in *Food Science and Techniques: Reports of the Scientific Committee for Food (thirty-sixth series)*. European Commission, Brussels, pp. 31–33.

European Commission (2001) Commission Regulation (EC) No 466/2001 of 8 March 2001 setting maximum levels for certain contaminants in foodstuffs (OJ L 77 16.3.2001, p12).

European Commission (2004) *Report of Experts Participating in Task 3.2.9*. Collection and collation of data on levels of 3-monochloropropanediol (3-MCPD) and related substances in foodstuffs, June 2004. Directorate-General Health and Consumer Protection, Brussels. Available from: http://ec.europa.eu/food/food/chemicalsafety/contaminants/mcpd_data_tables_en.htm [accessed November 2006].

Faesi, R.S., Werner, G. and Wolfensberger, U. (1987) Procédé de frabrication d'un condiment. European patent EP0226769.

Fellows, M. (2000) *3-MCPD: Measurement of Unscheduled DNA Synthesis in Rat Liver Using an In Vitro/In Vivo Procedure*, Report No. 1863/1-D5140. Covance Laboratories, Harrogate, UK.

Food Advisory Committee (FAC) (2000) Genotoxicity of 3-monochloropropane-1,2-diol. Food Advisory Committee paper for discussion, FdAC/Contaminants/48. Available from: http://archive.food.gov.uk/pdf_files/papers/fac_48.pdf [accessed November 2006].

Food Standards Agency (FSA) (2001a) *Survey of 3-Monochloropropane-1,2-diol (3-MCPD) in Soy Sauce and Related Products (Number 14/01)*. Food Standards Agency, London. Available from: http://www.food.gov.uk/science/surveillance/fsis2001/3-mcpdsoy [accessed November 2006].

Food Standards Agency (FSA) (2001b) *Survey of 1,3-Dichloropropanol (1,3-DCP) in Soy Sauce and Related Products (Number 15/01)*. Food Standards Agency, London. Available from: http://www.food.gov.uk/science/surveillance/fsis2001/13dcpsoy [accessed November 2006].

Food Standards Agency (FSA) (2001c) *Survey of 3-Monochloropropane-1,2-diol (3-MCPD) in Selected Food Groups (Number 12/01)*. Food Standards Agency, London. Available from: http://www.food.gov.uk/science/surveillance/fsis2001/3-mcpdsel [accessed November 2006].

Food Standards Agency (FSA) (2002) *Survey of 3-MCPD in Soy Sauce from Catering Outlets*. Food Standards Agency, London. Available from: http://www.foodstandards.gov.uk/news/newsarchive/2002/aug/soy_sauce [accessed November 2006].

Food Standards Australia New Zealand (FSANZ) (2003) *Chloropropanols in Food: An Analysis of the Public Health Risk*. Technical report series no. 15. Food Standards Australia New Zealand, Canberra. Available from: http://www.foodstandards.gov.au/_srcfiles/Final%20Chloropropanol%20Report%20-%2011%20Sep%2003b.doc [accessed November 2006].

Food Standards Agency (FSA) (2004) Chloropropanols in meat products. Press Release Thursday 20 May 2004. Available from: http://www.food.gov.uk/news/newsarchive/2004/may/chloropropanols [accessed November 2006].

Food Standards Australia New Zealand (FSANZ) (2006) *Australia New Zealand Food Standards Code, 2006*. Anstat, Canberra, Australia. Available from: http://www.foodstandards.gov.au/_srcfiles/FSC_Amendment_Cover_Page_v90.pdf [accessed January 2007].

Fromberg, A. (2001) *Survey of 3-Monochloropropane-1,2-diol (3-MCPD) in Food and Food Ingredients*. The Danish Veterinary and Food Administration, Søborg. Available from: http://www.foedevarestyrelsen.dk/FDir/Publications/2001901/Rapport1.asp [accessed November 2006].

Hamlet, C.G. (1998) Analytical methods for the determination of 3-chloro-1,2- propandiol and 2-chloro-1,3-propandiol in hydrolysed vegetable protein, seasonings and food products using gas chromatography ion trap tandem mass spectrometry. *Food Additives and Contaminants*, **15**(4), 451–465.

Hamlet, C.G. (2004) *Monochloropropanediols in Bread: Model Dough Systems and Kinetic Modelling*. PhD thesis, University of Nottingham.

Hamlet, C.G. and Sutton, P.G. (1997) Determination of the chloropropanols, 3-chloro-1,2-propandiol and 2-chloro-1,3-propandiol, in hydrolysed vegetable proteins and seasonings by gas chromatography ion trap tandem mass spectrometry. *Rapid Communications in Mass Spectrometry*, **11**(13), 1417–1424.

Hamlet, C.G., Jayaratne, S.M. and Matthews, W. (2002) 3-monochloropropane-1,2-diol (3-MCPD) in food ingredients from UK food producers and ingredient suppliers. *Food Additives and Contaminants*, **19**(1), 15–21.

Hamlet, C.G. and Sadd, P.A. (2002) Kinetics of 3-chloropropane-1,2-diol (3-MCPD) degradation in high temperature model systems. *European Food Research and Technology*, **215**(1), 46–50.

Hamlet, C.G., Sadd, P.A., Crews, C., Velisek, J. and Baxter, D.E. (2002) Occurrence of 3-chloro-propane-1,2-diol (3-MCPD) and related compounds in foods: A review. *Food Additives and Contaminants*, **19**(7), 619–631.

Hamlet, C.G., Sadd, P.A. and Gray, D.A. (2003) Influence of composition, moisture, pH and temperature on the formation and decay kinetics of monochloropropanediols in wheat flour dough. *European Food Research and Technology*, **216**(2), 122–128.

Hamlet, C.G. and Sadd, P.A. (2004) Chloropropanols and their esters in cereal products. *Czech Journal of Food Sciences*, **22**, 259–262.

Hamlet, C.G., Sadd, P.A. and Gray, D.A. (2004a) Chloropropanols and their esters in baked cereal products, in *Abstracts of the Division of Agricultural and Food Chemistry 227th Annual Meeting*, Anaheim, California, 28 March–1 April 2004 (ed. W. Yokoyama). American Chemical Society, Anaheim.

Hamlet, C.G., Sadd, P.A. and Gray, D.A. (2004b) Generation of monochloropropanediols (MCPDs) in model dough systems. 1. Leavened doughs. *Journal of Agricultural and Food Chemistry*, **52**, 2059–2066.

Hamlet, C.G., Sadd, P.A. and Gray, D.A. (2004c) Generation of monochloropropanediols in model dough systems. 2. Unleavened doughs. *Journal of Agricultural and Food Chemistry*, **52**, 2067–2072.

Hamlet, C.G., Jayaratne, S.M. and Morrison, C.M. (2005a) *Processing Contaminants in Bread from Bread Making Machines: A Continuing Project Under C03020*. A report prepared for the UK Food Standards Agency. RHM Technology Ltd, High Wycombe.

Hamlet, C.G., Baxter, D.E., Sadd, P.A., Slaiding, I., Liang, L., Muller, R., Jayaratne, S.M. and Booer, C. (2005b) *Exploiting Process Factors to Reduce Acrylamide in Cereal-Based Foods, Report No C014*. A report prepared for the UK Food Standards Agency. RHM Technology Ltd, High Wycombe.

Hamlet, C.G. and Sadd, P.A. (2005) Effects of yeast stress and pH on 3-MCPD producing reactions in model dough systems. *Food Additives and Contaminants*, **22**(7), 616–623.

Hasnip, S., Crews, C., Brereton, P. and Velíšek, J. (2002) Factors affecting the formation of 3-monochloropropanediol in foods. *Polish Journal of Food and Nutrition Sciences*, **11**(52), 122–124.

Hasnip, S., Crews, C., Potter, N. and Brereton, P. (2005) Determination of 1,3-dichloropropanol in soy sauce and related products by headspace gas chromatography with mass spectrometric detection: Interlaboratory study. *Journal of AOAC International*, **88**(5), 1404–1412.

Health Canada (2005) *Canadian Guidelines ("Maximum Limits") for Various Chemical Contaminants in Foods*. Health Canada, Ottawa, Canada. Available from: http://www.hc-sc.gc.ca/fn-an/securit/chem-chim/contaminants-guidelines-directives_e.html [accessed January 2007].

Hirsbrunner, P. and Weymuth, H. (1989) Procédé de frabrication d'un condiment. European patent EP0380371B1.

Huang, M., Jiang, G., He, B., Liu, J., Zhou, Q., Fu, W. and Wu, Y. (2005) Determination of 3-chloropropane-1,2-diol in liquid hydrolyzed vegetable proteins and soy sauce by solid-phase microextraction and gas chromatography/mass spectrometry. *Analytical Sciences*, **21**(11), 1343–1347.

Jin, Q., Zhang, Z., Luo, R. and Li, J. (2001) Survey of 3-monochloropropane-1,2-diol (3-MCPD) in soy sauce and similar products. *Wei Shang Yan Jiu*, **30**, 60–61.

Joint FAO/WHO Expert Committee on Food Additives (JECFA) (2002) *Evaluation of Certain Food Additives and Contaminants: Fifty-seventh Report of the Joint FAO/WHO Expert Committee on Food Additives*, Rome, 5–14 June 2001. WHO technical report series; 909. WHO, Geneva.

Joint FAO/WHO Expert Committee on Food Additives (JECFA) (2006) *Summary and Conclusions from the Sixty-seventh Meeting*, Rome, 20–29 June 2006, JECFA67/SC. Available from: http://who.int/ipcs/food/jecfa/summaries/summary67.pdf [accessed November 206].

Joint Food Safety and Standards Group (1999) Industry alerted to contaminant levels in soy sauce. Press Release FSA 18/99, 31 August 1999. Available from: http://archive.food.gov.uk/maff/archive/inf/ newsrel/fsa/fsa1899.htm [accessed November 2006].

Kasai, N., Suzuki, T. and Idogaki, H. (2006) Enzymatic degradation of esters of dichloropropanols: Removal of chlorinated glycerides from processed foods. *Food Science and Technology*, **39**(1), 86–90.

Konings, E.J.M., Ashby, P., Hamlet, C.G. and Thompson, G.A.K. (2007) Acrylamide in cereal and cereal products: A review on progress in level reduction. *Food Additives and Contaminants*, **24**, 47–59.

Kraft, R., Brachwitz, H., Etzold, H.G., Langen, P. and Zöpfl, H.-J. (1979) Halogenolipids. I. Mass spectrometric structure investigation of the isomeric halogeno-propanediols esterified with fatty acids (deoxyhalogenoglycerides). *Journal für Praktische Chemie*, **321**(5), 756–768.

Kuballa, T. and Ruge, W. (2004) Analysis and detection of 3-monochloropropane-1,2-diol (3-MCPD) in food by GC/MS/MS. Available from: http://www.varianinc.com/cgi-bin/nav?applications/gcms&cid= JLQOILQPFJ [accessed November 2006].

Kubec, R., Velíšek, J., Doležal, M. and Kubelka, V. (1997) Sulfur-containing volatiles arising by thermal degradation of alliin and deoxyalliin. *Journal of Agricultural and Food Chemistry*, **45**(9), 3580–3585.

Kuksis, A., Marai, L., Myher, J.J., Cerbulis, J. and Farrell, H.M., Jr. (1986) Comparative study of the molecular species of chloropropanediol diesters and triacylglycerols in milk fat. *Lipids*, **21**, 183–190.

Kuntzer, J. and Weisshaar, R. (2006) The smoking process – a potent source of 3-chloropropane-1,2-diol (3-MCPD) in meat products. *Deutsche Lebensmittel-Rundschau*, **102**(9), 397–400.

Laurent, H., Dreyfus, T., Poulet, C. and Quillet, S. (2002) Process for obtaining aminopolyamide-epichlorohydrin resins with a 1,3-dichloro-2-propanol content which is undetectable by ordinary means of vapor-phase chromatography. U.S. Patent No 6342580. Available from: http://gb.espacenet.com [accessed November 2006].

Leung, M.K.P., Chiu, B.K.W. and Lam, M.H.W. (2003) Molecular sensing of 3-chloro-1,2-propanediol by molecular imprinting. *Analytica Chimica Acta*, **491**(1), 15–25.

Lynch, B.S., Bryant, D.W., Hook, G.J., Nestmann, E.R. and Munro, I.C. (1998) Carcinogenicity of monochloro-1,2-propanediol (alpha-chlorohydrin, 3-MCPD). *International Journal of Toxicology*, **17**(1), 47–76.

Macarthur, R., Crews, C., Davies, A., Brereton, P., Hough, P. and Harvey, D. (2000) 3-monochloropropane-1,2-diol (3-MCPD) in soy sauces and similar products available from retail outlets in the UK. *Food Additives and Contaminants*, **17**(11), 903–906.

Mamma, D., Papadopoulou, E., Petroutsos, D., Christakopoulos, P. and Kekos, D. (2006) Removal of 1,3-dichloro2-propanol and 3-chloro1,2-propanediol by the whole cell system of pseudomonas putida DSM 437. *Journal of Environmental Science and Health Part A-Toxic/Hazardous Substances and Environmental Engineering*, **41**(3) 303–313.

March, J. (1983) *Advanced Organic Chemistry: Reactions Mechanisms, and Structure*. 2nd edn. McGraw-Hill, Tokyo, pp. 810–811.

Marshall, R.M. (2000) *3-MCPD: Induction of Micronuclei in the Bone-Marrow of Treated Rats*, Report No. 1863/2-D5140. Covance Laboratories, Harrogate, UK.

Masten, S.A. (2005) *1,3-Dichloro-2-propanol [CAS No. 96-23-1], Review of Toxicological Literature*. A report prepared for National Toxicology Program (NTP), National Institute of Environmental Health Sciences (NIEHS), National Institutes of Health, U.S Department of Health and Human Services. NTP/NIEHS, North Carolina. Available from: http://ntp-server.niehs.nih.gov/ntp/htdocs/Chem_Background/ExSumPdf/dichloropropanol.pdf

Meierhans, D.C., Bruehlmann, S., Meili, J. and Taeschler, C. (1998) Sensitive method for the determination of 3-chloropropane-1,2- diol and 2-chloropropane-1,3-diol by capillary gas chromatography with mass spectrometric detection. *Journal of Chromatography A*, **802**(2), 325–333.

Mingxia, Z., Jianke, Z. and Junhong, L. (2003) Determination of chloropropanols in expanded foods by gas chromatography. *Food and Fermentation Industries*, **29**(7), 52–54.

Ministry of Agriculture Fisheries and Food (MAFF) (1999a) Survey of 3-monochloropropane-1,2-diol (3-MCPD) in acid-hydrolysed vegetable protein. Food Surveillance Information Sheet, No. 181. Available from: http://archive.food.gov.uk/maff/archive/food/infsheet/1999/no181/181mcpd.htm [accessed September 2006].

Ministry of Agriculture Fisheries and Food (MAFF) (1999b) Survey of 3-monochloropropane-1,2-diol (3-MCPD) in soy sauce and similar products. Food Surveillance Information Sheet No. 187. Available from: http://archive.food.gov.uk/maff/archive/food/infsheet/1999/no187/187soy.htm [accessed September 2006].

Myher, J.J., Kuksis, A., Marai, L. and Cerbulis, J. (1986) Stereospecific analysis of fatty acid esters of chloropropanediol isolated from fresh goat milk. *Lipids*, **21**, 309–314.

Myszkowski, J. and Zielinski, A.Z. (1965) Synthesis of glycerol monochlorohydrins from allyl alcohol. *Przemysl Chemiczny*, **44**, 249–252.

Nyman, P.J., Diachenko, G.W. and Perfetti, G.A. (2003a) Determination of 1,3-dichloropropanol in soy sauce and related products by using gas chromatography/mass spectrometry. *Food Additives and Contaminants*, **20**(10), 903–908.

Nyman, P.J., Deininger, C. and Perfetti, G.A. (2003b) Survey of chloropropanols in soy sauces and related products. *Food Additives and Contaminants*, **20**(10), 909–915.

Payne, L.S. (1989) Process for preparing improved hydrolysed protein. European patent EP0361596.

Pesselman, R.L. and Feit, M.J. (1988) Determination of residual epichlorohydrin and 3-chloropropanediol in water by gas chromatography with electron capture detection. *Journal of Chromatography*, **439**, 448–452.

Plantinga, W.J., Van Toorn, W.G. and Van Der Stegen, G.H.D. (1991) Determination of 3-chloropropane-1,2-diol in liquid hydrolysed vegetable proteins by capillary gas chromatography with flame ionisation detection. *Journal of Chromatography*, **555**, 311–314.

Rétho, C. and Blanchard, F. (2005) Determination of 3-chloropropane-1,2-diol as its 1,3-dioxolane derivative at the μg kg^{-1} level: Application to a wide range of foods. *Food Additives and Contaminants*, **22**(12), 1189–1197.

Riehle, R.J. (Hercules Inc.) (2006) Treatment of resins to lower levels of CPD-producing species and improve gelation stability. U.S. Patent No 7081512. Available from: http://www.uspto.gov/patft/index.html [accessed November 2006].

Riehle, R.J., Busink, R., Berri, M. and Stevels, W. (Hercules Inc.) (2003) Reduced byproduct high solids polyamine-epihalohydrin compositions. U.S. Patent Application 20030070783. Available from: http://www.uspto.gov/patft/index.html [accessed November 2006].

Robert, M.-C. and Stadler, R.H. (2003) Studies on the formation of chloropropanols in model systems, in *Euro Food Chem XII Conference on 'Strategies for Safe Food: Analytical Industrial, and Legal Aspects; Challenges in Organisation and Communication'*, 24–26 September 2003 (eds. T. Eklund, H. De Brabender, E. Daeselerie, I. Dirinck and W. Ooghe). KVCV, Brugge, Belgium, pp. 527–530.

Robert, M.-C., Oberson, J.-M. and Stadler, R.H. (2004) Model studies on the formation of monochloropropanediols in the presence of lipase. *Journal of Agricultural and Food Chemistry*, **52**(16), 5102–5108.

Robjohns, S., Marshall, R., Fellows, M. and Kowalczyk, G. (2003) In vivo genotoxicity studies with 3-monochloropropan-1,2-diol. *Mutagenesis*, **18**(5), 401–404.

Rodman, L.E. and Ross, R.D. (1986) Gas–liquid chromatography of 3-chloropropanediol. *Journal of Chromatography*, **369**, 97–103.

Schlatter, J., Baars, A.J., DiNovi, M., Lawrie, S. and Lorentzen, R. (2002a) 3-chloro-1,2-propanediol, in *WHO Food Additives Series 48*. Safety evaluation of certain food additives and contaminants, prepared by the Fifty-seventh meeting of the Joint FAO/WHO Rome 2001. Available from: http://www.inchem.org/documents/jecfa/jecmono/v48je18.htm [accessed November 2006].

Schlatter, J., Baars, A.J., DiNovi, M., Lawrie, S. and Lorentzen, R. (2002b) 1,3-dichloro-2-propanol, in *WHO Food Additives Series 48*. Safety evaluation of certain food additives and contaminants, prepared by the Fifty-seventh meeting of the Joint FAO/WHO Rome 2001. Available from: http://www.inchem.org/documents/jecfa/jecmono/v48je19.htm [accessed November 2006].

Schumacher, R., Nurmi-Legat, J., Oberhauser, A., Kainz, M. and Krska, R. (2005) A rapid and sensitive GC-MS method for determination of 1,3-dichloro-2-propanol in water. *Analytical and Bioanalytical Chemistry*, **382**(2), 366–371.

Sim, C.W., Muhammad, K., Yusof, S., Bakar, J. and Hashim, D.M. (2004) The optimization of conditions for the production of acid-hydrolysed winged bean and soybean proteins with reduction of 3-monochloropropane-1,2-diol (3-MCPD). *International Journal of Food Science and Technology*, **39**(9), 947–958.

Svejkovská, B., Novotný, O., Divinová, M., Réblová, Z., Doležal, M. and Velíšek, J. (2004) Esters of 3-chloropropane-1,2-diol in foodstuffs. *Czech Journal of Food Sciences*, **22**(5), 190–196.

Vanbergen, C.A., Collier, P.D., Cromie, D.D.O., Lucas, R.A., Preston, H.D. and Sissons, D.J. (1992) Determination of chloropropanols in protein hydrolysates. *Journal of Chromatography*, **589**(1–2), 109–119.

Velíšek, J. (1989) *Organic Chlorine Compounds in Food Protein Hydrolysates*. DSc thesis, Institute of Chemical Technology, Prague, Czech Republic.

Velíšek, J., Davídek, J., Hajšlová, J., Kubelka, V., Janíeek, G. and Mánková, B. (1978) Chlorohydrins in protein hydrolysates. *Zeitschrift fur Lebensmittel-Untersuchung Und-Forschung*, **167**, 241–244.

Velíšek, J., Davídek, J., Kubelka, V., Bartošová, J., Tuěková, A., Hajšlová, J. and Janíček, G. (1979) Formation of volatile chlorohydrins from glycerol (triacetin, tributyrin) and hydrochloric acid. *Lebensmittel-Wissenschaft und-Technologie – Food Science and Technology*, **12**, 234–236.

Velíšek, J., Davídek, J., Kubelka, V., Janíček, G., Svobodová, Z. and Šimicová, Z. (1980) New chlorine-containing organic compounds in protein hydrolysates. *Journal of Agricultural and Food Chemistry*, **28**, 1142–1144.

Velíšek, J., Davídek, J., Šimicová, Z. and Svobodová, Z. (1982) Glycerol chlorohydrins and their esters – reaction products of lipids with hydrochloric acid. *Scientific Papers of the Prague Institute of Chemical Technology, Food, E*, **53**, 55–65.

Velíšek, J. and Davídek, J. (1985) Lipidy jako prekurzory organických sloučenin chloru v bílkovinných hydrolyzátech. *Potravinářské*, **1**, 1–18.

Velíšek, J., Davídek, T., Davídek, J. and Hamburg, A. (1991a) 3-Chloro-1,2-propanediol derived amino alcohol in protein hydrolysates. *Journal of Food Science*, **56**(1), 136–138.

Velíšek, J., Davídek, T., Davídek, J., Kubelka, V. and Víden, I. (1991b) 3-Chloro-1,2-propanediol derived amino acids in protein hydrolysates. *Journal of Food Science*, **56**(1), 139–142.

Velíšek, J., Ledahudcová, K., Hajšlová, J., Pech, P., Kubelka, V. and Víden, I. (1992) New 3-chloro-1,2-propanediol derived dihydroxypropylamines in hydrolysed vegetable proteins. *Journal of Agricultural and Food Chemistry*, **56**, 1389–1392.

Velíšek, J. and Ledahudcová, K. (1993) Problems of organic chlorine compounds in food protein hydrolysates. *Potravinářské vědy*, **11**, 149–159.

Velíšek, J., Doležal, M., Crews, C. and Dvořák, T. (2002) Optical isomers of chloropropanediols: Mechanisms of their formation and decomposition in protein hydrolysates. *Czech Journal of Food Sciences*, **20**, 161–170.

Velíšek, J., Calta, P., Crews, C., Hasnip, S. and Doležal, M. (2003) 3-chloropropane-1,2-diol in models simulating processed foods: Precursors and agents causing its decomposition. *Czech Journal of Food Sciences*, **21**, 153–161.

Wittmann, R. (1991) Determination of dichloropropanols and monochloropropandiols in seasonings and in foodstuffs containing seasonings. *Zeitschrift fur Lebensmittel-Untersuchung Und-Forschung*, **193**(3), 224–229.

Wong, K.O., Cheong, Y.H. and Seah, H.L. (2006) 3-Monochloropropane-1,2-diol (3-MCPD) in soy and oyster sauces: Occurrence and dietary intake assessment. *Food Control*, **17**(5), 408–413.

World Trade Organization (WTO) (2001a) Committee on Sanitary and Phytosanitary Measures – Notification – Republic of Korea – Acid-hydrolyzed vegetable protein and soy sauce containing acid, G/SPS/N/KOR/106, 3 December 2001. Available from: http://www.ipfsaph.org/En/default.jsp [accessed January 2007].

World Trade Organization (WTO) (2001b) Committee on Sanitary and Phytosanitary Measures – Notification of Emergency Measures – Malaysia – Foods containing acid-hydrolyzed vegetable protein, G/SPS/N/MYS/10, 26 July 2001. Available from: http://www.ipfsaph.org/En/default.jsp [accessed January 2007].

World Trade Organization (WTO) (2002) Committee on Sanitary and Phytosanitary Measures – Notification – Thailand – Vegetables and derived products, G/SPS/N/THA/88, 26 March 2002. Available from: http://www.ipfsaph.org/En/default.jsp [accessed January 2007].

Wu, H. and Zhang, G. (1999) Determination of 3-chloropropane-1,2-diol in soy sauce by GC-MS-SIM. *Fenxi Ceshi Xuebao (Journal of Instrumental Analysis)*, **18**, 64–65.

Xing, X. and Cao, Y. (2007) Determination of 3-chloro-1,2-propanediol in soy sauces by capillary electrophoresis with electrochemical detection. *Food Control*, **18**(2), 167–172.

Xu, X., Ren, Y., Wu, P., Han, J. and Shen, X. (2006) The simultaneous separation and determination of chloropropanols in soy sauce and other flavoring with gas chromatography-mass spectrometry in negative chemical and electron impact ionization modes. *Food Additives and Contaminants*, **23**(2), 110–119.

Yamamoto, S., Yoshida, Y., Kurumatani, M., Ota, M. and Asano, S. (2001) Water-soluble thermosetting resin and wet-strength agent for paper using the same. U.S. Patent Application 20010034406. Available from: http://www.uspto.gov/patft/index.html [accessed November 2006].

Yonetani, R., Ikatsu, H., Miyake-Nakayama, C., Fujiwara, E., Maehara, Y., Miyoshi, S. I., Matsuoka, H. and Shinoda, S. (2004) Isolation and characterization of a 1,3-dichloro-2-propanol-degrading bacterium. *Journal of Health Science*, **50**(6), 605–612.

Zelinková, Z., Svejkovská, B., Velíšek, J. and Doležal, M. (2006) Fatty acid esters of 3-chloropropane-1,2-diol in edible oils. *Food Additives and Contaminants*, **23**(12), 1290–1298.

13   Heterocyclic Amines

Mark G. Knize and James S. Felton

Summary

Heterocyclic amines (HAs) are a class of mutagens found at low ng/g (part-per-billion – ppb) levels in cooked food (meat and fish) and are derived from natural constituent precursors. Some nine HAs are described being primarily of an amino-imidazo structure suggesting that creatine or creatinine is an important precursor of the muscle meat reactions. There is considerable variation in the extent of formation of HAs depending on the temperature and mode of cooking and a knowledge of the formation conditions can suggest ways to cook meat (lower temperatures, among others) that greatly inhibit the formation and, thus reduce the human intake of HAs. A variety of analytical approaches have been successfully applied to heterocyclic amine analysis including GC/MS, LC/MS, LC/MS/MS and capillary electrophoresis. The compelling conclusion from meat consumption and cancer studies is that humans are inevitably exposed to these genotoxic rodent carcinogens over a lifetime. Intake levels are low and these HAs can be absorbed and then either activated and covalently bound to DNA and proteins, or detoxified and excreted in the urine. The consequences of such exposure in food safety terms are therefore difficult to assess.

13.1   INTRODUCTION

This chapter provides a historical account of the discovery of mutagens in cooked food and then reviews current knowledge of their precursors and formation. Model reactions simulating meat cooking clearly showed that these compounds are derived from natural constituents. Analytical methods have been developed and used to determine levels of these bioactive compounds in the human diet. Knowledge of the formation of these compounds has led to cooking advice to reduce exposure. The risks of exposure in humans have been evaluated in many laboratories.

The observation that cancer rates differ worldwide has implicated lifestyle and accompanying diet as important factors (Doll and Peto, 1981; Holmes, 1998). A biologically plausible hypothesis for at least part of this association was the discovery in the 1970s of mutagenic activity, as detected by bacterial test systems, in meats cooked for human consumption (Sugimura et al., 1977b; Commoner et al., 1978). The discovery of substances derived from cooked meats to be mutagenic in bacterial tests paralleled the well-known presence of mutagens in the smoke collected from cigarettes at that time (Sugimura et al., 1977a).

The discovery of mutagenic substances in cooked meats was followed by demonstrations in many laboratories worldwide that the mutagens were formed during high temperature cooking and that the formation process was both time- and temperature-dependent. A large range of foods was analyzed and it was determined that the cooked muscle-meats from any vertebrate source were the major sources of extractable mutagenic activity in the human diet (Bjeldanes *et al.*, 1982).

The precursors responsible for the mutagens were proposed when the chemical structures of the first compounds from cooked fish (Kasai *et al.*, 1980, 1981a) and beef (Kasai *et al.*, 1981b; Felton *et al.*, 1986) were determined to be heterocyclic amines (HAs). These mutagenic chemicals were primarily of an amino-imidazo structure suggesting that creatine or creatinine was a component of the muscle meat reactions.

Early work in adding creatine to meat before cooking showed that the addition increased mutagenic activity (Jägerstad *et al.*, 1983). Experiments relating creatine levels in fish, which varied over a range of 2.5-fold, showed mutagenic activity after cooking to be only somewhat correlated with the creatine concentration (Marsh *et al.*, 1990). A later study of cooked meat from 17 animal species also showed that creatine or creatinine levels do not explain differences in mutagenic activity. These results suggested that other components were also important in determining the mutagen levels in cooked meats. Other work showed free amino acids to be involved in the development of mutagenic activity (Övervik *et al.*, 1989), but not amino acids included in the polypeptide chain of proteins (Jägerstad *et al.*, 1983).

Analysis of the specific mutagenic compounds formed during cooking shows that amino acids are important and changes in these can affect the amount and types of mutagens found in the cooked meat. Importantly, knowledge of the formation conditions does suggest ways to cook meat (lower temperatures, among others) that greatly inhibit the formation, and thus reduce the human intake, of HAs.

13.2   FORMATION OF HETEROCYCLIC AMINES

Figure 13.1 shows the structure, common name, and chemical name of nine HAs found in cooked meats. All but one of these structures (AαC) contain the five-membered imidazo ring with an amino group and *N*-methyl group. Other HAs have been isolated from model systems based on heated or pyrolyzed protein, and can occasionally be found in foods, but this report focuses on these commonly reported HAs.

From Fig. 13.2, it can be seen that the structure of creatine resembles the imidazo ring, including the amino and *N*-methyl groups. Modeling experiments show that the four amino-imidazo heterocyclic amine mutagens shown can be pyro-synthesized in laboratory experiments from creatine and other small molecules, such as amino acids and glucose. It is easy to see that the *N*-methyl-amino-imidazo moiety could form intact from creatine, but the source of other rings are derived from other small molecules, and their sources are not apparent from the reactants. These HAs were isolated by following their mutagenic activity in *Salmonella*-based mutation tests during extraction and chromatographic purification (Kasai *et al.*, 1980, 1981a, b; Felton *et al.*, 1984; Knize *et al.*, 1990; Hayatsu *et al.*, 1991). Interestingly, in determining the exact chemical structure for the naturally produced HAs, many structural features were shown to greatly affect mutagenic potency. Such structure/activity relationships have been the subject of many research efforts worldwide (Nagao, 1981; Jagerstad, 1985; Hatch, 1997, 2001; Knize *et al.*, 2006).

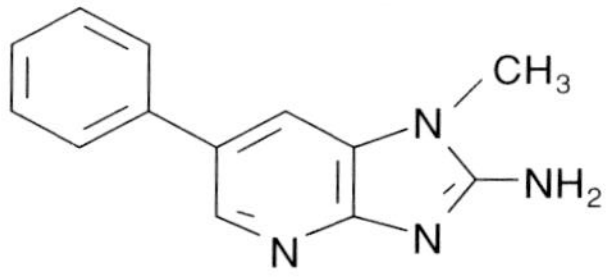

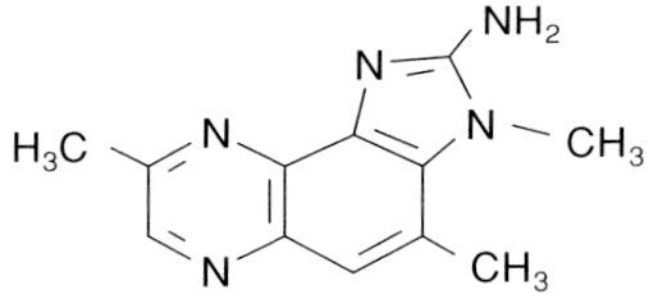

DiMeIQx
(2-Amino-3,4,8-trimethylimidazo[4,5-f]quinoxaline)

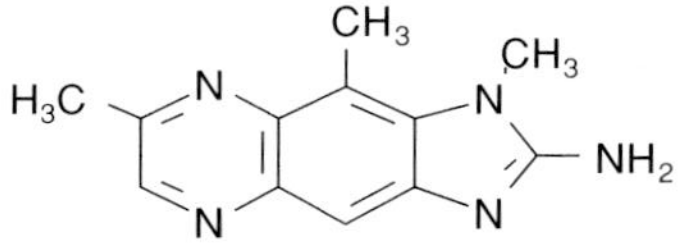

7,9-DiMeIgQx
(2-Amino-1,7,9-trimethylimidazo[4,5-g]quinoxaline)

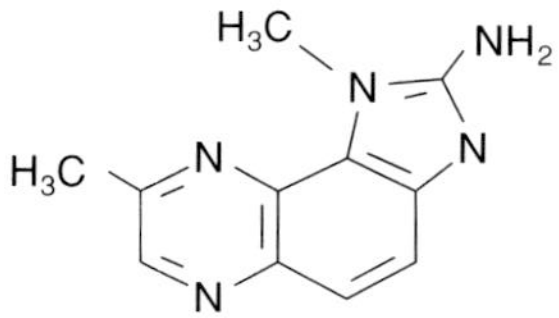

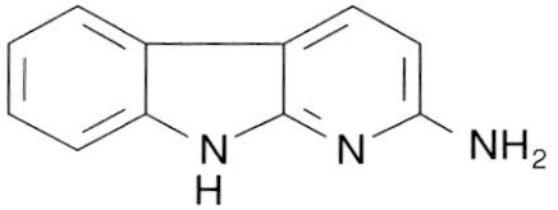

AαC
(2-Amino-9H-pyrido[2,3-b]indole)

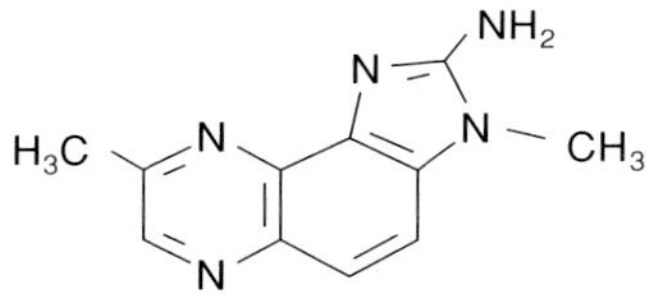

MeIQx
(2-Amino-3,8-dimethylimidazo[4,5-f]quinoxaline)

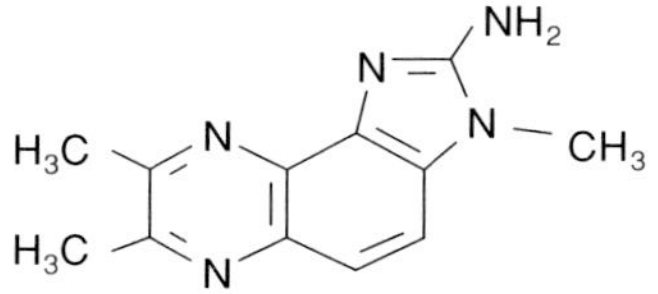

7,8-DiMeIQx
(2-Amino-3,7,8-trimethylimidazo[4,5-f]quinoxaline)

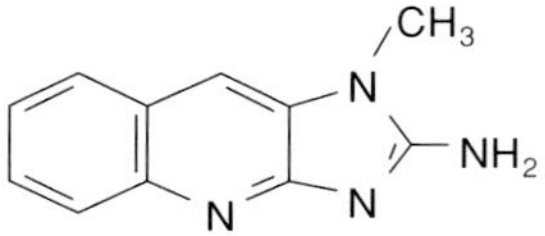

IQ[4,5-b]
(2-Amino-1,6-dimethylfuro[3,2-e]imidazo[4,5-b]pyridine)

PhIP
(2-Amino-1-methyl-6-phenylimidazo[4,5-b]pyridine)

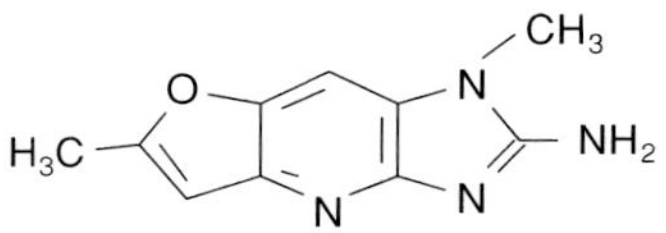

IFP
(2-Amino-1,6-dimethylfuro[3,2-e]imidazo[4,5-b]pyridine)

Iso-MeIQx
(2-Amino-1,8-dimethylimidazo[4,5-f]quinoxaline)

Fig. 13.1 Structures, common names, and chemical names of heterocyclic amines found in cooked meats.

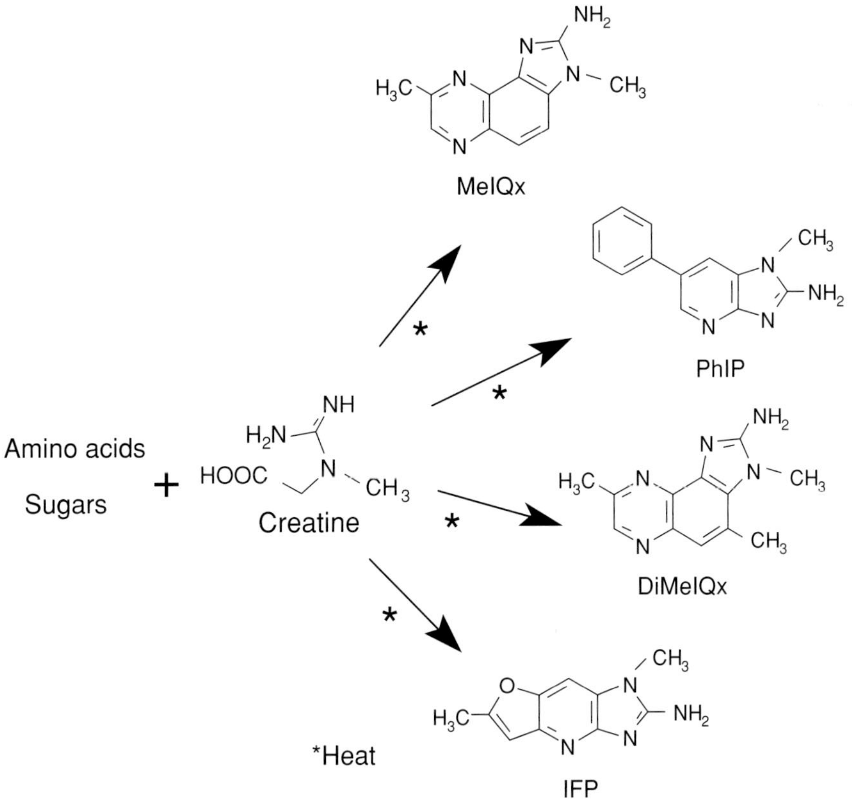

Fig. 13.2   Scheme of heterocyclic amine formation showing the precursors of amino acids and sugars in meat or model systems forming known heterocyclic amines.

13.2.1   Model systems

Model systems have been developed to understand the formation of the mutagenic/carcinogenic HAs, to help identify the foods and cooking conditions favoring their formation and to develop strategies to reduce their formation and, thus, human intake. Defined model systems composed of creatine or creatinine, amino acids, and sugars have been a good model for the trace level formation of these HAs. Jägerstad *et al.* developed a system for heating components in diethylene glycol (Jägerstad *et al.*, 1983) and the work was followed by many studies investigating HA precursors (Negishi *et al.*, 1984; Skog and Jägerstad, 1993; Johansson *et al.*, 1995) and kinetics (Arvidsson *et al.*, 1997) in a sealed-tube aqueous model. It was shown that 37°C is warm enough to produce 2-amino-1-methyl-6-phenylimidazo[4,5-*b*]pyridine (PhIP) from a mixture of phenylalanine with creatinine and glucose or 2-amino-3,8-dimethylimidazo[4,5-*f*]quinoxaline (MeIQx) from glycine with creatinine and glucose in aqueous buffers (Manabe *et al.*, 1992; Kinae *et al.*, 2002). But, no PhIP was found in a similar model system kept at room temperature for 2 weeks in another study (Zochling and Murkovic, 2002).

Simply dry-heating HA precursors produces HAs in the same relative amounts and molecule types which have been shown to form in cooked meats. Pais *et al.* (1999) showed that heating the amino acids, creatine, and glucose for 30 minutes at 225°C, in ratios mimicking

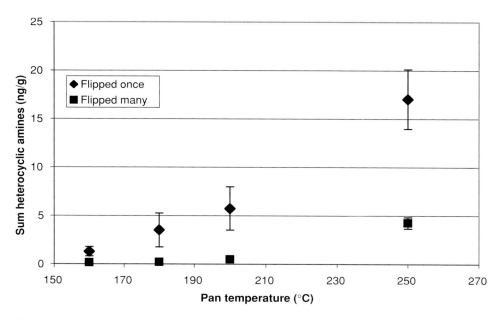

Fig. 13.3 Plot of heterocyclic amine formation during the frying of beef hamburgers to an internal temperature of 70°C with pan temperatures of 160, 180, 200, or 250°C. Diamond symbols show results from meat turned over once, at 5 minutes, during frying. Square symbols show results from meat turned over every minute until the final temperature is reached. (Reproduced from Salmon, C.P., Knize, M.G., Panteleakos, F.N., Wu, R., Nelson, D.O. and Felton, J.S. Minimization of heterocyclic amines and thermal inactivation of *Escherichia coli* in fried ground beef. *Journal of the National Cancer Institute*, **92**, 773–1778, copyright 2000 with permission of Oxford University Press.)

their content in beef muscle, chicken breast, or codfish, produced a family of HAs (Pais *et al.*, 1999). As is seen in the meats themselves, PhIP and MeIQx are formed in relatively large quantities in all three mixtures, but the model based on chicken meat produces the largest amount of PhIP. Surprisingly, IQ was only detected in the codfish model.

13.2.2 Conditions for formation in meats

The temperature dependence of the formation of HAs in beef patties cooked to 70°C, close to the U.S. Department of Agriculture/Food Safety and Inspection Services recommended internal temperature of 71.1°C, is shown in Fig. 13.3. The sum of the mass amounts of individual HAs identified, per gram of cooked meat, is shown for each of four pan temperatures, indicating a direct correspondence between increased cooking temperature and HA content (Salmon *et al.*, 2000). Also shown (square symbols) is the result of turning the meat over in the frying pan every minute, compared to turning the meat over once, after 5 minutes of cooking (diamond symbols). Heat flow simulations to understand HA formation during this pan-frying experiment were carried out by Tran *et al.* (2002). These simulations accurately modeled the experimental temperature increases, meat cooking times, HA spatial distribution, and total amount of HAs produced (Tran *et al.*, 2002), but the measured experimental result for HA content, based on turning the meat over every minute was greater than that expected from the computer simulation of meat cooking.

Together these data show the actual variation in formation (13-fold), for meat patties cooked to $70°C$. These results suggest the variation in HA formation, and, thus, the statistical range of the human HA dose.

13.2.3   Food sample analysis

The HAs isolated from cooked foods have stable multi-ring aromatic structures, and all have an exocyclic amino group. They all have characteristic UV spectra and high extinction coefficients, some of the compounds fluoresce (i.e., PhIP), and all can be electrochemically oxidized making electrochemical detection a suitable method. The aromatic structures of these HAs give little fragmentation and therefore show large base peaks, making mass spectrometry a good detection method (Knize *et al.*, 1992). With each of these detection methods, confirmation of peaks, for instance, by mass or UV spectra, is important to prevent false positives in the complex sample extracts.

There are several factors that make the analysis of HAs from foods a difficult problem. HAs are present in foods at very low ng/g levels (i.e., part-per-billion – ppb levels). Because of these low levels, chromatographic efficiency, detector sensitivity, and detector selectivity must be optimized. Several of the HAs are formed under the same reaction conditions, so the number of compounds to be quantified requires that the extraction, chromatographic separation, and detection be general enough to detect several of the HAs per sample. The complexity and diversity of food samples to be analyzed requires a rugged method not affected by the sample matrix. Several extraction and analysis methods were developed over the last 20 years, including solid-phase extraction (SPE), HPLC/UV, derivatization/GC/MS, and LC/MS/MS. These methods have made analysis cheaper and more accurate, enabling hundreds of samples from cooked meats and model systems to be investigated for HA content.

13.3   SAMPLE EXTRACTION

The detection of bacterial mutagens in cooked meats led to efforts to isolate these chemicals using a variety of extraction methods. Samples were extracted with acid or organic solvents (Sugimura *et al.*, 1977b; Felton *et al.*, 1981; Krone and Iowoka, 1981). The purity of the extracts obtained was generally not sufficient to measure specific HAs, and typically mutagenic potency was measured using the Ames/*Salmonella* test, an assay that performs well with crude mixtures.

Murray *et al.* (1988) devised an analysis method specific enough to be used on samples after a simple liquid/liquid extraction procedure. These were then derivatized and analyzed by GC/MS as described below.

The most widely used HA sample preparation scheme is the SPE method that uses liquid/liquid extraction on a solid support of diatomaceous earth as the first step. The inert diatomaceous earth carrier allows efficient and rapid organic solvent extraction without any risk of emulsion formation, a commonly encountered problem when extracting food.

SPE refers to procedures that use disposable cartridges typically containing 100–500 mg of a solid, often silica-based, sorbent. Gross developed a method to purify HAs from foods and related products with the goal of high sample throughput and high analytical sensitivity (Gross, 1990). The key to this method is the coupling of liquid/liquid extraction to a cation exchange resin column (propylsulfonic acid silica), concentrating the sample diluted from the large volume of organic solvent resulting from the first step in the extraction. Using the cation exchange properties of the column allows selective washing and elution to further purify the sample. This extraction method, followed by various detection schemes, was evaluated in interlaboratory comparisons in Europe (Santos *et al.*, 2004).

Another HA extraction method uses a solid support containing blue copper phthalocyanine trisulfonate linked to cotton (or rayon) and was developed for its ability to adsorb aromatic compounds having three or more fused rings. HAs can be adsorbed to the blue cotton from saline solutions and eluted with a methanol/ammonia solution (Hayatsu *et al.*, 1983), and this scheme has been used in many food analysis studies (Turesky *et al.*, 1988; Wakabayashi *et al.*, 1992).

A single SPE scheme was used to isolate three classes of genotoxic compounds from charcoal-grilled meat: HAs, polycyclic aromatic hydrocarbons, and nitrogen-containing polycyclic aromatic hydrocarbons (Rivera *et al.*, 1996). The method gave detection limits of 0.3–8.4 ng/g in a charcoal-grilled meat sample for the 12 compounds assessed. Very few studies have analyzed the same meat sample for more than one class of mutagen/carcinogen. We showed pan-fried hamburgers to have HAs only, and flame-grilled meat had both PAH and HAs, with a total of 37 ng/g of six PAH, and 17.2 ng/g for combined PhIP and MeIQx (Knize *et al.*, 1999).

13.3.1 Chromatography and detection

Baseline separation of all HAs is a prerequisite for the multi-compound analysis method. Chromatographic conditions were optimized by Gross *et al.* for the separation of 12 HAs in 1992 (Gross and Grüter, 1992). A more recent study evaluated HPLC columns for the separation of HAs for LC/MS, and the same column material favored by Gross, containing ODS-80 was found to be optimal for HAs in the newer study (Barcelo-Barrachina *et al.*, 2004b).

Peak confirmation is a crucial problem when working with such low levels of HAs since co-elution with other co-extracted compounds can occur. The most convenient and accessible instrument to identify HAs on-line during an HPLC separation is the UV photodiode array detector. Most instruments allow the recording of UV spectra even at a 0.1 ng level. A photodiode array detection system efficiently prevents most false peak identifications and is essential to prevent the interpretation of false positive results. Even with modern photodiode array detection with spectral library matching by computer, human interpretation is still needed for confirmation in many cases. Changes in separation selectivity were shown to further improve peak identification for complex samples (Pais and Knize, 1998). In addition, fluorescence detection is typically used in-line as a complement to photodiode-array detection. Not all HAs fluoresce, however, but PhIP a common HA in foods shows about a 9-fold greater peak signal with fluorescence than with the photodiode-array UV detection method.

Gross determined the standard error of repeated extractions to range from 6 to 11% for cooked fish, very acceptable for low-level analysis (Gross, 1990). Reproducibility of the SPE and HPLC method over time was determined in a blind study. Two hamburger samples, one cooked rare and one cooked "well-done," were repeatedly sent in separate aliquots for quality control determination. For the two samples analyzed repeatedly during the 24 months, MeIQx, PhIP, and DiMeIQx were consistently detected in the sample cooked "well-done," but not in the sample cooked rare. Relative standard deviations (coefficients of variation) for the amounts determined were 34% for MeIQx, 22% for PhIP, and 38% for DiMeIQx for the "well-done" cooked samples (Knize *et al.*, 1995a). No sample degradation was seen in the samples stored frozen at −4°C.

13.3.2   Mass spectrometry

Mass spectrometry has many desirable features as a detector of HAs. Mass spectrometry offers the selectivity of mass detection with the possibility of adding heavy-isotope-labeled internal standards to determine extraction recovery and act as chromatographic standards simultaneously.

13.3.3   GC/MS

A very sensitive approach for HA analysis was devised using gas chromatography/mass spectrometry (GC/MS). Most of the HAs exhibit poor chromatographic behavior for gas chromatography, and thus require derivatization prior to injection. For this method, extracts of food from a liquid/liquid extraction scheme were derivatized with 3,5-bistrifluoromethylbenzyl bromide at room temperature, washed with hexane and extracted with ethyl acetate. The total yield for these extraction steps was reported to be about 40% as determined by [^{14}C]MeIQx tracer (Murray *et al.*, 1989). PhIP can give multiple products upon derivatization with this method, and another derivatizing agent, pentafluorobenzyl bromide, was reported to give a single product (Friesen *et al.*, 1994). Additional derivatization methods for GC analysis of HAs were also reported (Kataoka and Kijima, 1997; Reistad *et al.*, 1997).

With mass spectrometry, recovery for extraction and derivatization can be calculated by the use of heavy-isotope-labeled internal standards as mentioned above. The chemical ionization and negative-ion detection gave a reported 1 pg detection limit using selected-ion monitoring (Murray *et al.*, 1988). GC/MS methods have not been optimized for as many different HAs as HPLC/UV methods, and IQ reportedly could not be detected by one GC/MS method (Vainiotalo *et al.*, 1993). Several groups reported GC/MS analysis gives limits of detection at least 20-fold lower than HPLC/UV methods (Murray *et al.*, 1989; Vainiotalo *et al.*, 1993).

13.3.4   LC/MS

The HAs are more easily separated by liquid chromatography (LC) and there are many more examples of groups successfully using LC/MS for food analysis. Gross *et al.* (1993) reported LC/MS-single-ion plots for 11 HAs and showed extracts of bacon to be free of interfering peaks at the masses monitored for the HAs (Gross *et al.*, 1993). Pais *et al.* separated 14 HAs and related compounds using LC/MS, reporting instrument detection limits of 10–600 pg depending on the HA (Pais *et al.*, 1997).

Three LC/MS systems with electrospray interfaces, an ion trap, a single quadrupole and a triple quadrupole, were compared for the analysis of HAs. The results clearly showed that the ion trap instrument was least sensitive for detection of HAs spiked into meat extract; the single quadrupole instrument was usually 2- to 7-fold better, and the triple quadrupole instrument, operated in the multiple reaction monitoring mode, was typically 10-fold more sensitive than the single quadrupole instrument, with levels of detection of 0.5–5 pg, depending on the HA (Barcelo-Barrachina *et al.*, 2004a). Multiple reaction monitoring with a triple quadrupole mass spectrometer was used to evaluate food samples by Klassen *et al.* (2002) and Turesky *et al.* (2005). These three studies, in three separate laboratories, confirm that triple quadrupole LC/MS/MS can detect multiple HAs concurrently at the picogram level.

13.3.5 High pressure LC-MS/MS

A promising new liquid chromatography system, using higher than normal pressure to provide high efficiency and resolution, was optimized for the mass spectral analysis of HAs. Sixteen HAs were separated and limits of detection were determined, showing this system to be sensitive and reproducible for routine analysis of meat extracts (Barcelo-Barrachina *et al.*, 2006).

13.3.6 Capillary electrophoresis

Capillary electrophoresis with photodiode-array detection was developed for the analysis of 12 HAs in foods (Wu *et al.*, 1996) and meat extracts (Olsson *et al.*, 1997; Puignou *et al.*, 1997). Capillary electrophoresis with field-amplified sample injection, coupled to a mass spectrometer used in either the MS or MS/MS mode, was evaluated for 16 HAs by Sentellas *et al.* (2003). They showed extraction of the amines from spiked urine, however, rather than the more difficult food analysis. Levels of detection are in the range of LC/UV for this separation method.

13.4 INCIDENCE AND OCCURRENCE

Table 13.1 shows typical amounts of PhIP and MeIQx for cooked beef, poultry, and pork meats cooked in a variety of ways. Cooking in the laboratory setting allows for determination of temperature and time, yet commercially cooked meats (labeled "Com.") are perhaps a better sample of foods from which to determine human exposure since these reflect meats cooked specifically for human consumption.

Beef cooked as steaks or ground and formed into patties, show levels in the low ng/g range, many samples containing both PhIP and MeIQx, with the highest sample totaling over 30 ng/g for these HAs. Commercially cooked "fast food" hamburgers and roast meat have low or undetectable levels of HAs.

Poultry samples differ from the beef or pork in that several labs show levels of PhIP greater than 50 ng/g, although MeIQx values are in the range of beef or pork samples. Roasted chicken, though, has very low or undetectable levels of HAs (Sinha *et al.*, 1995; Skog *et al.*, 1997), as does commercially fried chicken (Tikkanen *et al.*, 1993; Knize *et al.*, 1995). Chicken has notably more phenylalanine, tyrosine, and isoleucine compared to beef and pork, which appears to explain the high amounts of PhIP reported.

In pork samples cooked a variety of ways, PhIP and MeIQx are seen in the most "well-cooked" samples. Bacon appears to have the highest amount of HAs per gram. Commercially grilled pork or pork sausage have low levels. Pan residues from pork chops and bacon can also contain PhIP and MeIQx (Gross *et al.*, 1993; Johansson *et al.*, 1994; Skog *et al.*, 1995). Cajun-style pork (not shown), although very black in color, has low levels of HAs, a combined total of 3.5 ng/g (Knize et al., 1996). The HAs in pork are dependent on the preparation: hot dogs, having a casing, and thus, low surface area, are much lower in HAs than the thinly sliced and fried bacon.

Studies of the amounts of HAs produced in foods as a result of regional cooking practices are reported for Great Britain (Murray *et al.*, 1993), Sweden (Johansson and Jägerstad, 1994; Skog *et al.*, 1997), Switzerland (Zimmerli *et al.*, 2001), Spain (Busquets *et al.*, 2003), Japan (Wakabayashi *et al.*, 1993), and the United States (Knize *et al.*, 1995b, 1998).

Table 13.1 Heterocyclic amine amounts (ng/g) in beef, poultry, and pork.

Food	Cooking method	Surface temperature (°C)	Total time (min)	PhIP	MelQx	Reference
Beef						
Ground patty	Fried	—	—	16.4	2.2	Murray et al., 1993
Ground patty	Fried or grilled	—	12–20	<0.1–1.2	0.03–2.8	Johansson and Jägerstad, 1994
Ground patty	Com. fried	—	—	0.1–0.6	<0.1–0.3	Knize et al., 1995a
Ground patty	Com. fried	150–180	—	0.62–1.8	0.75–1.64	Klassen et al., 2002
Ground patty	Fried	—	10	0.16–15.2	1.4–5.3	Turesky et al., 2005
Roast	Roast	160	96–182	<0.1	<0.1	Sinha et al., 1998b
Steak	Broiled	—	—	16	2.1	Wakabayashi et al., 1992
Steak	Fried	150–225	7	0.1–1.8	0.02–1.6	Skog et al., 1995
Steak	Grilled	—	—	14	6	Fay et al., 1997
Steak	Griddled	—	—	4.8	2.9	Busquets et al., 2003
Steak	Fried	180–191	6–20	<0.1–23	1.3–8.2	Sinha et al., 1998b
Poultry						
Chicken	Barbecue	—	—	—	0.3	Murray et al., 1993
Chicken	Broiled	—	—	38	2.3	Wakabayashi et al., 1992
Chicken breast	Fried	197–211	14–36	12–70	1–3	Sinha et al., 1995
Chicken breast	Grilled	180–260	10–43	27–480	<0.1–9	Sinha et al., 1995
Chicken breast	Pan fried	140–220	12–34	<0.1–38.2	0.1–1.8	Solyakov and Skog, 2002
Chicken	Fried	175–225	15	0.5–10	0.4–0.5	Skog et al., 1997
Chicken	Fried	—	—	44.9	nd	Busquets et al., 2003
Chicken	Fried	—	—	<0.06–2.1	<0.06–1.8	Klassen et al., 2002
Chicken	Fried	150–180	—	10	0.34	Turesky et al., 2005
Chicken	Com. various	—	—	<0.1–1.75	<0.1–1.26	Wong et al., 2005
Turkey	Fried	—	20	3.8	1.4	Murkovic et al., 1997
Pork						
Bacon	Fried	170	12–16	<0.1–53	0.9–27	Gross et al., 1993
Bacon	Fried	225	12	1.6–2.7	0.9–1.2	Murray et al., 1993
Bacon	Fried	150–225	4–8	0.3–4.5	N < 0.03–23.7	Skog et al., 1995
Bacon	Fried	—	—	<0.1–36	1.0–27	Knize et al., 1997
Chop	Fried	150–225	8–9.5	<0.01–4.8	<0.03–2.6	Skog et al., 1995
Chop	Fried	175	5–15	<0.1	<0.1–3.8	Sinha et al., 1998a
Ham	Grilled	—	—	1.5	2.3	Knize et al., 1997
Pork	Barbecued	—	—	4.2	0.4	Murray et al., 1993
Pork loin	Fried	—	—	2.5	1.9	Busquets et al., 2003
Pork	Com. various	—	—	<0.1–5.41	<0.1–0.94	Wong et al., 2005
Sausage	Com. fried	—	—	<0.1	<0.1–0.3	Knize et al., 1995a
Sausage	Fried	160	6	0.1	0.7	Johansson and Jägerstad, 1994

In most cases, 2-amino-1-methyl-6-phenylimidazo[4,5-*b*]pyridine (PhIP) and 2-amino-3,8-dimethylimidazo[4,5-*f*]quinoxaline (MeIQx) tend to be the most mass-abundant of the HAs. Their concentrations in cooked meats typically range from nearly undetectable levels (typically 0.1 ng/g) to tens of ng/g for MeIQx and up to a few hundreds of ng/g for PhIP, depending on the cooking method.

13.4.1 Pan residues and food flavors

A source of HAs related to the meats themselves are pan residues and process flavors. Pan residues are sometimes consumed after being made into gravy and can be a source of HAs equivalent to or greater than that of the meat itself (Johansson and Jägerstad, 1994; Skog *et al.*, 1997; Pais *et al.*, 1999).

Process flavors are commercially produced flavors derived from heated mixtures of proteins, fats, and carbohydrates. These are added to foods in amounts up to a few percent by weight to improve the food's taste and color, and they can also be used as a base for soups. Because of their chemical complexity, specific sample preparation methods for HA analysis have been developed for these types of samples (Gross *et al.*, 1992; Perfetti, 1996; Pais *et al.*, 1997). Although most process flavors have undetectable levels of HAs, some samples contain as much as 20 ng HAs per gram of solid or liquid flavor (Jackson *et al.*, 1994; Stavric *et al.*, 1997; Solyakov *et al.*, 1999). However, since process flavors are greatly diluted and then consumed as only a tiny percentage in the food, we believe the bulk of HA exposure is not from process flavors, but is from "well-cooked" meats.

13.4.2 Modifying cooking practices to reduce the formation of heterocyclic amines

As shown in Fig. 13.3, the formation of HAs is related to pan temperature, even when meat is cooked to the same final internal temperature. Surprisingly, the time needed to reach the 70°C USDA-recommended internal temperature is nearly the same at 250°C (7 min) as it is at 160°C (9 min) (Salmon *et al.*, 2000). This is due to the slow heat transfer through the meat, suggesting that simply using lower pan temperatures is a practical way to reduce HA formation, without greatly increasing cooking time.

As shown above, turning pan beef patties over every minute, compared to turning the meat over once during the cooking period, and cooking at moderate pan temperatures until the target internal temperature of 70°C is reached, seems to be the most effective way to reduce HA content when frying. Monitoring the temperature avoids undercooking, which is defined as cooking to a final temperature below 70°C, the internal temperature needed to kill harmful bacteria in the meat (Salmon *et al.*, 2000).

Minor changes in recipes for preparing different meat dishes may provide another way to reduce the HAs formed. Schemes for reducing mutagenic activity or HAs by adding substances to ground meat have been reported. Additives such as soy flour or antioxidants, (Wang *et al.*, 1982) or glucose or lactose (Skog *et al.*, 1992) were shown to lower mutagenic activity when added to ground meat prior to frying.

The heat flow and mass transport in meat during frying is very complex. Water is important for the transport of water-soluble precursors for formation of HAs within the food. The transport of precursors from the inner parts of the food to the surface can be restricted by the addition of water-binding compounds, such as salt, soy protein, or starch to ground meat, thus reducing the formation of HAs. It has been shown that there is a significant effect with the addition of sodium chloride/sodium tripolyphosphate (Persson *et al.*, 2003). Enzyme treatment with creatinase was also used to reduce the available creatine in meat (Vikse and Joner, 1993).

HA formation can also be affected by meat surface treatment. The application of a 7-component marinade to chicken breast meat before grilling can greatly decrease PhIP, although MeIQx is increased at the longest cooking time. This increase is probably due to sucrose in the marinade (Salmon *et al.*, 1997). No change in HAs was seen after marinating chicken in another study (Tikkanen *et al.*, 1997), possibly due to differences in marinating or cooking conditions. Conversely, an HA reducing effect was seen when sugar was mixed with ground meat formed into patties before frying (Skog *et al.*, 1992). Experiments with eggs, bean cake and pork show that added sugar and soy sauce increases most of the HAs (Lan and Chen, 2002).

A microwave pre-treatment method reduced the amount of HAs formed during the frying of ground beef (Felton *et al.*, 1994). Beef patties received microwave pre-treatment for various times before frying. Microwave pre-treating for 2 minutes, then pouring off the resulting liquid and frying at either 200°C or 250°C for 6 minutes per side, reduced HAs. The liquid released by the microwave pre-treatment contained creatine, creatinine, amino acids, glucose, water, and fat, and discarding these precursors presumably reduced the concentration of reactants resulting in less HAs being formed. The sum of the HAs present decreased 3-fold following microwave cooking (1.5 min) and frying at 200°C (6 min per side) or 9-fold following microwave cooking and then frying at 250°C, compared to controls (non-microwave pre-treated beef patties fried under identical conditions).

13.5   EXPOSURE ASSESSEMENT

The primary method for assessing dietary consumption of different food types is through food frequency questionnaires. Exposures to HAA have been assessed using questionnaires for frequency of meat consumption, estimates of amounts consumed, and meat "doneness" preference, often using photographs to aid in assessing "doneness" (Sinha *et al.*, 1998a; Augustsson *et al.*, 1999; Keating *et al.*, 2000; Le Marchand *et al.*, 2002; Sinha, 2002). Worrisome is that validation of the "doneness" classification by analyzing foods consumed by the study population is typically not carried out, yet exposure estimates derived from these questionnaires and reported with great certainty often fail to consider meat preparation and cooking factors known to be important in HA formation. Nevertheless, numerous studies have used this data to correlate HA exposure with risk of cancers at many sites.

Despite the common endpoint temperature for the samples in Fig. 13.3, the sum of the four HAs varied from 0.14 ± 0.08 ng/g total to 17.05 ± 4.28 ng/g of cooked meat depending on pan temperature. This greater than 100-fold difference shows that the internal temperature of the meat (which was the same in all cases) may be meaningless for detecting an exposure gradient over the expected range in the human population. This is despite the reported use of endpoint temperature for classification. Asking the correct questions to get accurate HA exposure information is difficult, but clearly the cooking temperature and cooking time are the important parameters.

An investigation of HA amounts in meats obtained from survey participants demonstrates the limitations of food frequency questionnaires. For a study of foods cooked under typical household conditions, grilled meat samples were obtained from volunteers in households in the Midwestern United States as a part of a published study on pan-fried meats (Keating *et al.*, 2000). Participants were volunteers responding to an initial survey that they preferred their meat "well-done" or "very well-done." The participants were surveyed a second time several years later, and surprisingly, 46% of the participants changed their stated meat "doneness" preference. To correlate the stated "doneness" preference with HA levels, 92 samples of cooked meat were obtained and analyzed by SPE and reverse phase HPLC with photodiode-array detection, using published methods (Knize *et al.*, 1995a). Surprisingly, in this collection of meat samples thought to be cooked to a "well-done" or "very well-done" state, approximately 20% of the samples had undetectable levels of HAs. The standard questionnaires would categorize meats with these undetectable levels of HAs as high exposure samples. In addition, the quantified HA content of the meat samples spanned over two orders of magnitude, and the difficulty predicting quantities based on "doneness" assessments, the variation in HA formation, and the changing of stated meat "doneness" preference indicate that exposure estimates using diet surveys need to be made with caution. If misclassifications were reduced, the power of the epidemiology studies could be greatly improved. Additional information about the cooking method, temperature, and time should be coupled with the doneness results.

The bioavailable dose is an even better exposure measurement than the dietary intake, as it reflects the absorption of the carcinogens through the digestive system. Pioneering work was done in 1982 showing that exposure to mutagen-containing fried pork or bacon resulted in a spike of detectable mutagenic activity in urine (Baker *et al.*, 1982). It has also been shown that the ingestion of fried ground-beef hamburgers resulted in mutagenic activity in urine that was not detected in urine from the same individuals collected before the hamburgers were consumed (Hayatsu *et al.*, 1985).

Later, using oral or intra-peritoneal dosing of synthetic HAs, it was shown that in rodents, most of the dose was excreted in urine as metabolites (Turteltaub *et al.*, 1989; Alexander *et al.*, 1991; Buonarati *et al.*, 1992; Turesky *et al.*, 1993), suggesting it would be possible to monitor the exposure in humans as urinary metabolites.

The presence of specific HAs in human urine after a cooked-meat meal was shown by using a sensitive GC/MS detection method (Murray *et al.*, 1989). They showed that only a low percentage of the parent compound was in the urine. Other studies showed four HAs were detected in urine from volunteers on their normal diet (Ushiyama *et al.*, 1991) and PhIP and MeIQx conjugates detected in urine led to conclusions about Phase II conjugation reactions (Stillwell *et al.*, 1997). Another study showed PhIP metabolites were detected in a racially diverse population (Kidd *et al.*, 1999). One drawback of the analysis of urine for HAs and their metabolites is the short duration of the detectable exposure signal. It has been shown in many studies in humans and rodents that most of the dose is excreted within 12 hours, so detecting urinary HAs or their metabolites is not the desired long-term marker of dose or bioavailable dose, but only really useful for studying exposure after a single meal.

In 1992, it was discovered that PhIP has a high binding to pigmented tissues (Brittebo *et al.*, 1992). This finding led to detection of PhIP bound to human hair as a marker of dietary exposure and bioavailability (Reistad *et al.*, 1999). Interestingly, PhIP incorporation appears to vary with hair color and be dependent upon the eumelanin concentration in hair (Hegstad *et al.*, 2002). Because a hair sample may provide a record of exposure over a period of several months or longer, hair analysis may be a promising avenue for determining human exposure and internal dose.

13.6   RISK

Epidemiological studies relating cancer outcome and meat cooking "doneness" suggest effects in the breast, colon and prostate, plus esophagus, larynx, lung, lymphocytes, stomach, and pancreas. There are 31 such studies compiled by Knize and Felton (2005). Most of these studies show positive associations, but some are negative, as would be expected given the variety of cancer sites evaluated. Improved classification of exposure levels, however, would tend to elevate odds ratios and improve statistical significance if the hypothesis of HA involvement is true. Thus, improved determination of exposures would refine the risk assessment considerably.

Based on these observations it is apparent that quantifying human HA exposure is not a simple task. Formation of HAs in meat during cooking is highly dependent upon cooking method (time and temperature) and "doneness" levels. Individual exposure depends upon meat consumption patterns. The compelling conclusion for these meat and cancer studies is that humans are exposed to genotoxic rodent carcinogens over a lifetime. Intake levels are low; 1 mg of MeIQx (the total from a 200 g steak with 5 ng/g) consists of 2.8×10^{15} molecules that can be absorbed and then either activated and covalently bound to DNA and proteins, or detoxified and excreted in the urine.

An alternate route of HA exposure is through the respiratory tract via the fumes and smoke generated when meat is cooked. Although the HAs are not volatile chemicals, collected particles generated during cooking were shown to have mutagenic activity and to contain specific HAs (Nagao *et al.*, 1977; Rappaport *et al.*, 1979; Vainiotalo *et al.*, 1993; Thiebaud *et al.*, 1994; Thiébaud *et al.*, 1995). The total exposure to HA appears to be much greater through consumption of "well-done" meat than breathing of cooking fumes, but the metabolism and types of tissue exposed differs between the two routes. Occupational exposure to HA is not expected to be a large factor in human exposure, although the inhalation exposure noted above may be important in some occupational circumstances experienced by chefs.

13.7   REGULATIONS

There are no regulations of HA content in foods or process flavors. The State of California, using its Proposition 65 law to warn consumers about carcinogens, has listed PhIP as a carcinogen, but a no significant risk level (NSRL) has not been established for PhIP.

Conclusion

Human exposure to HAs can now be estimated with a degree of accuracy because HAs are easily measured in meat and human exposure to them can be estimated from food consumption surveys. The difficulty in assessing exposure arises from the number of different HAs of concern, variation in their formation depending on meat cooking conditions, and the lack of validated long-term biomarkers for either intake or biological consequences

The compelling part of HA research for population studies is that there are more than a 100-fold variations in human exposure, and, therefore, most study populations have enough range in variation to test the hypothesis that HA exposures are involved in human cancer etiology. Large cohort studies would be best for these types of studies. Also, compelling is the fact that these exposures can be modified through changes in cooking practices, if changes are warranted from risk assessment, without having to abstain from meat intake.

In most cases, 2-amino-1-methyl-6-phenylimidazo[4,5-*b*]pyridine (PhIP) and 2-amino-3,8-dimethylimidazo[4,5-*f*]quinoxaline (MeIQx) tend to be the most mass-abundant HAs. Their concentrations in cooked meat typically range from nearly undetectable levels (typically 0.1 ng/g) to tens of ng/g for MeIQx and up to a few hundreds of a ng/g for PhIP, depending on the cooking method and food source. Because of their prevalence in cooked meat, and their carcinogenic potency in rodents, these two compounds were used in most studies to understand the mechanisms of mutagenesis and carcinogenesis of HAs. The effects of these two compounds have been explored in model systems using biological endpoints related to mutagenesis in bacteria, cultured cells, and rodents.

Acknowledgments

This work was performed under the auspices of the U.S. Department of Energy by the University of California, Lawrence Livermore National Laboratory under Contract No. W-7405-Eng-48 and supported by NCI grant CA55861.

References

Alexander, J., Wallin, H., Rossland, O.J., Solberg, K.E., Holme, J.A., Becher, G., Andersson, R. and Grivas, S. (1991) Formation of a glutathione conjugate and a semistable transport glucuronide conjugate of N2-oxidized species of 2-amino-1-methyl-6-phenylimidazo[4,5-b]pyridine (PhIP) in rat liver. *Carcinogenesis*, **12**, 2239–2245.

Arvidsson, P., vanBoekel, M.A.J.S., Skog, K. and Jagerstad, M. (1997) Kinetics of formation of polar heterocyclic amines in a meat model system. *Journal of Food Science*, **62**, 911–916.

Augustsson, K., Skog, K., Jagerstad, M., Dickman, P.W. and Steineck, G. (1999) Dietary heterocyclic amines and cancer of the colon, rectum, bladder, and kidney: A population-based study. *Lancet*, **353**, 703–707.

Baker, R.S.U., Arlauskas, A., Bonin, A.M. and Angus, D. (1982) Detection of mutagenic activity in human urine following fried pork or bacon meals. *Cancer Letters*, **16**, 81–89.

Barcelo-Barrachina, E., Moyano, E., Puignou, L. and Galceran, M.T. (2004a) Evaluation of different liquid chromatography–electrospray mass spectrometry systems for the analysis of heterocyclic amines. *Journal of Chromatography A*, **1023**, 67–78.

Barcelo-Barrachina, E., Moyano, E., Puignou, L. and Galceran, M.T. (2004b) Evaluation of reversed-phase columns for the analysis of heterocyclic aromatic amines by liquid chromatography–electrospray mass spectrometry. *Journal of Chromatography B-Analytical Technologies in the Biomedical and Life Sciences*, **802**, 45–59.

Barcelo-Barrachina, E., Moyano, E., Galceran, M.T., Lliberia, J.L., Bago, B. and Cortes, M.A. (2006) Ultra-performance liquid chromatography-tandem mass spectrometry for the analysis of heterocyclic amines in food. *Journal of Chromatography A*, **1125**, 195–203.

Bjeldanes, L.F., Morris, M.M., Felton, J.S., Healy, S.K., Stuermer, D.H., Berry, P., Timourian, H. and Hatch, F.T. (1982) Mutagens from the cooking of food II. Survey by Ames/Salmonella test of mutagen formation in the major protein-rich foods of the American diet. *Food and Chemical Toxicology*, **20**, 57–363.

Brittebo, E.B., Skog, K.I. and Jagerstad, I.M. (1992) Binding of the food mutagen PhIP in pigmented tissues of mice. *Carcinogenesis*, **13**, 2263–2269.

Buonarati, M.H., Roper, M., Morris, C.J., Happe, J.A., Knize, M.G. and Felton, J.S. (1992) Metabolism of 2-amino-1-methyl-6-phenyli idazo[4,5-b]pyridine (PhIP) in mice. *Carcinogenesis*, **13**, 621–627.

Busquets, R., Bordas, M., Torbino, F., Puignou, L. and Galceran, M.T. (2003) Occurrence of heterocyclic amines in several home-cooked meat dishes of the Spanish diet. *Journal of Chromatography B*, **802**, 79–86.

Commoner, B., Vithayathil, A.J., Dolara, P., Nair, S., Madyastha, P. and Cuca, G.C. (1978) Formation of mutagens in beef and beef extract during cooking. *Science*, **201**, 913–916.

Doll, R. and Peto, R. (1981) The causes of cancer: Quantitative estimates of avoidable risks of cancer in the United States today. *Journal of Nature Cancer Institute*, **66**, 1191–1308.

Fay, L.B., Ali, S. and Gross, G.A. (1997) Determination of heterocyclic aromatic amines in food products: Automation of the sample preparation method prior to HPLC and HPLC-MS quantification. *Mutation Research*, **376**, 29–35.

Felton, J.S., Healy, S.K., Stuermer, D.H., Berry, C., Timourian, H., Hatch, F.T., Morris, M. and Bjeldanes, L.F. (1981) Mutagens from the cooking of food (1). Improved isolation and characterization of mutagenic fractions from cooked ground beef. *Mutation Research*, **88**, 33–44.

Felton, J.S., Knize, M.G., Wood, C., Wuebbles, B.J., Healy, S.K., Stuermer, D.H., Bjeldanes, L.F., Kimble, B.J. and Hatch, F.T. (1984) Isolation and characterization of new mutagens from fried ground beef. *Carcinogenesis*, **5**, 95–102.

Felton, J.S., Knize, M.G., Shen, N.H., Lewis, P.R., Andresen, B.D., Happe, J. and Hatch, F.T. (1986) The isolation and identification of a new mutagen from fried ground beef: 2-amino-1-methyl-6-phenylimidazo[4,5-b]pyridine (PhIP). *Carcinogenesis*, **7**, 1081–1086.

Felton, J.S., Fultz, E., Dolbeare, F.A. and Knize, M.G. (1994) Reduction of heterocyclic amine mutagens/carcinogens in fried beef patties by microwave pretreatment. *Food and Chemical Toxicology*, **32**, 897–903.

Friesen, M.D., Kaderlik, K., Lin, D., Garren, L., Bartsch, H., Lang, N.P. and Kadlubar, F.F. (1994) Analysis of DNA adducts of 2-amino-1-methyl-6-phenylimidazo[4,5-b]pyridine in rat and human tissues by alkaline hydrolysis and gas chromatography/electron capture mass spectrometry: Validation by comparison with 32P-postlabeling. *Chemical Research in Toxicology*, **7**, 733–739.

Gross, G.A. (1990) Simple methods for quantifying mutagenic heterocyclic amines in food products. *Carcinogenesis*, **11**, 1597–1603.

Gross, G.A. and Grüter, A. (1992) Quantitation of mutagenic/carcinogenic heterocyclic aromatic amines in food products. *Journal of Chromatography*, **592**, 271–278.

Gross, G.A., Grüter, A. and Heyland, S. (1992) Optimization of the sensitivity of high-performance liquid chromatography in the detection of heterocyclic aromatic amine mutagens. *Food and Chemical Toxicology*, **30**, 491–498.

Gross, G.A., Turesky, R.J., Fay, L.B., Stillwell, W.G., Skipper, P.L. and Tannenbaum, S.R. (1993) Heterocyclic amine formation in grilled bacon, beef, and fish, and in grill scrapings. *Carcinogenesis*, **14**, 2313–2318.

Hayatsu, H., Matsui, Y., Ohara, Y., Oka, T. and Hayatsu, T. (1983) Characterization of mutagenic fractions in beef extract and in cooked ground beef. *Gann*, **74**, 472–481.

Hayatsu, H., Hayatsu, T. and Ohara, Y. (1985) Mutagenicity of human urine caused by ingestion of fried ground beef. *Japanese Journal of Cancer Research*, **76**, 445–448.

Hayatsu, H., Arimoto, S. and Wakabayashi, K. (1991) Methods for separation and detection of heterocyclic amines, in *Mutagens in Food: Detection and Prevention* (ed. H. Hayatsu). CRC Press, Boca Raton; Ann Arbor, Boston, pp. 101–112.

Hegstad, S., Reistad, R., Haug, L.S. and Alexander, J. (2002) Eumelanin is a major determinant for 2-amino-1-methyl-6-phenylimidazo[4,5-b]pyridine (PhIP) incorporation into hair of mice. *Pharmacology and Toxicology*, **90**, 333–337.

Holmes, S. (1998) Food, nutrition and the prevention of cancer: A global perspective. *Health Policy*, **45**, 169–171.

Jackson, L.S., Hargraves, W.A., Stroup, W.H. and Diachenko, G.W. (1994) Heterocyclic amine content of selected beef flavors. *Mutation Research*, **320**, 113–124.

Jägerstad, M., Laser-Reutersward, A., Oste, R., Dahlqvist, A., Grivas, S., Olsson, K. and Nyhammer, T. (1983) Creatinine and Malliard reaction products as precursors of mutagenic compounds formed in fried beef, in *The Maillard Reaction in Foods and Nutrition* (eds. G.R. Waller and M.S. Feather). American Chemical Society, Washington, DC, pp. 507–519.

Johansson, M.A.E. and Jägerstad, M.I. (1994) Occurrence of mutagenic/carcinogenic heterocyclic amines in meat and fish products, including pan residues, prepared under domestic conditions. *Carcinogenesis*, **15**, 1511–1518.

Johansson, M.A.E., Fay, L.B., Gross, G.A., Olsson, K. and Jägerstad, M. (1995) Influence of amino acids on the formation of mutagenic/carcinogenic heterocyclic amines in a model system. *Carcinogenesis*, **16**, 2553–2560.

Kasai, H., Yamaizumi, Z., Wakabayashi, K., Nagao, M., Sugimura, T., Yokoyama, S., Miyazawa, T., Springarn, N.E. and Weisburger, J.H. (1980) Potent novel mutagens produced by broiling fish under normal conditions. *Proceedings of the Japan Academy*, **56**, 278–283.

Kasai, H., Yamaizumi, Z., Nishimura, S., Wakabayashi, K., Nagao, M., Sugimura, T., Spingarn, N.E., Weisburger, J.H., Yokoyama, S. and Miyazawa, T. (1981a) A potent mutagen in broiled fish. Part 1. 2-amino-3-methyl-3H-imidazo[4,5-f]quinoline. *Journal of Chemical Society*, 2290–2293.

Kasai, H., Yamaizumi, Z., Shiomi, T., Yokoyama, S., Miyazawa, T., Wakabayashi, K., Nagao, M., Sugimura, T. and Nishimura, S. (1981b) Structure of a potent mutagen isolated from fried beef. *Chemical Letters*, 485–488.

Kataoka, H. and Kijima, K. (1997) Analysis of heterocyclic amines as their N-dimethylaminomethylene derivatives by gas chromatography with nitrogen–phosphorus selective detection. *Journal of Chromatography A*, **767**, 187–194.

Keating, G.A., Sinha, R., Layton, D., Salmon, C.P., Knize, M.G., Bogen, K.T., Lynch, C.F. and Alavanj, M. (2000) Comparison of heterocyclic amine levels in home-cooked meats with exposure indicators (United States). *Cancer Causes Control*, **11**, 731–739.

Kidd, L., Stillwell, W., Yu, M., Wishnock, J., Skipper, P., Ross, R., Henderson, B. and Tannenbaum, S. (1999) Urinary excretion of 2-amino-1-methyl-6-phenylimidazo[4,5-b]pyridine (PhIP) in white, African-American, and Asian-American men in Los Angeles county. *Cancer Epidemiology, Biomarkers and Prevention*, **8**, 439–445.

Kinae, N., Kujirai, K., Kajimoto, C., Furugori, M., Masuda, S. and Shimoi, K. (2002) Formation of mutagenic and carcinogenic heterocyclic amines in model systems without heating. *International Congress Series*, **1245**, 341–345.

Klassen, R.D., Lewis, D., Lau, B.P.Y. and Sen, N.P. (2002) Heterocyclic aromatic amines in cooked hamburgers and chicken obtained from local fast food outlets in the Ottawa region. *Food Research International*, **35**, 837–847.

Knize, M.G., Roper, M., Shen, N.H. and Felton, J.S. (1990) Proposed structures for an aminodimethylimidazofuropyridine mutagen in cooked meats. *Carcinogenesis*, **11**, 2259–2262.

Knize, M.G., Felton, J.S. and Gross, G.A. (1992) Chromatographic methods for the analysis of heterocyclic amine food mutagens/carcinogens. *Journal of Chromatography*, **624**, 253–265.

Knize, M.G., Sinha, R., Rothman, N., Brown, E.D., Salmon, C.P., Levander, O.A., Cunningham, P.L. and Felton, J.S. (1995) Heterocyclic amine content in fast-food meat products. *Food and Chemical Toxicology*, **33**, 545–551.

Knize, M.G., Sinha, R., Salmon, C.P., Mehta, S.S., Dewhirst, K.P. and Felton, J.S. (1996) Formation of heterocyclic amine mutagens/carcinogens during home and commercial cooking of muscle meats. *Journal of Muscle Foods*, **7**, 271–279.

Knize, M.G., Salmon, C.P., Hopmans, E.C. and Felton, J.S. (1997) Analysis of foods for heterocyclic aromatic amine carcinogens by solid-phase extraction and high-performance liquid chromatography. *Journal of Chromatography A*, **763**, 179–185.

Knize, M.G., Sinha, R., Brown, E.D., Salmon, C.P., Levander, O.A., Felton, J.S. and Rothman, N. (1998) Heterocyclic amine content in restaurant-cooked hamburgers, steaks, and ribs. *Journal of Agricultural and Food Chemistry*, **46**, 4648–4651.

Knize, M.G., Salmon, C.P., Pais, P. and Felton, J.S. (1999) Food heating and the formation of heterocyclic aromatic amine and polycyclic aromatic hydrocarbon mutagens/carcinogens. *Advances in Experimental Medicine and Biology*, **459**, 179–193.

Knize, M.G. and Felton, J.S. (2005) Formation and human risk of carcinogenic heterocyclic amines formed from natural precursors in meat. *Nutrition Reviews*, **63**, 158–165.

Krone, C.A. and Iwaoka, W.T. (1981) Mutagen formation during the cooking of fish. *Cancer Letters*, **14**, 93–99.

Lan, C.M. and Chen, B.H. (2002) Effects of soy sauce and sugar on the formation of heterocyclic amines in marinated foods. *Food and Chemical Toxicology*, **40**, 989–1000.

Le Marchand, L., Hankin, J.H., Pierce, L.M., Sinha, R., Nerurkar, P.V., Franke, A.A., Wilkens, L.R., Kolonel, L.N., Donlon, T., Seifried, A., Custer, L.J., Lum-Jones, A. and Chang, W. (2002) Well-done red meat, metabolic phenotypes and colorectal cancer in Hawaii. *Mutation Research-Fundamental and Molecular Mechanisms of Mutagenesis*, **506**, 205–214.

Manabe, S., Kurihara, N., Wada, O., Tohyama, K. and Aramaki, T. (1992) Formation of PhIP in a mixture of creatinine, phenylalanine and sugar or aldehyde by aqueous heating. *Carcinogenesis*, **13**, 827–830.

Marsh, N.L., Iwaoka, W.T. and Mower, H.F. (1990) Formation of mutagens during the frying of Hawiian fish: Correlation with creatine and creatine content. *Mutation Research*, **242**, 181–186.

Murkovic, M., Friedrich, M. and Pfannhauser, W. (1997) Heterocyclic aromatic amines in fried poultry meat. *Zeitschrift fur Lebensmittel-Untersuchung und-Forsch A*, **205**, 347–350.

Murray, S., Gooderham, N.J., Boobis, A.R. and Davies, D.S. (1988) Measurement of MeIQx and DiMeIQx in fried beef by capillary column gas chromatography electron capture negative ion chemical ionisation mass spectrometry. *Carcinogenesis*, **9**, 321–325.

Murray, S., Gooderham, N.J., Boobis, A.R. and Davies, D.S. (1989) Detection and measurement of MeIQx in human urine after ingestion of a cooked meat meal. *Carcinogenesis*, **10**, 763–765.

Murray, S., Lynch, A.M., Knize, M.G. and Gooderham, N.J. (1993) Quantification of the carcinogens 2-amino-3,8-dimethylimidazo[4,5-f]quinoxaline, 2-amino-3,4,8-trimethylimidazo[4,5-f]quinoxaline, and 2-amino-1-methyl-6-phenylimidazo[4,5-b]pyridine in food using a combined assay based on capillary column gas chromatography negative ion mass spectrometry. *Journal of Chromatography (Biomedical Applications)*, **616**, 211–219.

Nagao, M., Honda, M., Seino, Y., Yahagi, T. and Sugimura, T. (1977) Mutagenicities of smoke condensates and charred surface of fish and meat. *Cancer Letters*, **2**, 221–226.

Negishi, C., Tsuda, M., Wakabayashi, K., Sato, S., Sugimura, T. and Jargerstad, M. (1984) Formation of MeIQx and another mutagen by heating a mixture of creatinine, glucose and glycine. *Mutation Research*, **130**, 378–379.

Olsson, J., Dyremark, A. and Karlberg, B. (1997) Determination of heterocyclic aromatic amines by micellar electrokinetic chromatography with amperometric detection. *Journal of Chromatography A*, **765**, 329–335.

Övervik, E., Kleman, M., Berg, I. and Gustafsson, J.-Å. (1989) Influence of creatine, amino acids and water on the formation of the mutagenic heterocyclic amines found in cooked meat. *Carcinogenesis*, **10**, 2293–2301.

Pais, P., Moyano, E., Puignou, L. and Galceran, M.T. (1997) Liquid chromatography-atmospheric-pressure chemical ionization mass spectrometry as a routine method for the analysis of mutagenic amines in beef extracts. *Journal of Chromatography A*, **778**, 207–218.

Pais, P. and Knize, M.G. (1998) Photodiode-array HPLC peak matching in complex thermally processed samples. *LC/GC*, **16**, 378–384.

Pais, P., Salmon, C.P., Knize, M.G. and Felton, J.S. (1999) Formation of mutagenic/carcinogenic heterocyclic amines in dry-heated model systems, meats, and meat dripping. *Journal of Agricultural and Food Chemistry*, **47**, 1098–1108.

Perfetti, G.A. (1996) Determination of heterocyclic aromatic amines in process flavors by a modified liquid chromatographic method. *Journal of AOAC International*, **97**, 813–816.

Persson, E., Sjoholm, I. and Skog, K. (2003) Effect of high water-holding capacity on the formation of heterocyclic amines in fried beef burgers. *Journal of Agricultural and Food Chemistry*, **51**, 4472–4477.

Puignou, L., Casal, J., Santos, F. and Galceran, M. (1997) Determination of heterocyclic aromatic amines by capillary zone electrophoresis in a meat extract. *Journal of Chromatography A*, **769**, 293–299.

Rappaport, S.M., McCartney, M.C. and Wei, E.T. (1979) Volatilization of mutagens from beef during cooking. *Cancer Letters*, **8**, 139–145.

Reistad, R., Rossland, O.J., Latva-Kala, K.J., Rasmussen, T., Vikse, R., Becher, G. and Alexander, J. (1997) Heterocyclic aromatic amines in human urine following a fried meat meal. *Food and Chemical Toxicology*, **35**, 945–955.

Reistad, R., Nyholm, S., Haug, L., Becher, G. and Alexander, J. (1999) 2-Amino-1-methyl-6-phenylimidazo[4,5-b]pyridine (PhIP) in human hair as a biomarker for dietary exposure. *Biomarkers*, **4**, 263–271.

Rivera, L., Curto, M.J.C., Pais, P., Galceran, M.T. and Puignou, L. (1996) Solid-phase extraction for the selective isolation of polycyclic aromatic hydrocarbons, azaarenes and heterocyclic aromatic amines in charcoal-grilled meat. *Journal of Chromatography A*, **731**, 85–94.

Salmon, C.P., Knize, M.G. and Felton, J.S. (1997) Effects of marinating on heterocyclic amine carcinogen formation in grilled chicken. *Food and Chemical Toxicology*, **35**, 433–441.

Salmon, C.P., Knize, M.G., Panteleakos, F.N., Wu, R., Nelson, D.O. and Felton, J.S. (2000) Minimization of heterocyclic amines and thermal inactivation of *Escherichia coli* in fried ground beef. *Journal of the National Cancer Institute*, **92**, 1773–1778.

Santos, F.J., Barcelo-Barrachina, E., Toribio, E., Puignou, L., Galceran, M.T., Persson, E., Skog, K., Messner, C., Murkovic, M., Nabinger, U. and Ristic, A. (2004) Analysis of heterocyclic amines in food products: Interlaboratory studies. *Journal of Chromatography B – Analytical Technologies in the Biomedical and Life Sciences*, **802**, 69–78.

Sentellas, S., Moyano, E., Puignou, L. and Galceran, M.T. (2003) Determination of heterocyclic aromatic amines by capillary electrophoresis coupled to mass spectrometry using in-line preconcentration. *Electrophoresis*, **24**, 3075–3082.

Sinha, R. (2002) An epidemiologic approach to studying heterocyclic amines. *Mutation Research-Fundamental and Molecular Mechanisms of Mutagenesis*, **506**, 197–204.

Sinha, R., Rothman, N., Brown, E., Levander, O., Salmon, C.P., Knize, M.G. and Felton, J.S. (1995) High concentrations of the carcinogen 2-amino-1-methyl-6-imidazo[4,5-b]pyridine (PhIP) occur in chicken but are dependent on the cooking method. *Cancer Research*, **55**, 4516–4519.

Sinha, R., Kulldorff, M., Curtin, J., Brown, C.C., Alavanja, M.C. and Swanson, C.A. (1998a) Fried, well-done red meat and risk of lung cancer in women (United States). *Cancer Causes and Control*, **9**, 621–630.

Sinha, R., Rothman, N., Salmon, C.P., Knize, M.G., Brown, E.D., Swanson, C.A., Rhodes, D., Rossi, S., Felton, J.S. and Levander, O.A. (1998b) Heterocyclic aromatic amine content of beef cooked by different methods to varying degrees of doneness and beef gravy made from meat drippings. *Food and Chemical Toxicology*, **36**, 279–287.

Skog, K., Laser-Reuterswärd, A. and Jägerstad, M. (1992) The inhibitory effects of carbohydrates on the formation of food mutagens in fried beef. *Food and Chemical Toxicology*, **30**, 681–688.

Skog, K. and Jägerstad, M. (1993) Incorporation of carbon atoms from glucose into the food mutagens MeIQx and 4,8-DiMeIQx using 14C-labelled glucose in a model system. *Carcinogenesis*, **14**, 2027–2031.

Skog, K., Steineck, G., Augustsson, K. and Jägerstad, M. (1995) Effect of cooking temperature on the formation of heterocyclic amines in fried meat products and pan residues. *Carcinogenesis*, **16**, 861–867.

Skog, K., Augustsson, K., Steineck, G., Stenberg, M. and Jägerstad, M. (1997) Polar and non-polar heterocyclic amines in cooked fish and meat products and their corresponding residues. *Food and Chemical Toxicology*, **35**, 555–565.

Solyakov, A., Skog, K. and Jägerstad, M. (1999) Heterocyclic amines in process flavours, process flavour ingredients, bouillon concentrates and a pan residue. *Food and Chemical Toxicology*, **37**, 1–11.

Solyakov, A. and Skog, K. (2002) Screening for heterocyclic amines in chicken cooked in various ways. *Food and Chemical Toxicology*, **40**, 1205–1211.

Stavric, B., Lau, B.P.-Y., Matula, T.I., Klassen, R., Lewis, D. and Downie, R.H. (1997) Mutagenic aromatic amines (HAAs) in "processed food flavor" samples. *Food and Chemical Toxicology*, **35**, 185–197.

Stillwell, W.G., Kidd, L.C.R., Wishnok, J.S., Tannenbaum, S.R. and Sinha, R. (1997) Urinary excretion of unmetabolized and phase II conjugates of 2-amino-1-methyl 6-phenylimidazo[4,5-*b*] pyridine and 2-amino-3,8-dimethylimidazo[4,5-*f*]quinoxaline in humans: Relationship to cytochrome P4501A2 and *N*-Acetyltransferase activity. *Cancer Research*, **57**, 3457–3464.

Sugimura, T., Kawachi, T., Nagao, M., Yahagi, T., Okamoto, T., Shudo, K., Kosuge, T., Tsuki, K., Wakabayashi, K., Litaka, Y. and Itai, A. (1977a) Mutagenic principles in tryptophan and phenylalanine pyrolysis products. *Proceedings of the Japan Academy*, **53**, 58–61.

Sugimura, T., Nagao, M., Kawachi, T., Honda, M., Yahagi, T., Seino, Y., Sato, S., Matsukura, N., Matsushima, T., Shirai, A., Sawamura, M. and Matsumoto, H. (1977b) Mutagen-carcinogens in foods with special reference to highly mutagenic pyrolytic products in broiled foods, in *Origins of Human Cancer* (eds. H.H. Hiatt, J.D. Watson and J.A. Winsten). Cold Spring Harbor, New York, pp. 1561–1577.

Thiebaud, H.P., Knize, M.G., Kuzmicky, P.A., Felkton, J.S. and Hsieh, D.P. (1994) Mutagenicity and chemical analysis of fumes from cooking meat. *Journal of Agricultural and Food Chemistry*, **42**, 1502–1510.

Thiébaud, H.P., Knize, M.G., Kuzmicky, P.A., Hsieh, D.P. and Felton, J.S. (1995) Mutagen production in airborne cooking products emitted by frying muscle and non-muscle foods. *Food and Chemical Toxicology*, **33**, 821–828.

Tikkanen, L.M., Sauri, T.M. and Latva-Kala, K.J. (1993) Screening of heat-processed Finnish foods for the mutagens 2-amino-3,4,8-dimethimidazo[4,5-f]quinoxaline, 2-amino-3,8-dimethylimidazo[4,5-f]quinoxaline, nad 2-amino-1-methyl-6-phenylimidazo[4,5-b]pyridine. *Food and Chemical Toxicology*, **31**, 717–721.

Tikkanen, L.M., Latva-kala, K.J. and Heiniö, R.-L. (1997) Effect of commercial marinades on the mutagenic activity, sensory quality and amount of heterocyclic amines in chicken grilled under different conditions. *Food and Chemical Toxicology*, **34**, 725–730.

Tran, N.L., Salmon, C.P., Knize, M.G. and Colvin, M.E. (2002) Experimental and simulation studies of heat flow and heterocyclic amine mutagen/carcinogen formation in pan-fried meat patties. *Food and Chemical Toxicology*, **40**, 673–684.

Turesky, R.J., Bur, H., Huynh-Ba, T., Aeschbacher, H.U. and Milon, H. (1988) Analysis of mutagenic heterocyclic amines in cooked beef products by high-performance liquid chromatography in combination with mass spectrometry. *Food and Chemical Toxicology*, **26**, 501–509.

Turesky, R.J., Stillwell, W.G.S., Skipper, P.L. and Tannenbaum, S.R. (1993) Metabolism of the foodborne carcinogens 2-amino-3-methylimidazo-[4,5-f]quinoline and 2-amino-3,8-dimethylimidazo[4,5-f]-quinoxaline in the rat as a model for human biomonitoring. *Environmental Health Perspectives*, **99**, 123–128.

Turesky, R.J., Taylor, J., Schnackenberg, L., Freeman, J.P. and Holland, R.D. (2005) Quantitation of carcinogenic heterocyclic aromatic amines and detection of novel heterocyclic aromatic amines in cooked meats and grill scrapings by HPLC/ESI-MS. *Journal of Agricultural and Food Chemistry*, **53**, 3248–3258.

Turteltaub, K.W., Knize, M.G., Healy, S.K., Tucker, J.D. and Felton, J.S. (1989) The metabolic disposition of 2-amino-1-methyl-6-phenylimidazo[4,5-b]pyridine in the induced mouse. *Food and Chemical Toxicology*, **27**, 667–673.

Ushiyama, H., Wakabayashi, K., Hirose, M., Itoh, H., Sugimura, T. and Nagao, M. (1991) Presence of carcinogenic heterocyclic amines in urine of healthy volunteers eating normal diet, but not of inpatients receiving parenteral alimentation. *Carcinogenesis*, **12**, 1417–1422.

Vainiotalo, S., Matveinen, K. and Reunanen, A. (1993) GC/MS determination of the mutagenic heterocyclic amines MeIQx and DiMeIQx in cooking fumes. *Fresenius Journal of Analytical Chemistry*, **345**, 462–466.

Vikse, R. and Joner, P.E. (1993) Mutagenicity, creatine and nutrient contents of pan fried meat from various animal species. *Acta Veterinaria Scandinavica*, **34**, 1–7.

Wakabayashi, K., Nagao, M., Esumi, H. and Sugimura, T. (1992) Food-derived mutagens and carcinogens. *Cancer Research (Supplement)*, **52**, 2092s–2098s.

Wakabayashi, K., Ushiyama, H., Takahashi, M., Nukaya, H., Kim, S.-B., Hirose, M., Ochiai, M., Sugimura, T. and Nagao, M. (1993) Exposure to heterocyclic amines. *Environmental Health Perspectives*, **99**, 129–133.

Wang, Y.Y., Vuolo, L.L., Springarn, N.E. and Weisburger, J.H. (1982) Formation of mutagens in cooked foods, V., The mutagen reducing effect of soy protein concentrates and antioxidants during frying of beef. *Cancer Letters*, **16**, 179–186.

Wong, K.-Y., Su, J., Knize, M.G., Koh, W.-P. and Seow, A. (2005) Dietary exposure to heterocyclic amines in a Chinese population. *Nutrition and Cancer, Nutrition and Cancer - An International Journal*, **52**, 147–155.

Wu, J., Wong, M.-K., Lee, H.-K., Lee, B.-L., Shi, C.-Y. and Ong, C.-N. (1996) Determination of heterocyclic amines in flame-grilled fish patty by capillary electrophoresis. *Food Additives and Contaminants*, **13**, 851–861.

Zimmerli, B., Rhyn, P., Zoller, O. and Schlatter, J. (2001) Occurrence of heterocyclic aromatic amines in the Swiss diet: Analytical method, exposure estimation and risk assessment. *Food Additives and Contaminants*, **18**, 533–551.

Zochling, S. and Murkovic, M. (2002) Formation of the heterocyclic aromatic amine PhIP: Identification of precursors and intermediates. *Food Chemistry*, **79**, 125–134.

14 Polycyclic Aromatic Hydrocarbons

Laura Cano-Lerida, Martin Rose and Paul Walton

Summary

Polycyclic aromatic hydrocarbons comprise a group of non-functionalised aromatic compounds containing either two or more CH-fused aromatic rings, and additionally their functionalised derivatives and heterocyclic analogues (totalling around 250 compounds). PAHs are ubiquitous as environment contaminants in the food chain, originating from a variety of pyrolysis sources, as well as directly contaminating foods from cooking and drying processes. Surveys have identified PAHs in vegetable oils, fish, shellfish, smoked and barbecued foods. Total diet studies have shown that the highest contributions to dietary intakes of PAHs come from miscellaneous cereals, fats and oils, and green vegetable food groups. PAHs are toxic, many of the compounds in this class are both genotoxic and carcinogenic. On the basis of risk assessments, the EU regulates PAHs in foods based on limits for individual compounds.

14.1 OCCURRENCE AND MECHANISM OF FORMATION

14.1.1 Definition

Polycyclic (Roy *et al.*, 2005) aromatic hydrocarbons (Baran *et al.*, 2003) or arenes form a large family of non-functionalised aromatic compounds containing either two or more fused aromatic rings made up of carbon and hydrogen atoms (Flowers *et al.*, 2002). This definition is strictly valid for unsubstituted parent PAHs and their alkyl substituted derivatives, and this is how the term is used in this chapter. The term PAH, however, can also be used for functionalised derivatives (Baran *et al.*, 2003) and heterocyclic analogues (i.e. indole, quinoline, benzothiophene, 9-cyanoanthracene, dibenzothiophene). There are about 250 compounds generally included in the term PAH and the structures of some of them are shown in Fig. 14.1.

14.1.2 Sources of PAHs

PAHs and their derivatives are widespread in the environment from both natural and anthropogenic sources, such as burning fossil fuels and processes shown in Table 14.1 (Larsson *et al.*, 1987; Mastrangelo *et al.*, 1996). They are created when substances such as coal, oil, gas and organic waste are burned incompletely (Phillips, 1983). At high temperatures organic compounds are partially cracked to smaller unstable fragments (Li *et al.*, 2003), mostly radicals, that recombine to give relatively stable PAHs (pyrosynthesis). Another source of PAHs

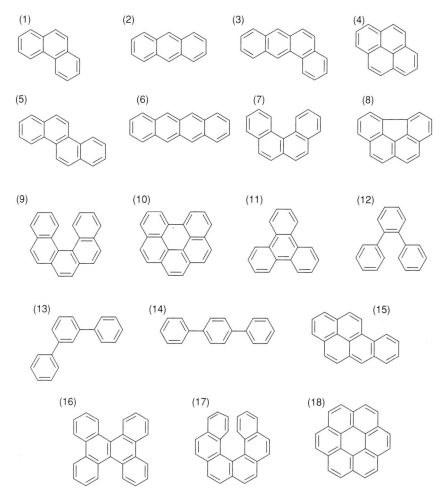

Fig. 14.1 Some members of the PAH family. (1) Phenanthrene, (2) anthracene, (3) benz[a]anthrecene, (4) pyrene, (5) chrysene, (6) naphthacene, (7) benzo[c]phenanthrene, (8) benzo[*ghi*]fluoranthene, (9) dibenzo[c, g]phenanthrene, (10) benzo[*ghi*]perylene, (11) triphenylene, (12) o-tephenyl, (13) o-terphenyl, (14) p-terphenyl, (15) benzo[a]pyrene, (16) tetrabenzonaphthalene, (17) phenanthro[3,4-c]phenanthrene, (18) coronene.

is aromatisation that occurs at lower temperatures (100–150°C), but this requires much more time and produces large quantities of alkylated PAHs (Moret and Conte, 2000).

PAHs are ubiquitous, being present in the atmosphere, surface water, sediments and soil, food and lipid tissues. For the general population, exposure to PAHs is mainly from food and inhaled air, although there have been cases of high level occupational exposure (European Commission, 2002).

14.1.3 Sources of PAHs in foods

14.1.3.1 *Environmental sources*

PAHs of lower molecular mass (generally those with three or fewer aromatic rings) are numerous in the atmosphere and those of higher molecular mass (the majority) enter the

Table 14.1 Sources of PAHs.

Anthropogenic sources

- *Industry*
  - Cokeries
  - Petroleum catalytic cracking
  - Iron and steel foundries
  - Aluminium works
  - Wood preservation using carbolineum and creosote
  - Road building
  - Roofing
  - Industrial incinerators

- *Domestic combustion*
  - Wood and kerosene stoves
  - Oil burners
  - Barbecues
  - Coal
  - Other fuels
  - Tobacco smoke

- *Traffic*
  - Diesel vehicles
  - Gasoline vehicles
  - Other vehicles

Natural sources
  - Forests fires and volcanoes
  - Lignite, coal and crude oil

European Commission (2002).

environment adsorbed onto particulate matter. The hydrosphere and geosphere are affected by dry and wet deposition of PAHs.

Solubility in water is a crucial characteristic for determining distribution patterns of PAHs in food (Lerario *et al.*, 2003). For example, Table 14.2 shows water solubility values (Leonard *et al.*, 2000) for some PAHs. Phenanthrene, for example, has a relatively high solubility when compared with other PAHs, and is therefore likely to have a higher environmental mobility; in fact, phenanthrene is the PAH found in highest concentrations in aquatic samples. As PAHs are lipophilic compounds, uptake by absorption by plants from contaminated soils is low. The waxy surfaces of vegetables and fruits can concentrate low molecular mass

Table 14.2 Examples of PAH water solubility.

PAH	MW[a]	S_W(mg/L)[b]
Phenanthrene	178.24	1.290
Anthracene	178.24	0.073
Pyrene	202.26	0.135
Fluoranthene	202.26	0.265

[a] MW, molecular weight.
[b] Solubility in water.
Reproduced from Paschke, A., Popp, P. and Schuurmann, G. Solubility and partitioning studies with polycyclic aromatic hydrocarbons using an optimized SPME procedure. *Fresenius Journal of Analytical Chemistry*, **363**, 426–428. Copyright 2004 with kind permission from Springer Science and Business Media.

PAHs through surface adsorption (European Commission, 2002). In marine sediments, PAHs become strongly bound and constitute a pollutant reservoir. In reality, only sediment dwelling organisms are likely to be adversely affected by these PAHs. Bivalves like mussels or oysters may accumulate high levels because they filter large quantities of water and unlike most other living organisms, they cannot metabolise the PAHs (Wootton *et al.*, 2003; Oros and Ross, 2005). Contamination can spread over the food chain. However, bio-magnification has not been reported in aquatic systems because of metabolism of PAHs by organisms (e.g. vertebrates) at higher trophic levels of the food chain (Narbonne *et al.*, 2005).

Other animals consumed by humans, such as cattle or poultry, may be exposed as a result of ingestion of other contaminated animals or plants or by breathing in contaminated air or by drinking contaminated water (Ciganek *et al.*, 2000, 2002; Schaum *et al.*, 2003). But most animals will metabolise the PAHs before any food products reach humans.

14.1.3.2 *PAHs originating from processing and cooking*

Despite PAHs present in foodstuffs due to environmental exposure, processing of food (such as smoking, industrial drying, barbecuing and other cooking methods) is thought to be the major source of PAHs in food (Mottier *et al.*, 2000). Charcoal grilling can generate high levels of PAHs, quantities produced being related to factors such as fat content, temperature and cooking time. For example, levels as high as 320 μg/kg of PAHs in charcoal-grilled duck breast steaks have been measured.

Vegetable oils such as coconut, sunflower, olive and grape-seed oils and fats contribute significantly to the presence of PAHs in the diet. Their contamination is mainly due to the industrial drying processes to which the seeds are subjected to when the seeds come into contact with combustion gases (Moret *et al.*, 2000). It is common practice to use open fires to heat seeds in industrial processes used to extract oils. Fortunately, refining can reduce the amount of these contaminants. Deodorisation methods remove the most volatile of these PAHs whereas charcoal treatment can remove higher PAHs (as a general rule, those with more than four rings) (Moret and Conte, 2000; Moret *et al.*, 2000). In coffee and tea, the roasting procedure can contribute to additional production of PAHs (European Commission, 2002).

Reports of the presence of PAHs in smoked food are numerous (Guillen and Sopelana, 2004; Jira, 2004; Stolyhwo and Sikorski, 2005). There are many variables intervening in this process: type of wood, type of generator (internal or external), oxygen accessibility, temperature and smoking time can all make the final concentration of PAHs very different in similar types of food.

14.1.4 Environmental fate and degradation

Environmental fate of PAHs is very varied; in terms of environmental distribution, partitioning between water and air, between water and sediment and between water and biota are the most important processes. Airborne emissions of PAHs from coal coking, exhausts of vehicles, combustion of fuels in general can release PAHs into the atmosphere. PAHs of higher molecular mass, entering the environment via the atmosphere are adsorbed onto particulate matter and are transferred into soils or watercourses by dry and wet deposition. PAHs bind strongly to soil organic matter (Laor and Rebhun, 2002) with an unusually slow rate of degradation (Siddiqi *et al.*, 2002). They reach watercourses and are rapidly transferred to sediments (Baran *et al.*, 2003).

Meat, poultry, fish, milk, eggs and their products are considered to be the major source of human exposure to PAHs (as well as to other lipophilic contaminants such as polychlorinated biphenyls (PCBs), polybrominated diphenylethers (PBDEs) and polychlorinated dibenzo-dioxins (PCDDs). Sources of food contamination are from direct intake from contaminated water and/or soils where plants and animals are produced and also from some methods of food preparation such as smoking, frying, baking, roasting, barbecuing and seed drying processes (vegetable oils and cereals). PAHs can also contaminate foodstuffs by diffusion from some packaging materials such as recycled polyethylene film (Moret and Conte, 2000; Tamakawa, 2004).

PAHs may be broken down as a result of photodegradation, microbial degradation and metabolism in higher biota (e.g. vertebrates). Photodegradation or photooxidation is a relatively facile process that occurs on exposure of many PAHs to air and light in air and water in the presence of sensitising radicals like OH, NO_3 and O_3 (Mallakin et al., 2002). PAHs are microbiologically degraded under aerobic conditions and the biodegradation rate decreases drastically with the number of aromatic rings on the PAH. Anaerobic degradation is much slower (Demaneche et al., 2004; Amir et al., 2005).

Metabolism in higher biota is a route of transformation of minor importance for the overall fate of PAHs in the environment, but it is important for biota and therefore for the food chain and for the equilibrium of ecosystems (Jones et al., 2000; Dreij et al., 2005). Biomagnification (the increase in the concentration of a substance in animals in successive trophic levels of food chains) of PAHs has not then been observed and would not be expected, because most organisms have a high metabolic potential for PAHs. Organisms at higher trophic levels in food chains show the highest potential for biotransformation. Bioaccumulation (the accumulation of material over a life time) is nevertheless a problem for organisms that lack the microsomal oxidase enzyme, which allows for the breakdown of PAHs to more water-soluble products for excretion (European Commission, 2002).

As PAHs are usually chemically stable, with no reactive groups, hydrolysis plays no role in their degradation. Calculations based on physicochemical and degradation parameters indicate that PAHs with four or more aromatic rings persist in the environment (IPCS, 1998). Table 14.3 shows some examples of some PAHs persistence in air, water, soil and sediment (based on model calculations) (Mackay and Shiu, 1992).

Interest in the biodegradation mechanisms and environmental fate of PAHs is prompted by their ubiquitous distribution and their dangerous effects on human health. PAHs are persistent pollutants with long life cycles in the environment, and there is evidence that suggests a direct relationship between PAH size and biodegradation rate. For example, reported half-lives in soil and sediment of phenanthrene (three rings) may range from 16 to 126 days while for the five-ring molecule benzo[a]pyrene (B[a]P) the reported range is from 229 to >1400 days (Kanaly and Harayama, 2000).

14.2 METHODS OF ANALYSIS

14.2.1 Current status

Due to the ubiquity of PAHs and their potentially deleterious effects on human health, these compounds are often analysed in a variety of matrices such as foods, oils, waters, soils etc. Being such a large and diverse family, standard methods, based on field sampling and laboratory analysis, are often used (Fisher, 2000). The following section does not provide an

Table 14.3 Suggested half-life classes of PAHs in various environmental compartments.

Class	Half-life (h) Mean	Half-life (h) Range
1	17	10–30
2	55	30–100
3	170	100–300
4	550	300–1000
5	1700	1000–3000
6	5500	3000–10 000
7	17 000	10 000–30 000
8	55 000	>30 000

Compound	Class in air	Class in water	Class in soil	Class in sediment
Acenalphthylene	2	4	6	7
Anthracene	2	4	6	7
Benz[a]anthracene	3	5	7	8
Benzo[a]pyrene	3	5	7	8
Benzo[k]fluoranthene	3	5	7	8
Chrysene	3	5	7	8
Dibenz[a,h]anthracene	3	5	7	8
Fluoranthene	3	5	7	8
Fluorene	2	4	6	7
Naphthalene	1	3	5	6
Perylene	3	5	7	8
Phenanthrene	2	4	6	7
Pyrene	3	5	7	8

Reproduced from Mackay, D.P. and Shiu, W.Y. Generic models for evaluating the regional fate of chemicals. *Chemosphere*, **24**, 695–717. Copyright 1992 with permission from Elsevier.

exhaustive list of analytical methods, but aims to identify well-established methods that are used as standard methods of analysis.

There are several well-established analytical procedures for the analysis of PAHs. Most of them involve pre-treatment of the sample, where the PAHs are extracted from the complex mixtures in which they are present, into a new matrix by liquid/liquid, solid-phase and/or ultrasonic extractions. Pre-treatment methods often serve the purpose of being a pre-cleaning of interfering compounds and also a pre-concentration step (Mastral *et al.*, 2004). The application of ultrasound for accelerating or assisting PAH (and other inorganic and organic compounds), extraction from solid materials is sometimes used (Rodriguez-Sanmartin *et al.*, 2005). Extracts are measured by either HPLC or GC (depending on the nature of the sample and its volatility) with the detectors being mass spectrometer (MS), ultraviolet (UV) or fluorescence spectrophotometers (Li *et al.*, 2003).

The conventional method for dust and soils consists of sample extraction, sample clean up (if needed) and sample analyses by GC-MS (Weisshoff *et al.*, 2002). With a few exceptions PAHs are analysed by HPLC with fluorescence or UV detection, or sometimes by GC-MS or GC-FID (Sluszny *et al.*, 2002). Thin-layer chromatography (TLC) is an inexpensive, quick analytical technique, but has low separation efficiency and is commonly used only for identification purposes of individual compounds or for very simple sample types.

With the exception of TLC, these techniques are accurate and can handle complicated mixtures, but at the same time they are expensive and require highly qualified personnel to operate. Their widespread use is justified by the high complexity of 'real' samples mostly containing a large number of PAHs. As an example, a standard GC column containing 3000 plates/metre allows good separation of mixtures of about 100 PAHs (IPCS, 1998). The limits of detection of these analytical methods for PAHs need to be extremely low as they are normally present at trace amounts (<ppb), especially in water, and are potentially toxic at these low concentrations (Manoli, 1999).

Immunoassays have been extensively applied to evaluate environmental contamination and analysis of biological fluids. Molecules detected by immunoassays vary widely in size, chemical and physical properties and biological activity. PAHs are enzymatically converted to highly reactive metabolites that bind covalently to macromolecules such as DNA, thereby causing mutagenesis and carcinogenesis in experimental animals (Section 14.3). For example, B[a]P is activated by microsomal enzymes to 7β,8α-dihydroxy-(9α,10α)-epoxy-7,8,9,10-tetrahydrobenzo[a]pyrene (Pavanello *et al.*, 1999) and binds covalently to DNA, resulting in formation of BPDE-DNA adducts (Pavanello *et al.*, 1999; Wani *et al.*, 2002; Smith and Hurtubise, 2004). Immunoassays are sensitive methods and able to detect PAH-DNA adducts in the blood and tissues of humans and animals and include, for example, enzyme-linked immunosorbent assays (ELISA) (Chuang *et al.*, 1998), radioimmunoassay (Schneider *et al.*, 1995; Baran *et al.*, 2003), dissociation-enhanced lanthanide fluoroimmunoassay (DELFIA) (Divi, 2002) and ultrasensitive enzyme radioimmunoassay (USERIA) (Ovrebo, 1992, 1994); [32]P- and ^{35}S-postlabelling with radioactivity counting (Gorelick and Reeder, 1993); surface-enhanced Raman spectroscopy (Olson, 2004); and synchronous luminescence spectroscopy (SLS) (Matuszewska, 2000).

The US EPA has developed a standard immunoassay method (number 4035) for screening of soils for PAHs. This method is a semi-quantitative enzyme immunoassay and correctly identifies 95% of samples that are PAH-free and those containing 1 ppm total PAH (as phenanthrene). The immunoassay is based on the use of antibodies immobilised on the walls of the test tubes that bind either PAH or PAH-enzyme conjugate. When PAH is present in the sample, it competes with the PAH-enzyme conjugate for a limited number of binding sites on the immobilised antibodies.

Table 14.4 lists some of the most representative methods in general use for PAHs analysis. The most commonly used are chromatographic methods like GC and LC with different detectors depending on the nature of the samples.

Solid-phase extraction (Biernoth and Rost, 1968) and solid-phase microextraction (Paschke *et al.*, 1999; Guillen and Sopelana, 2005; Hawthorne *et al.*, 2005; Hu *et al.*, 2005; Ter Laak *et al.*, 2005) are also attracting some attention in the field of PAH analysis. Supported liquid membrane extraction (SLM) (Zabiegala *et al.*, 2000), supercritical fluid extraction (SFE) (Rigou *et al.*, 2004; Chiu *et al.*, 2005; Librando *et al.*, 2005), online capillary microextraction (Bigham, 2002) and membrane extraction with sorbent interface (MESI) have also been used although less frequently (Roy *et al.*, 2005). Recent work by Roy *et al.* demonstrates the coupling of stir bar sorptive extraction to a new generation of gas chromatography mass spectrometry (GC-MS), using the field apparatus EM 640 S from Bruker (Roy *et al.*, 2005).

14.2.2 Analytical methods – forward look

The available standard methods have a series of drawbacks:

- They are very time-consuming
- They require highly trained operators

Table 14.4 Conventional analytical methods for the determination of PAHs (ATSDR, 1995; European Commission, 2002).

Analytical method	Sample	Detection limit	Per cent recovery
GC/MS	Biological tissues	5–50 ng/g	52–95
	Air	1 pg/sample	No data
		1×10^{-5} ppm	No data
		0.5–100 ng/m^3	74–90
	Water	1.9–7.8 µg/L	No data
		Sub-ng/L range	80
	Sediments	2.8–7.3 µg/g	91–97
	Coal-fly ash	No data	No data
	Seafoods	1–5 µg/kg	73–144
	Cigarette smoke	1–6 µg/cigarette	No data
LC-GC/MS	Vegetable oil	1 pg/sample	No data
	Aerosol samples	pg/m^3	No data
GC/FT-IR	Water and sediments	0.01–0.06 µg/g	No data
	Soil	0.025–0.25 µg/sample	0.998–0.85 µg/sample
GC/GPFD	Air	1×10^{-6} ppm	100.5
HRGC	Air	0.001–0.03 ng/m^3	No data
GC/FID	Biological tissues	50 ng/g	83–95
	Urine	1.2–6.48 µg PAH/mmol creatine	No data
	Air	0.05 ng/m^3	No data
GC/FID	Sea water	0.024–0.045 µg/L	No data
	Sediments	0.014–0.093 µg/g	76–110
HPLC/UV	Lungs	20 ng/g	93.7 (fluoroanthene); 65.3 (pyrene)
	Blood	20 ng/mL	107 (fluoroanthene); 108.6 (pyrene)
	Urine	5×10^{-12} mol/10 µg of labelled B[a]P	No data
	Water	0.01–3 µg/L	No data
	Coal and petroleum	0.1–7.1 µg/L	45–95
	Barley malt	2.5–5 ng/g	78–97
HPLC/fluorescence detector	Blood	5×10^{-12} g BPDE/sample	No data
	Urine	<1 µg PAH/mmol creatine	10–85
	Air	<0.01 ng/sample	No data
	Water	180×10^{-15} g/sample	89–100
		Low ng/L	86–87
	Cigarette smoke	3 pg/sample	89–108
HPLC/fluorescence detector	Coal and petroleum	3.5–46 µg/L	56–100
	Smoked foods	2–27 pg/sample	28–142
		0.1 µg/kg	75–90
	Charcoal-boiled beef	20–50 ng/g	No data
	Fat products	0.1–0.5 ppb	76–85
HPLC/SFS	Urine	0.01 pmol/mL	>30
HPLC/MS	Sediments	pg range	No data
ELISA	Blood	$0.38–2.2 \times 10^{-15}$ mol BPDE/µg DNA	No data

(continued)

Table 14.4 (continued)

Analytical method	Sample	Detection limit	Per cent recovery
		$0.06–0.2 \times 10^{-15}$ mol BPDE/µg DNA	No data
	Water	ng/L range	No data
	Soil	100–10 000 ppm range	140 B[a]P
USERIA	Blood	$0.38–2.2 \times 10^{-15}$ mol BPDE/µg DNA	No data
TLC	Blood	0.3×10^{-15} mols adduct/µg DNA	No data
	Sediments	0.2–2.7 µg/g	86–89
	Cooking oil fume	0.11–0.41 ng	96–99
	Coal and petroleum	µg/L range	No data
Time-related fluorometry (PAH-DNA adduct)	Blood	<1 adduct/10^8 nucleotides	No data
Spectrofluorometry	Sediments	0.008–4.5 ng/mL	80–95

- They are expensive to run and buy the equipment
- There is a strong dependence on the extraction conditions, which can sometimes produce lack of reproducibility
- There is a high risk of contamination during sampling prior to analyses
- There is also a lack of *in situ* and fast method of analyses

The major disadvantage of classical chromatographic analysis of PAHs, especially for heavily contaminated samples, is that it cannot be used as a tool to quickly identify total PAH content of a sample. As PAH identification is based on retention times, unambiguous compound identification requires either PAH separation (which might not be achieved in the HPLC column) or complete chromatographic resolution of sample components (Environmental Protection Agency, 1991). Because PAH separation might not be achieved, the use of a support analytical technique such GC-MS is preferable. In chromatographic analysis of PAHs, long elution times are typical (Larsson *et al.*, 1987), and standards must be run periodically to verify retention times, along with reference materials and other quality control procedures. If the concentrations of target species are found to lie outside the detector's response range, the sample must be diluted or concentrated and the process repeated.

Current investigations focused on direct on-site measurement of PAHs in soil and water described below, could address the problems of sample contamination prior to analysis and save time in the analyses (Sluszny *et al.*, 2004). Unfortunately, natural materials, such as humic acids found in water and soil, greatly reduce sensitivities of such techniques. Thus, at present, these methods are only capable of a semi-quantitative analysis when applied to 'real world' samples. There is some work reported that investigates the use of novel materials as selective sorbents for organic pollutants; various applications of polymer films have been suggested (Ensing and de Boer, 1999; Patel *et al.*, 2000; Sluszny *et al.*, 2004). An example of this includes a polymer film sensor for sampling and remote analysis of PAHs in aqueous environments (Sluszny *et al.*, 2004). A novel material for SPE with improved properties with respect to the normally used in conventional SPE methods (such as the C-18-reverse phase material) (Sluszny *et al.*, 2004) using a polymeric film as extraction medium for the direct analysis by fluorescence of PAHs has been developed. The polymer film is selective for pyrene

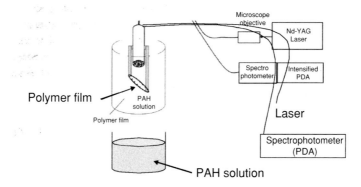

Fig. 14.2 Experimental set-up of Sluszny *et al*. (Reprinted from Sluszny, C., Gridin, V.V., Bulatov, V. and Schechter, I. Polymer film sensor for sampling and remote analysis of polycyclic aromatic hydrocarbons in clear and turbid aqueous environments. *Analytica Chimica Acta*, **522**, 145–152. Copyright 2004 with permission from Elsevier.)

and anthracene with respect to 'interfering substances' (Laor and Rebhun, 2002). The polymer is made out of a 1:2 ratio of bis(2-ethyl-hexyl) phthalate as a softening agent and poly(vinyl chloride-co-vinyl acetate) containing approximately 90% vinyl chloride in tetrahydrofuran (THF). After leaving thin films of this polymer into contact with samples spiked with PAHs for 45 minutes, the remote laser-induced fluorescence (Oliferova *et al*., 2005) of these films was measured with an experimental set-up like the one depicted in Fig. 14.2.

A dual-branch fibre-optic fluorescence probe was used for direct excitation of the PAH-containing polymer, as well as for the measurement of the emitted fluorescence. A special holder coupled the polymer film to the fibre-optic assembly at an angle of 55° with respect to the fibre-optic head and a laser light was focused onto the optical fibre by a microscope objective. The back-emitted fluorescence was transmitted through the optical fibres and the spectra were obtained by a PC-interfaced spectrometer, equipped with a photodiode array detector. The advantage of polymeric film sampling over direct fluorescence analysis in solution is supported by measurements carried out in the presence of humic acids as interfering substances. A quantitative evaluation of the method was given with fluorescence quenchers present at various concentrations (represented by humic substances) and when micro-particulates were suspended in the water (represented by various clay suspensions).

Environmental samples often contain high amounts of these natural materials including humic substances and clay suspensions. A major drawback for the wide application of this method is the fact that environmental samples also contain a wide range of different compounds that may became absorbed in the polymer films and can contribute to the total fluorescence or act as quenchers. In either case they would interfere with the quantification of PAHs. Work on novel materials for SPE applied to PAHs analysis has been reported by Oliferova *et al*. (2005). This consisted of an automated on-line method, which included both preconcentration (Biernoth and Rost, 1968) and analysis (Bystol and Campiglia, 2003) of water samples spiked with PAHs.

Hydrophobic organic substances are normally determined by on-line SPE-HPLC systems with microcolumns packed with non-polar sorbents. The most popular of these sorbents is octadecyl-bound silica gel (Booth and Gribben, 2005). These materials present several improvements such as the existence of residual polar groups (i.e. in ODS silanol groups, -Si-OH) that provide strong binding points for analytes making difficult their later desorption (Thurman and Mills, 1998). Due to these silanol groups ODS does not possess a high selectivity towards hydrophobic substances, it extracts not only hydrophobic but also hydrophilic substances (Hennion, 1999). Oliferova *et al.* (2005) proposed the use of fluorocarbon polymers (FPs) as sorbent materials to overcome these problems. FPs are highly hydrophobic, thus being ideal sorbents for recovery of hydrophobic substances from aqueous solutions. According to Oliferova *et al.* (2005) application of FP sorbents resulted in better extraction selectivity towards PAHs in comparison with several other sorbents.

14.3 PAHs IN FOOD, EXPOSURE ASSESSMENT AND HEALTH EFFECTS

14.3.1 PAHs in food and dietary intake

A risk assessment carried out by the EC Scientific Committee for Food (SCF) recommended further investigation and monitoring of 15 PAHs. The Joint FAO/WHO Expert Committee on Foods Additives (JECFA) in 2005 highlighted an additional PAH, benzo[*c*]fluorine as a compound of concern. Because of the continued development of European legislation in this area, analytical methodology for multiple PAHs in food is required in order to get a clearer picture of human exposure (Wenzl, 2006).

In view of uncertainties about the levels of PAHs in foods, especially those possessing both genotoxic and carcinogenic properties as identified by the former EC SCF, a review of regulations will be undertaken by early 2007. Information is required to support this review, and the European Food Safety Authority (EFSA) has been asked to initiate and coordinate PAH-data collection in various food categories. In response, EFSA has established an on-line analytical database in collaboration with EU Member States in order to collect data. This information should also show whether or not B[*a*]P is a suitable marker for more general PAH exposure (Wenzl, 2006).

There have been a large number of studies on PAHs in vegetable oils (Biernoth and Rost, 1968; Tiscornia, 1982; Larsson, 1987; Sagredos, 1988; Speer, 1990; Dennis *et al.*, 1991; Menichini, 1991; Gertz, 1994; Swetman, 1999; Moret *et al.*, 2000) and some studies of PAHs in fish and smoked and barbecued foods (Speer, 1990) and in fish and shellfish (Jones, 2001; Storelli, 2001; Mostafa, 2002; O'Connor, 2002).

A few estimates of dietary intakes of PAHs have been carried out, and these are summarised in Table 14.5. In the UK, the total diet survey (TDS) is one method used to provide information on dietary exposures of the general UK population to chemicals such as nutrients and contaminants. It allows a check to be made that intakes of these chemicals do

Table 14.5 Estimated dietary intakes by adults of PAHs estimated in other countries.

Type of samples analysed	Country	Year	Estimated dietary exposure (ng/kg bw/day)		Reference
			B[*a*]P	Total PAHs[a]	
Total diet study	UK	2000	1.6	69	FSA, 2002
Total diet study	Milan, Italy	ns	ns	50	Lodovici *et al.*, 1995
Market basket	Netherlands	1984–86	4.8	284	de Vos *et al.*, 1990
Retail foods	USA	ns	0.3–1.3[b]	n/a	Kazerouni *et al.*, 2001

[a] The sets of PAHs analysed were not the same in all the surveys.
[b] Range of dietary intakes by 90% of the consumers.
ns, not stated.

not give any general cause for health concern (Pufulete *et al.*, 2004) and also allows an evaluation of time trends in exposure. The design of the TDS is described more fully elsewhere (Peattie, 1983; Ministry of Agriculture, 1994), but it involves the preparation of a number of food group samples, each of which reflects the relative importance of different foods in the diet within 20 major food groups, e.g. fish, carcass meat or green vegetables etc. Average and high level (97.5 percentile) adult *lower bound* (assumes compounds not detected are at zero concentration) dietary intakes of B[*a*]P were estimated using the *lower bound* averages of the concentrations in the 24 composites of each food group. The average and high level adult *lower bound* dietary intakes of B[*a*]P in 1979 were estimated at 2.4 and 4.4 ng/kg body weight/day. The results showed that the highest contributions to dietary intakes of PAHs came from the miscellaneous cereals, fats and oils, and green vegetables food groups, and a follow-up survey in 1984 showed that vegetable oils were responsible for the presence of PAHs in the fats and oils food groups (Dennis *et al.*, 1991). The same work also showed that the concentrations of some PAHs were reduced significantly during the refining process.

Dietary intakes by UK consumers of PAHs were estimated from the fresh weight concentrations in the 2000 TDS samples using consumption data from the Dietary and Nutritional Survey of British Adults (Gregory *et al.*, 1990), the National Diet and Nutrition Survey (NDNS) of young people aged 4–18 years (Gregory *et al.*, 2000) and the NDNS of children aged $1^1/_2$–$4^1/_2$ years (Gregory *et al.*, 1995). For comparison purposes, the dietary exposures of UK consumers to the PAHs of greatest concern (Class A as described in Section 14.3.3.1 below) in 1979 were re-estimated. The estimated *upper bound* (assumes 'not detected' compounds are present at the limit of detection and is therefore the precautionary approach) average and high-level dietary intakes by adults of B[*a*]P in 2000 were 1.6 and 2.7 ng/kg body weight/day, and this compared with approximate *lower bound* values of 2.4 and 4.4 ng/kg body weight/day in 1979. Note that *upper bound* estimates from 1979 are not possible since limits of determination were not reported. The estimated dietary intakes by schoolchildren and toddlers of B[*a*]P in 2000 were higher than for adults, with average and high level (97.5 percentile) intakes of up to 3.8 and 6.2 ng/kg body weight/day respectively for toddlers, largely due to the greater amount of food that children eat per unit of body weight. The estimated *upper bound* average and high level dietary intakes by adults to the sum of the PAHs analysed in the samples from year 2000 were 69 and 116 ng/kg body weight/day. These figures cannot be compared directly with the results for 1979 as different sets of PAHs were analysed. Since certain PAHs were detected at concentrations exceeding the limit of determination (LOD) in only a few of the food groups, the *upper bound* estimates are likely to be significantly higher than the true values.

B[*a*]P was detected at concentrations exceeding the limit of determination (Hu *et al.*, 2005) in only four food groups (bread, miscellaneous cereals, oils and fats, and nuts). Five of the PAHs (acenapthalene, fluorine, fluoranthene and phenanthrene), all ranked relatively low by the scheme shown below in Section 14.3.2.1 (Department of Health, 1998) were detected at concentrations exceeding the LOD in all or all but one of the food groups, whilst two of the PAHs (dibenz[*ah*]anthracene and anthanthrene) were not detected at concentrations exceeding the LOD in any of the food groups. Only one of the PAHs ranked as amongst those of greatest concern, benz[*a*]anthracene, was detected in the majority of food groups. In general, the PAHs with the lower rankings in terms of health risk were detected more frequently, and at higher concentrations, than those with higher health rankings. The highest concentrations of PAHs were found in the bread, miscellaneous cereals, fats and oils, and fish, food groups. This is to be expected as fish tend to accumulate PAHs from the marine environment, and the direct drying of vegetable oils (Dennis *et al.*, 1991; Moret *et al.*, 2000) tends to introduce

PAHs into products made from them. The concentrations of the PAHs of greatest concern have fallen in the fish, and the fats and oils food groups since then. Concentrations in the other food groups cannot be compared directly as the TDS food groups have changed since 1979 (Peattie, 1983). Although the dietary exposures from the beverages, milk and milk product food groups appear to have risen, in 1979 the PAHs were seldom detected at concentrations above the limits of determination in these food groups. The available *lower bound* concentrations would therefore have been considerable underestimates of the true concentrations. Further details of the total diet study including the measured concentrations in different food groups can be found in the Food Standards Agency information sheet (FSA, 2002).

14.3.2 Health effects and risk assessments

14.3.2.1 *Health effects*

In 1992, the U.S. National Academy of Sciences identified PAHs and other semi-volatile organic compounds amongst the highest priorities for research in terms of exposure to chemicals through the diet. The control of their presence and reduction of concentrations is of great and growing importance in the context of both food safety and health and safety regulations (Fent, 2003).

Much work has been done in assessing the risks of PAHs to human health. A comprehensive review by the European Commission on 4th December 2002 deals with the key aspects involving PAHs risks and effects on humans ('Opinion of the Scientific Committee on Food on the risks to human health of Polycyclic Aromatic Hydrocarbons in food'). A brief review will be on health risks and effects caused by PAHs are given below, although an in depth review of this aspect is not given here.

PAHs are toxic, many of the compounds in this class are both genotoxic and carcinogenic. They are actively involved in enzyme induction, immunosuppression and teratogenicity (Flowers *et al.*, 2002; Ritchie *et al.*, 2003). Many PAHs are complete carcinogens as well as initiators, thus, in addition to their capacity of inducing mutations (via genotoxic mechanisms) they can also promote tumour formation (Baird *et al.*, 2005).

Modern research on PAH-induced carcinogenesis began with the isolation by Cook *et al.* of B[*a*]P from coal tar in 1930 and the demonstration that the compound initiates tumours when repeatedly painted on mouse skin. A review article published by *Nature* covers the first 50 years since the isolation of B[*a*]P and the subsequent progress in unravelling the metabolic fate and mechanisms of chemical carcinogenesis (Phillips, 1983). Because of the knowledge accumulated on B[*a*]P toxicity, it has by default become the reference chemical for PAH mixtures. The concept of toxicity equivalency factors for each PAH as an order of magnitude estimate of the relative toxicity of each PAH compared to B[*a*]P. This is similar to the scheme widely used and accepted for dioxins, although is less advanced and not so widely accepted for PAHs (see below under Risk Assessments – The Relative Potency Factor).

The binding capacity of PAHs to DNA was thought to be responsible for the PAH-induced carcinogenic potency (Fig. 14.3). This relationship was first discovered by Brookes and Lawley in 1964. However, recent results have shown that relative to B[*a*]P, a PAH mixture forms low levels of DNA adducts, yet is highly tumourigenic in mouse skin (Baird *et al.*, 2005). This is indicative of factors contributing to PAH-induced carcinogenesis beyond DNA binding ability.

Evidence that PAHs are carcinogenic to humans comes primarily from occupational studies of workers following inhalation and dermal exposure. No data are available for humans for

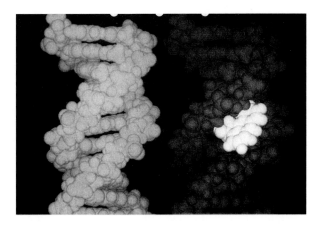

Fig. 14.3 Picture showing how a PAH will adhere to DNA (white patch). (Reprinted from Leonard, S.S., Wang, S. and Shi, X.L. Wood smoke particles generate free radicals and cause lipid peroxidation, DNA damage, NFKB activation and TNF-a release in macrophages. *Toxicology*, **150**, 147–157. Copyright 2000 with permission from Elsevier.)

other routes of exposure. However, there is a vast collection of data for animals such as mice (Rizova *et al.*, 2005).

Human exposure to PAHs occurs principally by direct inhalation, ingestion or dermal contact, as a result of their widespread presence and persistence in the environment (mostly urban ones). This exposure to ambient PAHs is usually in combination with other PAHs and other substances. These other substances may account for a more significant portion of the carcinogenicity of some mixtures, such as cigarette smoke, diesel emissions and urban aerosol (IPCS, 1998). For example, iron and steel foundry workers are exposed to PAHs and other potentially carcinogenic substances such as nickel, chromium, silica, soot, asbestos and benzene. Isolating the health effects associated with exposure to specific PAHs is, therefore, a complicated task and has been limited to experimental studies involving volunteers and accidental exposures (Rizova *et al.*, 2005).

For certain groups of workers there is an excess risk of lung, breast and urinary bladder cancer. These workers are those at coke ovens, coal gasification plants, petroleum refineries, aluminium smelters, iron and steel foundries and those working with bitumen, diesel and asphalt (Zmirou *et al.*, 2000; Jeffy *et al.*, 2002; Booth and Gribben, 2005). Chronic respiratory abnormalities risks are also higher for people submitted to PAHs contaminants, such as silicosis, asthma-like symptoms, lung function abnormalities and chronic bronchitis, especially, in aluminium plant workers (Mastrangelo *et al.*, 1996).

There are a number of schemes that have been used to rank PAH toxicity. One such scheme (Department of Health, 1998) uses the following categories:

A There is a high level of concern about a carcinogenic hazard for humans.
B There is concern about a carcinogenic hazard for humans, but the data are incomplete or the mechanism is unclear.
C The compound is a non-genotoxic carcinogen.
D The data are inadequate for assessment.
E There is no concern about carcinogenic hazard.

Three Class A PAHs were considered to be of greatest concern. These were B[*a*]P, benz(Amir)anthracene and dibenzanthracene.

14.3.2.2 *Risk Assessments*

Currently, there are at least three different approaches to conduct a health risk assessment of PAHs: (a) the surrogate approach; (b) the comparative potency approach; and (c) the relative potency factor.

The surrogate approach. The surrogate approach estimates the potency of the PAH fraction of a complex mixture of concern, based on the assumption that the health risk of this fraction is proportional to the level of an indicator or index chemical (typically B[*a*]P) in the sample. This approach treats the problem sample as a conceptual 'dilution' of a 'surrogate' mixture of PAHs. The 'surrogate' being a potent PAH-containing mixture, the chemical and toxicological properties of which are well known. An assumption must be made that the composition of the PAH mixture of concern is sufficiently similar to a surrogate PAH mixture. The extent of the 'dilution' is worked out examining the ratios of common PAHs formed in the problem sample and the surrogate (Costales and Bruya, 2005).

The comparative potency approach. The comparative potency approach was initially developed by EPA (Environmental Protection Agency) in the 1980s to evaluate adverse health effects of diesel fuels. The main assumptions made under this approach were similar mixtures (i.e. combustion mixtures) act in a similar way toxicologically, and the relative potency of a mixture in an in vivo or in vitro bioassay is directly proportional to its relative potency in humans (Amir *et al.*, 2005), the proportionality constant being represented by *k* (Pufulete *et al.*, 2004).

The relative potency factor. In this approach a marker compound is chosen as a reference for a toxicity scale (e.g. B[*a*]P). The carcinogenic and toxic properties of selected PAHs are determined with respect to that of the marker. The toxic equivalency factor (TEF) relates the toxicity of individual PAHs to that of the marker (Table 14.6). Individual PAH risks are then summed to yield a cancer risk estimate of the whole mixture. The key assumption of this approach is that carcinogenic risks for individual PAHs are additive and no cancellation or synergistic effects of these carcinogenic effects exist when in a complex mixture (Pufulete *et al.*, 2004).

All three approaches have strengths and limitations and supporters and critics. The US EPA sponsored a 2-day peer consultation workshop in October 2001 to examine these alternative approaches to the health assessment of PAH mixtures. More recently, a symposium held in 2005 on 'Toxicology of Chemical Mixtures' by the US National Capital Area Chapter Society of Toxicology (NCAC-SOT) dealt with this issue.

The Relative Potency Factor approach is the one most commonly used but even so it is agreed that it should be employed only 'as a last resort'. The reason it is seen as the favoured option is because it is the one relying most on scientifically feasible assumptions, although there are still significant data gaps resulting in considerable uncertainty. Another disadvantage to this approach is the fact that it may underestimate risk due to all PAH by considering only a few compounds. It may be that other compounds such as substituted PAHs contribute to the total toxicity and also it depends on extrapolation from animal models to humans which always adds to the uncertainty.

The comparative potency approach's main disadvantages are that it does not define the contribution of PAH to estimated overall risk, and it is therefore difficult to use for assessing speciated components of a mixture. Also, the assumption that mixtures from the same source are associated with similar risks may not be supported by the available data, and the levels

Table 14.6   TEFs based on B[a]P for 19
PAHs (Data from Burger, 1992).

PAH	TEF
Acenaphthylene	0.001
Acenaphthene	0.001
Fluorene	0.001
Phenanthrene	0.001
Anthracene	0.01
Fluoranthene	0.001
Pyrene	0.001
Benzo(b)naphtho(2,1-d)thiophene	0
Cyclopenta(c,d)pyrene	0
Benz(a)anthracene[a]	0.1
Chrysene	0.01
Benzo-(b)-fluoranthene[a]	0.1
Benzo-(k)-fluoranthene[a]	0.1
Benzo(e)pyrene	0
Benzo-(a)-pyrene[a]	**1**
Indeno-(1,2,3-cd) pyrene[a]	0.1
Dibenz-(ah)-anthracene[a]	5
Anthanthrene	0
Benzo-(g,h,i) perylene	0.01

[a] Toxicity relative to benzo-(a)-pyrene.

of compounds extractable in organic solvents are not usually reported, and the analytical methods are not standardised.

The main disadvantages of the surrogate approach are that it may result in an overestimate of the risk of PAH within a mixture and that there is no reason to believe that B[a]P is a good indicator of the potency of non-PAH compounds such as dioxins and dibenzofurans or volatile organic compounds such as benzene and 1,3-butadiene, which may be present in some complex samples. This is the reason why, in general, the surrogate approach does not predict the potency of a real complex mixture as a whole but only it's PAH component. The contribution of non-PAH to the overall risk of exposure to complex mixtures must thus be assessed separately.

Other more general concerns are that there are no human toxicity data on any of the individual PAHs (U.S. Environmental Protection Agency, 2002). As there is a fairly extended consensus within the scientific community on the fact that any approach that utilises the toxicity of a mixture as a whole is preferable to the use of a relative potency factor. This topic was reviewed in 2004 in *Regulatory Toxicology and Pharmacology* with a UK perspective (Pufulete *et al.*, 2004). There is a clear need for a more systematic collection of data on PAHs.

The US EPA identifies another challenge facing PAH risk assessments as the complexity of the scientific literature regarding PAHs and the different approaches for handling these assessments: surrogate, comparative potency and relative potency factor as discussed above (U.S. Environmental Protection Agency, 2002).

14.4   REGULATIONS

Given the long-term evidence suggesting PAHs elevate the risk of various cancers, immuno-toxic and respiratory problems, the EU has issued a number of Directives with an overall aim for the protection of human health. Directive 76/769/EEC harmonises Member States' controls over the marketing and use of certain dangerous substances. The Directive imposes restrictions on the market and use of the polycyclic aromatic hydrocarbon B[a]P as a substance or constituent of preparations in extender oils and tyres in a concentration equal or higher than 1 mg/kg, or more than 10 mg/kg for the sum of all polycyclic aromatic hydrocarbon. Directive 2455/2001/CE includes PAHs as priority contaminants and establishes a maximum acceptable concentration of 0.2 g/L in drinking water for the sum of six PAHs including fluoranthene, benzo(3,4)fluorantene, benzo(11,12)fluoranthene, benzo(1,12)perylene, benzo(3,4)-pyrene and indeno(1,2,3-cd)pyrene). Within the recommendations given by the European Union are directions for eliminating or minimising emissions in occupational settings, improved monitoring of urban air pollution and public education.

Early in 2005, three Commission documents relating to PAHs were published. They include a Regulation limiting levels of B[a]P in foods [Commission Regulation (EC) No 208/2005, 'Amending Regulation (EC) No 466/2001 as regards polycyclic aromatic hydrocarbons', *Official Journal of the European Union*, L34/3, 8.2.2005], a Directive covering sampling and analysis [Commission Directive 2005/10/EC, 'Laying down the sampling methods and the methods of analysis for the official control of the levels of benzo[a]pyrene in foodstuffs', *Official Journal of the European Union*, L34/15, 8.2.2005], and a Recommendation for further work to be carried out by Member States [Commission Recommendation, 'On the further investigation into the levels of polycyclic aromatic hydrocarbons in certain foods', *Official Journal of the European Union*, L34/43, 8.2.2005], prior to a review of the Regulation due in 2007.

There has also been publication of a Joint FAO/WHO Expert Committee on Foods Additives (JECFA) opinion on the risk from PAHs in food, which is summarised in the report: JECFA/64/SC Sixty-fourth meeting, Rome, 8–17 February 2005 (pp. 32–38).

Future legislation, including EU Regulations, Directives and other programmes are expected to toughen control and to review maximum permitted levels of PAHs in the environment and especially in foodstuffs.

References

Amir, S., Hafidi, M., Merlina, G., Hamdi, H. and Revel, J.C. (2005) Fate of polycyclic aromatic hydrocarbons during composting of lagooning sewage sludge. *Chemosphere*, **58**, 449–458.

ATSDR (1995) *Toxicological Profile for Polycyclic Aromatic Hydrocarbons (PAHs)*. Agency for Toxic Substances and Disease Registry, Department of Health and Human Services, Public Health Service, Atlanta, GA.

Baird, W.M., Hooven, L.A. and Mahadevan, B. (2005) Carcinogenic polycyclic aromatic hydrocarbon-DNA adducts and mechanism of action. *Environmental and Molecular Mutagenesis*, **45**, 106–114.

Baran, S., Oleszczuk, P. and Baranowska, E. (2003) Degradation of soil environment in the post-flooding area: Content of polycyclic aromatic hydrocarbons (PAHs) and S-triazine herbicides. *Journal of Environmental Science and Health Part B – Pesticides Food Contaminants and Agricultural Wastes*, **38**, 799–812.

Biernoth, G. and Rost, H.E. (1968) Vorkommen polcyclischer aromatischer Kohlenwasserstoffe in Speiseölen und deren Entfernung. *Archiv für Hygiene und Bakteriologie*, **152**, 238–250.

Bigham, S., Medlar, J., Kabir, A., Shende, C., Alli, A. and Malik, A. (2002) Sol-gel capillary microextraction. *Analytical Chemistry*, **74**, 752–761.

Booth, P. and Gribben, K. (2005) A review of the formation, environmental fate, and forensic methods for PAHs from aluminum smelting processes. *Environmental Forensics*, **6**, 133–142.

Burger, J., Nisbet, I.C.T. and Gochfeld, M. (1992) Metal levels in regrown feathers – assessment of contamination on the wintering and breeding grounds in the same individuals. *Journal of Toxicology and Environmental Health*, **37**, 363–374.

Bystol, A.J., Yu, S. and Campiglia, A.D. (2003) Analysis of polycyclic aromatic hydrocarbons in HPLC fractions by laser-excited time-resolved Shpol'skii spectrometry with cryogenic fiber-optic probes. *Talanta*, **60**, 449–458.

Chiu, K.H., Yak, H.K., Wai, C.A. and Lang, Q.Y. (2005) Dry ice-originated supercritical and liquid carbon dioxide extraction of organic pollutants from environmental samples *Talanta*, **65**, 149–154.

Chuang, J.C., Pollard, M.A., Chou, Y.L., Menton, R.G. and Wilson, N.K. (1998) Evaluation of enzyme-linked immunosorbent assay for the determination of polycyclic aromatic hydrocarbons in house dust and residential soil. *Science of the Total Environment*, **224**, 189–199.

Ciganek, M., Raszyk, J., Kohoutek, J., Ansorgova, A., Salava, J. and Palac, J. (2000) Polycyclic aromatic hydrocarbons (PAHs, nitro-PAHs, oxy-PAHs), polychlorinated biphenyls (PCBs) and organic chlorinated pesticides (OCPs) in the indoor and outdoor air of pig and cattle houses. *Veterinarni Medicina*, **45**, 217–226.

Ciganek, M., Ulrich, R., Neca, J. and Raszyk, J. (2002) Exposure of pig fatteners and dairy cows to polycyclic aromatic hydrocarbons. *Veterinarni Medicina*, **47**, 137–142.

Costales, M. and Bruya, J. (2005) Discussion of the error associated with polycyclic aromatic hydrocarbon (PAH) analyses. *Environmental Forensics*, **6**, 175–185.

Demaneche, S., Meyer, C., Micoud, J., Louwagie, M., Willison, J.C. and Jouanneau, Y. (2004) Identificaion and functional analysis of two aromatic-ring-hydroxylating dioxygenases from a Sphingomonas strain that degrades various polycyclic aromatic hydrocarbons. *Applied and Environmental Microbiology*, **70**, 6714–6725.

Dennis, M.J., Massey, R.C., Macweeny, D.J., Knowles, M.E. and Watson, D. (1991) Factors affecting the polycyclic aromatic hydrocarbon content of cereals, fats and other food products. *Food Additives and Contaminants*, **8**, 517–530.

Department of Health (1998) *1996 Annual Report of the Committees of Toxicity, Mutagenicity, Carcinogenicity of Chemicals in Food, Consumer Products and the Environment*. The Stationery Office. Available from: http://www.archive.officialdocuments.co.uk/document/doh/toxicity/toxic.htm

de Vos, R.H., van Dokkum, W., Schouten, A. and de Jong-Berkhout, P. (1990) Polycyclic aromatic hydrocarbons in Dutch total diet samples (1984–1986) *Food and Chemical Toxicology*, **28**, 263–268.

Divi, R.L., Beland, F.A., Fu, P.P., Von Tungeln, L.S., Schoket, B., Eltz Camara, J., Ghei, M., Rothman, N., Sinha, R. and Poirier, M.C. (2002) Highly sensitive chemiluminescence immunoassay for benzo[*a*]pyrene-DNA adducts: Validation by comparison with other methods, and use in human biomonitoring. *Carcinogenesis*, **23**, 2043–2049.

Dreij, K., Seidel, A. and Jernstrom, B. (2005) Differential removal of DNA adducts derived from anti-diol epoxides of dibenzo a,l pyrene and benzo alpha pyrene in human cells. *Chemical Research in Toxicology*, **18**, 655–664.

Ensing, K. and de Boer, T. (1999) Tailor-made materials for tailor-made applications: Application of molecular imprints in chemical analysis. *Trac-Trends in Analytical Chemistry*, **18**, 138–145.

Environmental Protection Agency (1991) *Methods for the Determination of Organic Compounds in Drinking Water*, EPA 600 ± 4-88 ± 039. US Government Printing Office, Washington, DC.

European Commission (2002) *Opinion of the Scientific Committee on Food on the Risks to Human Health of Polycyclic Aromatic Hydrocarbons in Food*, SCF/CS/CNTM/PAH/29. Peer Consultation Workshop on Approaches to Polycyclic Aromatic Hydrocarbon (Pah) Health Assessment.

Fent, K. (2003) Ecotoxicological problems associated with contaminated sites. *Toxicology Letters*, **140**, 353–365.

Fisher, M. and Schechter, I. (2000) *Encyclopedia of Analytical Chemistry, Polynuclear Aromatic Hydrocarbons Analysis in Environmental Samples*. Wiley, Chichester, pp. 3143–3172.

Flowers, L., Rieth, S.H., Foureman, G.L., Hertzberg, R., Nesnow, S., Murphy, D.L., Cogliano, V.J., Schoeny, R.S. and Hofmann, E.L. (2002) Health assessment of polycyclic aromatic hydrocarbon mixtures: Current practices and future directions. *Polycyclic Aromatic Compounds*, **22**, 811–821.

FSA (2002) *Food Survey Information Sheet 31/02 December 2002 PAHs in the UK diet: 2000 total diet study samples*. Available from: http://www.food.gov.uk/science/surveillance

Gertz, C.H. and Kogelheide, H. (1994) Untersuchung und Beurteilung von PAK in Speisefetten und-ölen. *Fett-Wissenschaft-Technologie (Fat Science Technology)*, **96**, 175–180.

Gorelick, N.J. and Reeder, N.L. (1993) Detection of multiple polycyclic aromatic hydrocarbon DNA adducts by a high-performance liquid-chromatography P-32 postlabeling method. *Environmental Health Perspectives*, **99**, 207–211.

Gregory, J., Foster, K., Tyler, H. and Wiseman, M. (1990) *Dietary and Nutritional Survey of British Adults.* HMSO, London, UK.

Gregory, J., Collins, D.L., Davies, P.S.W., Hughes, J.M. and Clarke, P.C. (1995) *National Diet and Nutrition Survey: Children aged 1½ to 4½ years. Vol 1: Report of the Diet and Nutrition Survey.* HMSO, London, UK.

Gregory, J., Lowe, S., Bates, C.J., Prentice, A., Jackson, L.V., Smithers, G., Wenlock, R. and Farron, M. (2000) *National Diet and Nutrition Survey: Young People aged 4–18 years. Vol. 1: Report of the Diet and Nutrition Survey.* The Stationery Office, London, UK.

Guillen, M.D. and Sopelana, P. (2004) Occurrence of polycyclic aromatic hydrocarbons in smoked cheese. *Journal of Dairy Science,* **87,** 556–564.

Guillen, M.D. and Sopelana, P. (2005) Headspace solid-phase microextraction as a tool to estimate the contamination of smoked cheeses by polycyclic aromatic hydrocarbons. *Journal of Dairy Science,* **88,** 13–20.

Hawthorne, S.B., Grabanski, C.B., Miller, D.J. and Kreitinger, J.P. (2005) Solid-phase microextraction measurement of parent and alkyl polycyclic aromatic hydrocarbons in milliliter sediment pore water samples and determination of K-DOC values. *Environmental Science and Technology,* **39,** 2795–2803.

Hennion, M.C. (1999) Solid-phase extraction: Method development, sorbents, and coupling with liquid chromatography. *Journal of Chromatography A,* **856,** 3–54.

Hu, Y.L., Yang, Y.Y., Huang, J.X. and Li, G.K. (2005) Preparation and applicaion of poly(dimethylsiloxane)/ beta-cyclodextrin solid-phase microextraction membrane. *Analytica Chimica Acta,* **543,** 17–24.

IPCS (1998) *Selected Non-heterocyclic Polycyclic Aromatic Hydrocarbons.* Environmental Health Criteria 202. World Health Organisation, Geneva.

Jeffy, B.D., Chirnomas, R.B. and Romagnolo, D.F. (2002) Epigenetics of breast cancer: Polycyclic aromatic hydrocarbons as risk factors. *Environmental and Molecular Mutagenesis,* **39,** 235–244.

Jira, W. (2004) A GC/MS method for the determination of carcinogenic polycyclic aromatic hydrocarbons (PAH) in smoked meat products and liquid smokes. *European Food Research and Technology,* **218,** 208–212.

Jones, J.M., Anderson, J.W. and Tukey, R.H. (2000) Using the metabolism of PAHs in a human cell line to characterize environmental samples. *Environmental Toxicology and Pharmacology,* **8,** 119–126.

Jones, S.H., Chase, M., Sowles, J., Hennigar, P., Landry, N., Wells, P.G., Harding, G.C.H., Krahforst, C. and Brun, L. (2001) Monitoring for toxic contaminants in mytilus edulis from New Hampshire and the Gulf of Maine. *Journal of Shellfish Research,* **20,** 1203–1214.

Kanaly, R.A. and Harayama, S. (2000) Biodegradation of high-molecular-weight polycyclic aromatic hydrocarbons by bacteria. *Journal of Bacteriology,* **182,** 2059–2067.

Kazerouni, N., Sinha, R., Hsu, C.H., Greenberg, A. and Rothman, N. (2001) Analysis of 200 food items for benzo[a]pyrene and estimation of its intake in an epidemiologic study. *Food and Chemical Toxicology,* **39,** 423–436.

Laor, Y. and Rebhun, M. (2002) Evidence for nonlinear binding of PAHs to dissolved humic acids. *Environmental Science and Technology,* **36,** 955–961.

Larsson, B.K., Eriksson, A.T. and Cervenka, M. (1987) Polycyclic aromatic hydrocarbons in crude and deodorizen vegetable oils. *Journal of the American Oil Chemists' Society,* **64,** 365–370.

Leonard, S.S., Wang, S., Shi, X.L., Jordan, B.S., Castranova, V. and Dubick, M.A. (2000) Wood smoke particles generate free radicals and cause lipid peroxidation, DNA damage, NF kappa B activation and TNF-alpha release in macrophages. *Toxicology,* **150,** 147–157.

Lerario, V.L., Giandomenico, S., Lopez, L. and Cardellicchio, N. (2003) Sources and distribution of polycyclic aromatic hydrocarbons (PAHs) in sediments from the Mar Piccolo of Taranto, Ionian Sea, southern Italy. *Annali Di Chimica,* **93,** 397–406.

Li, S., Olegario, R.M., Banyasz, J.L. and Shafer, K.H. (2003) Gas chromatography-mass spectrometry analysis of polycyclic aromatic hydrocarbons in single puff of cigarette smoke. *Journal of Analytical and Applied Pyrolysis,* **66,** 155–163.

Librando, V., Tomaselli, G. and Tringali, G. (2005) Optimization of supercritical fluid extraction by carbon dioxide with organic modifiers of polycyclic aromatic hydrocarbons from urban particulate matter. *Annali Di Chimica,* **95,** 211–216.

Lodovici, M., Dolara, P., Casalini, C., Ciappellano, S. and Testolini, G. (1995) Polycyclic aromatic hydrocarbon contamination in the Italian diet. *Food Additives and Contaminants,* **12,** 703–713.

Mackay, D., Paterso, S. and Shiu, W.Y. (1992) Generic models for evaluating the regional fate of chemicals. *Chemosphere* **24,** 695–717.

Mallakin, A., Babu, T.S., Dixon, D.G. and Greenberg, B.M. (2002) Sites of toxicity of specific photooxidation products of anthracene to higher plants: Inhibition of Photosynthetic activity and electron transport in *Lemna gibba* L. G-3 (duckweed). *Environmental Toxicology*, **17**, 462–471.

Manoli, E. and Samara, C. (1999) Polycyclic aromatic hydrocarbons in natural waters: Sources, occurrence and analysis. *Trac-Trends in Analytical Chemistry*, **18**, 417–428.

Mastral, A.M., Garcia, T., Lopez, J.M., Murillo, R., Callen, M.S. and Navarro, M.V. (2004) Where are the limits of the gas-phase fluorescence on the polycyclic aromatic compound analysis? *Polycyclic Aromatic Compounds*, **24**, 325–332.

Mastrangelo, G., Fadda, E. and Marzia, V. (1996) Polycyclic aromatic hydrocarbons and cancer in man *Environmental Health Perspectives*, **104**, 1166–1170.

Matuszewska, A. and Czaja, M. (2000) The use of synchronous luminescence spectroscopy in qualitative analysis of aromatic fraction of hard coal thermolysis products. *Talanta*, **52**, 457–464.

Menichini, E., Bocca, A., Merli, F., Ianni, D. and Monfredini, F. (1991) Polycyclic aromatic hydrocarbons in olive oils on the Italian market. *Food Additives and Contaminants*, **8**, 363–369.

Ministry of Agriculture, Fisheries and Food (1994) *The British Diet: Finding the Facts 1989–1993*. Food Surveillance Paper, 40. HMSO, London, UK.

Moret, S. and Conte, L.S. (2000) Polycyclic aromatic hydrocarbons in edible fats and oils: Occurrence and analytical methods. *Journal of Chromatography A*, **882**, 245–253.

Moret, S., Dudine, A. and Conte, L.S. (2000) Processing effects on polyaromatic hydrocarbon content of grapeseed oil. *Journal of the American Oil Chemists Society*, **77**, 1289–1292.

Mostafa, G.A. (2002) Monitoring of polycyclic aromatic hydrocarbons in seafoods from Lake Timsah. *International Journal of Environmental Health Research*, **12**, 83–91.

Mottier, P., Parisod, V. and Turesky, R.J. (2000) Quantitative determination of polycyclic aromatic hydrocarbons in barbecued meat sausages by gas chromatography coupled to mass spectrometry. *Journal of Agricultural and Food Chemistry*, **48**, 1160–1166.

Narbonne, J.F., Aarab, N., Clerandeau, C., Daubeze, M., Narbonne, J., Champeau, O. and Garrigues, P. (2005) Scale of classification based on biochemical markers in mussels: Application to pollution monitoring in Mediterranean coasts and temporal trends. *Biomarkers*, **10**, 58–71.

O'Connor, T.P. (2002) National distribution of chemical concentrations in mussels and oysters in the USA. *Marine Environmental Research*, **53**, 117–143.

Oliferova, L., Statkus, M., Tsysin, G., Shpigun, O. and Zolotov, Y. (2005) On-line solid-phase extraction and HPLC determination of polycyclic aromatic hydrocarbons in water using fluorocarbon polymer sorbents. *Analytica Chimica Acta*, **538**, 35–40.

Olson, L.G., Uibel, R.H. and Harris, J.M. (2004) C18-modified metal-colloid substrates for surface-enhanced Raman detection of trace-level polycyclic aromatic hydrocarbons in aqueous solution. *Applied Spectroscopy*, **58**, 1394–1400.

Oros, D.R. and Ross, J.R.M. (2005) Polycyclic aromatic hydrocarbons in bivalves from the San Francisco estuary: Spatial distributions, temporal trends, and sources (1993–2001). *Marine Environmental Research*, **60**, 466–488.

Ovrebo, S., Haugen, A., Phillips, D.H. and Hewer, A. (1992) Detection of polycyclic aromatic hydrocarbon-DNA adducts in white blood-cells from coke-oven workers–correlation with job categories. *Cancer Research*, **52**, 1510–1514.

Ovrebo, S., Haugen, A., Fjeldstad, P.E., Hemminki, K. and Szyfter, K. (1994) Biological monitoring of exposure to polycyclic aromatic hydrocarbon in an electrode paste plant. *Journal of Occupational and Environmental Medicine*, **36**, 303–310.

Paschke, A., Popp, P. and Schuurmann, G. (1999) Solubility and partitioning studies with polycyclic aromatic hydrocarbons using an optimized SPME procedure. *Fresenius Journal of Analytical Chemistry*, **363**, 426–428.

Patel, R., Zhou, R.N., Zinszer, K., Josse, F. and Cernosek, R. (2000) Real-time detection of organic compounds in liquid environments using polymer-coated thickness shear mode quartz resonators. *Analytical Chemistry*, **72**, 4888–4898.

Pavanello, S., Favretto, D., Brugnone, F., Mastrangelo, G., Dal Pra, G. and Clonfero, E. (1999) HPLC/fluorescence determination of anti-BPDE-DNA adducts in mononuclear white blood cells from PAH-exposed humans. *Carcinogenesis*, **20**, 431–435.

Peattie, M.E., Buss, D.H., Lindsay, D.G. and Smart, G.A. (1983) Reorganisation of the British Total Diet Study for monitoring food constituents from 1981 *Food and Chemical Toxicology*, **21**, 503–507.

Phillips, D.H. (1983) 50 Years of benzo(alpha)pyrene. *Nature*, **303**, 468–472.

Pufulete, M., Battershill, J., Boobis, A. and Fielder, R. (2004) Approaches to carcinogenic risk assessment for polycyclic aromatic hydrocarbons: A UK perspective. *Regulatory Toxicology and Pharmacology*, **40**, 54–66.

Rigou, P., Saini, S. and Setford, S.J. (2004) Field-based supercritical fluid extraction and immunoassay for determination of PAHs in soils. *International Journal of Environmental Analytical Chemistry*, **84**, 979–994.

Ritchie, G., Still, K., Rossi, J., III, Bekkedal, M., Bobb, A. and Arfsten, D. (2003) Biological and health effects of exposure to kerosene-based jet fuels and performance additives. *Journal of Toxicology and Environmental Health – Part B – Critical Reviews*, **6**, 357–451.

Rizova, V., Nikodinovski, M. and Kendrovski, V. (2005) PAHs – the sources of pollution of the environment and their toxicological impact on human's health. *Journal of Environmental Protection and Ecology*, **6**, 1–7.

Rodriguez-Sanmartin, P., Moreda-Pineiro, A., Bermejo-Barrera, A. and Bermejo-Barrera, P. (2005) Ultrasound-assisted solvent extraction of total polycyclic aromatic hydrocarbons from mussels followed by spectrofluorimetric determination. *Talanta*, **66**, 683–690.

Roy, G., Vuillemin, R. and Guyomarch, J. (2005) On-site determination of polynuclear aromatic hydrocarbons in seawater by stir bar sorptive extraction (SBSE) and thermal desorption GC-MS. *Talanta*, **66**, 540–546.

Sagredos, A.N., Sinha-Roy, D. and Thomas, A. (1988) The determination, occurrence and composition of polycyclic aromatic hydrocarbons in oils and fats. *Fett-Wissenschaft- Technologie (Fat Science Technology)*, **90**, 76–81.

Schaum, J., Schuda, L., Wu, C., Sears, R., Ferrario, J. and Andrews, K. (2003) A national survey of persistent, bioaccumulative, and toxic (PBT) pollutants in the United States milk supply. *Journal of Exposure Analysis and Environmental Epidemiology*, **13**, 177–186.

Schneider, U.A., Brown, M.M. and Logan, R.A. (1995) Screening assay for dioxin-like compounds based on competitive-binding to the murine hepatic Ah receptor 1. Assay development. *Environmental Science and Technology*, **29**, 2595–2602.

Siddiqi, M.A., Yuan, Z.X. and Honey, S.A. (2002) Metabolism of PAHs and methyl-substituted PAHs by sphingomonas paucimobilis strain EPA 505. *Polycyclic Aromatic Compounds*, **22**, 3–4.

Sluszny, C., Gridin, V.V., Bulatov, V. and Schechter, I. (2002) Analysis of polycyclic aromatic hydrocarbons in heterogeneous samples. *Reviews in Analytical Chemistry*, **21**, 77–165.

Sluszny, C., Gridin, V.V., Bulatov, V. and Schechter, I. (2004) Polymer film sensor for sampling and remote analysis of polycyclic aromatic hydrocarbons in clear and turbid aqueous environments. *Analytica Chimica Acta*, **522**, 145–152.

Smith, B.W. and Hurtubise, R.J. (2004) New methodology for the characterization of (+/−)-anti-BPDE-DNA adducts an tetrol I-1 with solid-matrix phosphorescence. *Analytica Chimica Acta*, **502**, 149–159.

Speer, K., Steeg, E., Horstmann, P., Kühn, T.H. and Montag, A. (1990) Determination and distribution of polycyclic aromatic hydrocarbons in native vegetable oils, smoked fish products, mussels and oysters, and bream from the River Elbe. *Journal of High Resolution Chromatography and Chromatography*, **13**, 104–111.

Stolyhwo, A. and Sikorski, Z.E. (2005) Polycyclic aromatic hydrocarbons in smoked fish – a critical review. *Food Chemistry*, **91**, 303–311.

Storelli, M.M. and Marcotrigiano, G.O. (2001) Polycyclic aromatic hydrocarbons in mussels (*Mytilus gallo-provincialis*) form the Ionian Sea, Italy. *Journal of Food Protection*, **64**, 405–409.

Swetman, A., Head, S. and Evans, D. (1999) Contamination of coconut oil by PAH. *Inform*, **10**, 706–712.

Tamakawa, K. (2004) Handbook of Food Analysis: *Food Science and Technology*. Vol. 2. (ed. L. Nollet). Marcel Dekker, Inc., New York, pp. 138.

Ter Laak, T.L., Mayer, P., Busser, F.J.M., Klamer, H.J.C. and Hermens, J.L.M. (2005) Sediment dilution method to determine sorption coefficients of hydrophobic organic chemicals. *Environmental Science & Technology*, **39**, 4220–4225.

Thurman, E.M. and Mills, M.S. (1998) *Solid-Phase Extraction: Principles and Practice*. Wiley/Interscience, New York, 372 pp.

Tiscornia, E., Forina, M. and Evangelisti, F. (1982) Composizione chimica dell'olio di oliva e sue variazioni indotte dal processo di rettificazione. *La Rivista Italiana delle Sostanze Grasse*, **59**, 519–556.

U.S. Environmental Protection Agency (2002) O. O. R. A. D., National Center for Environmental Assessment, Washington Office, Washington, DC.

Wani, M.A., El-Mahdy, M.A., Hamada, F.M., Wani, G., Zhu, Q., Wang, Q. and Wani, A.A. (2002) Efficient repair of bulky anti-BPDE DNA adducts from non-transcribed DNA strand requires functional p53 but not p21waf1/cip1 and pRb. *Mutation Research*, **505**, 13–25.

Weisshoff, H., Preiss, A., Nehls, I., Win, T. and Mugge, C. (2002) Development of an HPLC-NMR method for the determination of PAHs in soil samples – a comparison with conventional methods. *Analytical and Bioanalytical Chemistry*, **373**, 810–819.

Wenzl, T., Simon, R., Kleiner, J. and Anklam, E. (2006) Analytical methods for polycyclic aromatic hydrocarbons (PAHs) in food and the environment needed for new food legislation in the European Union. *Trends in Analytical Chemistry*, **25**, 716–725.

Wootton, E.C., Dyrynda, E.A., Pipe, R.K. and Ratcliffe, N.A. (2003) Comparisons of PAH-induced immunomodulation in three bivalve molluscs. *Aquatic Toxicology*, **65**, 13–25.

Zabiegala, B., Kot, A. and Namiesnik, J. (2000) Long-term monitoring of organic pollutants in water – application of passive dosimetry. *Chemia Analityczna*, **45**, 645–657.

Zmirou, D., Masclet, P., Boudet, C., Dor, F. and Dechenaux, J. (2000) Personal exposure to atmospheric polycyclic aromatic hydrocarbons in a general adult population and lung cancer risk assessment. *Journal of Occupational and Environmental Medicine*, **42**, 121–126.

Index

Food Science and Technology

Blackwell Publishing

GENERAL FOOD SCIENCE AND TECHNOLOGY

Title	Author	ISBN
Handbook of Fermented Meat and Poultry	Toldra	9780813814773
High Pressure Processing of Foods	Doona	9780813809441
Water Activity in Foods	Barbosa-Canovas	9780813824086
Food and Agricultural Wastewater Utilization and Treatment	Liu	9780813814230
Biotechnology in Flavor Production	Havkin-Frenkel	9781405156493
Frozen Food Science and Technology	Evans	9781405154789
Applications of Fluidisation in Food Processing	Smith	9780632064564
Encapsulation and Controlled Release Technologies in Food Systems	Lakkis	9780813828558
Accelerating New Food Product Design and Development	Beckley	9780813808093
Multivariate and Probabilistic Analyses of Sensory Science Problems	Meullenet	9780813801780
Handbook of Meat, Poultry and Seafood Quality	Nollet	9780813824468
Chemical Physics of Food	Belton	9781405121279
Handbook of Organic and Fair Trade Food Marketing	Wright	9781405150583
Sensory and Consumer Research in Food Product Design and Development	Moskowitz	9780813816326
Sensory Discrimination Tests and Measurements	Bi	9780813811116
Food Biochemistry and Food Processing	Hui	9780813803784
Handbook of Fruits and Fruit Processing	Hui	9780813819815
Nonparametrics for Sensory Science	Rayner	9780813811123
IFIS Dictionary of Food Science and Technology	IFIS	9781405125055
Kosher Food Production	Blech	9780813825700
Managing Food Industry Waste	Zall	9780813806310
Food Processing – Principles and Applications	Smith	9780813819426
Flavor Perception	Taylor	9781405116275
Food Supply Chain Management	Bourlakis	9781405101684
Plant Food Allergens	Mills	9780632059829
Welfare of Food	Dowler	9781405112451
Food Flavour Technology	Taylor	9781841272245
Adverse Reactions to Food	Buttriss	9780632055470
Advanced Dietary Fibre Technology	McCleary	9780632056347
Tastes and Aromas	Bell	9780632055449
Industrial Chocolate Manufacture and Use	Beckett	9780632054336

INGREDIENTS

Title	Author	ISBN
Sweeteners and Sugar Alternatives in Food Technology	Mitchell	9781405134347
Emulsifiers in Food Technology	Whitehurst	9781405118026
Technology of Reduced Additive Foods	Smith	9780632055326
Water-Soluble Polymer Applications in Foods	Nussinovitch	9780632054299
Food Additives Data Book	Smith	9780632063956
Enzymes in Food Technology	Whitehurst/Law	9781841272238

FOOD SAFETY, QUALITY AND MICROBIOLOGY

Title	Author	ISBN
Nondestructive Testing of Food Quality	Irudayaraj	9780813828855
Microbiological Safety of Food in Health Care Settings	Lund	9781405122207
Control of Food Biodeterioration	Tucker	9781405154178
Phycotoxins	Botana	9780813827001
Advances in Food Diagnostics	Nollet	9780813822211
Advances in Thermal and Nonthermal Food Preservation	Tewari	9780813829685
Biofilms in the Food Environment	Blaschek	9780813820583
Food Irradiation Research and Technology	Sommers	9780813808826
Preventing Foreign Material Contamination of Foods	Peariso	9780813816395
Aviation Food Safety	Sheward	9781405115810
Food Microbiology and Laboratory Practice	Bell	9780632063819
Listeria 2nd Edition	Bell	9781405106184
Preharvest and Postharvest Food Safety	Beier	9780813808840
Practical Food Microbiology 3rd Edition	Roberts	9781405100755
Metal Contamination of Food 3rd Edition	Reilly	9780632059270
Microbiological Risk Assessment of Food	Forsythe	9780632059522
Food Safety in Shrimp Processing	Kanduri	9780852382707
Shelf Life	Man	9780632056743
HACCP	Mortimore	9780632056484
Salmonella	Bell	9780632055197

For further details and ordering information, please visit

www.blackwellfood.com

Join our free e-mail alert service to receive journal tables of contents with links to abstracts and news of the latest books in your field.

Food Science and Technology from Blackwell Publishing

Microbiology of Safe Food	Forsythe	9780632054879
Clostridium Botulinum	Bell	9780632055210
E. coli	Bell	9780751404623
Environmental Contaminants in Food	Moffat/Whittle	9781850759218
Natural Toxicants in Food	Watson	9781850758624

FOOD LAWS AND REGULATIONS

Food Labeling Compliance Review 4th Edition	Summers	9780813821818
BRC Global Standard – Food	Kill	9781405157964
Guide to Food Laws and Regulations	Curtis	9780813819464
Regulation of Functional Foods and Nutraceuticals	Hasler	9780813811772
Concept Research in Food Product Design and Development	Moskowitz	9780813824246

DAIRY FOODS

Advanced Dairy Technology	Britz	9781405136181
Milk Processing and Quality Management	Tamime	9781405145305
Cleaning in Place	Tamime	9781405155038
Structure of Dairy Products	Tamime	9781405129756
Brined Cheeses	Tamime	9781405124607
Fermented Milks	Tamime	9780632064588
Manufacturing Yogurt and Fermented Milks	Chandan	9780813823041
Handbook of Milk of Non-Bovine Mammals	Park	9780813820514
Probiotic Dairy Products	Tamime	9781405121248
Mechanisation and Automation of Dairy Technology	Tamime/Law	9781841271101
Technology of Cheesemaking	Law	9781841270371

BAKERY AND CEREALS

Whole Grains and Health	Marquart	9780813807775
Baked Products - Science, Technology and Practice	Cauvain	9781405127028
Bakery Products Science and Technology	Hui	9780813801872
Bakery Food Manufacture and Quality	Cauvain	9780632053278
Pasta and Semolina Technology	Kill	9780632053490

BEVERAGES AND FERMENTED FOODS/BEVERAGES

Brewing Yeast Fermentation Performance 3rd Edition	Smart	9781405119085
Microbiology and Technology of Fermented Foods	Hutkins	9780813800189
Carbonated Soft Drinks	Steen	9781405134354
Brewing Yeast and Fermentation	Boulton	9781405152686
Food, Fermentation and Micro-organisms	Bamforth	9780632059874
Wine Production	Grainger	9781405113656
Chemistry and Technology of Soft Drinks and Fruit Juices	Ashurst	9781405122863
Technology of Bottled Water	Senior	9781405120388
Wine Flavour Chemistry	Clarke	9781405105309
Beer: Health and Nutrition	Bamforth	9780632064465
Brewing Yeast Fermentation Performance 2nd Edition	Smart	9780632064984

PACKAGING

Food Packaging Research and Consumer Response	Moskowitz	9780813812229
Packaging for Nonthermal Processing of Food	Han	9780813819440
Packaging Closures and Sealing Systems	Theobald	9781841273372
Modified Atmospheric Processing and Packaging of Fish	Otwell	9780813807683
Paper and Paperboard Packaging Technology	Kirwan	9781405125031
Food Packaging Technology	Coles	9781841272214
PET Packaging Technology	Brooks/Giles	9781841272221
Canmaking for Can Fillers	Turner	9781841272207
Design & Tech of Pack Dec for the Consumer Market	Giles	9781841271064
Materials & Development of Plastics Packaging for the Consumer Market	Giles/Bain	9781841271163
Technology of Plastics Packaging	Giles/Bain	9781841271170
Handbook of Beverage Packaging	Giles	9781850759898

OILS AND FATS

Trans Fatty Acids	Dijkstra	9781405156912
Chemistry of Oils and Fats	Gunstone	9781405116268
Edible Oil Processing	Hamm	9781841270388
Fats in Food Technology	Rajah	9781841272252
Rapeseed and Canola Oil – Production, Processing, Properties and Uses	Gunstone	9781405116251
Vegetable Oils in Food Technology	Gunstone	9781841273310

For further details and ordering information, please visit

www.blackwellfood.com

Join our free e-mail alert service to receive
journal tables of contents with links to abstracts
and news of the latest books in your field.